For Trish and Sue

Preface

Depending on one's perspective, the evolutionary study of family structure either began 3-5 decades ago (Lack 1947, 1954, Hamilton 1964a, Trivers 1974) or it has been a research area since 1859. Certainly the key conceptual roots took to the soil in the 1960s and 1970s, when it became more widely recognized that most of natural selection's potency exists at the level of individuals and genes. This clarification had many effects, none more counterintuitive than the need to re-evaluate that monolithic module of stability, the Family. Under scrutiny it became clear that the genetic interests of different family members were distinct. Imagine that.

This book is about what happens in families when demand outstrips supply. That occurs whenever (i) parents create more offspring than they can support fully, (ii) parents create more offspring than they *choose* to support, or (iii) offspring predict that they *might* be under-supported at some future time.

We had initially intended for the book to be a rich and well-mixed stew of theory and empiricism, so that readers would see how these two scientific traditions fuel each other. (Indeed, our own collaboration began and has been sustained by a fervent belief that theoreticians and empiricists both benefit from struggling to understand one another.) In our early vision of the book, each chapter was to blend both elements seamlessly, moving easily from models to data and on to verbal explanations and back. This has not come to pass, and we really should have known better. In this research area, as in most others, the lack of communication between empiricists and mathematical theorists has led to the construction of separate edifices, with rare connections between them. Too few theoreticians pay serious heed to the data; most field workers do not take the time necessary to penetrate the glittering but intimidating models. The lamentable result is that the former can waste much time exploring phenomena not of this (or any known) world, while the latter may fail to recognize exciting aspects of their study systems, and may end up gathering massive data sets that shed light on general problems only by chance.

This book cannot demonstrate how marvellously these approaches work in concert on the problems of intra-family conflict for the simple reason that they do not work in concert. Not on a regular basis. Not yet. But that is a worthy goal and we offer what follows as a progress report. Partly because of this reality and partly because we wrote different sections while separated by an ocean, the book is an amalgam of three elements: theory, egret siblicide studies, and sibling rivalry in all non-egret life forms. We have endeavoured to keep the mathematics as user-friendly as possible; virtually all of it should be comprehensible to readers with a semester of differential calculus under their belts. We have also tried to restate the essence of most formulae in English to make those points more accessible.

A second caveat pertains to taxon chauvinism. Although the extant sibling rivalry literature is biased toward birds (but perhaps not to the degree originally imagined!), this book is even more so. Within the ornithological literature reviewed, the egret work is given more space than it deserves because much of it is ours. It is a lot easier to acknowledge this skew than to provide a fairer balance! For what appears to be a slight to the invertebrates, we point to the fact that most of the sibling rivalry documented for them takes the form of cannibalism, which has been treated to an extensive series of excellent reviews (Elgar and Crespi 1992). By the same token, the existence of a forthcoming book, *Brood Reduction in Plants* (by Uma Shaanker *et al.*), was made known to us exactly one day after our first draft was completed, but we are counting on it to rectify our thin treatment of botanical topics. Thus we happily and sincerely hope that readers of this book will pursue those topics in these related volumes. Our primary goal all along has been to attract hungry and energetic postgraduate students to the topic of intra-family dynamics. If we can attract even two to the field, we stand to double our own contributions. And if the two we manage to con are ambidextrous with theory-building and theory-testing, perhaps the book we originally wanted to write can be produced in the near future.

Norman, Oklahoma   D.W.M.
Liverpool   G.A.P.
March 1997

Acknowledgements

This book was written with the generous support of fellowships from the Harry Frank Guggenheim Foundation (to D.W.M.) and Science and Engineering Research Council (to G.A.P.). Our own research programmes have been supported mainly by National Science Foundation and NERC grants. Special thanks are due to the Department of Psychology, University of Washington, where the writing began. We received great encouragement and much-needed suggestions and criticisms from many friends and colleagues, including the following who read the entire manuscript at least once: Eric Charnov, Scott Forbes, Charles Godfray, Paul Harvey, Catherine Kennedy, Tim Lamey, Robert May, Judith May, Colleen St Clair and Trish Schwagmeyer. In addition, we thank the following for reading chapters or parts thereof and suggesting improvements: Ingrid Ahnesjö, Per Alberch, Dave Anderson, George Barlow, Marc Bekoff, Andy Blaustein, Gary Bortolotti, Marty Burd, John Byers, Janalee Caldwell, Mertice Clark, Scott Creel, Curtis Creighton, Hernan Dopazo, Hugh Drummond, Philip Dziuk, Marilyn East, John Fauth, Lawrence Frank, David Fraser, Jeff Galef, Robert Gibson, David Grow, David Haig, Lynn Haugen, Peggy Hill, Heribert Hofer, Jim Krupa, Cammie Lamey, Jim Moodie, Christine Nalepa, Bart O'Gara, David Pfennig, Dave Queller, David C. Smith, Jim Stewart, Fritz Trillmich, Amanda Vincent, Laurie Vitt, Marvalee Wake, Susan Walls, Peter Waser and Karen Wiebe. Many of these provided access to unpublished manuscripts and references, as did Janis Dickinson, Mark Elgar, John Fauth, Yosiaki Ito, Marlene Machmer and Mary Reid. Jaqui Shykoff bravely attempted to explain endosperm genetics. Expert artwork was provided mainly by Coral McAllister. Debugging of the finished manuscript was aided by Kathy Farris. Finally, the many students and colleagues who have helped over the years in collecting egret field data are thanked collectively (they numbered 27 at last count), as are the members of the 1988 University of Washington graduate seminar on sibling rivalry.

Contents

1

General introduction

Through the animal and vegetable kingdoms, nature has scattered the seeds of life abroad with the most profuse and liberal hand; but has been comparatively sparing in the room and nourishment necessary to rear them.

(T.R. Malthus, 1798, An Essay on the Principle of Population)

1.1 Sibling rivalry and the evolution of selfishness

One fundamental tenet of Darwinism–Wallaceism, anticipated clearly by Malthus, is that organisms compete for critical resources. Individuals relatively adept at such games will tend to contribute (genes and their associated traits) disproportionately to future generations. For well over a century, biologists have examined the phenotypic attributes that foster such proficiency, and there is little disagreement today that the selfish promotion of personal welfare should be generally favoured by natural selection.

The special problem of 'altruistic' traits, i.e. those seeming to benefit other individuals, has attracted considerable attention precisely because such features appear incompatible with the classical perspective. In the past three decades (i.e. since Hamilton 1964a, b), inclusive fitness theory has come to occupy a central position in theoretical biology, offering as it does a compelling explanation for how traits (especially behavioural ones) that confer a net cost on the performer can increase within populations through the compensating benefits received by collateral or non-lineal kin (the so-called 'indirect component' of inclusive fitness: Brown and Brown 1981). The simple and fundamental principle, commonly known as Hamilton's rule, can be expressed as $br - c > 0$ (where c = net cost to the performer, b = net benefit to the recipient of the act's effects, and r = coefficient of relatedness between performer and recipient). Hamilton's rule thus neatly specifies the conditions, in terms of gains and losses to inclusive fitness, under which an individual should behave unselfishly on behalf of a genetic relative (Chapter 2).

Much of the excitement in behavioural ecology and sociobiology stems from this key insight, and armies of students have rushed to re-examine all proffered examples of altruism. The textbooks of animal behaviour are now brimming with examples from a rich and rapidly growing literature sweepingly in support of the Hamiltonian prediction. As a result, most of the erstwhile cases of altruism have either been re-classified as cryptically selfish (i.e. the performer actually incurs negligible costs and/or achieves unforeseen gains) or as 'nepotism' (the typical beneficiaries turn out to be close kin and the rule is satisfied). At the very least, suspected generosity among genetic relatives is in little danger of being overlooked.

Rather less consideration has been given to the fact that Hamilton's rule simultaneously specifies the evolutionary *limits on selfish behaviour* in social contexts: if the conditions for altruism are *not* met, the individual is expected to behave in ways that are self-promoting, i.e. selfish. Because many or most social interactions are not between close kin (i.e. $r \to 0$), such an expectation is of no particular interest; selfishness is unchecked. However, selfish behaviour often exacts a heavy toll on the very closest of genetic relatives, including offspring and siblings. While Hamilton's rule taught us that an individual should view its sibling as a glass half full, the same glass is also half empty.

Here, we explore this more sombre side of sibling relations, which is revealed vividly when critical resources are scarce and the competitors are, inescapably, close kin. Then the individual must weigh the direct vs. indirect components of its inclusive fitness (albeit not consciously). Sometimes it will pay to sacrifice some of the latter on behalf of the former. Thus, the theoretical significance of sibling rivalry as a topic stems from its position astride the evolutionary fulcrum between selfishness and kin-selected altruism.

We contend that the essential ecological squeeze required for siblings (even full siblings) to view each other as rivals is more pervasive than is generally appreciated. The daily needs of a young bird or mammal, for example, are often steep and much like those of its nest-mates. At times, the brood's net demands outstrip the parents' provisioning abilities to such a degree that the adults do best by cutting their losses, perhaps ending investment altogether (e.g. for reviews of parental infanticide see Hausfater and Blaffer Hrdy 1984). As David Lack emphasized half a century ago, family size is often adjusted downward via the incremental competitive process known as *brood reduction* (Lack 1947, Ricklefs 1965, O'Connor 1978). If each sibling's 'fair share' can be envisioned as simply its proportion of brood size multiplied by the available resources, there must be milder levels of resources shortage in which each sibling tries to sequester somewhat more than its share of the whole. Such sub-lethal rivalry may or may not project into future problems for the deprived brood-mates.

1.2 Manifestations of sibling rivalry

Sibling rivalry consists of a widespread and diverse family of phenomena, but it is by no means ubiquitous. It correlates with a number of ecological conditions and phylogenetic precursors. Its many forms depend partly on the nature of the critical limiting resource(s) and partly on the competitors' own abilities. Among birds, for example, we find competitive food-begging by songbird nestlings as well as the dramatic and fatal attacks by a 20-day-old black eagle (*Aquila verreauxi*), which commence soon after its lone nest-mate emerges from the haven of its eggshell (Figure 1.1).

Fig. 1.1 A dominant black eagle nestling enlarges a wound it has created on it 1-day-old sibling's back. Such attacks are relentless and the victim virtually always dies within a few days. (Photograph by Peter Steyn.)

Although most sibling rivalry necessarily occurs early in the life cycle (i.e. prior to dispersal), examples can be found at virtually all stages from gametogenesis (if one wishes to count the ovum and associated polar bodies as 'sibs'), to fertilization (intra-ejaculate scramble competitions among sperm 'sibs' and the ensuing chemical barrier created around the new zygote to prevent multiple fertilizations[1]), and on through embryonic stages, infancy, adolescence, and even, in some cases, to full adulthood.

Most cases of orthodox, post-zygotic sibling rivalry occur when brood-mates share a limited space and pool of resources. We shall call that restricted space a ***nursery***, which has meant 'a room set aside for young children with their nurse' since approximately AD 1300 (Barnhart 1988).[2] (The presence of someone in the 'nurse' role is presumably optional since the room remains a nursery even when empty.) We employ ***nursery*** as a conceptual device for distinguishing between families likely to have important competition among sibs and those of a more amorphous nature whose young drift apart (some-times literally, as when plankton broadcast vast numbers of offspring into ocean currents). Members of a diffusing sibship have no particular likelihood of encountering one another in the future, and are thus unlikely to compete with brood-mates for resources. In this book, the precise nature of the ***nursery*** will change vividly across taxa (a mammalian uterus with embryos, a den with cubs, a nestful of bird eggs, a puddle with toad egg mass, a pod with seeds, etc.), but the essential property is that the sibship remains intact for a time, and often exclusive of other sibships (Figure 1.2).

Once restricted to nursery systems, the problem of sibling rivalry can be seen as variations on the theme of mismatched supply (resource availability) and demand (family size). Whereas profligate reproduction in ***non-nursery*** taxa is basically a parental lottery (the more tickets, the better), nurseries are inherently finite. So why do ***nursery*** parents also overproduce? What benefits accrue to parents for creating more offspring than they typically support? There are at least three general categories of parental pay-offs for initial over-production in nursery taxa, even though the practice routinely requires a downward correction later and that adjustment may prove costly (e.g. Kozlowski and Stearns 1989, Forbes 1991a, b, c, Godfray *et al.* 1991, Konarzewski 1993, Mock and Forbes 1995). By creating extra progeny at the outset, parents may be (i) better able to capitalize on unpredictably favour-able ecological conditions, (ii) equipped with a backup offspring to stand in for one that is flawed (and/or dead), and (iii) fielding a team of propagules capable of assisting one another in various ways.

---

[1] This is probably too far-fetched. Sperm performance is determined by design features (mitochondrial packing, flagellum structure, etc.) assembled by the male's (that is, the paternal soma's) genome, not their own self-contained haploid genomes. Thus, a newly arising mutant can usually only affect the next generation's spermatogenesis (for an exception see Hecht *et al.* 1986), not its immediate gametic vehicle (see review by Sivinski 1984).

[2] Its parallel application in plant husbandry dates back to c. AD 1565.

Fig. 1.2 Diversity of 'nursery' types. **(a)** Each half-litter in the uterine horn of a pregnant mouse shares limited space and receives nutrients from a common source, the uterine artery. **(b)** The back of a male giant water bug serves as the female's oviposition site and its breadth may limit clutch size. **(c)** The renovated corpse of a small vertebrate accommodates the eggs and larvae of a pair of burying beetles, which attend the young closely. **(d)** Poison-dart frog parents carry newly hatched tadpoles on their backs, either temporarily as transport to a small water catchment or for longer periods. **(e)** Altricial avian nestlings are incapable of leaving the nest for many days, requiring all food to be delivered to them. **(f)** Many plants sequester multiple offspring (seeds) in common vessels, like this apple.

Although the first of these is most familiar as David Lack's '*brood reduction hypothesis*', we shall usually use its more general name, the **Resource-tracking Hypothesis** (Temme and Charnov 1987) to minimize certain ambiguities (Mock 1994). The idea is usually expressed as parents creating an 'optimistic' family size in order to take advantage of a good food year when all offspring can be supported properly (for refinements of this argument, see Williams 1966b, Martin 1987, Pijanowski 1992, Konarzewski 1993, Lamey and Lamey 1994).

The **Replacement Offspring Hypotheses** rest on there being a *core* subset of the brood (numbers matching what parents can normally support) plus some supernumerary or *marginal* young that are relatively expendable. In one version of this argument, danger to *individual* core brood members takes the form of fatal extrinsic forces (a grab-and-run predator, accidents, etc.), so the presence of a marginal sibling serves as **Insurance** against parents having to settle for an undersized brood (Dorward 1962). In an alternative version, the emphasis is placed on underlying variation in offspring 'quality' and parental over-production is viewed as a way of assembling an array of offspring from which only the most promising will receive full parental investment. This idea also travels under many different names, among which we favour '**Progeny-choice Hypothesis'** to accentuate its similarities with mate-choice aspects of sexual selection (Kozlowski and Stearns 1989; see also Darwin 1876, Buchholz 1922).

The possibility that marginal offspring may have some capacity for helping core siblings to survive and/or reproduce, the **Offspring Facilitation Hypo-thesis**, is most commonly suggested within the context of sibling cannibalism. There are, however, many other services that siblings can provide one another, the extremes being exemplified in eusocial insects (Williams and Williams 1957, Hamilton 1963). (Whether exploiting a sibling that was created solely for such a purpose counts as 'rivalry' *sensu stricto* offers a mildly interesting semantic puzzle that we shall duck in the interest of keeping the topic broad. One can usually retreat to the position that the victim sib would prefer to reverse the tables, to usurp the core brood role.)

Finally, two or more of these over-production dividends often contribute simultaneously (Figure 1.3). That is, by creating marginal offspring parents may be better prepared for environmental stochasticity (the 'good year'), ready for accidents involving the core brood, and also alert to the possibilities of weeding out an ontogenetic disappointment and judiciously upgrading that slot. If none of these events arise, the extra kid can be lunch for somebody.

With so many incentives for parents to aim high, and so much that can go wrong with resources during the nursery period, it is hardly surprising that diverse rivalries have evolved. In this book, we shall attempt to explore both the squeezes and their solutions.

1.3 Two broad contexts for sibling rivalry

To the nursery-inhabiting individual, a close genetic relative such as a sibling is evolutionarily both good and bad. On the plus side, it is a carrier of genes identical by common descent, a potential agent of indirect fitness. Alternatively, it might be useful as a food item, contributing to Self's direct fitness. (These two possibilities clearly *are* mutually exclusive.) On the down side, a sibling commonly depends on the same resource base as does Self, and thus often poses a significant threat.

Accordingly, we can dichotomize two broad families of sibling rivalry: *cannibalistic* vs. *resource-based*. We need not dwell overmuch on cannibalism in this book (although see Elgar and Crespi 1992), but when a sib is more valuable as food than as a gene bearer, it should be consumed. Such sibling

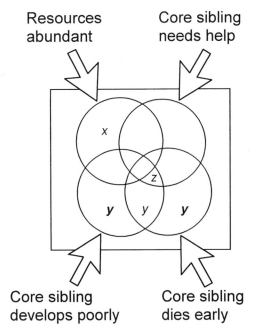

Fig. 1.3 Parents of many nursery taxa over-produce so as to gain potential fitness benefits from several mutually-compatible sources. Here we show four stochastic challenges (as overlapping circles) confronting a parent trying to set family size to an optimal integer value. By adding a 'marginal' offspring over and above the core brood it expects to raise, parents stand to gain one or more returns on investment. For example, a last-hatching egret chick has a non-zero chance of surviving along with all of its siblings, providing that its parents manage to hunt successfully (zone *x*). Even if parental deliveries fall short, that same marginal chick enjoys an improved probability of fledging should one of its elder nest-mates fail to survive or be stricken by disease (sum of all *y* zones). Lastly, in some brood-reducing birds the runt is consumed if it dies, in which case it may provide valuable nutrition (zone *z*). (From Mock and Forbes 1995.)

cannibalism should be most common (i) when alternative food levels are very low (and/or the acceleration in growth that Self can obtain from cannibalism is great), (ii) when ingestion of a sibling is a practical option (i.e. Self has the requisite anatomy and physiology to eat and digest its sib), (iii) when sibs are available, and (iv) when the victim's reproductive value is relatively low anyway. The degree to which selection should favour sib cannibalism depends, at least theoretically, on the impact such a trait has on both parties. If the intended victim has only a vanishingly small chance of surviving to breed in any case (as in many semelparous organisms), then its sib forfeits little inclusive fitness by eating it (Eickwort 1973, Milinski 1978, Charlesworth and Charnov 1980, Osawa 1992a). Sibling cannibalism is well documented for various newly pupated insects, anuran tadpoles and salamanders, and is probably generally common in taxa with clumped oviposition and relatively little post-zygotic care (see Elgar and Crespi 1992).

By contrast, *resource-based* sibling rivalry appears to be quite common in taxa such as birds, mammals, and plants that provide considerable parental care. Clutch size is often set initially at a high level and certain of the sibs' critical resources (e.g. food) are simultaneously non-shareable and limited by the bottleneck of parental delivery. What results is a classic lifeboat dilemma, with greater demand than guaranteed supply, plus some grave uncertainty about the likelihood of a timely deliverance. Resource-based rivalries also occur in animal taxa where dispersal is severely constrained by social or physical barriers (e.g. queen honeybees, co-operatively breeding birds and mammals, larval fig wasp brothers developing within a fig, etc.).

As with the over-production hypotheses, these two main sibling rivalry contexts are also not mutually exclusive. In principle, tight resource competition could be ameliorated by the expedient of cannibalism (i.e. if a sib must be sacrificed lest it consume too much food, its body tissues may be worth recycling). However, for reasons that are not well understood, such double-benefit patterns are far from universal in the studied cases of sibling rivalry. (For example victorious eaglets, and their parents, typically leave the dead victim to rot in the nest or drop the corpse over the rim.) Hence, most cases of sibling rivalry can be classified comfortably as either cannibalistic or resource-based.

1.4 Scope and biases

As mentioned above, it is our intention to keep this treatment of sibling rivalry as broad as possible. On the assumption that particular phenomena can always be jettisoned later, we wish to consider *any features of animals or plants that have the **effect** of promoting individual survival and/or reproduction at the expense of siblings (current and future)*. We acknowledge that many of these will not be sibling rivalry adaptations *sensu stricto* (see Williams 1966a), that they may well exist for various other reasons, but we want them all on the table, so to speak, for preliminary discussion. The point is that natural selection could have promoted (or contributed to) the evolution of them all.

The material for our treatment flows from three broad research traditions: descriptive natural history, theoretical modelling, and empirical hypothesis-testing (experimental and otherwise, field- and laboratory-based). Our goal is to weave these approaches around the central theme of sibling rivalry, with occasional digressions into ancillary topics.

Certain aspects of sibling rivalry have been treated theoretically far more than empirically. For example, sib–sib competition immediately abuts the issues of parental manipulation and parent–offspring conflict (e.g. Alexander

1974, Trivers 1974, Godfray 1995b). It is now clear that overt offspring selfishness *can* evolve even if it is detrimental to parental fitness (see Chapter 7). However, remarkably few satisfying data exist to show that such parent-offspring conflicts *have* evolved (see Chapter 11), suggesting either that the models are in dire trouble or that empirical progress has lagged far behind the theory (Mock and Forbes 1992).

The literature on real cases of sibling rivalry in the field has been scattered and anecdotal until quite recently. For example, early fatalities in nests of great birds of prey have been known for at least two millenia, but the dispersed and remote nature of eagle nests poses an obvious practical barrier to detailed study. The discovery that comparable siblicide also occurs in colonially nesting birds (particularly kittiwakes, pelicans, boobies, and herons) has helped bring many new facets to light. However, these new data represent only the beginning of exploration into sibling rivalry. It is our hope that a collation and discussion of the topic's current state, including some unabashed conjecture about where unstudied examples may reside, will help to direct attention to difficult or non-obvious areas where a reasonable investment of intensive study might pay substantial dividends.

We pay special attention to those cases among animals in which sibling rivalry is expressed through overt aggression (Table 1.1). Such systems not only offer vivid and easily quantified behavioural indications of selfishness, but may also shed light on some of the reasons for fighting *per se*. Perhaps if we can divine the ecological reasons for an individual's battling with its closest genetic relatives, we may better understand the depth of aggression in other contexts as well.

A different context in which sibling competition has been discussed at length concerns the evolution of sexual reproduction itself (e.g. Maynard Smith 1978, Young 1981, Bell 1982, Barton and Post 1986, Motro 1991). For example, it has been pointed out that the genetic similarity of very close kin is likely to make their ecological requirements more alike, and hence to intensify their rivalry over limited resources. When tested experimentally, this was shown not to be a very promising line of reasoning (Willson *et al.* 1987), so, in the interest of focusing on other matters, this usage lies beyond our purview.

1.5 Detecting sibling rivalry: logistics and semantics

If we step outside every morning at a given hour and record the status of a nestful of individually marked young songbirds in the hedge, we may notice one hatching later than the others and developing more slowly thereafter. If we go to the trouble of observing the brood in great detail (while suitably concealed, of course), it may become clear that the runt begs less effectively or tends to be ignored by the food-bearing parents. In addition, it may be physically intimidated by larger nest-mates. We document that its growth is

Table 1.1 A taxonomy and lexicon for sibling rivalry.

I. Resource-based sibling rivalry
  A. Lethal
    1. *Aggressive brood reduction*: fighting among siblings is used to establish a dominance hierarchy that renders the lowest ranking member(s) most likely to die
      a. **Obligate siblicide:** at least one sibling is virtually always killed (e.g. queen honeybees, certain predatory birds)
      b. **Facultative siblicide:** aggression is sometimes so severe that one or more siblings are killed (e.g. many predatory birds, fewer mammals)
    2. *Non-aggressive brood reduction*: the role of fighting, if any, is trivial compared to that of other fatal forms of competition. In some, the mortality may still be virtually guaranteed (e.g. pronghorns, polyembryony, crested penguins), but usually it is much less certain (e.g. songbirds, most mammalian suckling competition)
    3. *Filial infanticide*: brood reduction is effected by parents directly
  B. Non-lethal
    1. *Aggression-based*: relatively low levels of fighting are used to help skew resources to Self (e.g. piglets)
    2. *Scramble*: motor-skill competitions for limited parental resources (virutally all species with limited amounts of parental care, excluding those in the above categories)

II. Sibling cannibalism: consumption of potentially viable sibs (excluding unfertilized 'trophic' eggs or already dead sibs) for nutrients

---

slower; perhaps it becomes increasingly emaciated. Armed with such evidence, we could make reasoned causal inferences if that chick were eventually found dead or had vanished during the night. On the other hand, if we had made no observations other than our brief dawn records of 'present' or 'absent', such an inference after the death would require a greater leap of faith.

Considering the brevity of a census visit, it is unsurprising that most field studies of nestling birds are of the second variety. One chick ceases to exist and the familiar term 'brood reduction' seems ideally neutral and descriptive of that fact. Unfortunately, that term has now been used in a narrower sense for three decades (since Ricklefs 1965), and strongly connotes that the death was due to sibling competition *per se*. (The David Lack explanation was overwhelmingly persuasive and received nearly all of the attention for many years.) Accordingly, an attempt at this point to expand the definition of 'brood reduction' would probably confuse rather than clarify. Instead, it seems safer to retrofit 'brood reduction' with some operational criteria (as the quintessential model of resource-based sibling rivalry, it should be brood-size dependent and exacerbated by any factor that constricts supply), and then to adopt some benign and untainted term for those cases where family size diminishes because of unspecified or undetected causes. We have chosen *'partial brood loss'* for this nomenclatural niche (Mock 1994).

The term *siblicide* is used throughout this book as a particular subset of brood reduction, namely fatal sibling competition that includes *substantial overt aggression* (Mock *et al.* 1990). We prefer it to various alternatives because its meaning is obvious and unambiguous (it has not been used with other meanings), while also being gender-neutral (cf. 'fratricide' and 'sororicide', which add precision in cases where the victim's gender is non-random) and non-Biblical (cf. 'Cainism' or 'Cain-and-Abel battle'[3]). One can debate the precise meanings of *substantial*, *overt*, and *aggression*, but we leave those weighty matters to the future; this book describes many cases that we think qualify unambiguously. For the record, our case that egret brood reduction is often 'siblicidal' rests on two findings: (i) in many hundreds of hours of observation, one routinely sees chicks pecking each other repeatedly and with verve, causing losers to cower, bleed, flee (within the nest), avoid food and, occasionally, fall from the nest to die on the ground below; and (ii) from census visits at hundreds of otherwise unwatched nests, we find wounds on the smallest members of most broods that match the kinds of injuries suffered in our closely watched nests. Beyond that, it will be easier for us to show sceptics videotapes of the action than to define 'verve' scientifically.

1.6 Organization

We see two large bodies of theory as underlying intra-family conflict: that pertaining to sibling rivalry itself and that pertaining to parent-offspring conflict (Trivers 1974). We have organized most of the book around these two themes. Chapters 2–4 give a large dose of sibling rivalry theory (mainly as developed by O'Connor 1978 and Parker *et al.* 1989). This is followed by a more detailed introduction to avian brood reduction in general—and egret siblicide in particular—as an example of empirical field tests (Chapter 5), plus an extra chapter describing explorations into the ecological bases of egret aggression (Chapter 6). In Chapters 7–10, the fundamental and complex role played by parents takes centre stage, starting once again with a substantial review of the relevant theory (parent–offspring conflict), followed by a briefer treatment of the tepid empirical progress on that front (Chapter 11). These basic issues having been dealt with, five broad review chapters provide a cross-taxa sampling of sibling rivalry manifestations, starting with the relatively well-studied avian systems (Chapter 12), and then skimming more superficially across non-human mammals (Chapter 13), fishes, amphibia and reptiles (Chapter 14), invertebrates (Chapter 15) and plants (Chapter 16). Our goal in these reviews is to highlight the studies that have already been published, while prospecting for productive but unmined veins.

---

[3] This colourful appelation was apparently first applied to fatal sibling conflicts among golden eagles and continues to be popular in much of the raptor literature (e.g. Simmons 1988).

Summary

1. As with any form of ecological competition, the basic phenomenon of sibling rivalry centres on a mismatch in supply and demand, but here the participants are very close genetic relatives. The primary theoretical tool is therefore Hamilton's inclusive fitness concept, especially an inversion of Hamilton's rule for altruism.

2. Sibling rivalry can be characterized as a 'lifeboat dilemma' wherein shortages of key resources impinge critically on close genetic relatives, and hence on the indirect component of inclusive fitness. Because the shortages are often acute, the fitness consequences are sizeable. As a result, natural selection has shaped a great many phenotypic traits (especially of behaviour and life history) around these pressures.

3. In this book we focus on the nature of the limiting resources and on the proliferation of phenotypic responses to sibling rivalry across a broad array of life forms.

4. Most cases are considered within the context of parental over-production (the creation of more offspring than will be fully supported) and various constraints on post-zygotic parental care. The concept of **nursery** is introduced as a generic label for the places where dependent siblings are more or less entrapped spatially (usually by parental design), and thus where most sibling rivalry occurs. Diverse forms of nursery will be explored.

5. The parental over-production that sets most sibling rivalry dynamics in motion can usually be related to one or more of three general explanations. By creating supernumerary offspring, parents may accrue benefits through **resource-tracking**, possession of **replacement offspring** and/or various forms of **sibling facilitation** among the brood-mates.

6. In addition, sibling rivalry can be considered according to the classes of ecological competition displayed. Much of it is **resource-based**, hence subject mainly to *scramble* and *contest* (= *interference*) forms of competition. Alternatively, it may be primarily **cannibalistic**, hence representative of *exploitation* competition.

7. When referring to particular field studies where some but not all offspring die in the nursery, we adopt the broad non-committal term, 'partial brood loss'. Accordingly, we restrict our use of the term 'brood reduction' (and litter reduction, etc.) to cases where there is good reason to believe that the mortality was caused by sibling competition (or pre-emptive parental action). 'Siblicide' is reserved for the even narrower cases where overt aggression by nursery-mates is involved.

2

Theory I: Hamilton's rule and the evolutionary limits of selfishness

As a simple but admittedly crude model we may imagine a pair of genes g and G such that G tends to cause some kind of altruistic behaviour while g is null. Despite the principle of 'survival of the fittest' the ultimate criterion which determines whether G will spread is not whether the behaviour is to the benefit of the behaver but whether it is of benefit to the gene G; and this will be the case if the average net result of the behaviour is to add to the gene-pool a handful of genes containing G in higher concentration than does the gene-pool itself.

(Hamilton, 1963)

Relationships within families are complex mixtures of co-operation and selfishness; conflict abounds, but so do help and altruism. At one extreme a sib may kill its weaker nest-mate; at the other extreme an older offspring may devote its time, its energy, and even its life to the care of younger sibs. This richness and diversity of behaviour within the family seems inexplicable within a single general framework. Yet Hamilton's rule represents just that— a unified theory for family relationships. Here we attempt a simple introduction so that readers will not be distracted by the many rather complex issues surrounding it. The literature on kin selection is by now very extensive, and will be reviewed only briefly.

2.1 What is Hamilton's rule?

Hamilton's rule is a condition for the increase of an allele that has the phenotypic effect of helping a relative at a cost to Self. It is therefore a condition for altruism. Where the cost to Self's reproduction is c and the increment to the relative's reproduction is b, the allele will spread if

$$br > c \tag{2.1}$$

(Hamilton 1963, 1964a, b), where r is the probability that the relative shares the rare allele for the altruistic act.[1] The rule lies at the heart of his concept of *inclusive fitness*, which focuses attention on the social traits of an individual actor (here the altruist), and helps us see how the genes underlying the trait are able to increase in frequency. In short, an individual's total or 'inclusive' fitness is related to the sum of his/her personal fitness (e.g. proportional to reproductive success) plus a share of that achieved by kin, devalued by r. It should have been obvious from the beginning that Hamilton's approach can also be used to determine the conditions for increase in frequency of an allele for selfishness. Although our present interest is sibling selfishness, we shall follow tradition and give a simple explanation and derivation of the rule as it relates to an act of altruism.

To explain Hamilton's rule, we shall look at the specific case of sib altruism in a 'family structured' model.[2] This particular approach is chosen because it will be needed later (Chapters 7–10) for the models of parent–offspring conflict. Although our explanation is given in terms of single-locus genetics, readers must not get the impression that the main use of the rule is in population genetics. It is, in fact, a much more general paradigm for describing phenotypic evolution, with minimal genetic assumptions.

Imagine a diploid, sexual population in which a female produces a clutch of eggs after mating with one or more males. All the resulting offspring have the same mother, but may have different fathers; thus, some will be *full sibs* and others *half sibs*.[3] Alleles at a single locus control how an offspring behaves towards its sib. That locus is fixed initially (it exists at a frequency of 100%) for allele a. We consider whether a novel mutant allele A will spread. This allele causes its bearers to act in such a way as to *lose* reproductive prospects, while causing its sib to *gain* prospects of reproduction (i.e. altruistic behaviour).

Hamilton's rule can be viewed simply as a condition for the spread of the mutant allele A (although it is much more than this, and Hamilton was able to show that it applies whatever the frequency of allele A). For simplicity, however, we suppose that A is very rare initially, so we can ignore AA homozygotes (because they will hardly ever arise). Furthermore, we shall randomly tag one of the a alleles in the same population with an apostrophe, calling it a'. (This practice merely allows us to compare the future replication rate of our focal mutant, A, with that of an otherwise undistinguished non-mutant, a'.) Since both the mutant A and tagged a' alleles are exceedingly rare initially in this population fixed for aa, virtually all matings involving allele A will have

---

[1] We realize, of course, that little behaviour is controlled by just one gene, but use the convention of designating 'a gene *for* some phenotypic trait' as a convenient shorthand for 'a gene whose presence makes the difference between' developing that trait as opposed to some phenotypic alternative.

[2] That is, one in which the fate of a gene for altruism is examined within a given family composition.

[3] We use these terms to mean having two parents in common (full sib) vs. just one (half sib). In the botanical survey (Chapter 16) alternatives are discussed.

one *Aa* individual plus its *aa* partner (and hence will be *Aa* × *aa*); similarly, virtually all matings involving allele *a'* will be *a'a* × *aa*. (The possibility of an *Aa* × *a'a* mating is so slim that it can be ignored safely.) We imagine further that each female mates with *n* different males (where *n* is small); she then produces a large clutch of offspring, of which an average of 1/*n* were sired by each of her males.

How can we determine the fate of *A*? Will it increase or decrease? Allele *A* will spread if the number of copies of *A* in the grandchildren of our *Aa* mutant exceeds the number of copies of *a'* in the grandchildren of our *a'a* individual. That is, it spreads if it does better than a randomly chosen *a* allele[4].

We also assume for simplicity that there is a 1:1 sex ratio, and hence that the alleles *A* and *a'* have equal probabilities of occurring in males and females. Suppose that the mutant allele *A* occurs in a female, then each of her offspring has a probability of ½ of being *Aa*, and ½ of being *aa*. If *A* occurs in just one of the mother's mating partners (but *not* in her), then with random sperm-mixing the probability that her offspring is *Aa* is $1/2n$ (see footnote 2 below), and the probability that it is *aa* is $(1 - 1/2n)$. Similar probabilities apply for *a'a* and can also be calculated for *aa* progeny.

Imagine now that a unique opportunity arises every so often for one sib to help another. One sib can act as donor and the other as recipient of this action. If the potential donor is homozygous *aa*, it takes no action, and we shall assume that both donor and recipient go on with their lives and eventually produce one offspring each. In contrast, a sib bearing the *A* allele does perform the action, at some cost $-c$ in expected offspring. It therefore produces only $(1 - c)$ offspring, while its sib produces $(1 + b)$, where *b* is the increase in number of offspring due to receiving the help.

How often will altruistic acts occur? This depends on how often chance events generate the opportunities for altruism, and on the genotype of the potential donor (altruism occurs only if the donor carries *A*). Suppose that the opportunities for altruism arise with a fixed probability per sibship (= brood). Then if there is a probability *p* that altruism is shown in a sibship into which the **female** parent has transmitted the *A* allele, the probability that it is shown in a sibship in which the **male** parent has transmitted the *A* allele = p/n.

We can now marshall the accounts as the eight following points.

1. Probability that the altruistic event occurs per sibship, when the mutant allele is carried by the **female** parent = *p*.

---

[4] In a large population, the fate of the *a* alleles in the *Aa* × *aa* sibship is irrelevant; the ones in *this* sibship represent too small a fraction of the population's total *a* alleles to exert any significant effect on the average replication rate of *a*.

[5] To see this, focus on just one offspring. The chance that this individual was sired by the one male carrying the rare allele is 1/*n* and the chance that it received allele *A* (rather than *a*) from that male is ½, so the overall chance of having it is $(1/n)(1/2) = 1/2n$.

2. Number of sibships per parent in which the above ('point 1') occurs = 1.
3. Probability that the altruistic event occurs per sibship, when the mutant allele is carried by the **male** parent = p/n.
4. Number of sibships per parent in which 'point 3' occurs = n (because each male is assumed to mate with n different females).
5. Number of replicas of the A allele produced when the mutant allele is carried by the **female** parent = $p[(1 - c) + \frac{1}{2}(1+b)]$.
6. Number of replicas of the A allele produced when the mutant allele is carried by the **male** parent = $n(p/n)[(1 - c) + (1 + b)/2n]$.
7. Number of replicas of the a' allele produced when that labelled allele is carried by the **female** parent = $p(1 + \frac{1}{2})$.
8. Number of replicas of the a' allele produced when that labelled allele is carried by the **male** parent = $n(p/2)[1 + 1/2n]$.

Remember that Hamilton's rule is a condition for the increase of an allele for behaving altruistically towards a relative. Allele A spreads if there are more copies of A in the grandchildren of the Aa mutant than there are copies of a' in the grandchildren of the $a'a$ grandparent. Using the numbers to represent the eight points above, the condition for spread of A is:

$$(5 + 6) > (7 + 8),$$

which reduces to

$$\frac{b\left(\dfrac{1}{2} + \dfrac{1}{2n}\right)}{2} > c.$$

This conforms to Hamilton's rule (equation 2.1) if $r = (1/2 + 1/2n)/2$, where r is the probability that the recipient shares the allele A carried by the donor.

In the sibship to which the female parent transmitted the A allele, $r = \frac{1}{2}$ between sibs. In the n sibships to which the male parent transmitted the A allele, altruism occurs only $1/n$ times for every occasion in which the female parent carries the A allele. Sibship frequency and altruism frequency per sibship thus cancel. When altruism does occur, the probability that the recipient also carries the A allele is $1/2n$, hence the relative probability for r with 'male transmission' = $1/2n$. The overall probability that the recipient carries the A allele is therefore roughly $r = (1/2 + 1/2n)/2$, i.e. the mean of the effects via male and female parents. Our accounting system conforms to Hamilton's rule.

We have dwelt at some length on this example because it shows how careful one must be to calculate the exact value of r. Had we considered a simpler mating system in which each male and female mated only once per brood, all offspring would be full sibs and $r = (\frac{1}{2} + \frac{1}{2})/2 = \frac{1}{2}$.

Box 2.1   Formulations of Hamilton's rule

The above treatment of Hamilton's rule serves heuristic purposes only; serious derivations of the rule become mathematically complex, as indeed was the original analysis (Hamilton 1964a). A full review of the extensive theoretical literature on the topic is not appropriate here, although two important theoretical developments can be found in Grafen (1985) and Taylor (1988). The initial insight (Hamilton 1963, 1964a) used Wright's (1922) coefficient of relatedness (r), which is the probability that two individuals share a given gene by common descent. For our purposes, the two individuals are two players (donor and recipient) and the gene is the 'allele for' altruism. Various modifications to this have been suggested (e.g. Hamilton 1970, 1972, Charlesworth 1980, Michod and Hamilton 1980, Michod 1982), some of which aim specifically not to use an approach based on 'common descent (Orlove 1975, Orlove and Wood 1978). Hamilton (1975) used Price's (1970, 1972) covariance theorem for natural selection (see also Robertson 1966) to derive a form of r. Seger (1981) later used the same technique to produce an even more general coefficient. Various authors (e.g. Charnov 1977, Charlesworth 1980, Abugov and Michod 1981) derived the rule using a standard population genetics approach, and Charlesworth and Charnov (1980) extended the use of r to models of age-structured populations. Hamilton (1964a, 1970, 1972) showed that inbreeding does not alter the basic principle (see also Michod 1979).

Abugov and Michod (1981), Maynard Smith (1982a), and Grafen (1984) stressed that Hamilton's (1964a, b) approach and classical population genetics rely on two different methods of 'genetic bookkeeping' that nevertheless usually generate (exactly or approximately) equivalent conclusions. Standard population genetics relies on defining the fitness of individuals of given genotypes. According to this approach, we would first calculate the expected number of offspring produced by *Aa, aa,* etc. individuals, and would then use these genotype fitnesses to deduce the change in gene frequency of the allele *A*. The accounting consists essentially of summing the effects of all probabilities of social actions on an individual of each genotype, so as to predict its expected number of progeny. Much of Hamilton's (1964a) paper also related to establishing that his *inclusive fitness* technique gives the same solution as standard population genetics. Inclusive fitness is a measure of fitness that counts the effects of one individual's (the actor's) actions on its own and on other individuals' (the recipients') progeny. The numbers of copies of the *A* allele going through to the next generation are thus calculated as the sum of the changes to the actor's and recipients' progeny, with the effect on each recipient i being weighted by the appropriate r_i. The inclusive fitness approach is often much simpler to apply than the population genetics approach.

For a more detailed discussion of these calculations and some finer points, see Box 2.1. Additional background information can be obtained from two very readable comparisons of the population genetics vs. inclusive fitness approaches by Maynard Smith (1982b) and Grafen (1984). Dawkin's classic (1979) 'twelve misunderstandings of kin selection' is mandatory reading for all interested in the topic, and very clear explanations of inclusive fitness and Hamilton's rule are given by Grafen (1984, 1991a). For more advanced readers, we recommend Grafen (1985).

2.2 Selfishness between relatives

The most obvious way to convert Hamilton's rule into a rule for selfishness is to look at the condition under which the altruism allele A will **not** spread. From equation (2.1) this is when the inequality is simply reversed,

$$b_R r < c_A$$

where b_R is the benefit (subscript R designating 'to the recipient' for reasons that will become apparent soon) and c_A is the cost (to the actor). From Hamilton's original insight, we are to regard our non-mutant allele a (Hamilton's g in the quotation at the head of this chapter) as a 'null' gene. We shall return to the problem of what this may mean shortly (Section 2.4); suffice it to say that the condition for the spread of A defines an *'opportunity cost'*, that is, the pay-off necessary for performing one action instead of another,[6] measured in the currency of allelic replicas in the next generation. A positive opportunity cost $(b_R r - c_A > 0)$ means that the action pays in terms of inclusive fitness[7]; altruism is economically preferable to its null alternative and will be favoured by natural selection.

   However, if we are considering the spread of a selfish act between relatives, it makes sense to reverse b and c. The selfish actor now achieves a benefit $= b_A$ to its own prospects of reproduction, whilst the recipient sustains a cost $= c_R$. We now look at whether a rare allele S for selfishness will spread against its 'null' allele s (for doing nothing). For this we again require the opportunity costs of the action to be positive. The cost c_R to the recipient implies an expected loss of S alleles $= r c_R$, so the opportunity costs of the selfish action are positive for the actor if $(b_A - r c_R) > 0$, i.e. if

$$b_A > r c_R \tag{2.2}$$

R.M. Sibly (unpublished manuscript) has called expression (2.2) the 'selfishness rule' (although it is clear that the insight is due to Hamilton) in order to draw a distinction between this rule and its obverse for altruism. Both rules state that in order for a given action to spread, the effect on the actor must be weighed against the effect on the recipient diluted by the probability r that it shares the allele prescribing the action.

------

   [6] The idea of an 'opportunity cost' is crucial here. If an animal chooses to spend an hour courting members of the opposite sex, it cannot simultaneously be foraging under most typical circumstances. So the overall impact on its fitness must compare what it achieves, if anything, from the action it chose minus what it would have gained from the alternative open to it at the time.
   [7] It is slightly counterintuitive to think of a cost as something that 'pays', but the important thing is the net impact on fitness.

2.3 The mechanism behind Hamilton's rule

Much of the theoretical literature on kin selection is rather inaccessible to those with little mathematical training; it involves a detailed consideration of altruistic genes' mechanism of action (e.g. dominance, ploidy), the symmetry of gene frequencies in potential donors and recipients, and the exact selective effects on individuals (in terms of costs and benefits). In this section we give a very brief survey of some of these problems because they impinge on our later models of sib competition.

2.3.1 *The assumptions*

Grafen (1984) lists three assumptions essential to Hamilton's rule. We shall first look at these three, then consider a fourth, related complication: namely, the effects of random or genotypic allocation of the **roles** (actor vs. recipient) on the operation of the rule.

1. *Additivity of costs and benefits*

Several authors (e.g. Charlesworth 1978, 1980, Grafen 1984, 1991a, b, c, d) have pointed out that Hamilton's rule assumes fitness effects to be additive. In other words, the fitness of an individual that performs and receives one act of altruism is simply the sum of the costs and benefits, $1 - c + b$. This formulation of fitness seems perfectly plausible (see Maynard Smith 1982a). There are two cases where the additivity assumption is violated. If costs c include, say, a risk of death due to performing the altruistic action, it is more appropriate to assume (Charlesworth 1978) that the effects on this individual's fitness are multiplicative,[8] hence $(1 - c)(1 + b) = 1 - c + b - bc$. When fitness effects b, c, are small in magnitude (relative to '1' which really represents lifetime reproductive output; Grafen 1984), the two models of fitness become approximately equal because the second-order term, bc, in the multiplicative formulation approaches zero. (That is, if both b and c are very small relative to 1, their product approaches zero.)

A second problem with the additivity assumption occurs if, say, the benefits of each act of altruism decline with each successive act received (Seger 1981).

---

[8] The distinction between additive and multiplicative fitness consequences may be more apparent if we use some actual units. Running a mile for a candy bar might be reasonably modelled as an additive net effect because we might render both the costs (for running) and gains (ingested sugars) in calories, and determine which is greater. A multiplicative model is more appropriate in cases where one thing *must* be achieved in order for another to occur. For example, one *must* survive to reproductive age (say with probability x) in order to have y offspring, so the number of expected offspring would be xy. In the above example, a decrement to survival, c, is sliced off an otherwise unmarred expectation of survival (1.0) to produce a revised estimate, $1 - c$, and this is multiplied by the enhancement to that individual's baseline performance $(1 + b)$ to give the product.

If each individual cannot expect to receive or deliver more than one act of altruism in its lifetime, then there are no pay-offs from second, third, etc., acts to combine, so the additivity assumption is not a problem. For instance, sometimes altruism (or selfishness) only ever 'flows' in one direction within a given dyad, which simplifies the model. Such circumstances can arise if we are constructing models where the roles of actor and receiver are determined randomly with respect to genotype, but depend on some phenoytpic asymmetry. An example of this is the age determination of roles in a brood of two siblicidal birds: hatching order results from parental actions and is presumably arbitrary with respect to genotype, yet the first-hatched chick executes its nest-mate. Once carried out, the action cannot be repeated.

2. Weak selection

Now imagine a case where repeated acts of altruism are possible between pairs of relatives.[9] Because selection depends on the eventual summed effects of b and c, then if b and c affect survival, the relative proportions of the different genotypes change. For instance, if altruism involves a high probability of death by the altruist, the frequency of A-bearing genotypes gradually reduces by selection: in a sibship, they will no longer be predicted by Mendelian probabilities. Hence where r could be calculated from a knowledge of the relatedness between individuals *before selection occurs*, the subsequent changes in genotype frequencies by selection will cause the actual value of r (= the extent of genetic similarity at the A locus) to change. The difference between the 'ancestry-calculated' value for r and the 'true' value of r is significant only if the effects of selection are relatively strong. Grafen (1984) claims that if the opportunity cost $rb - c$ is relatively large in magnitude when r is calculated from a family tree, then the deviation from the true value of r will not affect the sign of $rb - c$. (It will not change a positive opportunity cost into a negative one; it will therefore not alter the decision that the animal will be selected to make, though it may affect the strength of selection for or against a given action. Alternatively, if $rb - c$ is small in magnitude, selection is weak and ancestry is a fair guide to the true r.) In practice, violation of the assumption of weak selection is unlikely to pose serious problems often.

3. Recipients are drawn randomly from the set of kin in which donors arise

If the first two assumptions seem to restrict the application of Hamilton's rule rather little, here we must be much more cautious. It is not so much that violation of assumption 3 causes *invalidation* of the rule, but rather that it

---

[9] Here we might envision two mammalian neonates suckling side by side. One individual may consume somewhat less than its share (½) of the limited current supply of milk, thereby allowing its sib to have a bit more. Their roles clearly *can* change freely, from meal to meal.

causes *errors in application* of the rule. An alternative way to state assumption 3 is that the *A* allele's frequency among *potential* donors and *potential* recipients must be the same (e.g. Seger 1981, Grafen 1984).

The problem once again concerns *r*, and is best illustrated by an example. Imagine that a dominant mutant gene *A* for altruism causes *all* of its bearers (*Aa*) to stop reproducing in order to help sibs, then all recipients would necessarily be *aa* . Such a gene could never spread, however great the benefits to the recipients, because an *Aa* never helps another *Aa* genotype. Although the ancestry-calculated *r* would be ½ (for sexually reproducing diploids), the true value for *r* = 0. Clearly, then, if we rely entirely on the ancestry method to calculate *r*, this will not equal the true *r* if altruistic (or selfish) actions are not directed randomly towards the set of relatives having the ancestry-calculated *r*.

Charlesworth (1978) pointed out that this sort of problem could have applied in the evolution of sterile castes in insects, and showed that it could be alleviated if the *A* allele has very low penetrance (a very low probability of being expressed). With full penetrance, *Aa* sibs help only *aa* sibs, but with very low penetrance, only a few of the *Aa* sibs are 'switched on'. The chances that one of the few 'switched on' *Aa* sibs helps another *Aa* sib become virtually equal to the proportion of *Aa* genotypes in the sibship. At this point the 'ancestry' *r* and the 'true' *r* also become virtually equal.

2.4 Hamilton's rule and conditionality

The implications of the last assumption must be considered further. Suppose, in a given interaction between two relatives (a dyad) we ascribe two potential **roles** to the two individuals. These roles, which may be called X and Y, may be any asymmetry between the pair of relatives, such as older/younger, stronger/weaker, etc. Let us assume that the potential roles are assigned purely randomly—in particular, they have nothing to do with an individual's genotype. Hamilton's rule applies strictly when an action is *conditional* upon the role, X or Y, that an individual currently occupies (Parker 1989). Thus consider an allele for an altruistic (or selfish) action. This gene must be expressed in one role only: its expression is conditional upon the individual's role, X or Y. On the other hand, if the gene's action is *unconditional* it *can be expressed in both roles* (by definition) and Hamilton's rule does not strictly apply. Essentially the same point was made by Seger (1981), who defined roles rather differently, in terms of whether an individual has the *potential* to be an actor or a recipient in an interaction. For Hamilton's rule to apply in its simplest form, potential actors and potential recipients must both have the same gene frequency for the allele (for altruism) under scrutiny. These gene frequencies will obviously be the same if the roles of potential actor and potential recipient are ascribed randomly with respect to genotype, and if the action which occurs when the potential actor carries the altruism gene occurs in only one role (that of the potential actor).

This principle turns out to be an important one because, depending on the mechanism of the gene action (conditional or unconditional), we may obtain different evolutionary results. Because it has a bearing on most of the models throughout the book, we give a detailed survey of how the conditionality effect arises, and what its implications are. Our treatment follows closely that of Parker (1989), and we begin with a couple of examples.

2.4.1 *A sib-selfishness model*

In this model there is a rare dominant allele S (selfish) in a population fixed for s. We examine whether S can spread; the relevant mating is therefore mutant parent Ss × wild-type parent ss. The progeny produced are Ss and ss, with equal probability.

Now consider the selfish action prescribed by S. Assume that a fixed amount of resource is provided by the parent(s). Offspring carrying S attempt to take $(1 + m)$ units of the resources, while ss offspring just take 1 unit each. The fitness f (= some measure of survival or reproductive success) of an individual offspring increases with the amount of resource gained, but with diminishing returns, as shown in Figure 2.1.

Suppose that there are just two full sibs in the sibship (this assumption is for simplicity only, it can be relaxed if we imagine that the interaction is between random dyads in a larger sibship). There are two phenotypic roles, X and Y, that correspond to an arbitrary asymmetry such as 'first to hatch,' 'second to hatch,' etc. We consider the effect of a gene when it is expressed in just one role, role X (conditional), or both roles, X and Y (unconditional).

There are four equally probable types of sibship (combinations of genotypes Ss and ss with roles X and Y), with specifiable offspring fitnesses (Table 2.1). For instance, consider sibship 4, where an Ss genotype occurs in both role

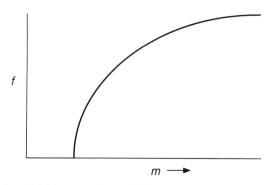

Fig. 2.1 Relationship between fitness f (prospects for survival and reproductive success) of an offspring and m, the amount of parental resources it receives.

Table 2.1. The sib competition model. There are four equiprobable sibships. Fitnesses of S-bearing sibs are displayed in bold type. In sibship 4, the fitnesses can be either $f(1)$ or $f(1 - c)$, depending on whether there is a cost (see text). (After Parker 1989.)

Genotypes		Fitnesses				
		Conditional		Unconditional		
	X	Y	X	Y	X	Y
1	ss	ss	$f(1)$	$f(1)$	$f(1)$	$f(1)$
2	Ss	ss	$\boldsymbol{f(1 + m)}$	$f(1 - m)$	$\boldsymbol{f(1 + m)}$	$f(1 - m)$
3	ss	Ss	$f(1)$	$\boldsymbol{f(1)}$	$f(1 - m)$	$\boldsymbol{f(1 + m)}$
4	Ss	Ss	$\boldsymbol{f(1 + m)}$	$\boldsymbol{f(1 - m)}$	$\boldsymbol{f(1)}$ or $\boldsymbol{f(1 - c)}$	$\boldsymbol{f(1)}$ or $\boldsymbol{f(1 - c)}$

X and role Y. When gene action is *conditional* (expressed only in role X) then fitnesses are $f(1 + m)$ to X and $f(1 - m)$ to Y. In the *unconditional* case, gene action is expressed equally in both sibs. Both attempt to take $(1 + m)$ units, but the resource is fixed and so each can only obtain 1 unit. Fitnesses are either $f(1)$, or perhaps $f(1 - c)$ where c is an energetic cost (equivalent to some loss of resources) due to the selfish interaction between the two sibs.

Under what conditions will S spread in an ss population? We again use the 'labelled allele' approach from Section 2.1: a wild-type s allele is arbitrarily labelled as s'. In a large population where S is a very rare mutant, s' will occur in an ss' parent, which will mate with an ss parent. All ss' progeny will attain a wild-type fitness of $f(1)$. S will spread if an Ss parent achieves more copies of S in its grandchildren than the number of copies of s' achieved by an ss' parent. From Table 2.1, we can see that for *conditional* gene action S can spread if

$$f(1 + m) + f(1) + f(1 + m) + f(1 - m) > 4 \cdot f(1),$$

The most useful rearrangement shows the dominant sib's perspective, namely

$$2[f(1 + m) - f(1)] > [f(1) - f(1 - m)],$$

which can be translated as

$$2[\text{benefit to Self}] > [\text{cost to sib}].$$

This is exactly rule (2.2), and so there is no violation of Hamilton with conditional action of the S allele. Note that if m is large, the S allele may not spread because of the diminishing returns relationship shown in Figure 2.1. We can calculate the optimal or ESS value of m (called m^*) for a selfish sib to take by finding the maximum value of this inequality. Mathematically, the maximum is found by setting the first derivative of the fitness function equal to zero, and then solving for that value of m (which is m^*). The second derivative must be negative for this to be a maximum. So

$$\frac{d}{dm} \{2[f(1 + m) - f(1)] - [f(1) - f(1 - m)]\} = 0,$$

which gives

$$\frac{d}{dm}[f(1 + m^*)] = \frac{1}{2}\frac{d}{dm}[f(1 - m^*)].$$

This selfish optimum is shown graphically in Figure 2.2. Stated verbally, individual X should continue to take extra resource up to m^*, at which point its marginal fitness gains from the resource become exactly half those available to sibling Y from consuming that same resource. This is again strictly Hamiltonian (Hamilton 1964b, Parker *et al.* 1989). We shall elaborate on this theme later to calculate an ESS m^* for how selfish stronger sibs should be to their weaker sibs (see Section 3.3).

However, if gene action is *unconditional*, we obtain very different answers. First, suppose there are no costs of two selfish Ss sibs' interactions. Applying the same analysis, the condition for spread of S is

$$f(1 + m) + f(1 + m) + f(1) + f(1) > 4 \cdot f(1)$$

if fitnesses in sibship 4 from Table 2.1 are $f(1)$. S will spread if

$$f(1 + m) > f(1),$$

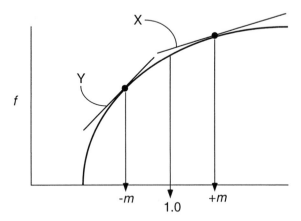

Fig. 2.2 Optimal amount of resource to take in role X in the conditional model of selfishness outlined in Table 2.1. Sibs that behave identically would receive one unit each, giving a fitness of $f(1)$. The optimal amount of resources to take in role X is $(1 + m^*)$ where m^* is such that the slope of line X is half the slope of line Y (see text). (After Parker 1989.)

which is always true if fitness increases with resources gained. Thus, whatever amount m is taken, S will spread. In fact, it is easy to show that the optimal amount of resources to take, m^*, has

$$\frac{d}{dm}[f(1 + m^*)] = 0.$$

In Figure 2.1 the slope of $f(m)$ becomes 0 only as m approaches infinity; in this case there can be no optimum, it clearly pays to take everything and to kill one's sib. m^* can be constrained only if the $f(m)$ relationship reaches a maximum—in this case the best thing to do is to take the amount of m^* that yields maximum personal fitness.

Why does the unconditional case obviously not obey Hamilton's rule? Note that in both cases where selfishness can occur (sibships 2 and 3, Table 2.1), the harmed sib is *not* a carrier of the S gene. From the point of view of the S allele, there is a relatedness of $r = 0$ and there is a violation of assumption 3 in the preceding section. In sibship 4 (rather bizarrely) both X and Y attempt to express the selfish action, but no fitness effects are felt because both sibs are equally prevented from achieving their 'aims'. The effects are therefore neutral relative to the null gene s. However, it makes little difference if we assume that there is an energetic cost to such an interaction (Table 2.1), so that X and Y in sibship 4 each gain $f(1 - c)$. This results in the following condition for spread:

$$[f(1 + m) - f(1)] > [f(1) - f(1 - c)],$$

which can be translated as

[benefit to Self when sib is ss] > [cost to Self when sib is Ss].

Stated this way, the spread of S is clearly unconstrained by kin selection; it is constrained only by personal costs. An optimal strategy, in which c becomes a function of m, would simply maximize the difference between personal benefits and personal costs (see Parker 1989), although an ESS version (see Macnair and Parker 1978) that behaves in Hamiltonian fashion can be constructed (H.C.J. Godfray, personal communication). Is this apparently non-Hamiltonian effect a result of the specific model, or is it genuinely the result of the unconditional gene expression? Before answering this question more fully, we shall consider another example.

2.4.2 A sib-altruism example

Consider the fate of a rare dominant mutant allele A (for altruism between full sibs) arising *de novo* in a wild-type aa population. Again (for simplicity) there are two sibs and these occupy roles X and Y, exactly as before. An A-bearing genotype behaves altruistically towards its sib, at a cost $-c$ to itself

Table 2.2. The sib competition model. There are four equiprobable sibships. Fitnesses of S-bearing sibs are displayed in bold type. In sibship 4, the fitnesses can be either $f(1)$ or $f(1 - c)$, depending on whether there is a cost (see text). (After Parker 1989.)

Genotypes		Fitnesses				
		Conditional		Unconditional		
X	Y	X	Y	X		Y
1 aa	aa	1	1	1		1
2 Aa	aa	**1 − c**	1 + b	**1 − c**		1 + b
3 aa	Aa	1	1	1 + b		**1 − c**
4 Aa	Aa	**1 − c**	1 + b	**(1 − c + b)** (1)		**(1 − c + b)** (1)
				(1 − c)/2 + (1 + b)/2 (2)		**(1 − c)/2 + (1 + b)/2** (2)
				1 (3)		1 (3)

and benefiting the recipient by $+b$. When neither sib performs any action, the fitness of each remains at unity. Once again, we shall examine the conditions for spread of the mutant A when its action is expressed in role X only (conditional) compared to when it is expressed in both roles X and Y (unconditional). The analysis is the same as before, and the fitnesses in relation to genotypes and roles are shown in Table 2.2.

When the action of A is conditional to role X, the A allele will spread if Hamilton's rule is met directly, i.e. if

$$\tfrac{1}{2} b > c,$$

$\tfrac{1}{2}$ [benefit to sib] > [cost to Self].

It is easy to calculate the optimal *amount* of altruism for X to show to Y if we allow benefits b to increase continuously with costs c (Figure 2.3). When b is a function of c, the optimal costs for X to pay, c^*, are such that $db(c^*)/dc = 2$, i.e. altruism should be increased until Y's marginal gains are reduced to twice the marginal costs of X. All is strictly Hamiltonian.

Now consider unconditional gene action. As in the selfishness model (Table 2.1), it is unclear what will happen in sibship 4 when both sibs now attempt to behave altruistically. We shall examine three possibilities (see Table 2.2):

1. Fitnesses are $(1 − c + b)$.

This assumes that two independent actions can flow between X and Y, and that fitness is *strictly additive* (each sib loses $-c$ and gains $+b$). There must be no interference between the two independent actions, each must be able to proceed in exactly the same manner and with the same effect as if only one

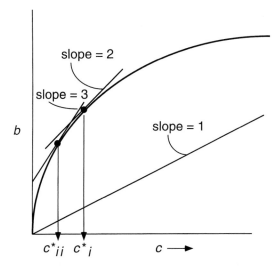

Fig. 2.3 Benefits b to Y, as a continuous function of the cost c to the donor X. In the conditional model, the optimal solution has c^*_i such that the slope of $b(c^*) = 2$; in the unconditional model the optimal slope can have c^*_{ii} such that the slope of $b(c^*) = 3$ (see text). (After Parker 1989.)

action had occurred. In a strict mathematical sense, it does not matter that the two actions proceed in opposite directions (X to Y, and Y to X); the results are equivalent if one sib gains a fitness of $(1 + 2b)$ and the other a fitness of $(1 - 2c)$, although two opposite actions are conceived in the model. Biologically, we would have to envisage a case where, in a given dyadic interaction, X and Y can either simultaneously give and receive altruism without restriction, or else the altruistic actions are exactly reciprocated, with one sib acting first as donor and then as recipient.

For unrestricted, reciprocal flow of altruism, the condition for spread of A is exactly Hamilton's:

$$b > 2c,$$

$$[\text{benefit to sib}] > 2[\text{cost to Self}],$$

and the optimal costs are the same as before with

$$db(c^*)/dc = 2;$$

(see Figure 2.3).

2. Fitnesses are $(1 - c)/2 + (1 + b)/2$
The most obvious possibility here is that each sib acts for half the time as donor, and for the other half as recipient, thus bisecting both the costs

sustained and the benefits received by each. Alternatively, just one complete altruistic action could flow between X and Y, with random probability of occurrence in either direction, so that one sib gains $+b$ and the other loses $-c$.

With this fitness in sibship 4, the condition for spread of A is

$$b > 3c,$$

[benefit to sib] > 3[cost to Self],

which clearly permits altruism, but is a more onerous condition to satisfy than Hamilton's. If benefits can be varied by continually increasing costs, the optimal cost, c^*, now has

$$\frac{db(c^*)}{dc} = 3;$$

which implies a reduced optimum for c^* (see Figure 2.3). The strict additivity of the effects of actions present in (1) have been destroyed.

3. Fitnesses are 1

This is the parallel of the selfish model examined in the previous section and Table 2.1. The assumption is that, because both sibs are actively attempting to give altruism to the other, neither is able to receive it. The spread of A is impossible since it requires that costs be

$$c < 0.$$

This result is unsurprising: there is neutral fitness in sibship 4, and in sibships 2 and 3 the altruism is 'wasted' because it flows to non-carriers of the A allele. Again, the strict additivity of effects has been lost. It is evident that the sib-selfishness example in Section 2.4.1 can also be interpreted in terms of destruction of additivity, brought about by the non-random allocation of effects and actions with respect to the receiver's genotype.

In summary, Hamilton's rule applies for the unconditional case only when two Aa sibs can give and receive altruism in such a way that the total fitness effects are exactly double those for a single Aa sib interacting with an aa sib. The most plausible way for this to occur biologically is for there to be strict independence of the altruistic events, so that there is a symmetrical pair of conditional events, X to Y and Y to X. In reality this particular form of un-conditionality converges towards conditionality. This way, strict additivity of costs and benefits is preserved. If there is some reduction in the fitness effects when two sibs act as altruists simultaneously, the additivity is lost and con-ditions for the evolution of altruism become more restrictive than Hamilton's. In contrast, with conditional (or role-restricted) gene action, the additivity is preserved and the conditions for simple cases concur exactly with Hamilton's rule. In many discrete games, the additivity can be lost for various reasons, as we have seen. Where we seek an ESS for continuous games, additivity automatically occurs near the ESS and thus Hamilton's rule usually applies, so long as marginal costs and benefits are used, as we shall see in later chapters. It is possible to construct more general versions of Hamilton's rule (Box 2.2).

Box 2.2 Generalized versions of Hamilton's rule

The original Hamilton's rule was shown to apply whatever the gene frequency of the allele A (Hamilton 1964a). Note that the equations in the previous sections have been expressed as conditions for spread of a rare allele A against its alternative allele a. It is easy to derive general versions (conditional and unconditional) of these spread conditions (Parker 1989). These can be used as stability conditions for kin-selected optima, since stability requires non-invasibility by rare alternative strategies.

Table 2.3 gives a generalized version of the model of the previous section. Fitnesses are given as costs and benefits in a Hamiltonian fashion. (An alternative version specifies fitnesses quite generally, but is more complex: see Parker 1989).

In each case, the fitness of a given individual is stated as a function of its role (X or Y) and of the genotype of the sib in the alternative role (Aa or aa). The convention adopted is to use upper case for costs and benefits when gene action is conditional, and lower case when gene action is unconditional. To obtain a feel for the notation, $C(X_{Aa}, Y_{Aa})$ is the cost sustained by an Aa sib in role X when it gives altruism to an Aa sib in role Y—which must behave exactly as would an aa sib, since gene action is conditional. Similarly, $b(Y_{Aa}, X_{Aa})$ is the benefit to an Aa sib in role Y that is behaving unconditionally, when its Aa sib is also behaving unconditionally in role X. From Table 2.3, Hamilton's rule becomes, for conditional action:

$$B(Y_{Aa}, X_{Aa}) > C(X_{Aa}, Y_{aa}) + C(X_{Aa}, Y_{Aa}), \qquad (2.3)$$

and for unconditional action:

$$b(X_{Aa}, Y_{Aa}) + b(Y_{Aa}, X_{Aa}) > c(X_{Aa}, Y_{aa}) + c(Y_{Aa}, X_{aa}) + \\ c(X_{Aa}, Y_{Aa}) + c(Y_{Aa}, X_{Aa}) \qquad (2.4)$$

Note that we are only ever concerned with costs and benefits to an Aa individual, since aa genotypes are irrelevant to the success of the A allele. Hamilton's rule is usually seen as being equivalent to the conditional form (equation 2.3), with costs to the donor being independent of the genotype of the recipient (i.e. $C(X_{Aa}, Y_{Aa}) = C(X_{Aa}, Y_{aa})$. Equation 2.4 seems less obviously Hamiltonian, but it becomes equal to equation 2.3 if:

$$b(Y_{Aa}, X_{Aa}) = b(X_{Aa}, Y_{Aa}) = B(Y_{Aa}, X_{Aa}); \qquad (2.5a)$$
$$c(X_{Aa}, Y_{Aa}) = c(Y_{Aa}, X_{Aa}) = C(X_{Aa}, Y_{Aa}); \qquad (2.5b)$$
$$c(X_{Aa}, Y_{aa}) = c(Y_{Aa}, X_{aa}) = C(X_{Aa}, Y_{aa}). \qquad (2.5c)$$

Here, the equivalence between equations (2.4) and (2.3) is achieved by a perfect doubling in the occurrence of the altruistic action, and by equivalence in fitness effects across X and Y. This is, in effect, strict additivity. There are other less obvious ways in which equation (2.4) can become strictly Hamiltonian, if additivity is restored by some balancing of the different effects.

Table 2.3. A general version of the altruism model. Fitnesses of A-bearing sibs are displayed in bold type. Costs or benefits may be a function of Self's role (X,Y), and of the genotype of the sib in the other role (see text). (After Parker 1989.)

Genotypes		Fitnesses			
		Conditional		Unconditional	
X	Y	X	Y	X	Y
1	aa	aa	1	1	1	1
2	Aa	aa	$\mathbf{1 - C(X_{Aa}, Y_{aa})}$	$\mathbf{1 + B(Y_{aa}, X_{aa})}$	$\mathbf{1 - c(X_{Aa}, Y_{aa})}$	$1 + b(Y_{aa}, X_{Aa})$
3	aa	Aa	1	$\mathbf{1}$	$1 + b(X_{aa}, Y_{Aa})$	$\mathbf{1 - c(Y_{Aa}, X_{aa})}$
4	Aa	Aa	$\mathbf{1 - C(X_{Aa}, Y_{Aa})}$	$\mathbf{1 + B(Y_{Aa}, X_{Aa})}$	$\mathbf{1 - c(X_{Aa}, Y_{Aa}) + b(X_{Aa}, Y_{Aa})}$	$\mathbf{1 - c(Y_{Aa}, X_{Aa}) + b(Y_{Aa}, X_{Aa})}$

2.4.3. *Two types of conditionality*

Finally, it is useful to draw a distinction between *enforced* conditionality and *evolved* conditionality. Enforced conditionality may be most easily explained through an example: if altruism is of the type where an onlooker saves a drowning person, the conditionality is **enforced**. Roles (onlooker, drowner) are set inexorably by force of circumstances, and it is impossible for the flow of action to proceed in the opposite direction (someone who is drowning cannot save an onlooker). Hence the gene action relates to the decision as to *what to do **when an opportunity arises*** to save a drowning person. On the other hand, if an action concerns, say, grooming or food-sharing between two siblings, the restriction of action to one particular role, or its magnitude in a given role, must evolve. It is likely to change gradually from an initial state of un-conditionality. If pay-offs do not differ between roles (equations 2.5a–2.5c above), there is no problem: simple additivity prevails. But if roles relate to, say, 'strong sib' and 'weak sib,' pay-off magnitudes are very likely to differ between roles. It will then be perfectly possible for an unconditional gene to have favourable effects if expressed in one role, but unfavourable effects when expressed in the other role (see Parker 1989 for a numerical example). If the summed effect satisfies equation (2.4), then we would expect the gene to spread, and later for modifiers to cause its action to be restricted only to the favourable role. However, if the overall effect is to violate equation (2.4), then in order to spread at all, such a gene must have some degree of conditionality right from the start, e.g. it must have a higher probability of expression in the role in which it is favourable.

2.5 Differences in prior states and loci—alternative actions

Suppose that there exists an array of possible altruistic actions A_1, A_2, A_3, etc., and an array of possible selfish actions S_1, S_2, S_3, etc. How can we predict the outcome of selection?

If actions A_1, A_2, A_3, etc. are the result of corresponding alternative alleles, A_1, A_2, A_3, at the same locus, Hamilton's rule defines which allele will increase in frequency against *a given alternative allele*. We need then to consider which allele (and hence what type of action) was fixed previously. The system's prior state can have a radical effect; it defines the standard against which we consider the fate of A_1. For instance, it is perfectly possible that allele A_1 can invade against A_2, but not against A_3.

In addition, because real genes do not operate in isolation, we must also distinguish between influential actions prescribed at *other loci* and those prescribed by alleles at the *same* locus. Suppose that we have loci *A, B, C, D, .. S*. A mutation occurs at locus *A*. For our purposes, all other loci are irrelevant *except in so far as they affect* the relative fitness of mutants bearing allele A *at the A locus.* We have been discussing altruistic (locus *A*) vs. selfish (locus *S*) genes, which could act to impel individual behaviour in opposite directions. Continuing on this theme, consider interactions between two nestling birds (full sibs). The *S*-locus is fixed as SS for acting selfishly: it causes the dominant sib to take a large share of the food resources from its weaker sib. However, over at the *A*-locus, aa is fixed. A mutant phenotype Aa arises that acts to assist its endangered nest-mate in some particular way. For example, A may cause its bearer to assist when the sib is about to topple out of the nest soon after hatching, or to curtail its sib-directed aggression when the victim appears to be nearing death (under such circumstances aa does nothing). Whether A will spread depends directly on Hamilton's rule as applied for the assisting action performed by Aa genotypes, *given* the current genetic background (i.e. that S is fixed). In short, the two loci (*S* and *A*) can be in conflict, depending on the alleles present[10].

Now suppose that there is an interaction between the two loci as follows. S* prescribes how much food to take from the total input (as in the dominance model on p. 43), and has reached its optimal level *given* that a is fixed. At this level, and assuming that falling out of nests or suffering fatal sib-inflicted injury is a rare event, it would not pay the dominant nestling to reduce its sib's fitness any further, even though this would mean extra food for itself. Hence allele A would spread; it must obey Hamilton's rule because we know that the threshold for the rule will be prescribed at the optimal level for S*, at which point the dominant chick's marginal benefits will be exactly half those of its sib. Any further food gain by the dominant chick at the expense of its weaker sib would be suboptimal. Thus we might expect to see the dominant chick saving its weaker sib from falling out of the nest, or curbing its attacks, yet later allowing it only minimal access to food brought in by the parents.

However, suppose now that there is no fixed dominance relationship between the two sibs. Each will now take half the available food. Call this state *S* (at the S locus). Whether allele A (for altruistic action) will spread now depends on a quite different condition; if SS is fixed then A may or may not spread depending on the form of the relationship between food gained and offspring success. Under a wide set of conditions, A will not spread. Thus in a population where stronger sibs vie to share food and cannot dominate their weaker bretheren, they may not attempt to save them from falling from the nest, even though they have the opportunity to do so.

The point is that although locus S and locus A control quite different actions, both affect an offspring's success through the effect each has on the probable amount of food gained during the nestling stage. The outcome of selection for or against the altruistic act depends on the prior state of alleles at the S locus—i.e. on whether one sib controls the other's access to food. The

---

[10] The *Ss-Aa* combination generates conflict, but the *Ss-aa, ss-aa,* and *ss-Aa* combinations do not.

history of mutations at the different interacting loci determines the outcome of selection, a feature common to many evolutionary studies.

To what extent will it matter if the alternative actions are prescribed at different loci rather than as alleles at the same locus? Our preceding explanations of kin selection and Hamilton's rule have been quite explicitly about competition between alleles at a given locus. However, we generally will not know whether the alternative actions relate to different alleles at the same locus, or to genes at different loci. Does this matter? If the answer is 'yes,' the implication is that the result we obtain from a hypothetical model may depend critically on the assumptions we make about the genetic mechanism. The answer is indeed 'yes'; we need, therefore, to be cautious. However, Hamilton's rule can actually be justified quite generally in a wide class of models, beyond single-locus genetics. A distinction can often be made between 'gene spread' models involving discrete strategies, which do often depend critically on assumptions about genetics, and ESS models where the assumptions are much weaker. The latter can be solved with a 'marginal Hamilton's rule' (Godfray and Parker 1991). This is simply a method of applying Hamilton's rule to continuously variable strategies, so as to derive an optimum or ESS. Like the direct version of Hamilton (equation 2.1), it is essentially phenotypic and non-genetic. Its general validity has yet to be proven mathematically, but so far it has given the same result as previous equivalent models in which the genetics has been explicit (Parker 1985). Assuming that it has the required generality, this version of the rule may prove very useful in future optimality studies of conflict between relatives. It is explained in relation to a particular model of sib competition in Section 3.3.

In later chapters we shall consider models of parent–offspring conflict. As modelled, this becomes equivalent to a form of conflict between genes at two loci. The different actions of the parent, and the different actions available to the offspring are modelled as alternative alleles at the two loci. Since both sets of strategies (parent, offspring) are continuous, it probably would not matter very much whether the two strategies were prescribed polygenically or by a hierarchy of alleles at two loci. However, the model would be affected very radically if the parental and offspring strategies were prescribed by the same genes at the *same* locus.

In reality, it would be rather unlikely that the very genes causing offspring to beg more or less vigorously would be equivalent to genes causing parents to provide more or less food. The assumption of two sets of loci for the two sets of strategies (parent, offspring) seems plausible. In the end, all we can do is to use our biological intuition to construct the simplest and most plausible genetic framework for the model we have in mind, in the hope that its conclusions will have enough generality to apply to a wide set of slightly more complex alternatives.

Some caution must be exercised concerning the choice of *complexity of action* prescribed by a single gene. In models, it seems sensible to allow complexity to build up gradually in a series of separate stages. Each step concerns the simplest modification of the existing action—possibly due to some independent modifier. Because the prior state can affect the outcome of the next step, one procedure is to establish the set of all possible stable states, and then to seek the most plausible stable state on the basis of gradually increasing complexity. The logic of such procedures is by no means fully formulated.

2.6 Kin recognition

The notion that animals may evolve kin recognition systems derives, unsurprisingly, from Hamilton (1964a, b), who suggested that 'the individual will seem to value his neighbours' fitness against his own according to the coefficient of relatedness appropriate to that situation'. It is important to note that kin selection can operate in two contexts. There may be no kin recognition, but animals are nevertheless tuned to behave in accordance with the circumstances that correlate with coefficients of relatedness. (Thus offspring produced typically in broods of full sibs need not 'recognize' that their immediate neighbours are close kin-yet their behaviour will be influenced by the fact that other offspring surrounding them are indeed full sibs.) Of course, such individuals would behave in exactly the same way towards unrelated 'artificially transplanted' offspring. Alternatively, if there is some way by which the individuals can recognize their genetic similarity at a locus regulating a particular altruistic action (e.g. from linked genes that show phenotypic cues), they are more likely to show the action when they are genetically similar. An example of this form of genetic matching is reported by Grosberg and Quinn (1986) for the planktonic larval stage of *Botryllus schlosseri*, a sessile colonial ascidian. Groups of siblings tended to clump together, while groups of unrelated larve settled at random. Closely packed colonies can grow into each other, fusing and (altruistically?) sharing a blood system. Whether two given colonies do fuse depends on their genotypes at one locus, a histocompatibility locus. If they share at least one allele, they fuse; otherwise, they do not. Nonrandom settlement occurred according to the genotype at that particular locus (or one tightly linked to it).

An extensive literature now exists on the phenomena of kin recognition in animals, and this has been the subject of both extensive reviewing (see Hepper 1986 and the volume edited by Fletcher and Michener 1987) and not inconsiderable debate (Grafen 1990b, 1991b, c, d, Blaustein *et al.* 1991, Byers and Bekoff 1991, Stuart 1991). We cannot attempt to cover the topic adequately here, so the reader is referred to these and related references. Connections with these issues and sibling rivalry are discussed in several of the taxon-based reviews (see Chapters 13–15).

2.7 The optimality approach

In later chapters, we shall outline a series of optimality and evolutionarily stable strategy (ESS, Maynard Smith 1982a) models for intra-familial conflict. The optimality approach has been criticized for its lack of consideration of underlying genetics and the problems that this may pose; in particular, the Stanford school of population genetics stresses the difficulties of attaining equilibria in *n*-locus population genetics models (e.g. Karlin 1975). However, although an optimum or ESS clearly cannot be attained if there is no genetic mechanism that can prescribe it, models of specific cases indicate that selection often takes a population as close to an ESS as the genetic mechanism can allow (Maynard Smith 1981). Furthermore, an exciting recent development (the 'streetcar' theory) now indicates that the optimality/ESS approach has a solid foundation within the framework of population genetics theory (Hammerstein, in press). The streetcar theory relates to interactions between non-relatives; special problems arise when, as here, the interactants are relatives. We must therefore await further developments before we can be certain that it will apply to ESSs in intra-familial conflict.

Summary

1. As is widely recognized, Hamilton's rule ($br > c$, where b is benefit to recipient's fitness, c is cost to performer, and r is Wright's coefficient of relatedness between the two players) formally specifies the condition for the evolutionary spread of altruism. A simple, single-locus model is developed along classical lines that illustrates the importance of calculating r accurately. Strictly, r is the probability that two individuals (e.g. altruist and recipient) share a rare allele by virtue of their common ancestry.
2. A simple modification of Hamilton's rule, reversing the inequality sign, equally specifies the evolutionary limits of selfishness. Whenever the regular rule's conditions for altruism are **not** met, selfish alternatives should prevail.
3. The assumptions underlying Hamilton's rule are (i) that costs and benefits vary independently in ways that allow them to be additive, (ii) that selection operates weakly on kin, and (iii) that aid is dispersed without bias among kin of differing genotypes (specifically, differing at the locus controlling altruism). The first two assumptions can often be ignored, but the third merits closer study because it can reverse the qualitative direction of a Hamiltonian prediction. In essence, (iii) raises the problem of how r is to be estimated. To be applied correctly, the rule requires r to be seen from the gene's perspective (i.e. r's 'true value'). In practice, it must generally be estimated from the whole individual's perspective (i.e. r

is inferred from the probabilities associated with an 'ancestry' calculation method). The latter requires considerable caution.

4. The implications of assumption (iii) are explored by considering individuals (whole bodies) playing distinct **roles** in a social interaction. If a gene's expression is contingent on its body being in a particular role, its mechanism is said to be **conditional** on the role.

5. In a simple model of conditionality, the spread of an allele for selfishness between siblings is shown to depend on whether its body is in control of the resources (role 'X'). Under these circumstances, a sib in role X does best by taking extra resources until its own marginal gain (its increment to personal fitness for consuming the next unit) is reduced to one-half of what its full sib would obtain from the same unit. At that point, sib X's inclusive fitness can be better served by letting sib Y have the next unit of resource.

6. However, if gene expression is **unconditional**, such that it is shown in roles X and Y equally, a different result is obtained; selfishness may then be totally unchecked by kinship. Such lack of constraint can arise if the costs of unconditional selfishness are always borne entirely by carriers of the competing allele.

7. When a similar model is applied to sibling altruism, a logical Hamiltonian result accompanies conditional gene action, but analysis of unconditional expression becomes confusing in those sibships where both players are trying to behave altruistically simultaneously. Working systematically through the possible alternatives, we find that unconditionality can fit with Hamilton's rule only when some restrictive additional requirements (concerning how symmetrical effects are resolved) have been met. In that case, unconditionality converges toward conditionality. Unconditional gene expression can violate requirements (i) and (iii) above for Hamilton's rule.

8. Generalized versions of Hamilton's rule are presented for both conditional (equation 2.3) and unconditional (equation 2.4) gene action.

9. We distinguish between *enforced* conditionality, in which expression is only possible in one role because of circumstances (a drowning sib cannot save its safe contemporary), and *evolved* conditionality, in which expression is possible in both roles, but evolves to be present only in one role. The latter may involve possible difficulties of an origin through unconditionality, which must satisfy different conditions.

10. Predictions of the nature of selfishness also depend on past history (i.e. which alleles already exist at various frequencies) and on other alleles in the genome that influence the phenotype at issue. For example, an allele favouring some form of altruistic assistance might be in substantial conflict with an allele that promotes selfishness from a different locus. In practice, we seldom know the necessary details about the underlying genetic mechanisms, even though these can greatly affect a model's predictions. Caution is indicated accordingly.

11. The fact that different loci within an individual can be in intra-genomic conflict in their expression will re-surface in later chapters, especially with regard *to parent–offspring conflict* and *genomic imprinting*.

3

Theory II: phenotypic models of sublethal sibling competition

It is theory which decides what we can observe.

(Einstein, 1926)

Thus far, we have sketched out the unifying concept of inclusive fitness, showing how it prescribes the boundaries of selfishness and altruism with respect to relatedness. Here and in later discussions (especially in Chapter 4 and Chapters 7–10), we shall examine explicit optimality models of sibling rivalry and other conflicts within the family.[1] First we look at the relatively mild forms of sibling rivalry, wherein selfishness is expressed as non-lethal skewing of limited resources toward Self, before moving on (in the next chapter) to fatal consequences.

3.1 Effects of gene action

In evolutionary biology, optimality models are essentially phenotypic; they make the implicit (but admittedly unrealistic) assumption that strategies[2] reproduce asexually and that the optima are achievable. Their function is to generate insights into the nature of adaptation. This is done by erecting and then testing predictions based on given adaptive interpretations (e.g. Maynard Smith 1978, Parker and Maynard Smith 1990). The difficulty with optimality models as applied generally to sib competition and intra-familial conflict is

---

[1] We are using the term 'family' in its simplest sense: one or more parent with one or more offspring. Alternative definitions can be used to build in such features as alloparental care by sexually mature siblings (e.g. Emlen 1995).

[2] As used in behavioural ecology, 'strategy' refers to a phenotypic option (from an array of alternatives often called a 'strategy set'). In this book, we are usually focused on behavioural traits (or occasionally life-history traits). It is especially important that this usage be recognized as a form of shorthand, a vehicle of convenience, and that it does not imply that the animal consciously plots its actions in the manner of, say, a human chess player. This is perhaps most easily appreciated by noting that plants exhibit numerous life-history traits that appear to be marvellously clever, although they have no brains with which to plot.

that individual fitness alone is not an appropriate optimization shared genes (i.e. the probability that the gene for a given action is pre relatives) must be accounted for exactly as part of the optimization proces Hamilton's rule can equally be applied in optimality models, it is in fact a method of determining the direction of phenotypic evolution without using formal population genetics. However, by now it will be clear that the nature of gene action may exert an influence on the outcome of the optimization. Hamilton's rule must be applied correctly (see Grafen 1984, 1991a), and we must be explicit about what happens when a gene for an action is expressed in different roles. That is, we must be precise in our assumptions about gene action and conditionality (see Box 3.1).

We next proceed to mark out the theoretical **battleground** for sibling rivalry. It exists within a nexus of conflicts within the 'nursery' family, all of which may interact simultaneously and in many subtle ways.

Box 3.1 Effects of gene action on optimality criteria

In a dyadic interaction in which one individual occupies role X and the other role Y, the most obvious approach is to assume that, say, two loci are involved, with gene expression limited to phenotypic role (X, Y), so that genes at one locus are expressed only when in role X, while genes at the other locus are expressed only when in role Y. We then require that each action (one for X and one for Y) should satisfy separately its own version of equation (2.3) above. However, 'one-locus-one-role' is not the only possible mechanism for **evolved conditionality**. A fuller discussion of how different gene actions may affect the optimality criterion is given in Parker (1989).

For continuously-variable strategies, we would use some version of equation (2.3) to determine the optimal strategy in each role, or their ESS equivalent (some explicit examples will be considered later; see Section 3.3). In addition, Parker (1989) discusses an alternative in which one locus defines the *absolute* magnitude of a given action (e.g. helping a sib) and another locus defines the *relative* expression of the action in the two roles. Although intuitively one would guess that the same optimum would be reached as with the 'one-locus-one-role' assumption, this need not be so unless special conditions are fulfilled. It seems likely that there are many further complexities to be discovered; the genetic mechanisms underlying possible dyadic actions are legion. Hamilton's Rule—in its simplest formulation applies most obviously to the 'one-locus-one-role' model, but is very robust if correctly applied.

For strategy sets that show continuous variation across alternatives, Godfray and Parker (1992) proposed a 'marginal Hamilton's Rule' as a method for deriving an optimum or ESS. Like the direct version of Hamilton (equation 2.1), it is essentially phenotypic and non-genetic. Its general validity has yet to be proven mathematically, but so far it has given the same result as many previous equivalent models in which the genetics have been explicit (e.g. Parker 1985). Assuming that it possesses the required generality, this version of the rule may prove very useful in future optimality studies of conflict between relatives. It is explained in relation to particular models of sib competition in Sections 3.3.4, 3.3.5, and 3.3.7.

ne family

imoderated bliss, and within-family conflict is simply
of life. Strife between parents, siblings, and parents and
more the special province of the biologist than of the
.t, solicitor, or whomever. Somewhat remarkably, in evolu-
'classical' view of the family tended to be one of harmony of
n was usually envisaged as maximizing family productivity.
A.. ne celebrated 'Lack clutch size,' which assumes that selection
maximi.. number of surviving offspring per brood (Lack 1947, 1948, 1954; see p. 205). However, if we look at other biological cases, there is usually no more evidence of intra-familial harmony than there is for humans.

Largely as a result of the works of Hamilton (1964a, b) and Trivers (1974), we now see the family as a unit rife with evolutionary conflicts of interests among its members. It is important to stress that conflict is not a direct alternative to co-operation. As we shall see in later chapters, there can be conflict over the extent and nature of co-operation (e.g. Emlen *et al.* 1995). In addition, there can be co-operative forms of conflict, as when team-work is advantageous (consider wartime alliances). Social conflict simply implies different interests of two or more individuals, in this context over Darwinian *fitness*. At a genetic level, conflicts may arise between the expression of genes at different loci and in different individuals (see Maynard Smith and Szathmáry 1995), and these levels can cross. Consider two interacting individuals that occupy two different roles, *A* and *B*, in a social interaction. Characteristics relating to one locus may be expressed when in the role of *A*, and those relating to another locus when in the role of *B*. Thus the conflict concerns the spread of genes at two or more loci.

For our discussion, we identify that there can be three essential and interacting social games: sexual conflict (i.e. parent–parent), sib competition and parent–offspring conflict (see Figure 3.1). Each game generally concerns parental investment (PI): its supply (by parents), its demand (by competing offspring) and its distribution (among competing offspring). As we shall see, the resolution of each conflict is influenced by the resolutions in the other dimensions: the eventual ESS requires the *simultaneous solution of all three games*.

Why are sexual conflict and parent–offspring conflict important to the understanding of sibling rivalry? The reason is that the intensity of sib competition typically depends on the magnitude of resources, which are supplied by parents, one way or another, in nursery taxa. Sexual conflict and parent–offspring conflict, in turn, affect the *supply* of these parental resources. In short, sib competition fuels parent–offspring conflict, and both are influenced by sexual conflict. All three dimensions interact.

Consider first the supply side. A parent may be (simplistically) viewed as having a fixed budget of parental investment (PI) that it allocates to separate

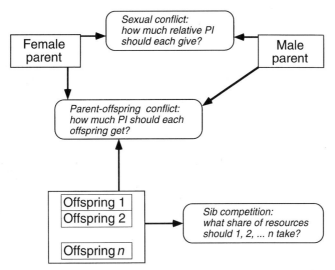

Fig. 3.1 General schema for intra-familial conflict over resources provided by parents. If there is biparental care, the two parents may have different interests concerning the amount of parental investment (PI) each should supply to the offspring (sexual conflict). Offspring and parental interest may differ over the amount of PI to be given (parent–offspring conflict). The offspring will also have different interests regarding the amount of PI that each one should obtain.

offspring (Smith and Fretwell 1974, Parker and Macnair 1978, see Figure 3.1). Just as dividing a cake into equal slices inevitably leads to a quantity-quality trade-off (the greater the number of slices, the smaller each one will be), increasing the amount m of PI in a given offspring may increase that offspring's fitness, $f(m)$, but it also reduces the number of offspring that the parent can produce. If offspring fitness shows diminishing returns with increasing PI (see Figure 3.2), it pays the parent to produce several offspring with an optimal amount of PI in each one. In Chapters 7 to 10 we shall see that sib competition disrupts the parent's optimal allocation of its PI, causing parental retaliation and changes in supply of resources: this disruption is the result of parent–offspring conflict. At that time we shall also explore how conflict between parents might affect this optimum, and how it might affect the supply of resources to progeny produced in clutches. In the remainder of this chapter and the next, we shall assume that the supply of resources is fixed so that the sib competition component of Figure 3.2 can be examined as an isolated issue. Sib competition can occur whether parents are present or absent, provided that the resource provided by parents is limited and that offspring remain together and compete actively for it. Thus, for now, parents simply provide a fixed resource, and we shall not here be considering alterations in parental strategy due to the effects of sib competition (but see Chapters 7–11).

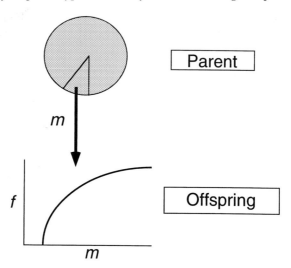

Fig. 3.2 Supply of resources from parents to offspring. Each parent can be considered as having a fixed amount of parental investment that it can allocate to its lifetime production of progeny (represented by the shaded circle). If it allocates a large portion, *m*, of this reproductive budget to a single offspring, this offspring achieves a high personal fitness, *f(m)*, although relatively few offspring can be produced in total.

We shall look first at graduated selfishness models of non-lethal sib competition, saving the all-or-nothing games of siblicide for Chapter 4.

3.3 Non-lethal sib competition: graduated selfishness

Here offspring compete over shares of the limited parental resources, whatever these may be. Parental interests are traditionally assumed to be best served by an equal apportionment of resource to each brood-mate (for further treatment of this point, see Chapters 7–11), but an allele regulating an offspring's degree of selfishness only has a probability of ½ of being carried by a full sib. Hence it pays a given offspring to take a greater share of PI than its nest-mate (Hamilton 1964a, b). This forms the basis of parent–offspring conflict (Trivers 1974).

There have been two types of models of non-lethal sib competition. In the first, there is a rigid dominance hierarchy such that individual offspring take shares of the resources according to their rank order (Parker *et al.* 1989). In the second, there is a begging scramble[3], and offspring are 'paid' in proportion

---

[3] 'Scramble' competitions are those in which rivals compete by hastening to consume the limited resources, e.g. vying to out-eat each other when food is scarce. 'Interference' competitions allow rivals to interfere with each other's performance more directly. The dominance relationships in our hierarchy model are of this latter class.

to their relative begging levels (Macnair and Parker 1979, Metcalf *et al.* 1979, Parker 1985, Harper 1986, Parker *et al.* 1989, Godfray and Parker 1991, 1992).

3.3.1 *Hierarchy models*

Imagine a strict linear hierarchy among full sibling avian nest-mates. Nestling *A* dominates nestling *B*; *A* and *B* dominate *C*; *A*, *B* and *C* dominate *D*; and so on. How much of the total input should each chick take? *A* takes first share, then *B* takes its cut from what remains, then *C*, etc. With three chicks, only *A* and *B* need make any decision about how much to take; *C* simply takes whatever is left (Figure 3.3). It is assumed that *C* gets less than it wants because resources have to be insufficient for a competition to exist.

To gain a feeling for how the model will be organized, consider first the case of just two chicks, *A* and *B*. A rare dominant allele causes *A* to take

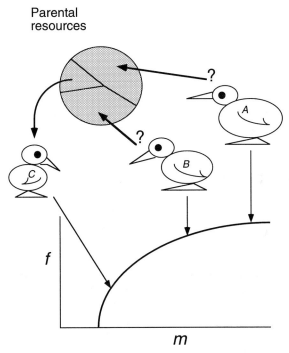

Fig. 3.3 The hierarchy model of sibling competition. The fixed total input of parental resources is represented by the shaded circle. The dominant chick *A* gains first access to the input and must decide how much to take, given that its sibs (*B* and *C*) are, in this example, full sibs. After taking this amount, the remainder of the input is left for the next most dominant chick (*B*). It has a similar decision to make, but has only the fate of its weaker nest-mate (*C*) to consider. C then simply consumes whatever is left. The share *m* of the resource that each chick takes determines its personal fitness, *f(m)*.

proportion p of the total food input, M. That is, A ingests a portion, $m_A = pM$. The fitness of A (its survival prospects multiplied by its relative future reproductive success) is described by $f_A(m_A)$. It can be seen that B ingests the remaining proportion, $1 - p$, of the food, so its share is $m_B = (1 - p)M$, and its consequent fitness is a function of that, i.e. $f_B(m_B)$. The relative replication rate of the rare allele is proportional to

$$f_A(m_A) + \tfrac{1}{2}f_B(m_B) - F$$

since there is a probability of $\tfrac{1}{2}$ that B also carries the rare allele characterizing p (and hence m_A, m_B). F is the fitness of the common alternative allele that prescribes what to do when in role A. Hence the rare allele spreads against the common allele if the above expression is positive. The optimal value for p, p^*, is found by setting

$$\frac{\partial}{\partial p}[f_A(m_A) + \tfrac{1}{2}f_B(m_B) - F] = 0$$

so that

$$f_A{'}(m_A) = \tfrac{1}{2}f_B{'}(m_B) \tag{3.1}$$

where, for example, the prime denotes the derivative of f_A with respect to m at m_A. Sib A should take parental resources up to the point where its marginal gains drop to half those of B (Parker *et al.* 1989). This solution is simply a restatement of Hamilton's own words: 'we expect to find that sibs deprive one another of reproductive prerequisites provided they can themselves make use of at least one-half of what they take' (Hamilton 1964a: 16).

There are two ways to extend this marginal-gains rule (equation 3.1). The simplest way is to assume that both sibs have equal gains or losses from the parental resources, as for example might be the case in the long term (e.g. over the entire period of parental care). If the fitness functions for each chick are the same, so that $f_A(m) = f_B(m)$, it is clear that for diminishing returns, p must be greater than $\tfrac{1}{2}$. A must take the larger share of resources ($m_A > m_B$). How much more? This depends critically on the shape of $f(m)$; that is, on how dramatically offspring fitness improves with greater provisioning. If resources are plentiful we expect $f(m)$ to rise steeply to its asymptotic value (see Figure 3.4a), and for there to be little difference between the shares (m_A, m_B) or fitnesses ($f_A(m_A)$, $f_B(m_B)$) of the two sibs. However, if resources are scarce we expect a slow rise to the asymptote (Figure 3.4b) and a consequently greater inequity between shares and fitnesses for A and B. To illustrate the effects of curve shape in more detail, we need to adopt an explicit mathematical form for chick fitness. There are many that would resemble Figure 3.4. Parker *et al.* (1989) used a form where $f(m)$ obeys an exponential law of diminishing returns:

$$f(m) = 1 - \exp[k(m - m_{min})] \text{ if } m > m_{min;} \tag{3.2}$$
$$f(m) = 0 \quad \text{if } m \leq m_{min}.$$

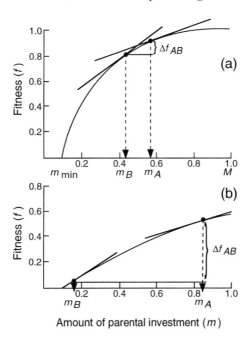

Fig. 3.4 The long-term hierarchy model with two sibs. The curves show the relationship between an offspring's personal fitness f (= survival prospects multiplied by reproductive success) and the amount m of parental resources it consumes throughout the period of parental care. $f(m)$ follows the form of equation (3.2) for exponentially diminishing returns, and no offspring can survive with less than m_{min} of resource (set here at 0.1). M is the total amount of parental resources and is standardized at one. **(a)** Resources are relatively plentiful ($k = 5$) and both sibs do well (i.e. achieve nearly maximal fitness). The fitness difference between the sibs (Δf_{AB}) is relatively small. **(b)** Resources are limited ($k = 1$) and the B-chick fares badly (Δf_{AB} is large). In both cases, the A-chick takes most resources ($m_A > m_B$), especially when resources are limited. (After Parker *et al.* 1989.)

Here, k is a 'shape constant': the effect of increasing k in equation (3.2) is to increase the rate at which $f(m)$ rises to its asymptote: increasing k effectively simulates increasing food abundance. If k is relatively low, $f(m)$ rises less steeply than if k is relatively high, so for a given parental effort m', an offspring's success, $f(m')$ is lower for low values of k. With this model, it can be calculated that the shares of A and B are

$$m_A = M/2 + (\ln 2)/2k;$$
$$m_B = M/2 - (\ln 2)/2k.$$

Thus A and B take virtually equal shares (half of the total M each) when k is huge (e.g. when conditions are bountiful), and their shares diverge as k gets smaller (when times are lean). At the point where k becomes equal to or less than $(\ln 2)/M$, it pays sib A to take **all** of the resources, thereby starving B to death (obligate brood reduction).

Increasing selfishness by A as resources become more scarce is an intuitively satisfying prediction from the marginal-gains rule, shown in equation (3.1). However, could we ever expect to see the A sib taking a *smaller* share of the resources than it allows B to take?

This is possible, at least in the short term. Imagine now that $f(m)$ represents the short-term increment to the survival prospects of an offspring that receives m units of food during a given feeding bout. It seems quite feasible that the two sibs could be affected quite differently by a given amount of food, so that the fitness relationship for sibling A, namely $f_A(m)$, is not the same as that for sibling B, $f_B(m)$. Although a difference between the forms of $f_A(m)$ and $f_B(m)$ can generate conditions where the dominant sib allows its subordinate to take more of the resource (Parker *et al.* 1989), this actually requires substantial differences between the two curves, either in their asymptotes (Figure 3.5a) or in their shape constants, k_A and k_B (Figure 3.5b). To state it another way, A's inclusive fitness might be better served in the short term by being generous to B if the resource is *sufficiently more valuable to B* than to A. In nestling birds, it is difficult to know whether sporadic lapses in A's gluttony occur for this reason (by whatever mechanism), but the possibility certainly cannot be ruled out. The general problem of state-dependent selfishness within sibships can be approached numerically using dynamic programming (e.g. Forbes 1991a, b, Forbes and Ydenberg 1992), especially for small broods. With large broods, the complexity of the models can quickly become cumbersome enough to consume very large amounts of computer time in deriving the ESS.

Indeed, life does become more complex if there are more than two siblings. Parker *et al.* (1989) analysed the case of just three chicks, each having the same $f(m)$ relationship (i.e. the long-term model). Here, A must first 'decide' how much of the total food to take, and B then decides what to take. Finally, C consumes whatever is left (see Figure 3.3). We know that once A has taken its share, the decision for B should obey the rule (equation 3.1) for two chicks, so that B's marginal gain will be half that of C. The complexity arises because the best decision for B depends on what A does, and vice versa; we now need an ESS approach. At the ESS, it turns out that A's marginal gains must lie between 0.5 and 1.0 times those of B; A's marginal gains are thus closer to those of B than B's are to C's. This means, in turn, that the shares and fitnesses of A and B will be closer than those of B and C (see Figure 3.6a). This suggests that for sibships greater than two, the differences in food shares and fitnesses will become greater down the dominance rank, so that the greatest disparity occurs at the bottom of the hierarchy. Using the same exponential law of diminishing returns for $f(m)$ as above, Parker *et al.* (1989) showed that the ESS shares for the three chicks are as follows:

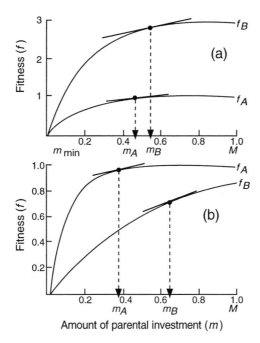

Fig. 3.5 The short-term hierarchy model with two sibs, showing two instances in which the *A*-sib should take less of the resource at a meal than it allows the *B*-sib to take (i.e. $m_A < m_B$). The relationship has a similar form to Figure 3.3, except that $f(m)$ here $= F[1-\exp(-km)]$ and represents a short-term increment (e.g. at a given feeding bout) to the personal fitness of a given chick. **(a)** The dominant chick, *A*, may allow *B* to take the greater share of food if *B* can experience a much greater maximum fitness increment in terms of the asymptote, *F*. Here $F_A = 1$, $F_B = 3$, although the shape constants *k* are the same ($k_A = k_B = 5$). **(b)** It may also allow *B* to take the greater share if *A* can attain its maximum much more slowly (i.e. if k_A is sufficiently higher than k_B). Here $F_A = F_B = 1$; $k_A = 10$, $k_B = 2$. (After Parker *et al.* 1989.)

$$m_A = \frac{M}{3} + \frac{5\ln 2 - 2\ln 3}{3k};$$

$$m_B = \frac{M}{3} + \frac{\ln 3 - \ln 2}{3k};$$

$$m_C = \frac{M}{3} - \frac{4\ln 2 - \ln 3}{3k};$$

(see Figure 3.6b). As might be expected, food shares converge towards equality (one-third of *M* each) when food is readily abundant (modelled as very high *k*), but become markedly disparate under conditions of chronic shortage (very low *k*), at which time there is an increasing chance that the *C*-

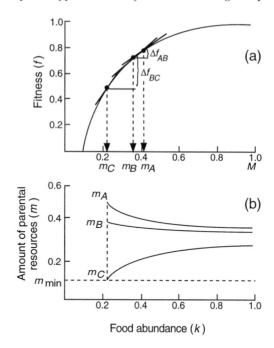

Fig. 3.6 The long-term model with three sibs. **(a)** $f(m)$ as in Figure 3.3, but with a sibship of three. Note that the fitness difference (Δf_{ij}) is greater between B and C (the two least dominant sibs) than between A and B. Similarly, the amount of resource taken decreases down the hierarchy ($m_A > m_B > m_C$). **(b)** Relationship between the shares (m_i) taken by each sib and k, a measure of resource abundance. Note that as resources become more plentiful (higher k-values), the shares of the three sibs converge toward one-third each. If resources are scarce, the A-sib takes a much greater share, ultimately at the expense of the C-sib. If C's share falls below the critical minimum, m_{min} (here shown arbitrarily at 0.1), then it has no prospect of survival. $M = 1$ throughout. (After Parker *et al.* 1989.)

chick will gain less than the critical minimum amount (m_{min}) necessary to ensure its survival. If C obtains less than m_{min}, it dies, after which the shares of A and B must then be derived by the two-chick model. A more subtle result is that the ESS share for the B-chick is much less sensitive to changes in the amount of food than are those of the A- and C-chicks. As food becomes more scarce, A effectively takes part of what would have been C's share.

It is more difficult to solve analytically the shares for a hierarchy of $n > 3$ chicks, using the exponential form for $f(m)$ in equation (3.2). Given that we know the relationship between the shares of the last three chicks, it is possible to iterate by computer the solutions for shares for $n = 4$ sibs. (So far, we have had little success in deriving solutions for sibships greater than $n = 4$.) Table 3.1 shows a numerical example of the shares (m_A, m_B, m_C, m_D) obtained when

Table 3.1. The shares (m_A, m_B, . . . etc.) of the total resources gained by the most dominant sib (A), the next most dominant (B), etc., in the hierarchy model. The total parental resource input increases in proportion to the number of chicks in the brood, so that the strategic choice of shares does not result from reduced per caput resources as the brood increases. The average amount of resource per chick is one unit, hence if a chick's share is greater than one it is gaining more than average resources. For details of the model, see text.

Number of chicks in the brood:	Shares of resources gained at the ESS:			
	m_A	m_B	m_C	m_D
2	1.35	0.65	—	—
3	1.42	1.14	0.44	—
4	1.45	1.27	0.98	0.29

the number of chicks in a nest varies from two to four. For this computation, we hold the total resource input per chick constant—it is $M = 2$ for two chicks, $M = 3$ for three and $M = 4$ for four chicks. Thus the difference in shares gained is not the result of decreasing total parental input. To calculate the results in Table 3.1, each chick's prospects obey equation (3.2), with k set as 1.0 and $m_{\min}$ as 0.1.

Again, with $n = 4$ sibs, the largest difference in shares (and personal fitness, f) occurs between the lowest and next-lowest ranking siblings. This difference seems to increase with number in the sibship. The highest ranking sib takes a larger share as sibship size increases, but the difference between the A- and B-sibs decreases as numbers increase. The general effect of increasing the size of the brood appears to be a bunching of shares and personal fitnesses at the top of the hierarchy, and a stretching out of these measures at the bottom. This can be interpreted in terms of the top-ranking sibs being able to approach asymptotic requirements, while competition down the hierarchy intensifies as a function of decreasing resource base (decreasing per-capita average remaining shares). Forbes (1993) has investigated a rather different explicit function for the personal fitness, f, of a chick in relation to the amount, m, of parental investment it receives. He used[4]

$$f(m) = \frac{m - m_{\min}}{m} \quad \text{for } m \geq m_{\min},$$

$$f(m) = 0 \qquad \text{for } m < m_{\min},$$

---

[4] Unfortunately, Forbes's paper generates a major clash of notation with our present (and past) versions of the model. He uses $\rho(m)$ in the sense of our $f(m)$ and he uses $f(M)$ to represent the future reproductive success of a parent that invests a total of M in the present brood.

(cf. equation 3.2). This gives a form very similar to the exponential form (equation 3.2), but proves more tractable—it allows the optimal food share to be derived for broods of any size. It has the following property: As the total amount M increases, the proportions of the total taken by each chick remain the same, for a given brood size. Thus, a changing food share (as a proportion of the total) with changing food levels is not an inevitable conclusion, but rather a property of the explicit function deployed in the model. However, as the total food supply increases, the disparity between the personal fitnesses of the chicks decreases, exactly as in the exponential model, and the rule persists that personal fitness disparities increase as we move down the hierarchy. We suspect that this conclusion about personal fitness is likely to be a general property of the hierarchy model; we have established its generality for the case of three sibs (Parker *et al.* 1989).

A word on the logic of hierarchy models is needed: we solve the best strategy for the dominant chick by maximizing the fitness function

$$f_A(m_A) + \tfrac{1}{2}\left[f_B(m_B) + f_C(m_C) + f_D(m_D)\right] - F$$

in which chick A 'counts' itself as one and all its sibs as $\tfrac{1}{2}$, i.e. the rare allele affecting the amount taken by A, m_A, is present in A but only has a probability of $\tfrac{1}{2}$ of occurring in any of the lower-ranking chicks. F is again the fitness of the common alternative allele for prescribing what to do when in role A. B's best strategy is found in a similar fashion, but chick A is excluded from B's formula, and so on for C and D until we reach the penultimate chick (call it Y and the runt Z), whose best strategy is found by maximizing

$$f_Y(m_Y) + \tfrac{1}{2}f_Z(m_Z) - F_Y$$

where F_Y is the fitness of the common alternative allele prescribing what to do when in role Y. This poses an apparent paradox. The best strategy for A takes into account all of the other sibs, but the best strategy for less dominant chicks becomes progressively less restricted by kin selection, so that penultimate sib Y takes into account only its subordinate Z. A rare gene affecting the share taken by Y, m_Y, is just as likely to occur in A, B, C, ... as it is in Z. So why are they not considered? The reason is that a sib's decision can only affect subordinates; Y's decision only affects Z, whereas A's decision affects the fitness of all the other chicks. Thus is there no paradox in calculating the ESS, even though this might seem to be the case at first sight.

This insight suggests an explanation as to *why* disparities between sibs increase down the hierarchy. The constraint on an individual's strategy due to kin selection becomes weaker down the hierarchy. The weighting of Self vs. full sibs is $1:0.5n_s$, where n_s is the number of subordinates. The second consideration is that the available per capita resource decreases down the hierarchy, because each chick will take rather more than its 'fair' share at each step of the process. The effect is to increase the disparity between shares (see Figure 3.5), especially between the last two chicks, so in large sibships we would expect the shares of most sibs to be rather similar, except for the lowest ranking one or two offspring, which will suffer badly.

This is a common finding, which is supported by the phenomenon of 'runting' in large litters of mammals, and by the many bird studies in which the most subordinate chick typically fares disproportionately badly, even relative to its immediate superior. Runting is not ubiquitous (the present model would not predict it with superabundant resources), but its occurrence fits the hierarchy model rather well.

Some further evidence that fits the predictions of this model comes from field studies of egrets, which we shall describe in Chapters 5 and 6. For the moment, however, we mention only that these birds exhibit a clear hatching asynchrony (1–3 days between successive chicks; Mock 1985) that imposes an age-related dominance hierarchy. Detailed observations of both natural three-chick broods and experimental broods whose food amount was doubled (via human provisioning; see Mock *et al.* 1987b) showed that the two senior chicks tend to ingest similar amounts of food and that the youngest obtains a much smaller share. Table 3.2 shows these results, with the exception of one nest where brood reduction occurred. As predicted generally, the shares ingested by *A* and *B* were more similar than those ingested *B* and *C*, and this difference is statistically significant in both natural and provisioned broods. However, a prediction of the explicit exponential model (equation 3.2) for *f(m)*—that food shares should become more similar as food becomes more plentiful—was not borne out. There was no evidence for any significant change in the shares between the supplemented and control treatments (Parker *et al.* 1989). Interestingly, constancy of food shares is predicted by the explicit function for *f(m)* used by Forbes (1994). Shares may change relatively little over a wide range of food availablities (as shown in Figure 3.6b), and this also fits Forbes's version rather neatly.

3.3.2 Begging models—The logic of begging scrambles

For many species, the hierarchy approach, in which sibs gain access to resources successively in strict dominance order, is too simplistic. Sib competition often involves scrambles of escalated begging, rather than effective

Table 3.2. Mean shares of parentally delivered food (SD) ingested by members of three-chick great egret broods. (After Parker *et al.* 1989).

Sib rank	Natural broods (n = 16)	Provisioned broods (n = 12)
A	43.2% (6.5%)	46.2% (4.9%)
B	37.0% (3.6%)	38.6% (11.4%)
C	19.8% (5.8%)	15.3% (11.9%)

despotism. When an adult passerine alights on the rim of its nest, the chicks rise instantly, each stretching towards its parent with the widest possible gape, and generating a very considerable tumult. Often it very much appears that there is a begging scramble in which the success of a chick presumably increases with its effort relative to that of its competitors.

In nature, it is probable that resource gains in most sib competitions are achieved by a mechanism lying somewhere between the extreme limits of perfect dominance hierarchy and chaotic scramble. In this section we shall examine the limit of non-aggressive competitive begging ('begging scramble') as our concept of sib competition. However, we anticipate that the same conclusions will apply whatever the exact form of the scramble between sibs for parental resources, provided that the scramble obeys the same assumptions. It can, for example, apply to the case where the costs of increased competition in the scramble are felt as a reduction in the quantity or quality of the nursery resources. Such a situation is analysed explicitly in the 'gregarious caterpillar' model described later in this chapter (see Section 3.3.7).

Models of competitive begging have been investigated by several authors (Macnair and Parker 1979, Metcalf *et al.* 1979, Parker and Macnair 1979, Parker 1985, Harper 1986, Parker *et al.* 1989, Godfray and Parker 1991, 1992), often in relation to parent–offspring conflict. An alternative approach is to view begging as an honest (but costly) signal to its parent of an offspring's need (Godfray 1991, 1995a), which we shall discuss later (see Chapter 9).

In a begging scramble (outlined in Figure 3.7), each offspring has a choice of begging level, x. For example, individual i chooses level x_i. Its gains (i.e. its share of the parental resources) are some increasing function of x; usually, it is assumed that the shares received match each offspring's relative begging level. The commonest assertion is that individual i obtains a share proportional to x_i/ϕ (where ϕ is the mean value of x for the entire brood). Note that i's begging level contributes to the mean begging level, so that a unilateral unit increase in x_i will have *less* effect in small broods because it affects the mean value to a greater extent. For example, in a brood of two, the unit increase in x_i is offset by half a unit increase in the mean level ϕ, whereas in large broods a unit increase in x_i has only a trivial impact on the mean level. Thus a given increase by i leads to a *greater* contrast with ϕ in a large brood than in a small brood. This effect—whereby a unilateral increase by one sibling will have less effect on its gains in small broods—has been termed the 'dilution effect' (Godfray and Parker 1992).

Alternatively, it could be that the share gained by individual i is proportional to x_i/ϕ', where ϕ' is the mean begging level for all sibs *other than i*. In this case, a unilateral unit increase in x_i will have the same effect regardless of brood size—there is no dilution effect. In reality there are many ways in which the share of individual i might alter with x_i. However, provided that i's gains increase with its *relative* (rather than *absolute*) begging level, then i's best strategy depends on the strategy played by other sibs. Thus our analysis must be based on ESS concepts.

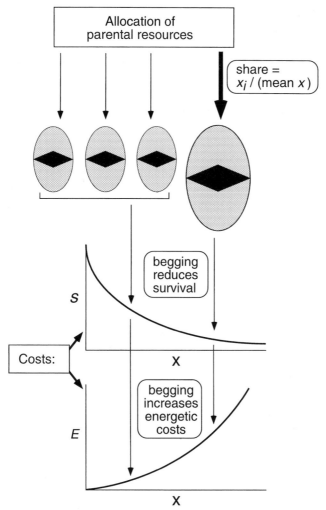

Fig. 3.7 A begging scramble among four chicks. The shaded ellipses are the open beaks as seen by the parent bird arriving to feed the brood. The begging effort is *x*, and in this instance three chicks have low begging efforts (schematically represented as small ellipses) and one has a high begging effort. The mechanism of payment to the offspring is by 'mean matching', in which the share of offspring i is equal to its begging level divided by the mean begging level. Two forms of begging cost are considered: increased begging may reduce survival prospects *S* or it may increase energetic costs *E* (see text).

Any approach which assumes that parents pay out investment in relation to an offspring's begging level relative to the mean level (termed 'mean-matching' by Harper 1986) makes specific assumptions about the parental behaviour. If *x* represents a probability of begging in unit time, or an 'attempted zone of parental attraction,' which is then squeezed down by the competing solicitations of others, and if the parent's probability of allocating the next food item is essentially random, we might expect i's share to follow the first type of mean-matching, that is x_i/ϕ. This form of mean-matching can be derived from first principles in 'competitive eating' games (see Section 3.3.7). The second form of mean matching (i's share is x_i/ϕ') could apply if the parent makes a specific comparison of i's level with that of the rest of the brood.[5]

In all of these models, begging is assumed to have costs to a chick in terms of energy or reduced survival (see Figure 3.7). The more effort spent on begging, the greater the cost. Thus if, say, the main cost of begging is a reduction in survival, we would expect survivorship S to decline as the begging level x increases. Field support for such costs is given by Haskell (1994). Alternatively, if the main costs are energy expenditure, E, we would expect E to increase with x. These different forms of cost lead to two different ways of expressing the fitness of an individual offspring, i. Remembering that the fitness, f, derived from gaining an amount, m, of the total resources, M, is $f(m)$, then if begging affects survivorship (e.g. because the noise tends to attract predators to the nest: Redondo and Castro 1992a, Haskell 1994), we can express individual i's personal fitness as

$$f(m_i(x_i,x^*)) \cdot S(x_i), \tag{3.3}$$

where $m_i(x_i,x^*)$ = the amount of resources gained by i when it plays x_i in the scramble against sibs that play x^*. Typically, $m_i \propto x_i/\phi$, where ϕ is the mean begging level for the entire brood. However, if begging costs are mainly energy expenditure, which is an unshared commodity, i's personal fitness is

$$f(y_i(x_i,x^*)), \tag{3.4}$$

where $y_i(x_i,x^*) = m_i(x_i,x^*) - E(x_i)$, the net resource gain (energy gains minus energy losses) in the scramble. Some authors (Macnair and Parker 1979, Parker and Macnair 1979, Parker 1985, Godfray and Parker 1992) have used the multiplicative form of equation (3.3), but others (Harper 1986, Parker *et al.* 1989, Godfray 1991, Godfray and Parker 1992) have used the additive form of equation, (3.4). For qualitative predictions, the results are unlikely to differ greatly, but if the models are to be applied quantitatively to real field

---

[5] A pathological form of mean-matching (for modelling purposes) would be to assume that the parent pays out only to the chick that begs most (i.e. all of the food is gained by i if $x_i > x_j$, for all j). However, an approach that makes few specific assumptions about the parental response was suggested by Harper (1986), and more recently by Godfray (1991). In Godfray's model, offspring occur singly (i.e. not in broods) and the parent simply monitors the offspring's begging and uses it as an honest signal of the offspring's condition. Provided that the signal is constrained to be honest (and if begging is costly as Godfray's analysis suggests that it will be) the parent responds to the signal essentially in accordance with parental interests. This model, which is applicable to brood production as well as the raising of single offspring, leads to some rather different conclusions about parent–offspring conflict (see Chapters 7–9).

observations, it will be necessary to make a correct choice of function, or to combine the two as

$$f(y_i)\,S(x_i) \tag{3.5}$$

in which there are both energetic and survival costs to begging.

3.3.3 *Gene expression: conditional or unconditional?*

Gene expression in the hierarchy model (see Figure 3.3) was assumed to be conditional: each chick makes the optimal decision depending on its position in the hierarchy. In begging games, gene expression may plausibly be either conditional or unconditional. If a mutant gene always causes its bearer to beg to a new level, its action would be unconditional. If a mutation affects only the behaviour of an individual when it occupies a given role (chick *A*, chick *B*, etc.), its action is conditional or role-dependent. From Section 2.4.1, we might anticipate that the mechanism of gene action may affect the solution to the model.

Hypothetical 'no-cost' begging

Consider a rare dominant allele *B* with unconditional action, which always causes its bearer to show increased begging (Figure 3.8). In a brood produced

Gene action unconditional

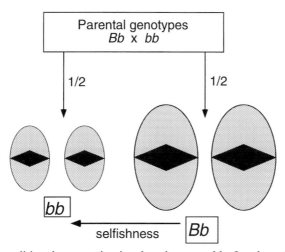

Fig. 3.8 Unconditional gene action in a begging scramble. In a brood produced by a mutant parent carrying the dominant mutant B for escalated begging, offspring bearing the mutation (segregating with a probability of ½) will take extra resources only at the expense of their wild-type (*bb*) sibs. Cost-free begging cannot here be moderated, because the selfishness harms only the *bb* sibs.

by a mutant parent, there is a probability of ½ that a given offspring will carry the allele. Imagine a vast sibship in which half of the offspring are *Bb* and half are *bb*. The *Bb* offspring beg more than the *bb* offspring, and hence take a greater share of the parental resources. For the sake of argument, suppose that there are no costs of begging. It is obvious that *B* will *always* spread, however escalated the begging, even if the *Bb* offspring effectively starve their full sibs to death (Macnair and Parker 1979). The selfishness is directed entirely by *Bb* genotypes towards *bb* genotypes, which are irrelevant to the fitness of the *B* gene. A dominant mutation of this type can always invade, even if it were to prescribe 'infinite' begging under our no-cost assumption. The same is true for small sibships. With just two sibs, unequal division of resources takes place only when *Bb* and *bb* occur together, and *Bb* can starve *bb* without cost to the *B* allele. (If two *Bb* genotypes share a nursery, they achieve equal gains and no losses because begging has no costs.)

This is not the case if gene action is conditional (Figure 3.9). Suppose that a rare, dominant, conditional mutant allele *C* acts to increase begging level *only* when its bearer is an *A*-chick. The increased begging causes the *A*-chick to gain resources at the expense of chicks in other roles (*B*-, *C*-, etc.), which have a probability of ½ of carrying the *C*-allele. The selfishness of the *A*-chick will now be moderated by kin selection. If begging has zero costs, it is clear that *C* will spread *only* if the amount taken from other chicks is justified in terms of the benefits to the *A*-chick. Consider a brood of two nestlings. Suppose that if both beg to the same level, each gains one unit of resource, but if chick *A* carries the *C*-allele it obtains $1 + m$ units. Following the arguments of Section 2.4.1, allele *C* will invade only if it causes a begging level in a *C*-allele bearing *A*-chick such that

$$2[f(1+m) - f(1)] > [f(1) - f(1-m)];$$

$$2[\text{benefit to } A] > [\text{cost to } B];$$

see p. 24.

Thus a conditional gene for highly-escalated begging will not invade because it is constrained by kin selection, although it is possible to show that begging will escalate by gradual steps to an infinite level if (hypothetically) it has no cost!

All this seems to imply that the genetic mechanism and the mode of gene action will exert an important influence on the solutions for begging games. To add to the complexity, a gene with unconditional action but that is *recessive* will be constrained by kin selection in much the same way as would the dominant conditional gene *C* just discussed (Macnair and Parker 1979). A rare recessive gene is usually expressed only when two heterozygotic parents meet; the recessive homozygotic offspring can hence harm the unexpressed recessive gene in their heterozygotic siblings. Again, escalation to an infinite level can proceed by gradual steps if begging has no cost. Furthermore, a

Gene action conditional
on role

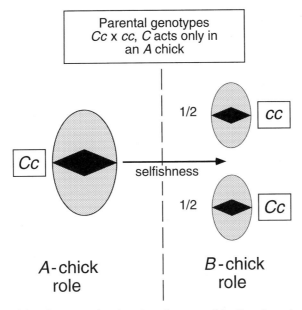

Fig. 3.9 Conditional gene action in a begging scramble. In a brood produced by a mutant parent carrying the dominant mutant *C* for escalated begging only when in the role of the *A*-chick, an offspring in role *A* which bears the mutation will take extra resources at the expense of its sib in role *B*. This sib has a probability of ½ of bearing the same mutant gene. A gene *C* for cost-free, highly escalated begging is now constrained by kin selection, since the selfishness harms bearers of *C*. However, if begging has zero costs, the ESS will still be for infinite begging (see text).

dominant *unconditional* gene that has low penetrance can also be constrained in the same way (Charlesworth 1978, Godfray and Parker 1991). When the gene is expressed, the low penetrance allows that the selfishness can damage bearers of the same gene in whom the gene is not expressed.

Costly begging

In fact, although it is important to consider genetic mechanisms, there is cause for optimism that the phenotypic results may be less complex than the above discussion might suggest. Whether gene action is conditional or unconditional, and whether it is recessive or dominant, the ESS solutions do not appear to be affected by the genetic mechanism. To demonstrate this (see Box 3.2), we

Table 3.3. Sib competition by costly begging in sibships of two, occupying roles X and Y (see text). There are four equiprobable sibshps (see Tables 2.1 and 2.2) in which the dominant mutant gene B causes its bearer to be to level x_j, whilst the alternative allele (b) prescribes the ESS begging level x^*. Fitness is the product of gains $f(m)$ through feeding and survival prospects $S(x)$, which reduce with begging. The share of food m is a function of self's begging level and the other sib's begging level. Thus $m(x_i, x^*)$ is the share gained when x_i is played against x^*. Fitnesses of B-bearing sibs are emphasised in bold type.

Genotypes			Gains $f(m)$ and survivorship $S(x)$			
			Conditional		Unconditional	
	X	Y	X	Y	X	Y
1.	bb	bb	$f(m(x^*,x^*))$ $S(x^*)$	$f(m(x^*,x^*))$ $S(x^*)$	$f(m(x^*,x^*))$ $S(x^*)$	$f(m(x^*,x^*))$ $S(x^*)$
2.	Bb	bb	$\mathbf{f(m(x_i,x^*))}$ $\mathbf{S(x_i)}$	$f(m(x^*,x_i))$ $S(x^*)$	$\mathbf{f(m(x_i,x^*))}$ $\mathbf{S(x_i)}$	$f(m(x^*,x_i))$ $S(x^*)$
3.	bb	Bb	$f(m(x^*,x^*))$ $S(x^*)$	$\mathbf{f(m(x^*,x^*))}$ $\mathbf{S(x^*)}$	$f(m(x^*,x_i))$ $S(x^*)$	$\mathbf{f(m(x_i,x^*))}$ $\mathbf{S(x_i)}$
4.	Bb	Bb	$\mathbf{f(m(x_i,x^*))}$ $\mathbf{S(x_i)}$	$\mathbf{f(m(x^*,x_i))}$ $\mathbf{S(x^*)}$	$\mathbf{f(m(x_i,x_i))}$ $\mathbf{S(x_i)}$	$\mathbf{f(m(x_i,x_i))}$ $\mathbf{S(x_i)}$

Box 3.2 ESSs for costly begging

First consider the unconditional case with just two chicks that are full siblings. There are four possible dyads, each having a probability of ¼ (Table 3.3), exactly as shown in Table 2.1. We look at the fate of a dominant mutant allele B that causes its bearer (Bb in the table) to beg at level x_i, and thus to deviate from the ESS value x^*. Because gene action is unconditional, all Bb chicks play x_i and all bb chicks play x^* (Table 3.3). We use the formulation in which personal fitness, w, is the product of gains, f, from uptake of an amount m of resources, and survival probability, S, through begging to level x_i. Thus in a sibship where one sib plays x_i and the other plays x^*, the personal fitness of the sib playing x_i is

$$w(x_i,x^*) = f[m(x_i,x^*)]\cdot S(x_i);$$

see equation (3.3). There are two dyads (2 and 3 in the table) in which one sib carries the rare allele for x_i, and one dyad (4 in the table) in which both sibs carry it, so the replication prospects, W_i, of the gene for x_i, are the sum of

$$W_i = \frac{2f[m(x_i,x^*)]\cdot S(x_i)}{4} + \frac{2f[m(x_i,x_i)]\cdot S(x_i)}{4}.$$

Assuming that shares match relative begging levels, then $m(x_i,x^*) = $ (*i*'s begging level)/(mean begging level) $= x_i/0.5(x_i + x^*) = $ the share gained by the chick playing x_i when its sib plays x^*; and $m(x_i,x_i) = x_i/x_i = 1 = $ the share gained by a chick playing x_i when its sib also plays x_i. To obtain the ESS, we set $\partial W_i/\partial x_i = 0$, and evaluate at $x_i = x^*$. (Technically, we also require that the second derivative of W_i be negative, which is normally satisfied automatically if $f(m)$ is monotonic increasing and $S(x)$ monotonically decreasing, eventually to zero.) Therefore, at the ESS, x^*,

$$\frac{f'(1)}{f(1)} = -4x^* \frac{S'(x^*)}{S(x^*)} \tag{3.6}$$

in which the derivative $f'(1)$ is positive, since the benefit in terms of resources gained increases with begging level x, and the derivative $S'(x^*)$ is negative, since survivorship S will decrease with begging.

Now consider the conditional case, where the deviant begging level x_i is expressed only, say, in role X but not in role Y (Table 3.3). X may be the dominant chick, and Y the subordinate. Note that in dyads 2 and 4, the gains to a B-bearing chick in role X are $m(x_i,x^*)$. However, where the B-bearing chick is in role Y, it actually plays x^*, and gains $m(x^*,x^*)$ when its sib is bb (dyad 3) and $m(x^*,x_i)$ when its sib is Bb (dyad 4). Our genetic accounting becomes

$$W_i = \frac{2f[m(x_i,x^*)] \cdot S(x_i)}{4} + \frac{f[m(x^*,x^*)] \cdot S(x_i)}{4} + \frac{f[m(x^*,x_i)] \cdot S(x_i)}{4},$$

where $m(x^*,x^*) = 1$ (undifferentiable in x_i), and $m(x^*,x_i) = x^*/0.5(x_i+x^*)$. Setting $\partial W_i/\partial x_i = 0$, and evaluating at $x_i = x^*$ to find the ESS gives exactly the same equation as (3.6). So in this model for just two chicks, there is no difference between the ESS for conditional or unconditional gene action.

Why should this be so? The solution becomes equal for the two cases because of an exact compensation of different effects in the accounting around the ESS. In the unconditional case, allele B is selectively neutral in benefits to b in sibship 4, but sustains additional costs if $x_i > x^*$. In sibships 2 and 3, B sustains the additional costs (if $x_i > x^*$), but receives additional benefits. The ESS is tuned by the mean effect of two units of cost bringing about only one change in benefit, which is then reduced because of the dilution effect (hence the '4' in equation 3.6). In the conditional case, allele B is selectively neutral in sibship 3, since it is not expressed. In sibship 2, circumstances are exactly as in the unconditional case: if $x_i > x^*$ allele B gets additional costs, but also additional benefits. In sibship 4, the B-allele in role X can gain additional benefits for additional costs, but it is constrained by the effect on the same B-allele in the sib occupying role Y (which loses some benefits although its costs remain the same). Thus two units of cost now bring about *three* changes in benefit, two of which are positive and one negative. Around the ESS, very tiny changes act additively, so that two positive changes plus one negative change become equivalent to one positive change. Hence, around the ESS, the conditional and unconditional cases give the same solution because the additivity required for Hamilton's rule is restored.

The calculations for sibships with many chicks are simpler. In the unconditional case, we can assume for a large brood of n sibs that proportions follow deterministic but familiar Mendelian ratios, so that for a rare dominant gene, half (i.e. $n/2$) the sibship are B-bearers, and half are not. Then

$$W_i = \left(\frac{n}{2}\right) f[m(x_i,x^*)] \cdot S(x_i),$$

in which $m(x_i,x^*)$ is the share of an x_i player in a sibship of equal numbers of x_i and x^* players, i.e. $= x_i/0.5(x_i+x^*)$. By the usual differentiation technique, the ESS is found as

$$\frac{f'(1)}{f(1)} = -2x^* \left(\frac{S'(x^*)}{S(x^*)}\right) \tag{3.7}$$

(Macnair and Parker 1979). Note that the '2' in equation (3.7) replaces the '4' in equation (3.6) because the dilution effect is lost in a very large sibship. We show later that equation (3.7) above implies a higher level of begging than does equation (3.6) for just two chicks, entirely because of the dilution effect (a unit of extra begging has less effect in small sibships).

In the case of conditional gene action, a sib deviates only if (i) it carries a copy of the deviant gene B and (ii) it occupies role I in the sibship (there are now n possible roles, corresponding to the n sibs). The mutant gene B differs in fitness from the wild type gene only when the mutation is expressed, i.e. when it occurs in a chick in role I. The probability that any of this chick's $(n-1)$ sibs also carry allele B is ½. The accounting now becomes

$$W_i = f[m(x_i,x_n^*)] \cdot S(x_i) + \left(\frac{n-1}{2}\right) f[m(x^*,x_{in})] \cdot S(x^*),$$

where $m(x_i,x_n^*) = x_i/\{[x_i + (n-1)x^*]/n\}$, the share gained by the chick in role I when playing x_i against the rest of the sibship $(n-1$ sibs) that are playing x^*; $m(x^*,x_{in}) = x^*/\{[x_i + (n-1)x^*]/n\}$, the share gained by one of $(n-1)$ sibs playing x^* when its sib in I plays x_i. The ESS becomes:

$$\frac{f'(1)}{f(1)} = -\left[\frac{2n}{(n-1)}\right] x^* \cdot \frac{S'(x^*)}{S(x^*)}. \tag{3.8}$$

which is the same as equation (3.7) if n is very large, and the same as equation (3.6) if $n = 2$.

shall look in detail at the ESSs when begging has costs. The conditional case has exactly the same ESS as the unconditional case at a brood size of two and at an infinite brood size, and probably (although we have not worked it all out) at each intermediate brood size (Box 3.2).

At least at the limits (the smallest and largest brood sizes), Box 3.2 shows that equation (3.8) above gives a general prediction for the ESS amount of begging for this model, and it appears that this is robust against assumptions about conditionality of gene action. Macnair and Parker (1979) derived the resulting equation (3.7), using a population genetics approach for very large broods of full sibs, and showed that the same ESS applies for a recessive gene. The general ESS result (equation 3.8 in Box 3.2) may therefore also be robust against assumptions about genetic dominance.

The case of n sibs has been analysed using direct applications of Hamilton's rule by Lazarus and Inglis (1986) and by Godfray and Parker (1991, 1992). Lazarus and Inglis did not use an ESS approach, and although their results are ESSs (H.C.J. Godfray, personal communication) they are not immediately comparable to the present analyses. However, Godfray and Parker's (1991, 1992) conceptual framework was similar to the analyses presented here in most respects, except that they developed a 'marginal' version of Hamilton's rule (see next section), rather than the genetic approaches used above. Encouragingly, this gives the same solution as equation (3.8), and appears to offer a general approach for continuous strategy games between sibs. We have shown at length how the ESS for the simplest conditional and unconditional begging games is the same, when calculated using a population genetics approach, although the discrete strategy version of the same game is often different (see Chapter 2). This is due to compensatory effects, and to additivity being restored for tiny changes around the ESS. The marginal Hamilton's rule appears to offer a much simpler approach for continuous games, and one that works. We shall now go on to outline the logic of the marginal Hamilton's rule and its conclusions concerning the effects of brood size.

3.3.4 *Begging costs and a marginal version of Hamilton's rule*

In the previous section, we defined the personal fitness of an offspring in terms of its benefits f through gaining m units of resource, and its costs S through begging to level x. Godfray and Parker (1991, 1992) were more general, defining the personal fitness of an offspring simply as $g(m,X)$, where m is again an individual's personal share of resources and X is a measure of the level of competition within a brood. If all offspring in a brood are begging at level x, and one individual increases its begging to a higher level, $x + \delta$, that individual is assumed to gain an extra amount of resources, Δ. Because resources are fixed, this means that all other brood members lose an average amount $\Delta/(n-1)$. The costs of competition are an increasing function of X.

Three types of costs of sib competition have been considered.

(i) *Individual costs (Macnair and Parker 1979, Parker and Macnair 1979, Parker 1985)*

The costs of competition are felt only by the competing individual, so that $X = x$. If the costs of begging are largely lost energy, they are likely to be experienced uniquely by the begging individual. Thus if a sib increases its begging unilaterally from x to $x + \delta$, its personal costs become a function of $X + \delta$. The costs felt by each of the other brood members remain as a function of $X = x$.

(ii) *Shared costs (Macnair and Parker 1979, Parker and Macnair 1979, Parker 1985)*

The costs experienced by each sib increase with the average intensity of begging for the brood, so that $X = \Sigma x/n$, where Σx is the sum of all begging for the entire brood. If one sib increases its begging unilaterally from x to $x + \delta$, the costs to all the sibs become a function of $X + \delta/n$. This formulation is appropriate if the costs of begging are risks of attracting predators, and the predation risk is related to the *average* intensity of begging for the brood.

Box 3.3   ESSs for different types of begging costs

Consider individual costs $(X = x)$. Assume that the strategy for begging at level x^* is an ESS. Thus the benefits of any mutant strategy that begs at level $x^*+\delta$ must be outweighed by its costs. If a chick begs at level $x^*+\delta$, it experiences a change in personal fitness of $g(m + \Delta, X+\delta) - g(m, X)$. The extra Δ of resources it gains is taken from the $(n-1)$ other brood members, so that the fitnesses of its sibs will drop by $g(m, X) - g(m-\Delta/(n-1), X)$. From Hamilton's rule, we know that at the ESS the personal benefits of tiny increases in begging must be offset by the costs via relatives. Thus

$$[g(m+ \Delta, X+\delta) - g(m, X)] = [r][n-1][g(m,X) - g(m-\frac{\Delta}{n-1}, X)] \text{ and}$$

$$[\text{benefits to Self}] = [\text{relatedness}]\cdot[\text{number of sibs}]\cdot[\text{cost to each}]$$

This equation can be Taylor-approximated (see Godfray and Parker 1992) to give the ESS:

$$\Delta g'_m = -\delta g'_X + \Delta rg'_m, \tag{3.9}$$

where g'_m and g'_X are the partial derivatives of $g(m, X)$, evaluated at the ESS and r is the coefficient of relatedness between sibs (Godfray and Parker 1992).

Although this technique for finding the ESS is formally equivalent to the usual technique of differentiating the fitness function with respect to the mutant value x (which cannot be equal to x^*), and then evaluating at $x = x^*$, where x^* is the ESS value, it allows a simple intuitive interpretation. Remember that a small change in the level of competition, δ, results in a change Δ in the resources gained by an offspring. The fitness consequences of these tiny changes can be expressed in terms of their marginal effects (the slopes, g'_m and g'_X), which for very small δ or Δ can be assumed to be linear. Equation (3.9) implies that at the ESS, the marginal benefits to Self of increased begging ($\Delta g'_m$) must be balanced by the marginal costs of the begging to Self ($-\delta g'_X$) plus the marginal costs to Self via relatives ($\Delta rg'_m$). Note that this latter cost is independent of clutch size, since each sib gets $\Delta/(n-1)$ less resources, but there are $(n-1)$ of them. For individual costs, then, the effects of clutch size cancel (see Godfray and Parker 1992) to give equation (3.9). However, the effects of clutch size do not cancel for the case of shared costs,

$$\Delta g'_y = -\{[1+(n-1)r]/n\}\delta g'_X + \Delta rg'_m, \tag{3.10}$$

and summed costs,

$$\Delta g'_y = -[1+(n-1)r]\delta g'_X + \Delta rg'_m \tag{3.11}$$

Note that brood size modifies only the scaling of the $\delta g'_X$ term (Self's marginal begging costs at the ESS) in these latter two cases. Brood size is felt as Self ('1') plus the $(n-1)$ sibs of relatedness r to Self. For shared costs, the costs of competition are divided among all the n members of the brood, so that as n becomes large, the scaling factor asymptotes at $r\delta g'_X$. For summed costs, the scaling factor continues to increase with n. All three equations clearly show that no ESS is possible unless begging costs increase with begging intensity (i.e. unless $g'_X < 0$), exactly as we have concluded in Section 3.3.3.

(iii) Summed costs (Harper 1986)

The costs experienced by each sib increase with the sum of the begging intensity for the brood, so that $X = \Sigma x$. If one sib increases its begging unilaterally from x to $x + \delta$, the costs of all of the sibs become a function of $X + \delta$. Summed costs are appropriate if begging not only attracts predators, but also that predation risk is related to the *total* begging carried out by the entire brood.[6]

Having explained how X (a measure of how begging translates into the costs experienced) differs from x (the individual's begging level), Box 3.3 works through ESS solutions for three different types of costs in terms of the general fitness function $g(m,X)$.

3.3.5 *Effects of brood size on begging levels*

Godfray and Parker (1992) examined the effects on the ESS level of begging of the various types of costs outlined in the last Section. Using the 'mean-matching' assumption that a chick's share of the resources is equal to its begging level relative to the mean begging level for the whole brood, they were able to calculate the extra resources gained (Δ) by an individual that unilaterally increases its begging from the normal level x to $x + \delta$. This could be derived eventually as

$$\Delta = [(n\text{-}1)/nx]\delta \qquad (3.12)$$

(Godfray and Parker 1992). The efficiency with which increased begging produces increased resources is Δ/δ; and this is influenced both by brood size and by the present norm of the begging level, x. The dependency on brood size is due to the dilution effect (the change in the average begging level caused by the unilateral increase from x to $x + \delta$ in one sib's begging). If there are few brood members, this increase has a strong effect in increasing the mean, and so the benefits of the unilateral increase (ratio of Self's begging level to the new mean level) are diluted. If there are many brood members, this effect is trivial. The dependency on the present norm of the begging level, x, is due to the fact that the difference between x and $x + \delta$ decreases as x becomes greater. As competitive begging increases, a given unilateral increase becomes progressively less rewarding. We can call this the 'swamping' effect. Both the dilution and swamping effects are, of course, specific products of the mean-matching assumption. This specific mechanism implies that escalation of begging is more profitable in large broods and at low absolute levels of begging. However, note that if costs are summed, they increase with brood size. Then the summed swamping effect—as we shall see—negates the advantages of dilution in large broods.

---

[6] Lest it be overlooked, the sole distinction between shared and summed costs concerns whether the brood's risk increases as a function of average begging (*shared costs*) or total begging (*summed costs*).

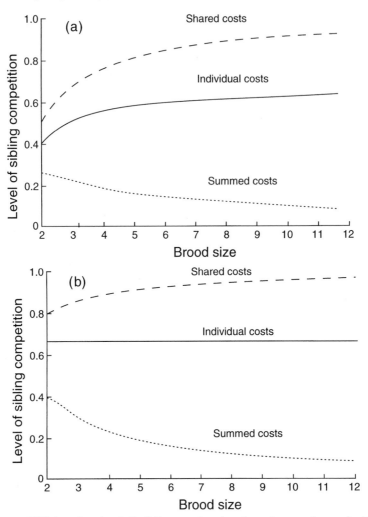

Fig. 3.10 ESS begging for full siblings under the three forms of cost (individual, shared and summed) in relation to brood size, and where survival prospects decrease linearly with begging level x. ($S(x) = 1-bx$; $b = 0.5$; see text.) **(a)** Case where an individual's share of the resource is proportional to its begging level divided by the mean level. **(b)** Case where an individual's share of the resource is proportional to its begging level divided by the mean level for the entire brood (Self's level excluded). (After Godfray and Parker 1992.)

By substituting equation (3.12) into equations (3.9) to (3.11), Godfray and Parker (1992) derived expressions for the ESS level of begging, x^*, in terms of n, r and the 'marginal' benefit: cost ratio ($g'_m/-g'_x$) at the ESS. To obtain some numerical solutions, they followed equation (3.3) in assuming that personal fitness, $g(m,X)$, is the product of two multiplicative components: benefits, $f(m)$, and survival prospects, $S(X)$. Benefits were assumed to follow the form of equation (3.2), and survival prospects decreased linearly with X as $S(X) = 1 - bX$, (where b is a positive constant). Thus, benefits $f(m)$ increase with diminishing returns (see Figure 2.1) above a fixed minimum amount of resource (m_{min}). An individual can gain more resources by a unilateral increase in its begging level. However, these benefits are reduced by the survival function $S(X)$; as an individual offspring increases its begging, the incentives for doing so diminish (literally, the reduction in benefits becomes greater, depending on whether costs are 'individual', 'shared' or 'summed', as explained in the last section).

The ESS levels of begging at different brood sizes, calculated using this 'marginal Hamilton's rule' approach, are highest with shared costs and lowest with summed costs (Figure 3.10a). In addition, the ESS level increases with brood size under shared and individual costs, but decreases under summed costs. For individual costs, this increase must be due to the mechanism of competition, shown in equation (3.12), because the marginal Hamilton's rule for this type of cost (equation 3.9) does not explicitly include a brood size term, n. The increase in ESS begging is here mainly due to the dilution effect, as explained above. A unilateral increase in begging becomes more profitable as brood size increases.

The dilution effect also applies for the other two forms of costs. However, with shared costs, as brood size increases, the costs are distributed amongst a greater number of siblings. Thus personal costs are increasingly distributed to sibs (which are intrinsically less valuable than Self), allowing a greater ESS level of begging. The reverse occurs with summed costs: the cumulative cost soon swamps the dilution effect and leads to a decline in the ESS begging level.

The model results (Figure 3.10a) thus depend on the specific mechanism of competition. Equation (3.12) applies if personal gains are proportional to Self's begging level relative to the brood average. However, suppose that the parent, instead of comparing an individual's level with the brood average, compares the individual's level with the average of all *other* brood members (i.e. excluding the individual in question). For this case, Godfray and Parker showed that if an individual makes a unilateral increase in begging from x to $x + \delta$, its resource share increases by an amount

$$\Delta = \delta/x. \tag{3.13}$$

This can be compared with equation (3.12). Note that there is no longer any dependency on brood size: the dilution effect is now absent, although the swamping effect remains. If equation (3.13) is substituted into the marginal Hamilton's rule equations (3.9) to (3.11), using the same multiplicative formulation of fitness as before, we obtain a somewhat different picture of ESS begging at low brood sizes (Figure 3.10b). With individual costs, sibling conflict

---

Box 3.4 Genetics and the marginal Hamilton's rule

In Section 3.3.3, we were very careful to define precisely the mechanism of gene action, and to consider the Mendelian segregation of progeny and hence the various possible sibships. In developing the marginal Hamilton's rule (see Section 3.3.4) and its implications (present section), we have abandoned explicit genetics in favour of inclusive fitness. This raises one very obvious cause for concern. The marginal Hamilton's rule considers the effects of a *unilateral* deviation by Self on Self's benefits, and on Self's personal and inclusive fitness losses. However, it would seem very likely that these benefits and losses will be altered if deviations are *multilateral* rather than *unilateral*. For instance, consider a large sibship. If a parent carries a rare mutant gene that causes its bearers to behave differently, then in a large sibship, approximately half of the progeny will deviate, not just one. Is the marginal Hamilton's rule approach therefore incorrect?

There are two obvious genetic conditions under which it will be correct. The first is the case of conditionality of action of mutant genes. Suppose that a gene acts only in a particular role in a sibship (there are n unique roles in a sibship of size n; see Section 3.3.3). Such conditional action will then always be unilateral. The second is the case of mutant genes with very low penetrance (i.e. the chances that the gene is expressed is very low). This, too, could preclude multilateral action. Remarkably, however, it appears that the marginal Hamilton's rule may work much more generally. It appears to give the same solution for brood sizes of two. Remember that for the genetic model of begging described in Section 3.3.3 , we obtained no difference between conditional and unconditional gene action in sibships of two (see p. 58). The deviant action can be *bilateral* if gene action is unconditional. However, it also gives the same solution for infinite brood sizes. The model of Macnair and Parker (1979) assumed Mendelian ratios and hence (implicitly) infinite brood sizes; they give the same results as Figure 3.10 for large brood sizes. The result is also unchanged at brood sizes of just three (H.C.J. Godfray, personal communication). Whether there is general concordance for finite brood sizes above four has yet to be determined (the algebra becomes complex because of the genetic permutations of sibships). There seems to be reason for optimism. It appears that for genes of very small effect within the neighbourhood of the ESS, the two approaches (phenotypic and genetic) converge, essentially because very tiny effects behave additively.

The genetic approach has greater realism, although it involves making explicit assumptions about the genetic mechanism. Macnair and Parker (1979) assumed single-locus genetics, as have the previous models described throughout this chapter. It also allows the investigation of evolutionary dynamics (e.g. Macnair and Parker 1979). However, it becomes algebraically complex (see e.g. the non-ESS models of Stamps *et al.* 1978), especially for sibships greater than $n = 3$. The phenotypic approach may (because of its much greater tractability) allow the analysis of more interesting and more complicated biological interactions.

(as measured by ESS begging level) is unaffected by brood size (equation 3.13 is independent of n). Shared and summed costs show similar general trends to before, but since only the costs of begging are now influenced by brood size, relatively higher ESS begging levels are now predicted at small brood sizes. Thus the two mechanisms of mean-matching (Self vs. whole-brood and Self vs. rest-of-brood) can lead to different begging levels, but only at low brood sizes.

Godfray and Parker (1992) also examined how parental fitness would be affected by escalated begging as the brood size increases in the above models. Interestingly, for both types of mean-matching mechanism, the reduction in parental fitness is the same for summed and shared costs, even though the latter results in much greater begging (Figures 3.10a and b). This is because, although the costs to parent (and offspring) are greater if they rise with the sum of the begging, the ESS begging level is reduced in order to compensate: the two effects balance to give the same reduction in parental fitness. With individual costs the reduction in parental fitness is less than with shared/summed costs. These results obviously have implications for parent–offspring conflict over clutch size (see Section 10.1).

But what, we may ask, became of genetics? This may not be a problem, as we discuss in Box 3.4.

The models in the present chapter have all concerned competition over the distribution of parental resources between sibs within the same brood. This has been termed intra-brood conflict to distinguish it from inter-brood conflict, which relates to the allocation of parental resources among successive broods of offspring, i.e. present vs. future broods (see Macnair and Parker 1978, 1979, Parker and Macnair 1978, 1979, Parker 1985). Lazarus and Inglis (1986) investigated inter-brood conflict, and used inclusive fitness as an optimality criterion to model sibling conflict and parent–offspring conflict in relation to brood size. Their analysis assumes no intra-brood conflict (any extra resources gained from the parent are distributed equally among the brood). They predict that conflict should decrease with brood size, essentially because the advantages to an individual offspring of increased investment beyond the parental optimum become diluted as brood size increases. This prediction remains controversial.

3.3.6 *Conditional gene action with unequal sibs*

A feature of all of the begging models discussed earlier (Sections 3.3.2 to 3.3.4) is that all sibs are considered to be equal. Yet we know from empirical studies that this is not necessarily so. Typically in many birds and plants, and in many other taxa, there are often strong competitive differences between siblings, which are likely to affect their relative competitive abilities, either in terms of differences in the costs of begging, or in the effectiveness of each unit of begging. In this section we shall attempt to analyse a simple begging scramble in which the sibs have unequal competitive ability.

Consider two unequal full-sib avian nestlings, A and B, in a nest. Gene action is conditional, i.e. a given gene acts only in a particular role (it acts either when in chick A, or in chick B, etc.) as shown in Figure 3.9. Ideally, the A-chick would take parental resources up to the point where its marginal gains drop to half those of B, as in the hierarchy model (see Section 3.3.1 and equation 3.1), but of course B 'wishes' to do the same to A. We have seen in the last section how this conflict between two equal chicks may be resolved when resources are obtained by a costly begging scramble. What happens if the two chicks are not equal in their begging ability?

Parker *et al.* (1989) examined the case of unequal chicks by assuming that the inequality between A and B acts to modify the effectiveness of their begging levels, x_A and x_B. The realized effectiveness is modelled by multiplying x_A and x_B by the positive constants, a and b, which express the degree of asymmetry between the chicks. Thus if A is the stronger chick, $a > b$ and A begs more effectively. In the begging scramble over the total resource M, the shares of the two sibs follow a form of mean-matching, so that the amount of resources obtained by A is

$$m_A = Max_A/(ax_A + bx_B),$$

and similarly, for B

$$m_B = Mbx_B/(ax_A + bx_B).$$

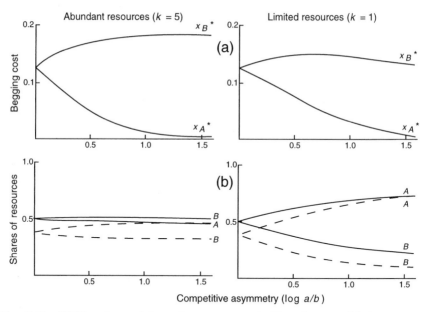

Fig. 3.11 ESS begging and shares in an asymmetrical begging scramble between two chicks in relation to the log ratio of their competitive abilities (a/b). Energetic costs rise linearly with begging level x ($E(x) = jx$, see text). Two levels of resource abundance are shown: abundant resources ($k = 5$) and limited resources ($k = 1$). $M = 1$. (a) ESS begging levels, x_A^* and x_B^*, for the two chicks, A and B. (b) ESS resource shares for the two chicks, A and B. Continuous curves are the total resource gained by each chick; broken curves are the net energy uptake (gain from recourse share minus the energetic cost of begging). (After Parker *et al.* 1989.)

Parker *et al.* (1989) assumed that begging costs were energy losses, $E(x)$, to be subtracted from the energy gained, m_A or m_B (see equation (3.4)). Hence the inclusive fitness of A in relation to changes in its begging strategy, x_A, is

$$f(m_A - E(x_A)) + \tfrac{1}{2} f(m_B - E(x_B)) - F \qquad (3.14)$$

where F is the inclusive fitness of the alternative allele to that prescribing x_A.

The ESS level of begging for A, x_A^*, is found by the usual technique of differentiating the fitness function (i.e. expression (3.14) above, for inclusive fitness) with respect to x_A, remembering that m_B is dependent on x_A. The same procedure can be applied to find the ESS level for B, x_B^*. This gives two equations, one containing x_A and x_B^*, and the other containing x_B and x_A^*. At the ESS, $x_A = x_A^*$, and $x_B = x_B^*$, so the two equations can be 'plugged together' to give a condition that is to be satisfied for the ESS (see Parker *et al.* 1989).

This ESS condition is expressed in terms of the general functions, $f(m)$ and $E(x)$. To proceed further, Parker *et al.* (1989) used an explicit exponential form for $f(m)$ similar to that of equation (3.2), in which shape-constant k scales the amount of resources available. When k is high, resources are plentiful, so that each chick can expect a personal fitness towards the asymptotic level, and vice versa (see Section 3.3.1; Figure 3.3). For simplicity, begging costs were assumed to rise linearly as $E(x) = jx$ (where j is a positive constant). Solutions for x_A^* and x_B^* could then be iterated using a suitable computer procedure.

Some results are illustrated in Figure 3.11, which shows ESS begging costs and shares of resources gained for the cases when resources are relatively abundant ($k = 5$) and relatively scarce ($k = 1$). Remember that the A-chick is competitively superior—it has a higher 'competitive weight' ($a > b$) and can therefore have a greater begging level for the same cost. In the calculations made for Figure 3.11, the competitive weight of the A-chick (a) was set arbitrarily at 1, and B's competitive weight (b) was varied; the competitive asymmetry is shown as $\log[a/b]$. The main conclusions are as follows. (i) The stronger sib A always spends less on begging, $x_A^* < x_B^*$, whether resources are abundant or not (Figure 3.11a). This is simply because A obtains whatever it wants more cheaply, owing to its initial competitive advantage. (ii) The difference in ESS begging levels increases as the difference in competitive weights (i.e. physical strengths of the rivals) increases (Figure 3.11a). B's begging costs increase initially as the inequality between the sibs increases, but as the inequality continues to increase, B's costs may actually decline (e.g. as further insistence becomes more and more pointless). Similarly, as A's basic advantage grows, it needs to pay out less and less to get what it 'wants', so A's costs always decline with the competitive asymmetry. (iii) With scarce resources ($k = 1$), A gets a greater share of the total resource input M, but with abundant resources ($k = 5$), B gets the greater share (Figure 3.11b). (iv) However, A always has the greater *net* energy return (energy uptake minus energy costs: Figure 3.11b). Thus although B may (paradoxically) obtain slightly more of the resource input, it must pay a much greater begging cost than *A* to do so, and as a result its net gains are less than those of *A*. (v) As would be expected intuitively, *A* achieves a higher personal and inclusive fitness than *B*. Fitness differences increase as the competitive asymmetry increases.

Although it is difficult to know as yet how many of the above results are specific to the exact form of the model (e.g. see Section 3.3.1), they do indicate that some complexity may be expected in natural systems. Some conclusions are intuitively obvious (e.g. that the stronger chick should fare better), and appear to be to some extent supported by observations. Others, such as the conclusion that, under abundant parental input, the weaker chick may take more of the incoming food and yet still fare less well, have not as far as we know ever been reported from nature. However, one study provides good evidence for a situation of the type predicted for limited resources (Figure 3.11). Redondo and Castro (1992b) showed that, in general, magpie parents feed chicks more if they beg more. However, although smaller chicks beg more than larger ones, the larger chicks obtained greater resource shares. An alternative explanation for this finding (see p. 193) is favoured by Redondo and Castro.

Parker *et al.* (1989) examined data for three-chick egret broods in relation to this begging model. When an adult enters the nest and lowers its head, the three chicks attempt to grasp the bill crossways (with a scissor-grip). Parents typically withdraw several times before producing food boluses. There is much about the behaviour which suggests a begging scramble between the sibs. The senior sibs (*A* and *B*) obtain significantly more scissor-grips of the parent's bill than the subordinate *C*-chick, and (see p. 49) achieve greater shares of the resource. However, it is not as yet possible to assess the costs of each attempted scissoring, since this action by *C*-chicks may involve direct attack by senior sibs. Parker *et al.* (1989) concluded that egrets more closely fitted the hierarchy model of sib competition (Section 3.3.1) than this begging model, although they could not be certain of this until more accurate methods for assessing costs have been devised.

3.3.7 A sib competition scramble without parental presence

If a group of individuals is exploiting a resource patch, the speed with which each competitor consumes the resources will determine the share that each obtains before the resources are exhausted. Harvesting the resource faster may mean that each unit of resource taken up is harvested less efficiently (e.g. through greater personal energy expenditure or greater resource wastage), but it may pay to sacrifice harvesting efficiency for harvesting speed in order to gain a greater share of the available resource. Parker (1985) (see also Parker and Maynard Smith 1990) proposed a simple model that derives the ESS level of efficiency (or speed) of harvesting under these types of condition.

The following account is based on the work of Godfray and Parker (1992), who applied similar assumptions to a model of sib competition.

As a possible example, Godfray and Parker (1992) envisaged a brood of herbivorous, gregarious insect larvae feeding together on a host plant. (An alternative scenario relates to chemical defences of sibling plants.) The mechanism of competition is the speed with which a larva eats the host plant. When all larvae feed at the same rate, each gains the same amount of resource. If a larva unilaterally increases its feeding rate, it can obtain more food, but this increased rate has costs. Two types of cost of a unilateral increase were considered (see also Section 3.3.4): (i) *Individual costs*, which are borne solely by the individual, e.g. because the measures necessary for increasing its uptake rate are energetically expensive, or because they render the individual's assimilation less efficient; and (ii) *Summed costs*, which are borne equally by each member of the brood and depend on the total amount of competition, e.g. because increasing the uptake rate causes some wastage, so that less resource is then left for all brood-members to feed on.

Once again, let us call the personal fitness of our focal offspring $f(m)$, where m is the amount of resource obtained by that individual. The amount m it gains depends on its own competitive effort, x. By increasing its effort unilaterally from x to $x + \delta$, the individual increases the amount of resource it gains from m to $m + \Delta$, whilst that for each of the other brood members drops from m to $m - \epsilon$. The ESS competitive effort was calculated using the marginal Hamilton's rule, as in Section 3.3.4. The unilateral increase will be favourable up to the point where the benefits are exactly balanced by the costs to the siblings devalued by the coefficient of relatedness r:

$$[f(m+\Delta) - f(m)] = [r][n-1][f(m) - f(m-\epsilon)]$$

[benefits to Self] = [relatedness][number of sibs][cost to each]

see p. 62. Godfray and Parker (1992) were able to show that the ESS has

$$\Delta = \epsilon r(n-1) \tag{3.15}$$

which is but a very simple restatement of Hamilton's rule in marginal form: the benefits to the actor of a tiny increase in personal gains must be offset by the costs to sibs, weighted for relatedness (see equation 3.9 in Box 3.3 for another version of marginal Hamilton's rule). The specific mechanism used to investigate this model was to standardize the fitness of each sibling as $f(1)$ for the case where all offspring have the same feeding rates, x. If one larva increases its feeding rate to $x + \delta$, it obtains $(x + \delta)/\phi$, where ϕ is the mean rate for the brood (see also Section 3.3.4).[7] Costs were assumed to be linear, so that

---

[7] Note that here we expect this form of mean-matching to apply from first principles: if a given larva has a feeding rate of x_i, its gains of the total resource will be proportional to x_i/ϕ, where ϕ is the mean for the whole brood. This is the form of mean-matching that generates the 'dilution effect' (see Section 3.3.2).

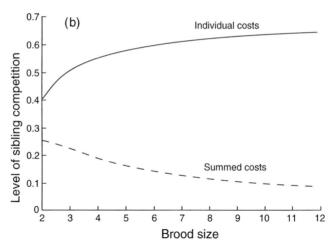

Fig. 3.12 ESS levels of sib competition (as measured by feeding rates) in relation to brood size, for the model of sib competition without parental presence (see text and cf. Figure 3.9). (After Godfray and Parker 1992.)

$$\text{for individual costs: } m = (x/\o)(1-bx); \text{ and}$$
$$\text{for summed costs: } m = (x/\o)(1-bn\o).$$

ESS values for feeding rate, x^*, could be derived by combining these equations with the marginal rule (3.15). The ESS feeding rate (level of sib competition) increases with brood size under assumptions of individual costs (see Figure 3.12). This is because benefits are constrained at low brood sizes by the dilution effect (an individual's increase in feeding haste contributes less to the mean rate as brood size increases; see Section 3.3.5), whereas those costs are unrelated to brood size. However, with summed costs, the ESS feeding rate declines with brood size. Although the dilution effect still operates, the costs themselves are now strongly influenced by brood size (competition becomes increasingly expensive as the number of sibs increases).

Although the present model is structurally very different, the results are very similar to those of the begging model described in Section 3.3.5 (see Figure 3.10a), in which gains follow the same pattern, $(x_i/\o)$.

High levels of sib competition (high feeding rates) mean that resources are lost that would otherwise have been available—either through each individual's lost energy (individual costs) or through resource wastage (shared costs). Figure 3.12 shows the percentage reduction in the resources available to the offspring as brood size increases. For both types of cost, the resources that are lost increase with brood size, but for full sibs the asymptote is 33% with individual costs, and 100% with summed costs. Thus, particularly with summed costs, the optimal clutch size for the parent will be strongly constrained by this form of scramble competition amongst sibs.

3.3.8 *Hierarchy vs. scramble in nature*

Our last example, competitive feeding, is a 'pure scramble,' and there are good reasons to expect gains to fit mean-matching of the 'Self vs. whole-brood' form. There are many other cases to which this may apply. For instance, if feeding gains depend on relative speed of searching for food in the nursery (e.g. non-cannibalistic tadpoles and aquatic insect larvae in small catchments might be prime candidates), we would expect *a priori* the same form to apply. Parental absence is not a necessity for a pure scramble, as we have explained in Section 3.3.2, so that if a parent simply delivers food randomly to each 'begging event unit,' the same form of mean-matching is expected. In particular, when the parent exerts little control over the flow of resources to offspring, the 'Self vs. whole-brood' type of mean-matching is entirely plausible. Biochemical scrambles should also follow this form, e.g. apples on a branch extracting nutrients from the maternal xylem (see Chapter 16), placental extractions by fetal mammals (Chapter 13) and analogous diffusion to pouch-dwelling pipefish eggs (Chapter 14).

Many natural cases consist of a mixture of scramble and hierarchy elements. The frantic competition of nestling egrets includes both hustle and jab, the balance of which differs sharply across hatching order. The first-hatching *A*-chick uses a fair amount of aggressiveness during its first week or so, and can then often get by on hustle and occasional threat. The *B*-chick, by contrast, keeps the attacks (on *C*) going for much longer, and the *C*-chick often eschews battle whenever possible. For nursing mammals, there are some cases where nursing scrambles are embellished by aggressive interference, and so on. Indeed, cases of 'pure hierarchy' may be quite rare, existing mainly where there are strong initial asymmetries, such as some extreme hatching asynchrony in birds and some cases of extended contact between siblings born in different seasons (marsupials and pinnipeds).

Summary

1. Both co-operation and conflict occur frequently within families. They are too intimately enmeshed to be regarded meaningfully as opposites: conflict exists between co-operating individuals (e.g. during biparental care), and rival factions often form mutually beneficial alliances in order to compete effectively.
2. Family structure is depicted in the form of three games or 'social dimensions', which share the common currency of parental investment. These are sexual conflict (between the two parents), parent–offspring conflict and sibling competition. Each dimension impinges on the other two.
3. This chapter focuses mainly on non-lethal forms of sibling rivalry, wherein an offspring's primary decision concerns its own degree of selfish consumption of parental investment, a continuous variable. Two major classes

of 'graduated-selfishness' models are presented: the *hierarchy models* (whose defining feature is a rigid linear dominance order) and the *begging models* (where success is achieved solely via scramble competition).

4. In a two-sib hierarchical brood, a simple optimality solution exists: the despotic *A*-sibling (one in total control of how much PI it consumes) should follow Hamilton's rule, taking PI until its marginal gain declines to half that of its full sibling, *B*. The actual difference between the two siblings' consumption depends on their fitness functions (whether these differ between the sibs) and on resource availability.

5. In a three-sib hierarchical brood, the best decision for *A* depends on *B*, and vice versa, so an ESS solution is required. The *C*-sibling consumes by far the least PI, while *A* and *B* take more similar shares and hence achieve more similar personal fitnesses. Surprisingly, *B*'s personal fitness is least affected by variations in total PI, but only *C* faces real hardship (because *A* essentially takes from *C*). Some field data from egrets are broadly consistent with this model.

6. A new four-sib hierarchical model shows that, as with the three-sib version, the lowest-ranking sibling's personal fitness lags far behind that of the others.

7. *Begging models.* In non-aggressive 'scramble' forms of sibling competition, each offspring chooses a level of effort ('begging') and is rewarded (by parental feeding) in proportion to some assessment of its relative signal. Begging is assumed to have personal *costs*, such as the energy expended in signalling (individual costs) or some increased risk or loss to the whole brood (summed or shared costs).

8. Gene expression in the begging models can be either *conditional* (shown in just one of two possible roles in a dyadic interaction) or *unconditional* (shown equally in either role). In the (unrealistic) case of cost-free begging that is based on unconditional gene action, the individual should escalate its signal to infinite levels. A dominant gene for such action would always gain the benefits and only share in the costless contingency of sometimes having hyperselfish nest-mates. Ironically, a conditional gene for escalated begging can be constrained by kin selection, but begging will gradually escalate to infinity as one allele replaces another by small steps.

9. However, although genetic mechanisms *can* influence begging model predictions for discrete strategy models, continuous strategy models appear to be much more robust because tiny deviations around the ESS tend to be additive in their effects. (A detailed explanation of this argument is given in Box 3.2.)

10. A 'marginal' version of Hamilton's rule (Godfray and Parker 1992) allows ESS solutions to be derived for continuous-strategy games, using an inclusive fitness approach (see Box 3.3). It gives the same ESS solutions for a large array of models (in fact, for all of the cases so far examined) as does the explicit genetic approach (see Box 3.4).

11. A unilateral increase in individual begging level can produce varying effectiveness (the efficiency with which an extra unit of effort attracts an extra unit of PI), depending on brood size. Most models assume that gains from begging follow some form of 'mean-matching'. Under the assumption that parents apply mean-matching by paying out to an individual's begging level in proportion to its magnitude *relative to that of the entire brood's mean*, a unit increase in begging level is less effective in small broods because it also pulls the mean level up (the dilution effect). In contrast, if parents pay out to an individual's begging level in proportion to its magnitude *relative to the mean for the rest of the brood*, there is no dilution effect.

12. The type of mean-matching used by parents (about which little is known empirically) can radically alter the predictions of begging models. With mean-matching of the 'Self vs. whole-brood average' type, ESS begging levels are highest with shared costs, followed by individual costs, and in both cases there is an increase in begging with brood size. ESS begging levels are lowest with summed costs, in which case they actually decrease with brood size. When mean-matching is of the 'Self vs. rest-of-brood average' type, ESS begging levels are similarly ordered, except that for individual costs the ESS is independent of brood size.

13. If siblings differ in their ability to solicit PI, such that some can afford to expend greater effort than others, the ESS levels must be adjusted via the 'competitive weights' of each. Analysis of an explicit model generates five results (Parker *et al.* 1989): (i) a stronger sib should always spend less on begging; (ii) as the ratio of competitive weights (stronger/weaker) increases, the weaker sib should initially increase its begging (but only up to a point beyond which it declines), while the stronger sib can afford to conserve by begging less and less; (iii) when resources become scarce, the stronger sib's share of the total increases; (iv) the stronger sib always gets the greater net energy return, although the weaker sib may (in some cases) ingest more food while paying higher begging costs; and (v) not surprisingly, the stronger sib always achieves higher fitness.

14. Sibling scramble competitions can also exist without parents, if offspring are deposited in a food-bearing nursery (e.g. phytophagous insects, parasitoids, puddle-dwelling tadpoles, etc.). Escalations in harvesting rate are likely to reduce the efficiency of converting food into offspring tissue, which can be modelled as individual or summed costs to yield a variant of the marginal Hamilton's rule. Pay-offs are expected (from first principles) to follow a 'Self vs. whole-brood average' form of mean-matching. The ESS feeding rates differ according to which costs are most relevant, rising with increased brood size under individual costs and falling with increased brood size under shared costs. Under summed costs, the wastage can be sufficient to reduce the optimal clutch size for parents.

4

Theory III: fatal sibling competition

... and it came to pass, when they were in the field,
that Cain rose up against Abel his brother, and slew him.
And the Lord said unto Cain, Where is Abel thy brother?
And he said, I know not: Am I my brother's keeper?

(Genesis, Chapter 4: 8–9)

We have seen how individual brood-mates (and, by extension, individual offspring with prospective future siblings) are expected to practise 'graduated selfishness,' the incremental consumption of limited resources that siblings need. One emphasis was on the continuous nature of such decisions, with the clear implication that they are relatively flexible, even reversible; high selfish consumption one day may be followed by low (generous) consumption the next. It is obvious that this can be taken to extremes, either through scrambles (wherein Self simply takes such a large portion that sibs cannot be sustained on what remains) or through the more direct application of aggressive acts. If an individual offspring can assess the ecological realities it faces with some accuracy, and if circumstances require that someone must die, then the choice is usually a simple one.[1] Less apparent is the possibility that individual family members occupying empowered roles (e.g. dominant sibling or parent) may be able to anticipate future shortfalls, a situation that has come to be labelled as *pending competition* (Stinson 1979), and to use that information to make similar decisions even before the crisis arises. For example, a bully nestling might sacrifice a sib while it is tiny, cheap and easily killed. Furthermore, parents presumably make some kind of guess about the nursery's carrying capacity when first setting initial family size; they may still choose over-production for various reasons, but the degree of that life-history decision should be tuned prudently by natural selection.

---

[1] Erstwhile US Senator and presidential hopeful Barry Goldwater, speaking in a very different context, made this point succinctly, 'Extremism in the defense of liberty is no vice!'

4.1 Brood reduction

Although earlier authors had mentioned tangentially that the loss of one or two hungry mouths would leave more food for the survivors, it was Lack (1947, 1954) who pointed out that such an eventuality might well be part of the parental gambit. It became part of his theory of clutch size and a key element in his ideas on why many avian parents effect an asynchronous hatching of their eggs. However, the term 'brood reduction' was coined by Bob Ricklefs as part of an undergraduate project testing some of Lack's predictions with songbirds (Ricklefs 1965). Over the next 25 years, the term was used with a vast diversity of meanings. To some it connotes a catch-all for any losses of dependent young; to others, including us, it is more usefully restricted to the original usage of within-brood partial mortality that is due to sibling rivalry *per se*. We use this narrower version (for further discussion, see Mock 1994).

4.1.1 *O'Connor's thresholds*

Conditions for brood reduction were first examined in a remarkable paper by O'Connor (1978), which combined development of a theoretical framework with an attempt to review the existing ornithological data with respect to the new predictions. O'Connor used a simple inclusive fitness argument to derive the condition under which it should pay a senior sib to eliminate a junior nestling. The tacit assumption was that this action can be achieved at no cost to the senior sib. Suppose that there are n nest-mates in competition for the limited resources provided by the parent. The survival (or other fitness) prospects, S, of each nestling are likely to decrease with the number in the brood, so we expect that $S(n)$ will be a declining function. If the senior chick allows a given junior nestling to survive, then the prospects for each sib will be $S(n)$; if it eliminates that junior chick, the prospects for all remaining chicks become $S(n-1)$. So it pays the senior sibling to eliminate a junior nest-mate whenever its own inclusive fitness is greater after one chick is removed. Assuming the nest-mates to be full sibs, inclusive fitness is a function of Self's fitness plus half the fitness of the remaining chicks:

$$S(n-1) + \tfrac{1}{2}[(n-2)\cdot S(n-1)] > S(n) + \tfrac{1}{2}[(n-1)\cdot S(n)] \qquad (4.1)$$

$$\frac{S(n-1)}{S(n)} > \frac{n+1}{n}. \qquad (4.2)$$

O'Connor pointed out that it can similarly pay a parent to eliminate one of the chicks, if the resulting gains to the rest of the brood are high enough. Using the same inclusive fitness approach, but this time from the parent's point of view, he calculated that filial infanticide would be in parental interests when

$$\frac{S(n-1)}{S(n)} > \frac{n}{n-1} \qquad (4.3)$$

Finally, he also realized that a victim chick's own inclusive fitness could theoretically be enhanced if, under extreme conditions, it effectively eliminates itself from competition with its nest-mates. For such suicide, it is required that

$$\frac{S(n-1)}{S(n)} > \frac{n+1}{n-1} \tag{4.4}$$

Of these three thresholds, the easiest one to satisfy is the first (condition 4.2, for siblicide); next easiest is the second (condition 4.3, for filial infanticide). The last condition (condition 4.4 for suicide) is the most difficult to satisfy. Of course, if condition 4.4 is satisfied, both the other conditions will already have been realized. Put another way, when all conditions are met, the strongest selection will be on the senior sib to perform siblicide. Thus O'Connor's interpretation of brood reduction differed from previous interpretations due to Lack (1954) and Ricklefs (1965), who envisaged brood reduction solely as an adult adaptation. O'Connor calculated the threshold probability of mortality—the difference in survivorship above which it becomes favourable to reduce the brood number from n to $(n-1)$. These thresholds, from the three different 'viewpoints,' are shown in Figure 4.1. The threshold for infanticide lies only just above that for siblicide at larger brood sizes; the suicide threshold is well above both at brood sizes that apply in most birds. All three thresholds decrease as brood size increases—it is obvious that the right-hand sides of conditions (4.2) to (4.4) are greatest at low values of n. This simply reflects the fact that eliminating one consumer has a greater impact when each individual represents a larger share of the original pie.

4.1.2 Large broods: 'optimal brood reduction'

When the difference in survivorship lies between the O'Connor thresholds (Figure 4.1), there will be conflicts of interest. For instance, if the survivorship difference between broods of n and $(n-1)$ chicks lies above the curve for siblicide but below that for infanticide, there will be parent–offspring conflict over brood reduction. This amounts to parent–offspring conflict over family size (anticipated by Hamilton 1964a, b and Trivers 1974; considered further in Chapters 7–10).

For large broods, such as are common in many insect species, we can consider the survival prospects, S, of each offspring to be a continuous function of n, the clutch/brood size. Parker and Mock (1987) used an inclusive fitness approach to deduce the optimal amount of brood reduction from the point of view of the senior sibs. The senior sibs may be the first larvae in a clutch to hatch. They have the unique opportunity of killing and/or cannibalizing their unhatched sibs without danger of retaliation. The details of the Parker and Mock model will be considered more fully in Section 10.1.3, but it

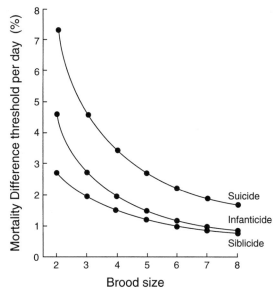

Fig. 4.1 O'Connor's thresholds for a family containing only full siblings. If the increase in nestling survival due to reducing the brood by one chick exceeds the siblicide threshold (lowest curve), it pays senior sibs to lose a junior sib. If the increase exceeds the infanticide threshold, it pays the parent to lose one chick. If the increase exceeds the suicide threshold, it pays any given chick to sacrifice itself. (After O'Connor 1978.)

predicted that senior sibs should reduce the clutch (by siblicide or some other mechanism) until inclusive fitness of a given senior sib is maximized. Parker and Mock also predicted that (on the simplest assumptions) there should be no conflict among the senior sibs over the number of junior sibs that are to be killed. The optimal amount of siblicide depends on the starting clutch size (i.e. on the parental strategy), on the form of $S(n)$, i.e. the function defining survival prospects in relation to brood size, and degree of relatedness among sibs.

4.1.3 *Siblicide thresholds from the hierarchy model*

The hierarchy model of sib-competition (Section 3.3.1; see Figures 3.1 to 3.4) predicts how selfish stronger sibs should be with respect to weaker sibs when there is a rigid dominance hierarchy (i.e. access to resources strictly follows rank). Suppose the sib hierarchy consists of several young birds in a nest and the resource is food brought by parents. For the two lowest-ranking chicks, the more dominant (penultimate) of the two has control of the remaining food (that left by its 'more senior' sibs). Parker *et al.* (1989) showed that the amount of this remaining food consumed by the more dominant chick is determined by the marginal gains rule (equation 3.1): it should take food until its marginal gains drop to half those of the junior full sibling (see Section 3.3.1). They also derived a similar marginal gains rule for the last three chicks in the hierarchy: the relationship between the last two chicks remains the same as in equation (3.1), but there is a further, more complex, marginal gains rule that determines how much food the most senior of the three chicks should take.

Thus in the hierarchy model, the decision concerning the amount of food to take is set entirely by marginal gains, that is the *gradients*, of the relationship between f (the personal fitnesses) of given chicks, and m, the amount of food each consumes out of the total food base. When $f(m)$ follows an exponential law of diminishing returns, as in equation (3.2) (see Figures 3.3 and 3.5), the amount of food to be taken by each of the three chicks can be calculated explicitly (see p. 47). It is possible, for the same model, to compute the ESS amounts to be taken by four chicks in a hierarchy. Again, the ESS food share for the most dominant of the four chicks is likely to be determined by some form of marginal gains rule, although this remains to be established analytically.

A hierarchy model with four siblings and progressively decreasing food abundance produces a complicated picture (Figure 4.2), which we shall now explore. As before, the effect of decreasing food supply can be modelled by decreasing shape parameter k in equation (3.2). Moreover, as before, we call the four chicks A (most dominant), B, C, and D (least dominant). The personal fitnesses of the chicks are shown in Figure 4.2. When resources are plentiful ($k = 3$), all of the chicks do well and even D has a personal fitness of 0.86 (maximum value $= 1.0$). In the range for k shown here (Figure 4.2), all chicks survive so long as $k > 0.79$, although the decline in D's personal fitness is steep. As expected (from Section 3.3.1), the personal fitnesses show a marked 'runting' effect: the greatest fitness difference is between the two most subordinate sibs. If and when food abundance drops to $k = 0.79$, D is sacrificed, and the personal fitness of A and B increase at the expense of a sharp drop in the personal fitness of C. However, at this threshold, the changes in inclusive fitness of chicks B and C and the parents are rather small (see also Figures 10.1 and 10.2). For that matter, so is the change in inclusive fitness of the victim sib, D! From D's point of view, its only loss in dying at the threshold (where its personal fitness is zero) relates to the change in distribution of resources among A, B, and C, which become less equal than if it were to continue to live. The same types of effects are visible at $k = 0.45$, the threshold at which chick C is eliminated. At $k = 0.18$, even chick B is sacrificed, and at this point chick A's personal fitness cannot experience an increase for the simple reason that it is already consuming all of the food. The fitness of A therefore continues to decline as k decreases.

In our numerical four-chick example (Figure 4.2), the total investment value of M was set at 4, and m_{min} was set at 0.1 (recall that m_{min} is the minimum resources needed for any survival prospects: see Figure 3.4). Although this

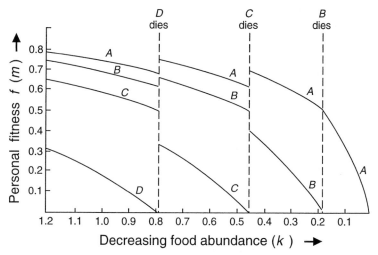

Fig. 4.2 Siblicide thresholds in the hierarchy model. The personal fitness, $f(m)$ of each of four chicks in a brood is plotted against decreasing k, a measure of decreasing food abundance in the exponential form for $f(m)$ given in equation (3.2). In this example, $M = 4$; $m_{min} = 0.1$. If a chick dies it is assumed not to have consumed any resources (M stays at 4 throughout). The thresholds are the levels of food abundance k at which one of the chicks is lost from the brood (see text).

model needs further study, it is clear that the effects shown in Figure 4.2 are not simply due to the fact that some resources (m_{min}) are freed up at each siblicide threshold (when $M = 4$ gets re-allocated across a brood of one less, thereby conserving m_{min}). This can be established unequivocally by setting m_{min} at zero: the same form of results is achieved, although the siblicide thresholds move to slightly lower values of k.

It seems counterintuitive that the ESS division of parental resources among surviving chicks should change as the siblicide threshold is crossed. This effect stems from the fact that the ESS shares are dependent only on the relative *gradients* on the personal fitness function $f(m)$, and not on the absolute value of personal fitness (see equation 3.1). The gradient $f'(m)$ is positive when $f(m)$ is zero (for the chick that is about to die), but as soon as this chick is removed, there is no $f'(m)$ for that chick. The ESS shares for the remaining chicks therefore alter. This result is paradoxical, and invites further investigation of the implications of behaviour around the siblicide thresholds. *A further complication in nature is that parent birds may cut M in what looks like a per-capita correction after siblicide (see Section 6.3.1). Cattle egret parents bring two-thirds as much food to the two senior sibs after the runt is removed, etc., paradoxically producing a 'pay cut' to the seniors, which were busily consuming C's share of the food before it died (Mock and Lamey 1991). Brown pelicans do much the same (Ploger, in press).

Whatever the value of k or m_{min}, the siblicide threshold occurs when the personal fitness of the next-to-lowest-ranking (penultimate) chick becomes 0.5. The inclusive fitness differences for parents and senior sibs across each siblicide threshold did not appear to exceed 3%, but that for the penultimate chick could drop by more than 10%. Not surprisingly, in a family on the cusp of brood reduction, the *least empowered survivor* is most affected by the victim's death and this change is twofold. On the logical 'plus' side, any resources the victim may have been consuming become available to the others, including the penultimate sib; on the 'minus' side, the penultimate chick moves into the unenviable 'next-victim' slot. Under the formal assumptions of the model, of course, the dying chick is not allowed to consume resources anyway, so the penultimate chick gains nothing.

One interesting feature of this model is that O'Connor's thresholds—at least in their original form—vanish completely. From either the parent's or the victim's point of view, the only interest in avoiding the brood reduction event rests entirely on the increasingly skewed shares that result from re-allocating PI after one sib is lost from the brood. This means that parental and victim inclusive fitnesses are affected only very weakly (see Chapter 10), a feature that may help to account for the seeming lack of parental interest in siblicide, and the apparent resignation sometimes seen in victims (see Chapter 11).

What about siblicide thresholds in the non-hierarchical begging models? This topic remains uninvestigated. Much would depend on whether begging serves as a true and honest signal to the parent of offspring condition (Godfray 1991), or whether it is the result of a scramble competition between unequal sibs in which the parental mechanism of resource allocation is fixed, as in mean-matching (see Section 3.3.5). In the former, we might expect brood reductions, like parental investment, to be geared to parental interests (see Godfray 1991). In begging scrambles of the type depicted in Section 3.3.5, the thresholds may be ordered rather similarly to those of the hierarchy model, especially if high levels of inequality exist between sibs.

4.1.4 *Population genetic models of siblicide evolution*

All of the models of brood reduction discussed so far have been based on inclusive fitness approaches. The earliest population genetics approach to the evolution of siblicide was that of Stinson (1979): it gives rather different results to O'Connor's inclusive fitness approach, as do the ensuing studies of Dickins and Clark (1987) and Godfray and Harper (1990), although the latter tends to support O'Connor under the assumption of conditionality.

An insight into the reasons for the differences between the inclusive fitness and population genetics approaches is best given by the work of Godfray and Harper (1990). Remember that one of the conditions for Hamilton's rule to operate is weak selection (Section 2.3.1), i.e. that the trait has relatively small effects on fitness. This is not likely to be so for brood reduction (unless it forms part of a graduated selfishness continuum, as in the last section). The conditions for spread of a gene for, say, siblicidal aggression are very much dependent on the presence of a hierarchy within the sibship, and on an identifiable runt that can be eliminated readily[2]. This prerequisite relates to conditionality; as we have seen, Hamilton's rule may not apply unless gene action is conditional (Section 2.4).

Godfray and Harper distinguish between two forms of siblicidal gene action.

1. Random-victim model

A siblicidal individual chooses a victim at random. If there is more than one potential aggressor, an aggressor is picked at random from this group in proportion to its siblicidal tendency (the penetrance of the siblicidal gene).

2. Runt model

An individual in a particular role ('runt') is pre-designated as the victim. If one or more members of the rest of the brood are siblicidal, the runt dies. The role of runt is determined environmentally, and a runt is not allowed to be an aggressor even if it carries the siblicide allele.

The difference between the two models can be seen most clearly in broods of two sibs. According to the runt model, there is strict conditionality, following the dyadic relationship shown in Table 2.1. Suppose that allele S in Table 2.1 is now 'siblicide'; its alternative allele is s. Siblicide occurs only when the chick in the non-runt role (X) carries the siblicide gene (sibships 2 and 4), in which case it kills a sib that has a 50% probability of bearing the siblicide allele as well. In the random victim model, Godfray and Harper assumed there to be no identifiable runt, although the model applies to this case provided that siblicide occurs when at least one chick (in roles X or Y) carries the allele (including sibship 3). In other words, gene action is unconditional in the random victim model, but conditional in the runt model.

Godfray and Harper investigated the conditions for spread of siblicidal genes with differing levels of penetrance. If a gene has very low penetrance, its effects are expressed with very low probability. Very low penetrance has essentially the same effect as conditionality: it ensures that the gene frequency of the allele is approximately equal in potential actor and recipient. (This can also be interpreted as ensuring additivity: H.C.J. Godfray, personal com-

---

[2] There are various ways, some quite subtle, by which one offspring can slide into the 'designated victim' slot. Nursery-mates may simply develop at differing rates, either because of intrinsic variations (e.g. more or less favourable match-ups of maternal and paternal haplotypes) and/or many kinds of external influences. The point is that the victim role is not always *imposed* by parents. In particular, though hatching asynchrony is one conspicuous way of creating a victim, it is not the only approach, as others have argued.

munication.) According to both models, low penetrance siblicide yields the same condition for spread as an inclusive fitness approach (invasion occurs in areas a and b, Figure 4.3a).

However, if there is complete penetrance, so that the siblicidal gene is always expressed (when the model conditions permit it), the condition for spread of the gene can differ from that predicted by inclusive fitness. As might be expected, the runt model gives the same result as inclusive fitness if there are two sibs. If the personal fitness of each sib is 1.0, a siblicide gene with full penetrance spreads if the benefit b to the remaining sib is greater than ½. But as the brood size increases, the benefit b of the brood reduction to the entire surviving brood must be increasingly greater than ½ to allow spread, and the benefit must approach unity for large clutches (see area a, Figure 4.3a). However, Godfray and Harper found in the runt model that once a gene of low penetrance has spread to fixation, a gene with slightly higher penetrance can invade, until the population consists of individuals fixed for an obligate siblicide gene (complete penetrance). Thus, whatever the clutch size, the predictions of the runt model become identical to those from inclusive fitness. Figure 4.3a shows only the zones of invasion of genes with differing levels of penetrance (complete, or very low) into populations in which the alternative allele is either non-siblicidal or fully siblicidal.

By contrast, in the random victim model with two sibs, a siblicide allele with complete penetrance will spread if $b > 1/3$, i.e. more easily than is suggested by inclusive fitness (as expected for unconditionality; see Section 2.4). Again, conditions become more restrictive as brood size increases, but now it is easier for a gene with complete penetrance to spread (regions a and b, Figure 4.3b) than one with low penetrance (only region a).

For both models, the condition for spread of a siblicide allele in a non-siblicidal population differs from the reverse condition (namely for the spread of non-siblicide alleles into a siblicidal population). In area d of both graphs (Figure 4.3a), neither a population fixed for siblicide nor one fixed for non-siblicide can be invaded by its alternative allele, and in this area the evolutionary fate depends on the starting conditions. In area d of the runt model, although siblicide is technically an ESS (*sensu* Maynard Smith 1982a), it is not a continuously stable strategy or CSS (*sensu* Eshel and Motro 1981; Eshel 1983). A CSS is a strategy that is stable at fixation against invasion by mutant alternative strategies (i.e. it is an ESS), and furthermore, if perturbed slightly from the stable value, it is also stable against invasion. In area d of Figure 4.3a, any small perturbation away from obligate siblicide allows the invasion of an allele with a smaller probability of siblicide (lower penetrance) to spread to fixation. This allele can itself be replaced by one with even lower penetrance, and so on, until the population becomes fixed for non-siblicide, which is both the eventual state and a CSS. Bearing in mind that, for similar reasons, obligate siblicide is expected to be the eventual state in both areas a and b (in b the penetrance will gradually 'shuffle up' to obligate siblicide),

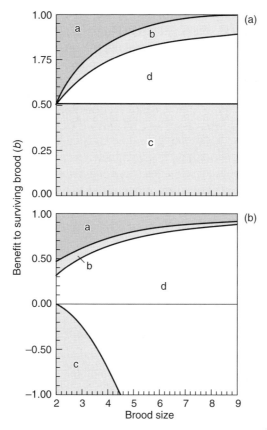

Fig. 4.3 The spread of siblicidal and non-siblicidal alleles in Godfray and Harper's models. Siblicide results in a benefit, b, that is shared among the remaining members of the brood. A value of $b = 1$ is equivalent to one brood member, so that the parent favours brood reduction when $b > 1$. **(a)** The runt model. If the population is fixed for non-siblicide, a siblicidal gene of very low penetrance will invade in areas a and b, and a gene with complete penetrance spreads only in area a. If the population is fixed for siblicide, a non-siblicidal gene (of any penetrance) can spread in area c. In area d, both siblicide and non-siblicide are ESSs, but only non-siblicide is a CSS (see text). **(b)** The random victim model. If the population is fixed for non-siblicide, a siblicidal gene of complete penetrance will invade in areas a and b, but a gene of very low penetrance can spread only in area a. Note the difference here from upper Figure **(a)**. If the population is fixed for siblicide a non-siblicidal gene (of complete penetrance) can spread in area c. In area d, both siblicide and non-siblicide are stable (see text). A negative value for b means that brood reduction is actually harmful to the remaining sibs. (After Godfray and Harper 1990.)

then the boundary between b and d is the unique one for the transition from siblicide to non-siblicide. This is the O'Connor threshold for siblicide; hence there is no discrepancy between the population genetics result and the inclusive fitness result in the runt model.

In area d for the random victim model (Figure 4.3b), if there is any per-turbation away from either full siblicide or non-siblicide, the only mutations that will spread are those that tend to return the population to its original state. Either strategy is thus both an ESS and a CSS in area d of this model.

Of course, there are many possible rules for siblicide other than those of the runt and random victim models. Each will have its own properties, depending on the exact rule for gene action. For example, Godfray (1987a) analysed the fate of a 'fighting' gene that causes its bearer to kill *all* other sibs, not just one. This is a plausible proposition for the evolution of siblicidal behaviour of parasitoid wasps (Godfray 1994, see Chapter 15). In his models, siblicide continues until only one larva per sibship survives: the chances of being the survivor are random for all bearers of the fighting gene. For full sibs, such genes are likely to spread easily if clutch sizes are small (e.g. < 4), and even more so if some members of the clutch are unrelated (see also Smith and Lessells 1985). The presence of solitary, fighting young appears to be a locally absorbing state; conditions for the spread of a fighting gene into a non-fighting population are much easier to satisfy than those for spread of a non-fighting gene into a fighting population.

This outline of Godfray and Harper's analyses gives some idea of the complexity of applying a population genetics approach rather than using inclusive fitness. It indicates once again that the exact mechanism of gene action can be important in determining the outcome of interactions between relatives. We should expect siblicide to obey the O'Connor threshold if there is an obvious runt that is typically the victim, and that cannot itself act siblicidally. Something very similar to this is common in birds with hatching asynchrony and finds clear parallels in many other taxa (see Chapters 12–16). The odds are generally weighted strongly against the last hatcher. However, with synchronous hatching, the random victim model could be appropriate, leading to very different predictions. The complexities of models with an adaptive basis to runting—as proposed in the siblicide threshold version of the hierarchy model (Section 4.1.3)—have yet to be analysed in terms of population genetics.

4.2 Overview: some additional considerations

As stressed earlier, in these three theoretical chapters we have simplistically regarded parents as resource providers, rather than as players in a dynamic game of supply and demand (Figure 3.1). Many investigations have analysed the interaction between supply and demand (e.g. Parker 1985, Harper 1986,

Hussell 1988, Godfray 1991). The balance between the two both influences and is influenced by the other games (sib competition and sexual conflict). Chapters 6–10 investigate the supply side and its interactions with sibling competition.

Graduated selfishness models (Chapter 3) generally predict an increase in selfishness as food abundance decreases, an effect that can ultimately result in brood reduction if conditions are harsh enough. This may create a false impression; there are cases where sib selfishness appears to be unmoderated, such that siblicide is obligate, without any obvious (or realized) stringency of resources. For example, in some parasitic wasps, siblicide nearly always proceeds swiftly until only one survivor remains (see Godfray 1987a), and in certain birds (Stinson 1979, Anderson 1990a, Mock *et al.* 1990; see also Chapter 12) two eggs are laid, and the stronger sib (the first to hatch) routinely kills the weaker one. Assuming that the conditions in these cases favour brood reduction rather than some more measured form of selfishness, the problem here is mainly to explain the laying of more than one egg by the female parent (see Parker and Mock 1987). For some raptors, boobies and others, this is perhaps most easily explained by Dorward's (1962) 'insurance-egg hypothesis', which predicts that if there is a significant risk that the offspring represented by the first egg may fail (through infertility, early embryonic death, hatching failure or immediate post-hatching mortality), the parents may optimally insure against reproductive failure by laying a second egg. Evolution of an insurance egg is expected if the cost of that egg is less than the benefit of the increased probability of producing a surviving offspring. As discussed earlier (Chapter 1), there are additional incentives for parental overproduction.

Some models of brood reduction tacitly assume that there are no costs or benefits to the action, other than through inclusive fitness. Direct costs (e.g. energy expenditure, risk of injury, etc.) may often be associated with siblicide. It may pay a given senior sib not to perform the act if there is a likelihood that another sib will do it instead. (Godfray and Harper (1990) suggest that the techniques suggested by Eshel and Motro (1988) and Motro and Eshel (1988) could be used to analyse such problems.) Thresholds for brood reduction are likely to become higher if such direct costs are non-trivial. Conversely, if there are additional benefits to siblicide (e.g. the aggressor cannibalizes the victim, as in many insects), then we would expect lowered thresholds and greater amounts of siblicide.

Most models assume that the fitness of all sibs is (at least potentially) equal. If offspring vary in their intrinsic quality, and are hence likely to have discernibly different reproductive values, this will certainly affect siblicide thresholds, and of course also infanticide and suicide thresholds (Temme 1986, Haig 1990). Note that here we are not considering differences that arise by chance variations in the amounts of resource gained (e.g. through environmental effects such as hatching asynchrony), but differences between offspring that would cause them to have different relationships between personal fitness, *f*, and the amount of resources gained, *m*. Much will depend on the stage at which differences in offspring quality become apparent.

We have stressed many times and in different contexts that the mode of gene action may be vital in determining the extent of sib selfishness. An inclusive fitness approach does not usually take such subtleties into consideration. Rather, it assumes certain features of the genetic mechanism. Nevertheless, if applied with understanding, Hamilton's rule has immense virtues as a model of phenotypic evolution. It usually proves to be much more robust than first anticipated.

Summary

1. When available resources are insufficient (or likely to be so) for all concurrent offspring to survive, sibling rivalry is expected to become fatal and 'brood reduction' results.
2. An early inclusive fitness model for brood reduction was developed by R.J. O'Connor (1978), deriving its basic logic from Hamilton's rule. Expressing the ecological squeeze in terms of reduced per capita survivorship, he found limits at which it pays (i) surviving siblings to kill one nest-mate (*siblicide threshold*), (ii) parents to kill one of their own offspring (*filial infanticide threshold*) and even (iii) the victim to die so as to free more resources for its viable siblings (*suicide threshold*). With gradually worsening conditions, these should be reached in the order given; that is, siblicide should be favoured under somewhat milder shortfalls than infanticide, which in turn should be favoured before suicide).
3. The discrepancies between thresholds of different family members are most exaggerated in small broods (where loss of even one offspring represents a relatively large change in demand level). In large broods (e.g. those of many insects) the thresholds become almost congruent.
4. Analogous thresholds can be derived from the hierarchy model (Chapter 3) of sibling selfishness, where dominance is rigidly enforced. A new four-sib model is presented that offers an inclusive-fitness explanation for the phenomenon of 'runting'. When the youngest (or weakest) member for each brood size dies (as resource levels worsen), the next-to-youngest sib's value takes a sharp drop in the eyes of other family members; it assumes the new designated-victim role. This model also shows that once a given chick is near death, the step over that irreversible line has surprisingly little impact on the inclusive fitness of itself, its parents or its sibs, which may help to explain their frequent disinterest.
5. A comparable analysis for sibling competition by begging has not yet been attempted, but is likely to depend substantially on whether there is a begging scramble (e.g. Chapter 3) or whether begging constitutes an honest signal to parents (Chapter 9).

6. Formal population genetics models of fatal sibling rivalry have been developed that draw attention to the presumed mechanisms of gene expression, particularly the degree of penetrance, underlying the phenotypic traits of interest. Models in which a designated-victim is readily identified to all family members ('runt' model) or not ('random victim' model) are described. These assumptions fit different taxa and generate sharply different predictions. In the runt model, gene expression (siblicide) is conditional upon not being a runt and ultimately obeys an inclusive fitness approach. In the random victim model, gene expression is unconditional, and predictions do not conform directly to Hamilton's rule.

7. Relatively little empirical attention has been paid to the actual costs of brood reduction, although the theoretical models usually make explicit assumptions about their existence. Another neglected area is the fitnesses of different siblings, especially those that appear to suffer substantially through the process of sibling rivalry.

5

Introduction to sibling rivalry in birds

'That lays three, hatches two, and cares for one . . .

This is the case in most instances, though occasionally a brood of three has been observed. As the young ones grow, the mother becomes wearied with feeding them and extrudes one of the pair from the nest . . . the phene[1] is said to rear the young one that has been expelled.

Aristotle *History of Animals* Book VI, Chapter 6

In the biological study of sibling competition, more scientific effort has been devoted to birds than to any other taxon. That bias stems historically from an enduring descriptive natural history tradition, begun and nurtured by true amateur bird enthusiasts (artists, bird-watchers, and other keen observers), substantially pre-dating any theoretical framework. Furthermore, many birds happen to be well suited to the subject by the virtues of being diurnal, nesting above-ground where their young can be observed (as opposed to giving birth in subterranean burrows), and various reproductive traits that suit them well to key life-history measurements (e.g. Lessells 1991). Finally, the central role played by birds in this, as in so many topics in evolutionary ecology, stems from the fact that David Lack was an ornithologist.[2]

Because avian young develop very rapidly, transforming from egg to full adult size and flight in a few short weeks, they require enormous quantities of food. In nidicolous species the young must rely on parental deliveries for that food, which is often insufficient even if both parents labour unremittingly (Ricklefs 1968, Ydenberg 1994). Thus, a severe ecological squeeze frequently serves as the backdrop to nestling life. Prior to hatching, young birds presumably cannot do much one way or the other about siblings: the yolk food base is individualized and a restrictive shell prevents all non-vocal interactions. On the other hand, a few weeks after hatching, the youngsters have often acquired the ability to fly, making themselves forbiddingly difficult for humans to observe well. Thus, most of the rivalry *that we study* unfolds during a brief window of research opportunity, namely the nestling period. This very

---

[1] A type of vulture.
[2] One imagines that he might well have focused on different biological issues had he studied, say, bats in the Oxford clock tower, rather than swifts.

brevity has the appeal of fitting neatly into the circannual cycle of academics (allowing field study after spring term, but leaving time for late-summer conferences!). Our point is not that this is an ideal focus for the scientific literature, but just that it provides some likely reasons for the skew that exists.

Because birds tend to be relatively monogamous (e.g. Verner and Willson 1969, Lack 1968, Black 1996), two additional features come into play. First, the nestlings frequently receive tangible post-zygotic investment from not one but both parents. Secondly, the chance that a given nest-mate is a full sibling is often quite high (but by no means guaranteed: see McKinney *et al.* 1984, Gowaty and Karlin 1985, Westneat *et al.* 1990, Birkhead and Møller 1992). Although these considerations should tend to discourage sibling rivalry, it is obvious that they are routinely overridden.

Ecological interest in avian sibling rivalry began to take shape with Lack's (1947, 1954) interpretation of hatching asynchrony. He proposed that the habit of hatching the brood over a period of two or more days can be viewed as a parental manipulation (effected simply by starting effective incubation while still laying) for dealing with unpredictable food supplies. Parents must commit themselves to an integer number of eggs well in advance of the brood's peak demands. By establishing an initial age/size hierarchy among the nestlings, parents can often improve their seasonal output. This modest early bias, in turn, produces asymmetries in the sibs' abilities to compete for potentially insufficient food deliveries.

Essentially, the parents are seen as covering multiple bets. In years of particularly rich food, even the least competitive nestling can thrive, but in poor (or even average) years, the system generates a self-tuning downward adjustment in family size. Thus, the parents efficiently produce the largest number of adequately fed, healthy young. In fact, Schüz (1943, cited in Lack 1947) had previously published a similar idea—that the last egg in white stork (*Ciconia ciconia*) clutches constitutes a 'reserve' and becomes a fledgeling only when food is abundant—but Lack's independent derivation and subsequent development of the principle (1954, 1966, 1968) has linked it tightly to his name. The partial mortality within families that is predicted to occur except in particularly good years, later dubbed 'brood reduction', has since been sought and documented in a great many bird species (see Chapter 12).

In recent decades Lack's suggestion that early incubation is the mechanism chiefly responsible for hatching asynchrony has also received support (e.g. Inoue 1985, Magrath 1992). However, his interpretation that the ultimate *cause* is to facilitate brood reduction has met some discursive challenges (e.g. Hussell 1972, 1985, Clark and Wilson 1981, Amundsen and Stokland 1988, Amundsen and Slagsvold 1991), owing to the emergence of viable alternative hypotheses for the phenomenon of hatching asynchrony (reviewed in Magrath 1990, Konarzewski 1993, Ricklefs 1993, Stoleson and Beissinger 1995). Whatever the underlying evolutionary reasons for hatching asynchrony, the *effects* on avian sibling rivalry are of primary interest to us here.

In a broader sense, Lack's ideas on avian reproductive rates were part of the emerging theoretical debate about the key units of selection (e.g. Wynne-Edwards 1962, Williams 1966a, Lewontin 1970, Alexander 1974, Dawkins 1976, Grafen 1984; see also Appendix I of Lack 1966). His premise that individuals typically behave so as to maximize breeding success (a position subsequently refined as the maximization of *lifetime* success, e.g. Williams 1966b, Charnov and Krebs 1974, Smith and Fretwell 1974) directly opposed the classical 'group selection' of Wynne-Edwards (1962).

The application of these perspectives to sibling rivalry *per se* can be seen clearly in the original writings of Hamilton (1964a, b), Alexander (1974), and others, but the light was really switched on directly by R.J. O'Connor (1978). That paper sewed Lack's brood reduction argument to both Hamilton's inclusive fitness theory and Trivers' notion of parent–offspring conflict, and then scoured a cumbersome ornithological literature for *post hoc* tests of his new models. This synthesis not only opened a theoretical front (Chapter 4), but it also opened the eyes of more than a few field-workers (e.g. Stinson 1979, Braun and Hunt 1983, Mock 1985, Drummond *et al.* 1986) who were in position to observe some especially vivid cases, such as siblicidal aggression.

O'Connor made explicit the conditions under which brood reduction by the elimination of one offspring would be advantageous from the point of view of different family members, as we have outlined in Section 4.1.1. All of these conditions require that the survival prospects, S, of an individual offspring increase as brood size, n, decreases; namely:

$$S(n) < S(n - 1) \tag{5.1}$$

where $S(n)$ is the per capita survivorship of chicks in an intact brood of size n, and $S(n-1)$ represents survivorship after one nestling has been eliminated. A slightly more general version of this inequality draws attention to the individualized roles of different siblings in the hierarchy,

$$S_i(n) < S_i(n-1) \tag{5.2}$$

where the fitness pay-offs to brood member of rank 'i' are compared. In a simple three-sib brood, for example, the youngest member may be executed primarily by the next youngest in a pre-emptive defensive strike that precludes role reversal. Such a unilateral assault may or may not affect the eldest sibling's security, which was never threatened by the victim's existence. To date, rather little empirical effort has been directed toward this point (but see Husby 1986, Mock and Parker 1986, Evans 1996). A few other general problems are only starting to attract research attention, including effects on long-term survival (e.g. Spear and Nur 1994), the possibility that the party gaining from brood reduction might not be a nest-mate but a parent (Hõrak 1995), and the selfish uses of aggression.

5.1 Life-history aspects

One critical assumption is that reproduction is a costly endeavour (in terms of time, energy and risk of parental death), so parents have to budget their activities for maximum lifetime success (e.g. Williams 1966b, Trivers 1972, 1974, Charnov and Krebs 1974). Such costs can be hard to document for birds, whose mobility and longevity make them inconvenient subjects for the study of this issue, but strong cases are now accumulating (e.g. Gustafsson and Sutherland 1988). Parents forced experimentally into making elevated expenditures tend to show reduced survival and/or productivity subsequently (Jarvis 1974, Bryant 1979, Nur 1984a, 1984b, Røskaft 1985, Gustafsson and Sutherland 1988: for reviews see Nur 1987, 1988, Winkler and Wilkinson 1988, Hochachka 1992, Roff 1992, Stearns 1992).

Less attention has been paid so far to the long-term fitness consequences of fatal sibling competition for the survivors. In the short term, one imagines that loss of one hungry mouth means diminished hassles for the parents and presumably slightly more food for the remaining sibs, either of which points intuitively to the expectation of lifetime benefits. Some theoretical models also support this view, both for surviving sibs (Temme and Charnov 1987, Pijanowski 1992) and for the parents (Konarzewski 1993, Mock and Forbes 1994). Pijanowski's model, for example, shows that the parental practise of hatching eggs asynchronously (with consequent early losses from sibling competition, including some that may not be strictly necessary) can be favoured by natural selection even if the so-called 'bad food' seasons, where brood reduction clearly *is* necessary, occur infrequently (this point is developed more strongly in a follow-up commentary; see Lamey and Lamey 1994). On the empirical side, one finds some very promising connections between brood size and fledging condition, and between that condition and subsequent recruitment into the breeding population (e.g. Perrins 1963, 1965, Garnett 1981, Krementz *et al.* 1989, Magrath 1990, 1991, cf. Nisbet and Drury 1972, Newton and Moss 1986), but little that directly links brood reduction in overtaxed families with enhanced survivorship of parents or siblings (see Husby 1986).

5.2 The element of sibling aggression

Even though Aristotle falsely impugned eagle mothers as the perpetrators of brood reduction, by Lack's day it had been known for decades, and probably for centuries, that the nestlings of certain raptors sometimes fight to the death (e.g. Salther 1904, Bent 1937). However, the early brood reduction literature paid scant attention to overt fighting among nest-mates, perhaps because of the enormous time investment required to document such events quantitatively. In discussing raptor chick losses, Lack (1947: 324) wrote simply that 'if food becomes short . . . the youngest and smallest chick gets a proportionately smaller share of the food and quickly dies.' His wording leaves open the issue of whether the distribution 'decisions' are made via parental favouritism (e.g. selectively giving food to certain nestlings), scramble competitions among nest-mates, and/or outright bullying. Documentation of overt sib fighting is provocative in part because the aggressive behaviour is an offspring phenotype and presumably evolved to serve offspring interests.

It is understandable that the sheer logistical difficulty of observing newly hatched eaglets must have discouraged the accumulation of many details and

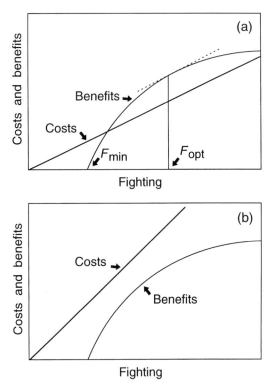

Fig. 5.1 The amount of aggression used in brood reduction is highly variable; frequently none is used at all, and a simple optimality model suggests why (Lamey and Mock 1991). **(a)** Although the benefits obtainable from fighting are limited (hence the decelerating curve), the costs are not (drawn here as a linear function for simplicity). If benefits ever exceed the costs, fighting can deliver a net gain that is maximized at F_{opt}, where the tangent to the benefits curve (= instantaneous gain rate) parallels the cost gradient (or its tangent if curvilinear). This level of fighting is the most cost-effective. **(b)** Similarly, if the two curves do not intersect, either because fighting is very costly or because the ability to control critical resources is impaired (so benefits are too low), selection should favour non-aggressive forms of competition, such as begging and jockeying for position.

robust sample sizes. More recently, several research programmes have focused on siblicidal species that are colonial in nesting habits (reviewed in Mock *et al.* 1990), thereby producing empirical dividends of both observational and experimental dimensions.

Here, we wish to emphasize that aggression is only one style of several available in sibling competitions and there are many brood-reducing taxa (probably the great majority) that seldom or never use it. The likeliest reason for such pacifism during a fatal rivalry is probably the simplest one: fighting can be expensive, especially for taxa lacking weapon-like morphology (see Chapter 6). A simple additive optimality model of sibling aggression's cost-effectiveness summarizes this argument. If we assume that the net benefit an individual can gain from abusing siblings follows a decelerating gain curve (e.g. that the food energy one stands to gain must asymptote as control of the food approaches hegemony) and that there are non-trivial costs for fighting (e.g. energy expended that might have been devoted to growth), there can be a maximum net gain that indicates an optimal amount of aggression (Figure 5.1a). According to the same logic, though, it can be seen that many possible benefit and cost curves may not cross, and thus have no positive net value for fighting (Figure 5.1b). Any factors that make the cost curve too steep (e.g. lack of weaponry or physical separation of siblings) or the gain curve too flat (e.g. unmonopolizable units of food) should, therefore, make aggression unsuitable for use against nursery-mates (Lamey and Mock 1991).

In the next two chapters, we shall focus first on nestling birds in which aggression does play substantial and varied roles in sib competition. For this purpose, we describe the facultative siblicide of egrets in considerable detail (Section 5.3), using them as a model system from which to examine *resource-based sibling rivalry* in general. After some intervening discussion of parent–offspring conflict, the competitive patterns discerned for birds will then be compared with other taxa, starting with mammals (for reasons of parental care similarity). Finally, we shall move away from the twin issues of fighting and limited resources to address other kinds of sibling competition (e.g. cannibalism) across a much broader array of life forms.

5.3 The precursors of sibling aggression and facultative siblicide in egrets

In resource-based sibling rivalry (and to some degree in cannibalism systems, too), the evolution of overt fighting among siblings appears to hinge on five general features (Table 5.1). At its most basic level, competition requires that at least one key resource be limiting during the period of need. For nestling birds this usually means a food shortage (the traditional brood reduction context) but, as we shall see in later chapters, it can involve adult reproductive

Table 5.1. Evolutionary precursors of aggressive behaviour in resource-based sibling rivalry.

1. Weaponry
2. Resource monopolizability
3. Resource limitation
4. Site topography
5. Intra-brood asymmetries

opportunities (if siblings remain in contact and must vie with one another for, say, breeding territories) or other commodities. In the traditional avian model, food shortages derive from the growth requirements of the young, their dependency on parental care, and local resource availability (constancy in space and time). In the next chapter, we shall explore some ways in which overt fighting may be related to both the supply and demand components of egret food competition.

A fourth general factor is the amount and quality of space that is; the size and architecture of the nest (more generally of the nursery). In particular, the availability of potential escape routes may affect how much damage losers sustain from participating in combat. In Chapters 6 and 12, we shall consider the proximate effects of crowding *per se*.

Our final precursor, competitive asymmetries, puts the main focus on avian hatching asynchrony (plus a few cases of egg size polymorphisms), although other variations can be found as well. Some emphasis will be placed on how parents appear to manipulate sibling disparities in ways that affect parental interests, especially with respect to the complications of sib fighting.

5.3.1 *Some background information on egret biology and logistics*

Many members of the family Ardeidae (especially herons and egrets) are well suited to the study of competitive sibling rivalry by virtue of being large-bodied, diurnal and colonial. Their chicks hatch asynchronously and are reasonably well armed by virtue of being equipped for a fish-catching lifestyle (they have long bills, an elongated sixth cervical vertebra that produces the characteristic kink in the neck, and associated neck muscles that allow a slingshot striking action with the bill). Accurate pecking develops very early (even houseflies may be snatched deftly by 1-month-old nestlings) and those strikes pack sufficient power to inflict considerable damage on their targets, including nest-mates. Although famously skittish in the field (e.g. Bent 1962), the adults habituate somewhat in the colony if observers are willing to remain quiet and inside a blind. With patience, one can reap a considerable harvest of data by observing the activities at dozens of nests from a single location.

In addition, at most latitudes these birds tend to lay only three or four eggs—the small family sizes that are expected to promote particularly acute sibling rivalry (O'Connor 1978). This general prediction is based on the usual assumption that brood size is set at or reasonably near the parents' ability to supply resources and on the logical fact that elimination of a competitor has a greater proportional impact on competition if there are few chicks initially.

Finally, ardeids are surprisingly tolerant subjects for various logistical procedures involving nest contents. Newly hatched chicks can be dye-marked on their crowns or backs for identification purposes. Parents tolerate having their eggs (and chicks up to about 1 week old) moved about in order to adjust brood size (Mock and Lamey 1991), brood age composition (Werschkul 1979, Fujioka 1985a, Mock and Ploger 1987) or even species (Mock 1984b). The amount of food the chicks receive can be augmented (Mock *et al.* 1987a). The birds themselves seem to be unaffected by these alterations.

On the negative side, adult ardeids are notably difficult to capture (and thus to mark permanently with bands), they fly considerable distances to forage, and they can live for at least 20 years (making lifetime success measurements less practical). Furthermore, young herons and egrets disperse widely, so virtually nothing is known about the fates of individual fledgelings after they have left the colony. Thus, we can only speculate about the possible long-term effects of position in the nest hierarchy; that is, of growing up well fed vs. half-starved. This is an important shortcoming of the system, which we shall discuss later. For the moment we shall focus on the fitness-conferring consequences of early siblicide.

As the topic of brood reduction began to attract field students, herons were an early choice. Owen (1955, 1960) showed that the last-hatched chick in grey heron (*Ardea cinerea*) nests tended to die first. Soon, other studies confirmed the general outlines of the Lack perspective (e.g. Blaker 1969, Jenni 1969, Milstein *et al.* 1970, Siegfried 1972, Werschkul 1979). Detailed observations of parental care and chick aggression began with Blaker's (1969) descriptions of cattle egrets (*Bubulcus ibis*). This approach became more theory-driven and quantitative when O'Connor's (1978) model inspired new field studies.

5.3.2 *The nesting cycle*

When the birds return from their wintering areas to our Texas and Oklahoma study colonies, males begin to defend nest sites, add sticks to old nests, perform elaborate courtship displays, and attract mates (for descriptions of signals and the pair-formation process see Blaker 1969, Mock 1976). Although extrapair copulations have been reported for these species (e.g. Fujioka and Yamagishi 1981), their role in producing broods of mixed paternity is unknown. For the time being, we shall assume that the great majority of nest-mates are full siblings (having both parents in common), although a recently completed DNA assay of cattle egret broods has shown there to be more half-sibs in the youngest/victim slot than in either of the two elder sibling ranks (J.A. Gieg, personal communication).

In all of these species, laying generally occurs at 1- or 2-day intervals, with incubation beginning soon after the laying of the first egg, hereafter referred to as the '*A*' egg (e.g. see Inoue 1985). Modal clutch size in all these populations is three. Hatching intervals correspond roughly to those of laying. In one sample of great egrets, for example, the mean interval between *A* and *B* was 1.3 days and between *B* and *C* was 1.7 days (and in a few four-chick families the *C–D* interval was 2.1 days; Mock 1985).

Our basic method has been simple observation from blinds, using spotting scopes at short range (usually 15 m or less) that provide an extremely close view of the details of nestling life. Up to 10 focal broods are typically kept under continuous daylight surveillance from hatching through the first month by rotating observers once every 24 h (for additional information on habituating parent ardeids, see Mock *et al.* in press). These detailed records are often supplemented with census studies of early survivorship in many additional nests, usually in a different but nearby colony.

5.3.3 *The early post-hatching period*

Newly hatched ardeids are semi-altricial (i.e. quite helpless, but possessing a full thatch of dorsal natal down). Great egret hatchlings eat very little at first— just a few tiny fish regurgitated by parents on to the nest floor (a delivery method we have termed *indirect* feeding)—and presumably they are sustained by the yolk's residuum, which is sometimes visible through the semi-transparent belly skin. The brood's early lethargy is interrupted only during meals, which become increasingly active episodes. In the Texas great egret nests, a typical food bolus consists of about eight tiny fish (each about 5 cm in total length and 2 g in wet mass) held together by parental saliva and fish mucus, readily falling apart into discrete items when pecked. Cattle egret boluses look similar, but are composed mostly of grasshoppers.

On the first day or two after hatching, prey items are seldom contested among the nestlings, in part because there is a surplus relative to the brood's collective gut capacity and in part because each chick is still so slow and clumsy that it is fully absorbed with the task of ingestion. A 1-day-old great egret chick, for example, grasps one of the small fish crosswise at mid-body and, with a series of short lunges, shifts its grip to either head or tail end before attempting to swallow. These lunges may cause it to lose its balance and topple, an indignity it resists by gripping the nest sticks with sharp-clawed toes. Whole fish, even tiny ones, are much easier to swallow when oriented head-first, which great egret chicks apparently must learn through trial and error. Within a few days, chicks orient their initial strike selectively toward the fish's anterior end (Figure 5.2), thereby greatly accelerating the whole handling procedure. Handling time drops from > 6 s per fish (on the day after hatching) to just over 2 s by day 7. By day 10, it reaches the apparent minimum

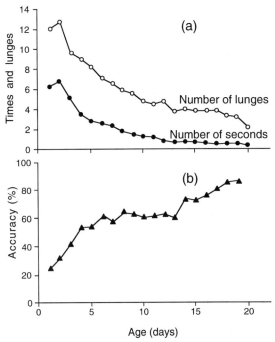

Fig. 5.2 Improvements in the great egret nestling motor patterns that affect success in scramble competitions for food. Captive chicks were presented with single, standard-sized single prey items (a fish 2 cm in total length and 5 g in wet mass). **(a)** Mean number of lunging motions (○) made with fish in chick's bill during swallowing attempts and total handling time (●) from first bill contact with a small presented fish until that prey item disappears in the chick's throat. **(b)** Percentage of initial strikes made toward the head end of the fish (a choice that facilitates swallowing for very young chicks). Once chicks are about 2 weeks old the throat lumen accommodates tail-first swallowing almost as well as head-first swallowing.

value of about 1 s per fish (Mock 1985). Cattle egret hatchlings show a similar, but faster, improvement in prey consumption.

We have detailed this motor skill development because of its relevance to sibling competition. It is during this first week, as small appetites enlarge, that resource limitation becomes an issue. Whereas newly hatched chicks seldom consume the entirety of even one parental bolus (the adult reswallows whatever remains), within a few days all boluses disappear very rapidly. Thereafter, the parent seldom reswallows anything and chicks almost always continue begging after the last bolus (Mock 1985).

Sibling aggression may commence at any time during the first week, sometimes as soon as the *B*-chick hatches. At first fighting is feeble, consisting of little more than one chick raising its head shakily to full neck extension and then falling bill-first on to its nest-mate. Soon, however, even the small act of one chick elevating its head comes to constitute an implied threat. Should a junior nest-mate rise taller than a more senior sib, the latter usually ascends quickly and challenges. The opponents face each other briefly, each straining for height with bills tilted slightly above horizontal (Figure 5.3), until pecking blows are exchanged or one drops down, conceding the air-space. Even during these first days, the superiority of age and size conferred by hatching asynchrony produces a noticeable advantage during bouts, although no physical damage is done. Indeed, there is no obvious immediate benefit from winning a

Fig. 5.3 Great egret sibs threaten each other by elevating their heads maximally, the position from which they peck an opponent. Note that there is a parent in attendance and that it is preening, apparently taking no notice of the sib fighting. (Photograph by D. Mock.)

given early fight. Specifically, early combats do not seem to influence food distribution.

In short, the first week is spent quietly, with intermittent scramble competitions for food whenever a parent chooses to regurgitate, and occasional fights that superficially appear to be inconsequential. Gradually, though, the rapid rise in prey-handling speed confers an increasingly clear edge to senior chicks. A 2-day age difference allows each item to be swallowed about 50% faster, which is highly significant once all of the food is being consumed. Moreover, the early fights turn out to be very important by establishing age-related dominance that has profound consequences in the weeks to come.

5.3.4 *Direct feeds*

Like early fights, the chicks' scissoring of the parental bill makes no immediate sense. As if ignoring the grasping actions, a providing parent delivers an *indirect* feed by unhurriedly lowering its bill tip to within a few centimetres of the nest floor and expelling the food. Chicks initially scissor weakly and near the tip of the parent's bill (Figure 5.4), but as they gain size and strength their grips become much firmer and migrate closer to the parent's gape. Nevertheless, at first virtually all of the food falls to the nest floor, from which it must be retrieved.

The value of scissoring becomes apparent when the *A*-chick starts to cheat on the original system, intercepting parts of boluses as they emerge from the parent's bill (for great egrets this usually occurs on or around *A*'s seventh day). The younger sibs soon acquire that skill, and by the end of *A*'s second week nearly half of all boluses are caught before they reach the nest floor, at least partially. After a third week, that proportion has reached 90% and most of these *direct* feeds involve little or no dropped food (Mock 1985). The smaller cattle egret chicks are slightly faster, with little food touching the nest floor after just 1 week (Ploger and Mock 1986).

From this point, the chicks clamour for position in front of the parent, each seeking to advance its scissoring ahead of the others, from which position it can slide unimpeded up to the parent's gape. Once there, it resists usurpations by nest-mates. Unless it is the *A*-chick, any scissoring chick is subject to being pecked harshly on its unprotected nape or of being dragged off backwards. Whoever holds that top scissor-grip at the time of bolus emergence (the 'pole position') has an excellent chance of monopolizing the offering. Another problem confronting the scissoring chick is that the parent may not deliver a bolus immediately or, for that matter, any time in the near future. In fact, most top scissor-grips gain no food whatever and the parent, which can use its vastly superior size and strength to extract its bill vertically from the chick's grasp, is another opponent of sorts. In dealing with uncooperative parents, chicks holding the top grip soon learn to grasp the parent's bill diagonally, such that the slightly serrated tomium of the chick's bill crosses firmly over the parent's eye!

Fig. 5.4 Great egret chicks scissoring parent, a key element in the control of food delivered in small, monopolizable boluses. Here an *A*-chick reaches quickly for the newly arrived parent's bill as a recently pecked and 'partially intimidated' *B* sibling hesitates, making no move to contest *A*'s priority of access. (Photograph by D. Mock.)

Although the latter's cornea is partially protected by the nictitating membrane, such a head-lock takes the starch out of much parental struggling—especially the brute force technique of vertical withdrawal—and thus extends the chick's tenure at the gape. As a counter-move, the parent sometimes escapes from an eye-scissor by 'faking a bolus'. It makes peristaltic convulsions and opens the gape slightly, as if delivering, which usually inspires the scissoring chick to release the eye and assume a bolus-catching position beneath the parent's gape. The parent then extricates itself swiftly.

Notwithstanding these fine points of grip-retention, obtaining the pole position becomes an increasingly essential part of obtaining food. In both great egrets and cattle egrets, the chick holding the highest grip *when the bolus comes forth* ingests, on average, 92% of that delivery. Of course, that also depends on sibling rank; because *C*-chicks are far more likely than *A*- or *B*-chicks to be hammered or hauled off backwards just as the food arrives, the pole advantage is on average about 10% lower for them (Mock 1985, Ploger and Mock 1986).

In summary, the transition to *direct* feeding signals a qualitative change in the nature of sibling food competition. Whereas the initial premium is on pure handling speed (while competing for a food source that is equally available to all nest-mates), it later becomes possible to improve one's share of the food by controlling *access* to the parent's bill. The coveted bill descends slowly and somewhat predictably, such that success hinges on height and position; a chick standing tallest and directly in front of the parent has a better chance. Repeated bludgeoning of rival nest-mates trains them to stay low.

5.3.5 *Sib-fights*

Every great egret and cattle egret brood we have studied has exhibited considerable fighting, averaging 4 to 5 bouts per day in typical three-chick great egret broods (Mock 1984b). Within a few days of hatching, the relatively mild early fights can escalate into something quite serious, as the chicks' hitting power increases. These are not symbolic or 'ritualized' battles, although they often assume a see-sawing rhythm as the combatants exchange blows. Most jabs are directed at the opponent's face. Simultaneous thrusts meet midway between the two bodies, where the bills sometimes grasp each other and hold for a moment (Figure 5.5). As with human prize-fighters, the chick coming off worse from an altercation actively tries to seize its opponent's punishing bill— it 'clinches.' Most fights are brief, but bouts containing well over 100 rapid exchanges without respite may occur. Remarkably linear dominance relationships usually develop within broods, corresponding to hatching order.

Sibling fights take many forms, depending mainly on how the loser concedes and how quickly it does so. The simplest fights, which usually occur while the participating dyad has had a series of increasingly one-sided battles, are those in which the attack inspires no retaliation. At the next level, return fire is brief until the loser is tagged with several unanswered shots and crouches low. From there, the severity of the beating is left largely to the victor's discretion. Sometimes it continues to jab at its opponent, causing the latter to screech and hide its face. As an alternative to jabbing, a dominant chick may seize the cowering victim by it head or neck, lift that part a few centimetres and then slam it forcefully against the nest cup (Figure 5.6a). If the attack persists for more than a few extra blows, the loser is likely to flee, sometimes squawking loudly and racing about the nest or dodging behind

Fig. 5.5 An egret chick getting the worst of a pecking exchange (usually the younger member of the dyad) often tries to catch the aggressor's bill and 'clinch'. This clearly reduces the number of blows the agggressor can land and may also fatigue the dominant chick as it struggles to free itself and renew the attack. (Photograph by D. Mock.)

other nest occupants while being hit (Figure 5.6b). During such chases, the primary target is the back of the head. Frequently bullied chicks soon develop a characteristic baldness, dotted with fresh and crusted blood, where the nape feathers have been plucked forcibly during fights. If fleeing within the nest fails to stop the assault, the loser may hurry to the nest rim and, if necessary, drape its neck as far as possible over the side, making the head harder to reach (Figure 5.6c). Beyond that, its only remaining option is to quit the nest, at least

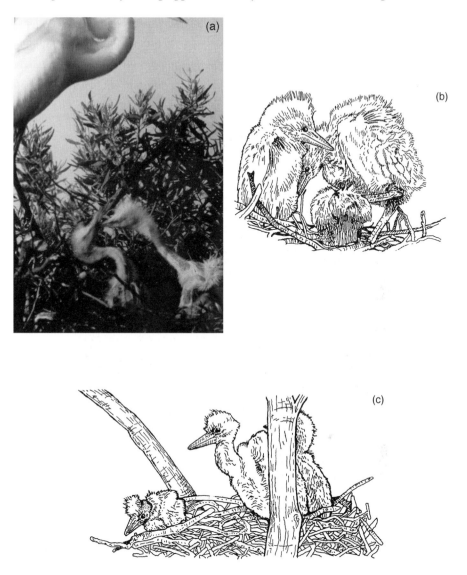

Fig. 5.6 Sibling fighting among egret nest-mates. **(a)** The initial face-to-face pecking with which early fights typically commence, shown here for great egret chicks, is simple and vigorous. **(b)** A badly beaten and completely-intimidated cattle egret *C*-chick concedes all dignity and here is used as a perch for its elder sib. **(c)** When all else fails, a beaten chick simply leaves the nest for a while. If a safe refuge exists in the surrounding vegetation, this can be a very useful method; otherwise, the victim risks falling from a dubious perch. (From photographs by D. Mock.)

temporarily, which carries some risk of falling to the ground and never returning.

Such serious fighting makes sense once the transition to direct feeding begins. By then, cumulative subjugations have produced an all-important *anticipation* of further abuse by each sibling dyad's junior member. The degree of social handicap conferred varies along a continuum. 'Partial intimidations' lead to slight but noticeable hesitations by a subordinate chick during meals, wherein they do not seek the tallest and most central nest posture possible, or they fail to reach as rapidly as they might to scissor the parent effectively. At the extreme, 'complete intimidations' involve total avoidance of bolus presentations, a tuning out. On average, *C*-chicks are eight times more likely to be completely intimidated than either senior nest-mate; partially intimidated *C*-chicks achieve only half as many scissor-grips and fewer than one-third as many poles as senior sibs (Mock 1985). Thus, the switch to direct feeding methods means that aggression pays *incremental* food dividends to the elder member of each dyad. Basically, it means that food becomes 'economically defendable' (Brown 1964) via fighting.

5.3.6 *Food division*

As in Lack's original size-hierarchy idea, the two senior members of three-chick egret broods typically each receive more than their 'fair share' of the total food (i.e. more than one third apiece). The elements of fighting and intimidation exaggerate the skew more sharply than would be achieved by motor/scramble competition alone (Mock 1985). The rich do indeed get richer and the poor, poorer. Great egret parents produce only a few boluses per meal (mean value ± SD, 2.6 ± 1.5; maximum 8), and it is not uncommon for one

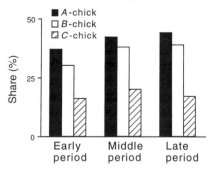

Fig. 5.7 Mean food consumption by sibs in three-chick great egret broods during the first month after hatching in Texas. The 'early' period is defined as that when >60% of the food is obtained through indirect feeding (off the nest floor), which in this species ends when the *A*-chick is about 13 days old. The 'late' period begins a week later, when >60% of the food is intercepted (see Figure 6.1). Throughout these changes, the mean share for *C*-chicks remains consistently low; each senior sib typically ingests twice as much food as *C*. (From Mock 1985.)

senior sibling to ingest two or more boluses. Through the first 3 weeks, the probability that a *C*-chick will receive no food at all during a meal (61%) is nearly double that risk for a senior sib (36%). As ability to catch boluses directly increases, the fraction consumed by the youngest chick declines (Figure 5.7).

Fig. 5.8 A cattle egret *C*-chick in the process of being evicted by its senior siblings. The victim's head was bald and scabbed from repeated beatings. Here it clings to a branch and tries to keep its head out of the seniors' reach; seconds after the photograph was taken, it was hit with a flurry of pecks from above and fell to the ground, 2 m below, from which it never returned. (From Mock *et al.* 1990.)

This food allocation pattern, in turn, creates disparities in sibling growth rates. In great egrets, the slope of the youngest chick's weight-gain curve ceases to parallel that of the seniors shortly after the second week, corresponding to the rise in direct feedings (Mock 1985). Similar bifurcations in the growth curves of senior and junior sibs have been documented for cattle egrets and other ardeid species (e.g. Owen 1960, Blaker 1969, Jenni 1969, Milstein *et al.* 1970, Werschkul 1979, Fujioka 1985b, Inoue 1985).

5.3.7 *Injuries and mortality*

Although a given attack seldom produces visible injuries, the youngest chick in an egret brood is often recognizable from its cumulative wounds. In addition to the conspicuous baldness, the repeated beatings from senior siblings frequently cause lesions on the neck, back, and rump, plus associated subdermal haemorrhaging. It is hard to assess the physiological contributions of wounding to subsequent mortality, but the victims certainly act as if they are in pain, and indeed, as if increasingly sore. This in turn makes effective bullying and food control easier for the senior sibs (Figure 5.8).

Despite food deprivation and physical abuse, the youngest member manages to survive in most three-chick nests of great egret (65% of 126 broods; Mock and Parker 1986) and cattle egret (67% of 100 broods; Mock and Chapman, unpublished data). Though *C* is *usually* the first to die (more than 60% of the first deaths in our sample of great egret three-chick broods), there is a very real chance that *A* or *B* may predecease it (in 11.7% and 18.3% of nests, respectively). When that happens, *C*'s survivorship leaps upwards, from 62% if neither senior predeceases *C* to 86% after one does. Thus, the 'marginal' *C*-chick represents two distinct kinds of *reproductive value* to its parents, one that is gained over and above the production of its two healthy nest-mates ('extra RV' or RV_e), and one that is achieved only when a senior's early death promotes the marginal chick's survival, such that it flourishes as a substitute/replacement[3] ('insurance RV' or RV_i). In our Texas great egret data, the initial investment in creating and nurturing a third and final egg yielded an average dividend equivalent to about two-thirds of a survivor, which breaks down as one part RV_e to three parts RV_i (Mock and Parker 1986). The total RV for *C*-chicks in Oklahoma cattle egret broods was quite comparable, but the contributions of RV_e and RV_i were reversed (Mock and Chapman, unpublished data) (Figure 5.9).

---

[3] This argument has recently been refined for facultative brood-reducers. Lamey *et al.* (1996) pointed out that, while the death of a core brood member is likely to facilitate the enhanced survival of a marginal nestmate, the latter's success is by no means guaranteed. Thus the chance that it, too, may fail—even in the face of brighter prospects—must be taken into consideration (e.g. see Evans 1996).

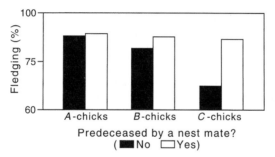

Fig. 5.9 The youngest members of three-chick great egret broods have much higher survival rates (% still alive after 1 month) in nests where one of the senior siblings dies early (open bars) than when brood size remains full (filled bars).

Partial brood loss takes various forms in egrets, with the first death per brood tending to occur around the end of the second week and usually involving an emaciated and often beaten *C*-chick (siblicidal brood reduction). The final *coup de grâce* sometimes takes the rather clear-cut form of a hostile eviction by senior sibs—with *C* assaulted so relentlessly that it tumbles to the ground and is unable to return (Figure 5.9)—and sometimes a more ambiguous pattern (e.g. the victim gradually succumbs on the nest floor or vanishes overnight after hours or even days of enforced food deprivation). It seems likely that dehydration, starvation, blood loss, increased susceptibility to nocturnal predators and lowered resistance to parasites interact synergistically in the mortality. *C*-chicks in broods that receive more food than average enjoy greatly improved prospects (Mock *et al.* 1987a).

5.4 Other ardeid species

Much of the above description for great and cattle egrets applies reasonably well to other ardeids. The developmental pace generally varies inversely with body size, such that small egrets acquire comparable motor skills such as direct feeding and various forms of locomotion (Blaker 1969, Milstein *et al.* 1970, Werschkul 1979) at earlier ages than the largest family members. In the following discussions of factors underlying avian sibling competition, we shall use ardeids frequently for illustrative purposes. Of course, some points are better understood in terms of other taxa, and shall be so presented. In addition, one of the taxon-based survey chapters (Chapter 12) is devoted entirely to birds other than ardeids, which raise a variety of points we simply have not studied (and/or cannot study) with egrets.

Summary

1. Avian nestlings grow rapidly while consuming impressive amounts of parentally delivered food. In many nidicolous taxa, parents lay more eggs than they can (or will) support adequately, creating an acute squeeze.
2. The history of ornithological sibling rivalry studies began in earnest with David Lack's ideas on optimal clutch size and the functional significance of hatching asynchrony. The area received a major boost from the review by O'Connor (1978) that connected it formally to Hamilton's rule. Empirical progress quickly accelerated with the extension of field studies of species that nest colonially and practise siblicide.
3. A difficult and relatively neglected aspect of the current literature concerns the long-term effects of sibling rivalry, specifically how the fitnesses of parents and surviving offspring are affected by brood reduction.
4. Overt sibling aggression is used during the brood reduction process by quite a few avian families, but is probably not a cost-effective aid to conflict resolution in most taxa. Its value appears to hinge on five 'precursors', namely (i) the possession of suitable 'weaponry' (usually bill morphology associated with predatory lifestyle), (ii) food delivered to nestlings in units small enough to be monopolized, (iii) resource limitation (usually food), (iv) a spatially limited 'nursery' (usually an elevated nest), and (v) intra-brood disparities in competitive ability.
5. Siblicidal brood reduction in egrets is described in some detail. These large, colonial birds have small families (often just three young) and hatch their eggs at 1- to 2-day intervals, creating a size/age hierarchy. At first, the senior *A*- and *B*- chicks simply use their strength and motor superiority to out-consume the youngest (*C*), but they change the strategy when they become large enough to intercept food directly from the parent's bill.
6. In egret broods, fighting creates a rather linear dominance hierarchy. Senior siblings use physical intimidation to enlarge the proportion of food they ingest, and usually grow at a faster rate than the youngest siblings. *C*-chicks thus suffer from both physical abuse and food deprivation; some are killed outright by their sibs. Nevertheless, most *C*-chicks manage to survive, at least through the first month. *C*-chicks apparently provide at least two fitness dividends to parents: they may boost total output (as proposed by the Resource-Tracking Hypothesis), or they may serve as a replacement for a senior sib that experiences early failure (Insurance Hypothesis).

6

Supply, demand, and defendability

When you ain't got nothin' you got nothin' to lose.

(Bob Dylan)

Were it not for the great blue heron neighbours, the importance of prey size in promoting egret sibling aggression (Chapter 5) might have been overlooked indefinitely. During regular nest censuses in multi-species colonies, we realized that the heron nestlings were less war-like, not pecking each other and not showing physical evidence of abuse. We began to observe a few great blue heron broods in the hope that interspecific differences might illuminate the egrets' violence. Two seasons of descriptive data collection confirmed the casual impression. The Lavaca Bay heron nestlings fought only about 5% as frequently as great egrets in adjacent nests (Mock 1985). No heron siblicides were observed directly, and only one case out of 46 censused broods could even be inferred from the presence of wounds. By comparison, 53% of the closely watched egret broods were classified as siblicidal, and a conservative figure of 36% for that species was derived from victims bearing external wounds in the census sample (Mock 1985). Despite scant aggression, partial brood losses in the heron nests none the less involved the youngest brood members (88% of these broods that lost any offspring lost their last-hatched individual first; Mock and Parker 1986) and were due to starvation. Heron brood reduction victims simply stopped growing, became emaciated, and expired on the nest floor.

Otherwise, the similarities between these two species are numerous. Clearly, they occupy the same breeding habitat (the same islands and the same nesting bushes) at about the same time (although the herons commence somewhat earlier in the spring, take slightly higher sites and construct bulkier nests). The two species feed in the same general habitats, although the egrets tend to be more gregarious when foraging (Mock et al. 1988). Taxonomically, they are closely related, with some revisions treating them as congeners (Payne and Risley 1976, Sheldon 1987, Sibley and Ahlquist 1990). Great blue herons are substantially larger (adults are about 125 cm tall and weigh about 3 kg compared to about 100 cm and 1.5 kg, respectively) and heron chicks are correspondingly slower to develop motor skills. Even so, young herons are perfectly capable of striking accurately and with sufficient force to break (human!) skin; obviously they have the armaments for internecine strife.

In these colonies, heron broods were about the same size at hatching as those of great egrets (2.7 vs. 2.5) and experienced comparable rates of partial brood loss (64% of heron 3- and 4-chick broods underwent partial mortality during the first month vs. 70.5% for the egrets; Mock and Parker 1986). In both species, brood reduction occurred during the first weeks (Mock and Parker 1986). The most dramatic difference between the two species was in the size of their respective food items. Heron fish tended to be very large, often 25 cm long and weighing on average 200 times more than a single egret fish, and were usually presented singly (Mock 1985). Whereas egret boluses (wads of several tiny fish) could be caught by scissoring chicks and thus monopolized effectively through fighting and intimidation, the size of heron prey seldom allowed this practise. Consequently, heron meals consisted almost entirely of *indirect* feeds, with one enormous fish deposited on the nest floor and lying there, exposed and available to all, while being struck and chiselled by the chicks. Heron meals consisted of just one or two such items (mean = 1.25), from which everyone fed.

Upon the arrival of a food-carrying parent, the behaviour of young herons differed in some important ways from that of their egret counterparts. Although the parent's bill was frequently scissored, this action seemed to be less urgent than than in the case of egrets. With hindsight, this makes sense: there was little to be gained from scissoring a heron parent because fish small enough to be caught in the *direct* manner were the exception, not the rule. A typical heron regurgitation consisted of a large, rather fresh fish, of which only the head and gill region had been sufficiently predigested in the parent's stomach acid for the chicks to knock off ingestible pieces. Just as egret chicks required trial and error to improve prey-handling, it took heron chicks a few days to discover the merits of directing pecks toward the more profitable anterior end of big fish. After a week, their striking power had increased appreciably and they could knock off larger hunks from each carcass. However, the same pattern remained: a large fish was presented, left available for a few minutes, and then re-swallowed. Depending on how long a particular parent remained at the nest, it often produced the same fish several times, allowing the chicks to whittle it down until one managed to swallow the tail. They commonly did not manage to finish consuming a particular fish before that parent was relieved by its mate and the whole process began again with a new fish. It is far more difficult to quantify the ingestion rates of heron chicks than that of egret chicks, simply because amorphous 'pieces' from a big fish are harder to assess visually than the discrete and uniform small prey of egrets. Consequently, less is known about the food distributions among asynchronously hatched heron chicks, but it seems that starvation in this species resulted from the junior sibs' inferior performance in this type of extractive foraging;

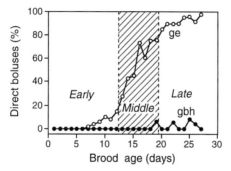

Fig. 6.1 The transition to direct feeding occurs smoothly for great egrets (ge) in Texas, but not in the neighbouring great blue heron nests (gbh). (From Mock 1985.)

that is, knocking off loose ingestible chunks before the limited amount of soft tissue was all consumed.

In short, the great blue heron nestling rivalry observed in Texas remained a scramble competition, never shifting to a resource-defence one. By the age of 4 weeks, although they had grown sufficiently to gulp down large, intact mullets, heron chicks could still not intercept them from the parent's bill (as *direct* feeds). Nearly every fish went to the nest floor (see Figure 6.1), where it was accessible to all nestlings for an average of about 15 s (and even this figure is for uncontested mullets—when two chicks tugged at the same fish the mean exposure time was nearly 2 min; Mock 1985). In great blue heron nests, accordingly, sibling aggression cannot yield continuous, incremental advantages of *partial intimidation* as in great egret nests.

6.1 Prey-size hypothesis

The prey-size hypothesis (that sibling aggression should be linked with direct feeds of offspring, i.e. with boluses small enough to monopolize) was derived *post hoc* from the above observations as both a proximate and an ultimate explanation for sib fighting. On the proximate level, it specifies a particular cue (prey must be delivered in units small enough to be caught, and hence economically defendable) that can promote the development of nestling aggression. On the ultimate level, the fact that the cue is food, which is often in fatally short supply, strongly suggests the fitness benefit gained from being responsive to that cue.

6.1.1 *Comparative applications of the prey-size hypothesis*

Both within the family Ardeidae and beyond it, the fit of the hypothesis is unidirectional. Birds whose nestlings are aggressive do feed them on diets of monopolizably small units. The converse holds less well; there are many taxa that feed very small prey (or pieces thereof) to their young, but the latter do not fight. Thus, small prey may be a necessary but not sufficient condition for the development of overt hostility between siblings.

Cattle egrets provide a good illustration. Sib fighting is common and effective in establishing stable dominance hierarchies that shape the food allocations among siblings (Fujioka 1985b, Ploger and Mock 1986, Mock and Lamey 1991). Among ardeids, this species is unusual in that it eats mainly insects (orthoptera), with only a few vertebrates—mainly amphibians and mice—added during the nestling period. The regurgitated boluses tend to be small and discrete (Telfair 1983, Ploger and Mock 1986) and there is a smooth transition to *direct* feeding. Two other ardeids that eat small prey and show siblicidal aggression are the reddish egret (*Egretta rufescens*) and little blue heron (*E. caerulea*). The former preys on small marine fishes, which it catches by remarkable active-pursuit foraging in very shallow water (Meyerriecks 1962, Kushlan 1978). These fishes are regurgitated to the young in small boluses via the *direct* method (D. Mock, personal observation) and the nestlings fight, sometimes to the death, among themselves (R. T. Paul, unpublished data). Little blue herons, by contrast, are much more passive 'wade-or-walk-slowly' hunters, mainly capturing anurans (Meyerriecks 1962, Kushlan 1978), but their young are also fed by the *direct* method (Werschkul 1979) and sometimes fight. Brood reduction in the latter species can also involve active evictions of the youngest sib (Werschkul 1979). In the Australian inter-mediate egret (*E. intermedia*), also, small prey are transferred via direct feeds and the young are aggressive, although siblicidal mortality has not been reported (McKilligan 1990). By contrast, little egrets (*E. garzetta*) in Japan have been reported to use *direct* feeding with nestlings, but exhibit virtually no fighting (Inoue 1985). Great egrets may show geographic variation in nestling aggression. Casual observations of great egret broods in Japan (M. Fujioka, personal communication) suggest that they may be less aggressive, despite a small-prey diet. The Texas-like combination of *direct* feeds and fighting has been noted for this species in Argentina (D. Boersma, personal communication).

The great blue heron's Old World sibling species, the grey heron, has been studied extensively. Its diet varies substantially (reviewed in Lowe 1954, Milstein *et al.* 1970, Cramp and Simmons 1977), which apparently leads to variations in the methods used to deliver food to nestlings. In a UK colony where the predominant prey were large eels (*Anguilla anguilla*), Milstein *et al.* (1970) observed only *indirect* feeds, some sib fighting (not quantified), and no siblicidal deaths during the first month (and only one thereafter). By contrast, in other colonies where the regurgitated food included many small prey items and brood reduction was common, intimidation at mealtimes apparently contributed to the unequal distribution of food among sibs (Owen 1960; see also Holstein 1927), and permanent evictions of the youngest brood members have been observed (Owen 1955). At the other end of Eurasia, Litvinenko (1982) reported that chicks of this same species engage in profuse fighting, both within and between broods, but the typical prey size for that population was not stated. Congeneric goliath heron nestlings (*A. goliathi*) probably receive only very large fish (Mock and Mock 1980) and have not been seen fighting, but they have been observed very little.

It appears that the nestlings of all other predatory birds that exhibit siblicidal aggression are fed in some variation of the *direct* method. Newly hatched raptors and skuas are offered shreds torn by the parent from larger carcasses. Booby and pelican chicks insert their heads into the gular pouch of the parent. Young cranes and oystercatchers (Safriel 1981, Groves 1984) are handed small individual prey or parts of larger ones. Finally, the kittiwake (*Rissa tridactyla*), the only gull species known to practise siblicide regularly (Braun and Hunt 1983, Dickins and Clark 1987), may also be the only one in which regurgitated food is intercepted by the chicks before it reaches the substrate (Walk 1978). Brown pelican (*Pelecanus occidentalis*) parents regurgitate small fish and predigested fish mush for their chicks, which is initially taken in an *indirect* fashion, but increasingly intercepted as the chicks age until, during their third or fourth week, more than half is taken by the *direct* method from the parent's pouch (Pinson and Drummond 1993). Both fighting and siblicides are frequent. Pinson and Drummond (1993) concluded that pelican senior sibs, like egrets, use aggression in a pre-emptive manner to establish dominance and control over junior nest-mates.

An experimental test with captive American kestrels (*Falco sparverius*) showed that the ability of larger siblings to consume disproportionate shares of limited prey depended on whether the food was small enough to be monopolized (Anderson *et al.* 1993), even though overt aggression is apparently not used. In nine four-chick test broods, female chicks were 10% heavier than their brothers and used their greater bulk to station themselves strategically at the point of parental delivery (the 'hole position'). When food items were small enough to be swallowed immediately, females ate more than their 'share' (an average of 31%, instead of 25%), and males ate less (19%). However, when the food items were too large to be monopolized, all offspring could tear off pieces and no skew resulted. Thus, prey-size effects are not limited to interference competitions, but may also be important in certain kinds of scrambles. The implications of these results for sex ratio will be discussed later (see Section 12.5).

Of all the comparisons, however, the most exceptional and compelling example concerns two species of waterfowl. The peculiar thing is, of course, that waterfowl chicks are highly precocial and normally feed themselves, but in the magpie goose (*Anseranas*) and musk duck (*Biziura*), the parents not only provide the food, but also deliver it by the *direct* method. In the goose, 'bill-to bill feeding occurs for about six weeks' (Kear 1970: 376) and this species 'is unusual in having a conspicuous dominance hierarchy within the brood' (Davies 1963). In musk ducks (see Lowe 1966), the mother alone feeds her brood 'strenuously' for about eight weeks, and brood size often falls to a single duckling, which '. . . may suggest that one bird receives most of the food and so survives' (Kear 1970: 378). There is also reported to be some 'much simpler form' of parental feeding in whistling ducks (*Dendrocygna* species) (Kear 1970), but whether that leads to sibling aggression is apparently unknown.

6.1.2 *Experimental test of the prey-size hypothesis*

To explore whether a small-prey diet can turn sibling aggression on and a large-prey diet can turn it off, we reversed the diets of great egrets and great blue herons in Texas. The logic of the hypothesis also requires that any changes in fighting be accompanied by a switch to the appropriate feeding techniques (and degree of food control). Embedded in this approach is the tacit assumption that the birds possess sufficient developmental flexibility to make such adjustments, that is, to modify their behavioural ontogeny radically as a function of the cue.

A simple cross-fostering design was used, wherein 10 three-chick broods of great blue heron hatchlings were placed in nests belonging to great egret parents, whose own broods went to the vacated heron nests. Adults of both species readily accepted their new families (Figure 6.2) and fed them the reversed diets. A rotating team of field observers maintained a continuous daylight vigil on the broods for 25 days, thus documenting the diets as delivered, the food transfer methods, aggressive activities, and fates.

The results were provocatively mixed (Mock 1984b). Relocated heron chicks changed in accordance with the predictions. They quickly acquired the direct feeding method and fought at egret-like rates (15 times more often than usual; Figure 6.3). Seven of these broods experienced the siblicidal loss of their youngest chick. By contrast, the fostered egret broods showed essentially no response to their unlikely situation. Far from curtailing their enthusiasm for scissoring, they tried to make the usual transition to *direct* feeding under what were clearly impossible circumstances. Week after week they fought for position, seized the adult bill (Figure 6.4), and were consequently overwhelmed by each emerging prey item. The chick holding the pole position was sometimes pinned beneath the huge fish while its sibs ate what they could hack from the fallen carcass. These egrets did not fare well, overall. They appeared to be undernourished generally (perhaps lacking the brute strength to dislodge enough mullet tissue) and were chronically caked with fish slime. Furthermore, the youngest chick in each brood had the usual assortment of bruises and cuts from its beatings. In all, the fighting rate was somewhat lower, but not significantly so, and siblicidal deaths occurred at six of the 10 nests. To summarize, the swap experiment revealed that great blue herons have some ability to switch their nestling development dramatically, turning on a whole

Fig. 6.2 An interspecific fostering experiment. Great blue herons and great egret hatchlings were exchanged in a Texas colony in order to determine the effects of a reversed diet on sibling aggression. (From Mock 1984b; see text for details.)

suite of food-monopolizing behaviour patterns when prey items are small, but great egrets showed no reciprocal capacity and apparently cannot extinguish those patterns when food items are large.

6.1.3 *Developmental plasticity*

This difference in ontogenetic flexibility between the two species, although not the intended focus of the swap experiment, can be accounted for in at least three different ways. First, the high levels of fighting might be due to egret parents *per se*, rather than to egret diet. Of course, this argument does not

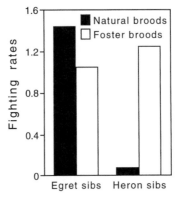

Fig. 6.3 In Texas, broods of great egrets fight at relatively high rates (shown here as bouts per day per sibling dyad), while broods of great blue herons fight relatively little when they receive their normal diets from conspecific parents (filled bars). Broods raised by heterospecific 'foster parents', and thereby receiving opposite-sized prey items, responded differently (open bars). Egrets receiving large fish decreased their fighting slightly, but not significantly so, while herons receiving small fish fought at very high levels and killed their youngest nest-mates (Mock 1984b).

account for egret chicks' persistent war-like behaviour in heron foster nests. Secondly, the herons' flexibility might be adaptively related to changes in their own local diet, specifically if there are sporadic shifts in prey size. Temporal variations in heron diet might result, for example, from transient periods of superabundance in younger cohorts of resident fish populations, such that small items are sometimes the most profitable type. Under such a recurring scenario, selection could favour adjustable nestling behaviour to fit any diet. Testing the plausibility of this argument is tricky. One might wait patiently for nature to provide at least one season's duration of this hypothetical prey-size shift (we were not that patient). Alternatively, one might devise a clever field experiment to induce certain pairs to consume small prey, presumably by provisioning targeted adults directly (we were not that clever). As an aside, certain individuals of this species nesting in the Florida Keys have adopted the habit of 'pan-handling'; that is, of soliciting fish hand-outs from generous human residents. They actively defend particular food patches (Powell 1983), which could be exploited by organizing a network of feeders to establish large-prey vs. small-prey parents (we have not been this organized!).

A third possibility recognizes that great blue herons, which are known to have a highly variable diet across their broad range, probably face the large vs. small prey situation every season in different areas. If the ability to turn food-monopolizing behaviour on and off is locally gainful, depending on its cost-effectiveness, gene flow between demes might allow the latent trait to exist everywhere. In particular, it may reside in the Lavaca Bay herons because it is routinely valuable elsewhere. Of course, this geographical argument is com-

Fig. 6.4 Great egret nestlings attempted to use 'direct' feeding methods with great blue heron foster parents. **(a)** The chicks can scissor the adult very well, but **(b)** they are incapable of catching the very large fish that emerge. (Photograph by D. Mock.)

patible with that of temporal variation, but much easier to test, requiring only the discovery of one or more populations that normally consume monopolizably small prey. Nestlings in such areas should show the combination of *direct* feeding and high sibling aggression. This logic can also be reversed to explain the lack of flexibility in great egrets: if egrets always take small prey, selection might never have the opportunity to favour an ontogenetic switch. This view has the satisfying feature of falsifiability. If a great egret population were found that *normally* consumes large prey and still has siblicide, the hypothesis would be in dire straits.

6.1.4 *Laboratory study of prey-size*

The inflexibility of Texas great egrets with respect to prey size was demonstrated once again when 40 captive broods were fed on identical amounts and contents of food, but in large vs. small packages (Mock *et al.* 1987a). These three-chick broods were collected near Lavaca Bay, and their laboratory diets were composed of the same small Lavaca Bay fish normally provided by their parents. However, in the laboratory boluses were presented via 'parent' egret hand-puppets, manipulated from behind a curtain, and experimenters controlled whether standardized amounts (24 g wet weight) of the tiny fish were presented either as 'small-food' (i.e. monopolizable) boluses or in a 'large-food' format that could not be monopolized. Specifically, in 20 'large-food' nests, the measures of food were packed inside sausage casings and accompanied by a 15-cm-long, 3-cm-diameter wooden dowel. The food end of this unit was slit, so as to allow removal of individual items (comparable to the pre-digested head end of a heron's mullet), but the dowel's bulk prevented swallowing of the whole bolus. As a result, *direct* feeding was feasible only for the 20 'small-food' broods (Figure 6.5).

As in the cross-fostering field experiment, laboratory broods receiving 'large' prey did not alter their attempted feeding methods or fighting levels. Senior sibs receiving 'small' boluses took them as *direct* feeds and thus controlled them. Broods receiving the 'large' boluses persisted in their efforts to do exactly the same thing, but failed, and had to peck the morsels out individually from off the nest floor. The fighting rate comparison showed the two groups to be virtually identical (Mock *et al.* 1987a).

6.1.5 *Geographical variation field studies*

Attempts were made to observe natural egret and heron populations that take opposite-sized prey to the Texas pattern, but a rumour that great egrets eat large fish at Isla Taboga, Bay of Panama, proved erroneous and to this day we remain unaware of any great egret population whose natural diet is composed mainly of items that are too large for nestlings to manage. The reciprocal

Fig. 6.5 Captive great egrets in the laboratory were fed artificially with their natural prey fishes, which were either hand-shaped into normal 'small' boluses or stuffed into 'large' boluses made with sausage casings and wood (see text). (From Mock *et al.* 1987a.)

study proved much more feasible: the requisite small-prey diet had been documented properly for a Québec population of great blue herons (DesGranges 1981). In the summer of 1984 a sample of 17 nests was observed from the top of a granite dome on Île Laval in the St Lawrence River. This population's diet consisted of fish just slightly larger than those eaten by egrets in Texas, mostly less than 10 cm in length (Figure 6.6), and these heron chicks showed the egret set of tricks—they scissored parents vigorously, switched from indirect to direct feeds, and fought. When the appropriate corrections were applied (matching the Texas and Québec heron samples by brood size, chick ages, and hours of observation), the fighting rate was on average 850% higher in the Canadian birds (Mock *et al.* 1987b).

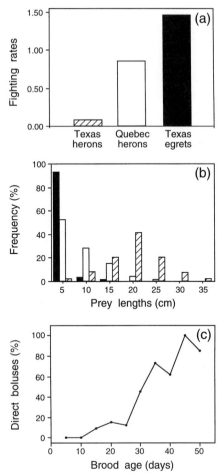

Fig. 6.6 A comparative study of great blue heron nestlings in Québec. **(a)** The Canadian herons fought 8–10 times as often as their Texas counterparts (units = bouts per day per sibling dyad), although still not as often as great egrets in Texas. **(b)** These Canadian herons ate mostly very small fish, more like those taken by egrets than those taken by herons in Texas. (Bar patterns as in **(a)**). **(c)** Like egrets, these northern herons also switch to the direct feeding method, but a few weeks later than the smaller and more agile Texas egrets. Compare this curve with that shown for egrets in Figure 6.1. (From Mock *et al.* 1987b.)

Geographical variation in great blue heron sibling aggressiveness as a function of prey size needs further attention. The species certainly offers an adequately diverse diet (see references in Mock *et al.* 1987b), and presumably participates in the same kind of postnatal dispersal that has been documented for other ardeids (e.g. Rydzewski 1956, Dusi 1967, Bruckhardt and Studer-Thiersch 1970), which could generate substantial rates of gene flow between populations. A brief report of great blue heron siblicide in a southern Ontario colony (David and Berrill 1987) adds some complexity to the above picture. The authors' verbal descriptions of the fights correspond closely to what has been seen elsewhere, and they conclude that many of the chick deaths were 'associated with siblicidal attacks.' The puzzle is that the diet observed changed from 'small and unidentifiable,' during the first two weeks, to single large fish (10–20 cm long) that could not be monopolized thereafter. Thus, it seems directly counter to the prey-size evidence summarized above. On the other hand, the actual fight rates that they report match the Texas heron data almost exactly (four-chick broods in Ontario fought 0.008 times per hour per dyad vs. 0.005 per hour per dyad for Texas; for three-chick broods the rates were 0.006 and 0.005 times per hour per dyad, respectively). The one siblicide that they actually observed (in which the loser of a fight left its nest and fell to the ground) is compelling evidence that fights can have fatal consequences, but they also inferred that other deaths were siblicides, without mention of wounding or data for aggressively enforced starvation.

In general, it seems fair to conclude at this point that prey size is probably one of several important factors that contribute to sibling aggression across bird taxa (Table 5.1). What these findings indicate about the ontogeny of sib fighting is far from clear, but the most parsimonious explanation is probably a simple operant conditioning mechanism. Individuals that are rewarded with food for intimidating nest-mate competitors may attack more frequently, and those not rewarded may elect not to waste the effort. If so, then the small size of the prey is merely one factor shaping the reinforcement schedule.

A more onerous possibility, that the developmental flexibility is genetically based (Smith-Gill 1983), suggests that local heron diets change with respect to prey size and/or that parents specialize in particular prey sizes. The latter would certainly be of interest if some broods naturally receive consistently small-prey diets, while neighbouring broods are fed large prey. The fact that individual parent birds of some species often specialize in particular classes of prey is well known, so this is not a hopeless wish. Indeed, different heron pairs in Texas showed notable diet fidelity (e.g. some took mostly mullet, while others seemed to be fixated on ribbonfish or rice rats), but none of these crossed the size threshold for monopolization. Black-crowned night-herons (*Nycticorax nycticorax*) in those Texas colonies might serve admirably for such a comparison, as many parents take mainly small brackish crustaceans, while others appear to be oriented heavily toward a diet of cattle egret nestlings too large for their own chicks to swallow whole.

6.2 Supply and demand in egret siblicide

The argument that a given trait (like being able to switch sibling aggression off) is adaptive because of its efficacy in reducing competition loses much of its steam if the competition itself is apocryphal. Sibling rivalry studies require good evidence that at least one key resource is clearly limiting, such that selection can work on competitive skills. For the nestlings of many avian taxa, widespread evidence of partial brood starvation has long focused attention on food amount as the most important limiting resource (Lack 1954, Ricklefs 1965). We shall now consider various aspects of how food supply, and then demand, affect the fact and form of avian sibling competitions. In particular, we shall re-examine the matter of how fatal sibling aggression *per se* relates to the participants' access to sufficient food.

6.2.1 *Proximate vs. ultimate roles of food amount*

Just because food shortage occurs sufficiently often to constitute a major selection pressure promoting competitive traits does not necessarily mean that it also constitutes the specific cue(s) that elicit aggressiveness or other responses. The issues of proximate and ultimate causation must be considered independently. The ultimate question is usually inferred from repeated observations that victims of partial brood loss showed progressive outward signs of food deprivation prior to their deaths. With or without additional problems from physical wounds (evidence of fighting), the youngest sib in an asynchronous brood may be only half the size of its senior nest-mates.

Sometimes O'Connor's (1978) more formal criteria of brood-size-dependent mortality (see p. 77) can be applied or approximated. Direct application requires that individual nest records be analysed so that only broods of a particular initial size are compared with the matched subset that underwent a loss of a single chick. For example, the survivorship of great egret nestlings that lost one sibling early was higher than that for broods which were not thus trimmed (Mock and Parker 1986). (Comparable examples for other taxa are given in later chapters.) More commonly, census data show only that smaller heron broods enjoy an overall higher per caput survivorship (Pratt and Winkler 1985, Mock and Parker 1986). Overall, there are good correlational reasons for believing that the loss of one chick from egret broods relaxes the major risks to its surviving sibs. In addition, experimental food supplementation diminishes egret chick mortality under field conditions (see below).

6.2.2 *The 'food amount hypothesis' for sibling aggression*

Where food supply is very restricted, a simple proximate relationship between current food levels and sib fighting is usually assumed. Lack (1966: 309) summarized this view in a confident tone: '. . . the predisposing cause of death is food shortage among the nestlings, since *a well-fed chick does not attack its nest-mate*' (our emphasis). This reasoning was subsequently dignified with a label, the 'food amount hypothesis,' and formally cast as 'a causal relationship between insufficient food and overt sibling aggression' (Mock *et al.* 1987b). The behavioural mechanism is presumed to operate via the internal signals of hunger, with chicks expected to fight more if they are hungry and less if surfeited. Such a system would have the obvious advantage of being both reversible (Procter 1975) and efficient; unnecessary beatings should cease if and when food conditions improve. (Note that the hypothesis is based on the tacit assumption that fighting carries a non-trivial cost, perhaps simply energy that might otherwise be diverted to some worthier purpose, such as growth.) In being both economical and conditionally flexible, it parallels our preceding discussion on applying the prey-size arguments to great blue herons.

Perhaps because of the idea's intuitive appeal, it received virtually no critical attention, while a few lines of indirect evidence (i.e. descriptive observations 'consistent with' the hypothesis) accumulated. For example, nestlings of various species tend to fight more *during meals* (Siegfried 1968, Gargett 1978, Braun and Hunt 1983, Jamieson *et al.* 1983, Groves 1984, Fujioka 1985a, David and Berrill 1987, McLean and Byrd 1991) and chicks replete with food typically seem drowsy and unwarlike (Lack 1954, Nuechterlein 1981, Fujioka 1985a). Similarly, periods of inclement weather have been reported to correlate with the demise of younger sibs in various brood-reducing species (Williams and Burger 1979, Poole 1982, Braun and Hunt 1983), the implication being that these victims were executed by their deprived and increasingly desperate senior nest-mates. However, none of those studies actually showed that *attacks* became more numerous or vigorous during low-food periods, leaving open alternative possibilities (e.g. that the tendency for victims to *succumb* is expedited by reduced food deliveries).

Two early attempts to test the idea experimentally proved inconclusive. Procter (1975) deprived South Polar skua (*Catharacta maccormicki*) chicks of food for various intervals and claimed that hunger increased attack rates. Unfortunately, that experiment's design was seriously confounded (e.g. different chick ages, re-use of some chicks but not others, testing some individuals against their real nest-mate while testing others against alien chicks, etc.), leaving his conclusion open to question. Subsequently, Braun (1981) force-fed kittiwake *A*-chicks, but concluded that her results, too, were unclear.

The first convincing demonstration that sibling aggression is a food-dependent mechanism was made with blue-footed boobies (*Sula nebouxii*). In two-chick broods on Isla Isabela (in the Mexican Pacific), the *A*-chick routinely directs a moderate, but non-injurious, level of pecking toward its one younger sib, which apparently escalates sharply into fatal attacks only when *A*'s personal condition is poor. Specifically, if the weight of the *A*-chick drops below a certain threshold (approximately 75–80% of the maximum for age-matched chicks), *B* vanishes permanently from the nest (Figure 6.7). Interestingly, these doomed *B*-chicks were not lighter than the neighbouring *B*-chicks. This correlation between the *A*'s weight and *B*'s demise proved to

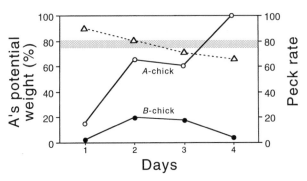

Fig. 6.7 When blue-footed boobies were experimentally deprived of food, the senior sib (*A*-chick) increased its attacks as its own mass declined. Here, the mean mass of *A*-chicks is expressed (broken line) as the percentage of its 'potential weight' (the mean for age-matched control *A*-chicks in the same season). By the second day, deprived *A*-chicks had dropped to 80% of their potential and their peck rate (solid line connecting open circles) had tripled while that of the subordinate (solid line connecting solid circles) had changed little. Previous correlational work had shown that most *A*-chicks show escalation when their mass falls by 20–25% (hatched zone). (From Drummond and Garcia Chavelas 1989.)

be consistent across several seasons (Drummond *et al.* 1986, 1991). This relationship was subsequently demonstrated experimentally (Drummond and Garcia Chavelas 1989). When *A*-chicks were temporarily prevented from feeding (by encircling their necks with cloth tape that did not expand to allow the passage of swallowed items), their attack rates increased several fold, relative to controls. This increase was especially noticeable as *A*-chick weights approached the predicted 75–80% 'threshold' (Figure 6.7). Most recently, it has been shown that relative hunger can affect fighting and dominance. By pairing non-sibs of equal sizes but differing need (time since previous meal), Rodríguez-Gironés *et al.* (in press) showed that this asymmetry can also affect the outcomes, as predicted by the logic of asymmetrical contests (e.g. Maynard Smith and Parker 1976, Enquist and Leimar 1983).

Similar food-sensitive aggression had been reported and debated for ospreys (reviewed in Mock *et al.* 1990; see also Forbes 1991d), and has now been demonstrated experimentally (Machmer 1992, Machmer and Ydenberg, submitted).

6.2.3 *Egrets and the 'food amount hypothesis'*

Although we also assumed initially that egret sibling aggression was predicated on food amount, something soon seemed amiss. For various reasons, the first summer (1979) found us hand-raising six captive great egret broods, which were given as much food as they wanted three times a day (we later calculated that these chicks greedily ingested six times more food than is normally provided by parents!). Despite this absurd supply, severe hostilities persisted among the siblings. Although we did not quantify the captives' aggressiveness, its very presence and intensity were puzzling. The following summer, it became clear that certain wild broods are fed far more lavishly than others (probably due to parental hunting skills, etc.), yet the relatively wealthy broods were not particularly peaceful. The simplest prediction of the 'food amount hypothesis,' namely a negative relationship between food delivered to the nest and brood fighting, has found little support in the ardeid data. Data from 19 great egret broods monitored through their first 25 days

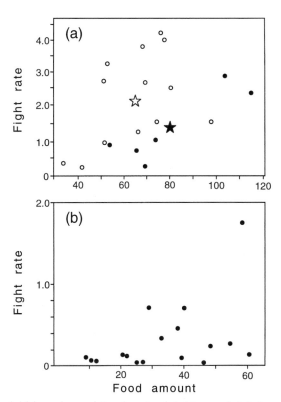

Fig. 6.8 Do ardeid broods receiving less food fight more? **(a)** A regression for 19 broods of Texas great egrets showed a weak and sloppy trend in the opposite direction. Food amount is expressed here as estimated grams of fish delivered per day, and fight rate is expressed bouts per dyad per day. Open circles represent broods that experienced partial brood losses (open star = mean for that sample), while closed circles represent broods that remained intact throughout the first month post-hatching (closed star = mean). **(b)** A similar regression for 17 great blue heron broods in Québec was also weakly positive. Here food amount is expressed as the hourly mean. (From Mock *et al.* 1987a.)

showed a weak (and not statistically significant), but *positive* regression that accounted for only 14% of the variance (Figure 6.8a). A similar analysis for cattle egret broods in Texas showed no slope at all (Ploger and Mock 1986). In Japan, Fujioka (1985b) reported a strong negative relationship between cattle egret fighting rates and an index he called the 'Success Rate' (the proportion of chick scissoring attempts that led to parental bolus production), but he did not test fighting against food levels *per se*. Indeed, his result may expose an opportunity cost for fighting if the attacker's own most profitable scissorings are precluded by its zeal to go on the offensive. Alternatively, it may reflect aggressive interruptions of scissorings that would have delivered food to the chick being attacked. Additional data on sib fighting and delivered food for 19 cattle egret broods in Oklahoma revealed another weakly positive regression (Mock and Lamey 1991), as did the aggressive great blue heron chicks in Québec (Mock *et al.* 1987b) (Figure 6.8b).

Descriptive analyses of food and egret fighting over shorter time scales have produced more mixed results. Great egret fighting was not negatively related to the amount of food received by the brood during the previous 3 days. In fact, only two of 22 sample days showed a significant relationship at all, and both were positive (Mock *et al.* 1987b). On the other hand, a recent study of cattle egrets has provided correlational evidence that underfed senior sibs may escalate their attacks in the last few days before the victim's death (Creighton and Schnell 1996).

Two experiments were undertaken in order to check this pattern more rigorously using our standard broods of three great egret chicks. The amount of food delivered to nests was manipulated in both directions, by adding extra food to experimental broods in the field and by creating low-food experimental broods in the laboratory. The basic result recurred: broods receiving more food tended to fight slightly (but not significantly) more, not less, and tended to enjoy higher survivorship relative to controls.

In the field experiment, the mean amount of food delivered by parents (from the previous seasons' data) was doubled for experimental broods by adding a wad of equivalent mass to their daily food supply. Each afternoon, that pre-measured supplement was simply dumped on to the floors of 50 nests by extending a plastic rain-gutter over the rims and inverting it. Fifty control broods were visited and sham-fed with the empty gutter. Of these 100 broods, 13 experimentals and 10 controls were kept under continuous vigil from blinds. The chicks co-operated by eating the supplementary food quickly in almost all cases, and the parents co-operated by continuing to bring their usual deliveries, but the fighting rates did not differ between treatments (Figure 6.9), and the mean bout lengths were identical (5.4 blows). The only significant responses were in growth and survival of the chicks. Overall, *C*-chicks grew much faster (they were 61% heavier on the tenth day) in the provisioned nests, where they also suffered significantly diminished mortality (Mock *et al.* 1987a).

Meanwhile, the 40 captive broods in the laboratory (already described in Section 6.1.4) were being fed either of two food amounts. The 20 'high-amount' broods received roughly one-third more than normal (the field average), while the 20 'low-amount' broods received only half that much (*c.* one-third less than the field average). Once again, a weak positive result emerged: 'high-amount' broods fought more, but not significantly so. Interpretation of this experiment was complicated by the exceptionally high general aggression rates, which led to an unusually high frequency of sib evictions.[1] Thus, all broods—including both 'high-amount' and 'low-amount' treatments—eventually experienced such mortalities. The 'low-amount' broods suffered significantly more *second* losses (Figure 6.10). In both field and laboratory experiments, then, the 'food amount hypothesis' was contravened. Great egrets seem either to ignore food levels when fighting, or perhaps to allocate slightly more energy to battle when they are well fed. In fact, the best predictors of great egret *C*-chick survival include one measure of that sib's competitive vigour (specifically, its ability to acquire the pole position; see Chapter 5) and the brood's overall fighting rate. In unmanipulated broods, *C*-chicks that survived their first month averaged 20% of their brood's pole positions, while those that died seldom achieved that grip (4% of their broods' totals). This ability is crucial, as it correlated strongly with *C*'s total amount of food ($r = 0.85$, $P < 0.01$). Not surprisingly, *C*'s ingestion rate was also related to how much food the parents provided for the brood as a whole ($r = 0.60$, $P < 0.01$). Therefore, parents that brought more food to everyone also managed to get more food into the *C*-chick, which apparently helped it to remain competitive and alive.

Fighting rates were on average significantly higher in wild broods whose victim died, both on an absolute basis and when calculated per unit food. In the light of the various lines of evidence showing that such aggression is not stimulated by food shortages *per se*, this suggests the obvious alternative—that the youngest chick tends to survive in nests where parental food deliveries are greater mainly because it is more likely to receive enough food to withstand the punishment it inevitably suffers. This explanation, in fact, seems to account adequately for all the great egret data up to this point.

6.2.4 *Food-insensitivity and sibling aggression*

What may seem surprising, at least initially, about this aspect of the great egret siblicide system is the sharp demarcation between the proximate and ultimate levels of causation. One expects behaviour to be elicited by cues that allow reliable prediction of the circumstances that directly affect survival (or

---

[1] Any chick driven out of its nest by bullying sibs was declared to be 'officially dead', removed to safer quarters, and fed to satiety for the duration.

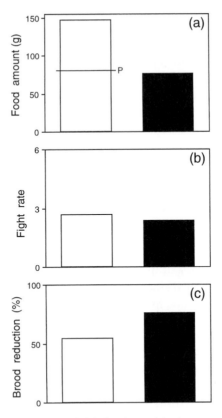

Fig. 6.9 Results of great egret field food-provisioning experiment. **(a)** Broods receiving extra food in the colony (open bars) continued to receive the same contributions from their parents (line marked 'P') as control broods (filled bars). **(b)** Provisioned broods did not reduce their fighting rates. **(c)** However, significantly fewer provisioned broods experienced chick losses. (From Mock *et al.* 1987b.)

other components of fitness). That is, proximate causes often link neatly to the underlying ultimate ones. Here, by contrast, food shortage seems to be clearly established as important to survival, yet one key intervening variable, namely sibling aggression, may pay it no heed. On the other hand, the aspect of food abundance that really matters is whether it will be *sufficient to meet the brood's needs* through the period of dependence, specifically whether a fatal supply bottleneck will occur at some point, relative to the sibship's future demands ('pending competition' of Stinson 1979). If current food levels (and hunger) shed light on that probability, then predicating food-control behaviour on such stimuli can make sense. Under such circumstances, sibling aggression might be viewed as a pre-emptive strike, which should be lethal if a life-threatening food shortage is inevitable and at least temporarily sublethal if adequate control can be gained through a policy of habitual intimidation. Conversely, if it is clear that favourable current conditions will continue, then fighting with nest-mates may be a worthless expense. However, if the correlation between today's food and tomorrow's (or next month's) needs is weak or non-existent, then current hunger would seem to be a poor proximate cue on which to base aggression, and any alternative cue that predicts future risks to senior sibs more accurately may be a better anchor for fighting. If *no* forecast can be made, a policy of relentless oppression may be the only realistic option.

Across siblicidal taxa, then, one might expect to find a range of such long-term probabilities, with 'food-sensitive' sibling aggression in the species with high correlation values and 'food insensitivity' in the others. Such comparisons are just beginning to be made. For example, preliminary analyses of the food deliveries by great egret parents show that even the short-term stability was quite low in Texas (Mock *et al.* 1987b). It seems possible, then, that egret senior sibs can do no better than oppress their juniors, sacrificing a (fairly small) unit of indirect fitness if food turns out to be so scarce that the junior sibling dies. This thesis also suggests that obligate brood reduction is appropriate when the outlook for full brood survival is consistently poor (Stinson 1979). Theoretical explorations of how variability in parental provisioning should affect the decision to execute a nest-mate have shown that the pending competition need not even be consistent. If there is even a modest chance that food will be insufficient for the entire brood, senior siblings often fare better by killing early on, even if this entails some unnecessary wastage of indirect fitness (Forbes and Ydenberg 1992). Such reasoning may well explain the paradox of eagle sibs fighting and killing each other while food is still plentiful (see Section 12.3).

6.3 The demand side: brood size and crowding

As summarized up to this point, the avian sibling aggression literature has dwelt mainly on the *supply* aspects of nestling competition, seldom considering that *demand* is also an integral component of any such dynamic. However, the evidence indicating that egrets ignore current food levels (and the less robust evidence suggesting that the instability of their supply may render such insensitivity the best behavioural policy) forces adoption of a broader perspective. Perhaps sibling aggression in some species is adjusted more to the number of consumers (i.e. brood size), rather than to the actual resources. Maybe sib-fighting occurs automatically, with the victim's fate being determined by the amount it gets to eat, until brood reduction trims demand to a safe, easily met level, whereupon further fighting becomes superfluous.

Although hardly revolutionary logic, this proposition may have escaped

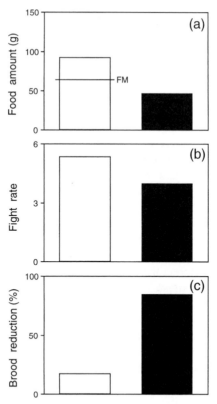

Fig. 6.10 Results of great egret laboratory food-provisioning experiment. **(a)** Broods in the 'high-food' treatment group (open bars) received roughly one-third more food than is normally provided by parents (line marked 'FM' is the estimated field mean), while 'low-food' broods (filled bars) received only half as much. **(b)** Fighting was not lower in the 'high-food' broods. **(c)** Mortality was much higher in the 'low-food' broods (see text). (From Mock *et al.* 1987b.)

consideration until now for one very good reason. The intuitive expectation that food amount *ought* to provide the primary proximate cue for sibling aggression recognizes implicitly that such a system, if reliable, would be inherently superior to one based on brood size, because it offers the potential of substantial savings on fighting costs (to all parties). If abundant current food allows solid prediction of a similarly ample future supply, then all of the energy that might otherwise be squandered on inflicting and recovering from injury could be invested in growth. Everybody wins. By comparison, a system that relies only on brood reduction garners none of that efficiency, responding only after an irretrievable unit of fitness (the victim's life) has been forfeited. In hindsight, then, it made sense to have searched for the more parsimonious and efficient explanation first and to seek secondary solutions only when and if necessary. In any case, this is how it unfolded; when the 'food amount hypothesis' appeared dubious for egrets, the search was broadened.

6.3.1 *Brood size experiments*

The number of chicks in a nest is surely the least ambiguous cue to prospective competition. For egrets, it normally changes only when someone dies, an event with obvious potential impact for the brood's commerce. Because the per capita share of parental food changes as the reciprocal of brood size, the consequences of a single death should be much greater for small broods than for large ones (O'Connor 1978). Of course, the net advantage available to Self for systematically eliminating *all* nest-mates is curbed by indirect fitness tariffs. Moreover, the direct fitness benefits from sequestering additional resources are likely to be asymptotic. These two elements fit together to produce the general predictions outlined in Chapter 1. There may also be additional direct fitness costs of siblicide if, for example, loss of a companion diminishes Self's thermal efficiency, increases its risk of predation, or tempts its parents to desert and re-nest (Mock and Parker 1986). Finally, if the parents do not turn the dead victim's share of the food over to the survivors, all that has been gained is a hypothetical margin of safety, a supply of resources that the parents might release in an emergency.

The idea of parents 'withholding' such food, and diverting that which would have gone to the now deceased victim chick to self-maintenance or some other purpose, is not wildly far-fetched. Egret parents spend quite a lot of time being inactive (even while a siblicide drama is in progress), they have a demonstrated ability to bring more food than they normally do (Fujioka 1985a, Mock and Ploger 1987; see Chapter 11), and they may adjust food deliveries to fluctuations in brood size (Mock and Lamey 1991). However this works out, one can imagine a *core brood* size (Mock and Forbes 1995) within a despotic sib's comfort zone, that suits the survivors' best interests. When that brood size has been reached, bullies should curtail aggressive rivalry.

For the egrets we have studied, nests typically start with three hatchlings and the *core brood* size appears to be two. Great egret individual survivorship rates are higher in nests with two chicks, where hatchlings have a probability of 0.85 of surviving the first month, than in nests with three, four or *even one* chick (Figure 6.11). Broods of three and four chicks exhibit all of the trappings of acute competition (starvation, high levels of aggression, etc.). On the other hand, single chicks tend to be abandoned by their parents (especially if their solitary condition is reached early in the breeding season, when the parents still have time to re-nest; Mock and Parker 1986).

Provocatively, great egret sibling aggression levels are also much lower in broods of two than when more chicks are present (even when corrected for the number of combatants; Mock and Parker 1986). This difference might reflect some recognition by those two chicks of their now more genial circumstances or some other covarying factor. This possibility was clarified by arranging artificial brood reductions in cattle egret nests (Mock and Lamey 1991). We wanted to determine whether reducing brood size from three chicks to two *per se* was sufficient to decrease fighting activities (again corrected for the number of dyads). If so, then restoring the test broods back to full strength should also renew aggression.

The basic protocol was to record natural nest activities in full-sized (three chicks) 1-week-old broods for 3 days, to remove one chick for a second 3-day period (maintaining it in captivity), and then to return it to its home nest for a final 3-day period of post-manipulation recording. Thus, each of the experimental broods served as its own within-group control ('A-B-A' or switchback design) in relation to brood size. The removal was intended to simulate the sudden disappearance that might occur as a result of a siblicidal death (e.g. being driven over the rim and falling to the ground) or predation (e.g. night-heron attacks typically result in just one chick being snatched). Although persecuted ambulatory chicks do occasionally leave their natal nests for a day or so of seeking food elsewhere in the colony, and may then return home, our replacement of the 'lost' chick was not meant to simulate anything in particular, but just to re-establish full brood size. (It was the Easter Miracle design.) In addition, a matched set of unreduced egret broods was observed as a second line of (between-group) controls. One member of each of these broods was sham-handled briefly on days 3 and 6, but remained with its nest-mates.

The full 9-day experimental cycle was performed twice, using entirely

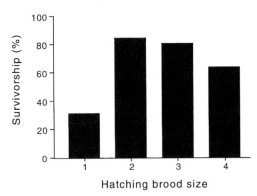

Fig. 6.11 Great egret mean survivorship (% alive at end of first month) for broods hatching 1, 2, 3, or 4 chicks initially. The surprisingly low value for singletons was probably due to parents abandoning one-chick broods that occurred early in the season, when the parents may have sufficient time to nest again and produce a larger number of offspring. (From Mock and Parker 1986.)

different sets of nests. In the first experiment, 10 C-chicks (the typical victims) were removed, so as to mimic the normal brood reduction phenomenon as closely as possible. As predicted, fighting in the reduced broods plummeted (from 2.4 daily fights per dyad during the pre-test to 0.1 during treatment). When C was returned 3 days later, fighting resumed at the normal high levels (1.5 daily fights per dyad). Aggression at the 10 sham-handled control nests showed only the temporal decline (Figure 6.12a).

To make sure that the apparent brood size effect had not been confounded by chick rank (for example, if our removal of C-chicks had quelled fighting merely because we had absconded with the favourite 'target' individual, the proverbial whipping-boy), the experiment was repeated exactly as before (using new nests and broods), except that this time the top-ranking A-chicks were removed instead. The same plunge in daily fights per dyad was observed in the nine manipulated broods and, once again, the nine control broods remained level. As in previous studies, the possible influence of large food amount could be discounted (see Figure 6.12b). While this would have complicated interpretation if the parents had delivered to the reduced broods at the original level (thus increasing consumption by 50% per capita), they apparently 'withheld' the missing chick's food, delivering a number of boluses that was neatly tailored to the smaller brood size. The important point here is that the test of brood size was not confounded by food amounts.

Thus, it appears that cattle egret chicks do monitor brood size and base part of their fighting decisions on that factor. Just how they make this assessment is not known. It could involve a simple form of visual counting (requiring only a discrimination between '2' and '>2'). Alternatively, it might be accomplished via vocal or tactile cues (e.g. frequency of bumping into a sib, receiving warmth from two sides of the body rather than just one, etc.). On a practical level, this indicates that fighting behaviour hinges on at least one proximate cue that provides an accurate forecast of the 'pending competition.' The cue of brood size may seem a clumsy and drastic choice for that vital role, but at least it has the virtue of being a very solid predictor and, as such, provides one aggression switch-off mechanism, which is better than none at all. Broods may have to make the best of a bad job.

6.3.2 *Escape space*

To date, all research on ardeid sibling relations has involved nests where the brood's spatial options are sharply limited. For the most part, a harassed junior sib can flee no further than the nest's rim without having to give up solid footing. Depending on its age and motor development, a persecuted chick that ventures beyond the nest risks falling to the ground. In the Texan colonies, such falls only involved a height of 1–2 m, but young nestlings were seldom capable of scaling back up to the nest. Not surprisingly, most chicks remain in the nest and accept the physical punishment for as long as they can.

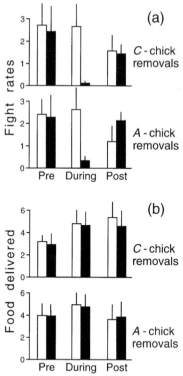

Fig. 6.12 Cattle egret brood size experiment. Broods were recorded for 3 days initially (PRE = pre-manipulation), then one chick was removed for 3 days (During) in the experimental nests (filled bars) (no removals from the control nests, open bars). Finally, the missing chicks were restored and all broods were recorded for 3 more days (Post = post-manipulation). **(a)** Fighting virtually disappeared with the simulated brood reduction, regardless of the removed chick's rank. **(b)** Parents adjusted their provisioning so as to keep the per-capita food amounts constant. (From Mock and Lamey 1991.)

In a few cases, however, the surrounding vegetation fortunately contained leafy nooks in which desperate junior sibs sought, and found, temporary asylum. One great egret *C*-chick adopted the useful strategy of leaving its home nest at the *beginning* of each meal, when the most vicious fights typically occur, and staying out of harm's way until both of its senior sibs had consumed exactly one bolus apiece, at which time it stormed back and joined the fray for any late boluses. With some food already in their necks, its sibs' pecks no longer intimidated at this stage. In a few other cases, close adjacent nests that happened to become vacant were used as havens for junior siblings. These anecdotes aside, however, most beaten chicks had no safe place to which to flee. It would be interesting to observe sibling relations in a ground-nesting ardeid population, where movement from and between nests poses less danger (e.g. Palmer 1962, Litvinenko 1982).

In general, the effects of nestling space on sibling competition have been little explored (but see Slagsvold 1985) and need to be investigated further, not only for ground-nesting taxa such as gulls, cranes, skuas, and owls, but also for the opposite extreme. Certain kinds of ***nurseries*** feature spatial con-figurations that narrow the options available to rival sibs much as a boxing ring constrains the mobility of prize-fighters. In hole and burrow-nesting birds, for example, the parent's one point of entry is perfectly predictable, and a bully's task in defending it is greatly simplified (see Lessells and Avery 1989, Bryant and Tatner 1990, Litovitch and Power 1992, Anderson *et al.* 1993). Finally, there is one escape route for a victimized egret chick that opts to leave the abusive 'haven' of the nursery, although its chances of success are lean. We have seen such individuals snuggle into a neighbour's brood, although they are usually repelled vigorously by the resident chicks. If one manages to get in, it crouches into a tight ball and accepts whatever pecking it can stand as submissively as possible. Occasionally, this works and the others eventually abandon their attempt to evict it; the residents may be too young to kill the intruder outright and the latter may be several days older than they are. Even the resident parents are sometimes fooled, although intruders are usually driven off and even severely wounded by the first returning adult. In short, there is a longshot gamble which desperadoes attempt—that of parasitizing another family. The prospects for survival at home may rapidly be approach-ing zero when it moves, so the probability of its unilaterally arranging to be adopted need not be especially high. A small window of opportunity is open because of high nest density, asynchrony of brood ages, and imperfections of kin recognition and nest defences.

Summary

1. Great blue herons nesting in the same Texas colony as the great egrets described in the previous chapter show almost no siblicidal aggression, despite having comparable rates of brood reduction. Part of this be-havioural difference was proposed to stem from the size of prey fish regurgitated for chicks. Heron meals consisted of single, partially digested fish too large to be taken straight from the parent's bill and swallowed whole, while egret chicks received discrete boluses of much smaller fish. The key element seemed to be that egret food is intercepted from parents in the form of *direct* feeds. Accordingly, egret meals change from being *indirect* scramble competitions (during the first week when chicks are too young to catch boluses) and become increasingly a resource-defence competition. The selfish benefits from fighting and resulting dominance are

available mainly after that transition to mostly *direct* feeds, which the herons never made.

2. This argument was set out as the **Prey-size Hypothesis**, that sibling aggression's cost-effectiveness should hinge on the monopolizability of the food. On a comparative basis, this appears to be unidirectional. Monopolizably small food units (and, specifically, *direct* feeds) may be a necessary, but not sufficient, condition for avian sibling aggression.

3. The hypothesis was field-tested by exchanging whole broods of herons and egrets so that each was given the opposite diet by heterospecific parents. The prediction that herons should adopt *direct* feeding and develop siblicidal aggression was supported; the reverse situation, that egrets should abandon *direct* feeds and adopt pacifism, was not. The difference in developmental plasticity was explored through intraspecific comparisons. Great egrets fed on 'large-prey' vs. 'small-prey' diets in the laboratory showed no behavioural flexibility, and no natural populations of 'large-prey' egrets have been located for a geographical comparison. A 'small-prey' population of great blue herons in Québec shows that this species faces different ecological conditions across its range, which may help to account for its flexibility.

4. The amount of food delivered by parents determines the severity of sibling competition in birds. Food amount has also been widely assumed to influence the intensity of fighting, in part because a logical mechanism ('hunger') exists that could mediate useful adjustments in aggression. In blue-footed boobies and ospreys it has now been shown experimentally that fighting varies inversely with food supplies: hungry chicks escalate their aggression.

5. In great egrets, however, attempts to detect hunger-sensitive adjustments to sibling aggression have failed. Unmanipulated broods receiving larger amounts of food did not fight less, nor did broods whose ingestion levels were experimentally doubled (in both laboratory and field settings). The youngest chick's ability to survive appears to be strongly related to both total food amounts (to the brood) and that individual's ability to remain active and competitive. Food-insensitivity in egrets may be due to inconsistency in parental deliveries.

6. A more conservative proximate regulator of sibling aggression might couple it only to brood size, and hence demand level. This was tested by temporarily removing one chick from three-chick cattle egret broods: in one trial the displaced sibling was the youngest, and in a repetition it was the eldest. As predicted, fighting fell off sharply when brood size was lowered, and then rebounded to control levels after it was restored. This effect could not have been due to shifts in food amount for the simple reason that parents compensated for the reduction in brood size by bringing one-third less food. In this experiment, then, the senior siblings that remained in the nest actually took a cut in food supply.

7. Some egret victims opt to leave the home nest, either briefly (to minimize their beatings at meal-time) or permanently. In the latter case, they may attempt to invade a nearby nest. Such intrusions are usually, but not always, futile. Sometimes the chick succeeds in overcoming the resistance of the resident nestlings and their parents, effectively arranging its own adoption. If it is larger than its new nursery-mates, it may dominate and eventually kill them.

7

Parent–offspring conflict I: introduction to theory

Parents are classically assumed to allocate investment in their young in such a way as to maximize the number surviving, while offspring are assumed to be passive vessels into which parents pour the appropriate care. Once one imagines offspring as *actors* in this interaction, then conflict must be assumed to lie at the heart of sexual reproduction itself—an offspring attempting from the very beginning to maximize its reproductive success (RS) would presumably want more investment than its parent is selected to give.

(Trivers, 1974)

Parents represent the 'supply side' and offspring the 'demand side' for the flow of parental resources from parent to offspring (see Ydenberg 1994). In Chapters 2–4, we saw that sibling rivalry is likely to depend on the amount of resources over which nursery-mates compete, but we simplified the picture by assuming that the resources from parents were fixed. Conflict between parent and offspring may affect the flow of parental resources to a brood of siblings (as shown in Figure 3.1). In the present chapter we shall consider how this conflict affects, and is affected by, sib competition.

7.1 How can parent–offspring conflict arise?

Although the origins of the idea of parent–offspring conflict (which we shall abbreviate as POC) can be traced back to Hamilton's classic insights (1964a, b), the concept and its logical foundation were first proposed in a remarkable paper by Trivers (1974). He used Hamilton's rule to formulate, for sexually reproducing diploids, a theory of conflict over the amount of parental investment (PI) that each offspring should receive from its parent; offspring are generally under selection to 'want' more PI than it is in parental interests to give. Trivers's argument was based on simple benefit/cost considerations (Figure 7.1). As an offspring receives more PI, its expected future reproductive success (and hence the benefit B to the parent) is likely to increase, up to some asymptotic maximum (see also Figure 2.1). However, the costs C to

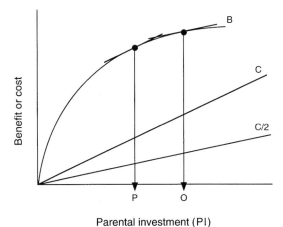

Fig. 7.1 Trivers' original formulation of parent–offspring conflict. The parent's in-clusive fitness maximizes the difference between benefits (B) and costs (C), occurring at the point where the benefit curve's slope is equal to that of the cost curve. The offspring's inclusive fitness is maximized at (B−C/2). Displaced offspring are assumed to be full siblings. (Modified from Trivers 1974.)

the parent increase with PI because, following the definition of PI, fewer future offspring can be produced if more is spent on the present offspring. Trivers argued that selection on parents would maximize $B-C$.

Suppose that offspring are produced successively by a given parent, and that they are all genetic full siblings. A given offspring receiving benefit B reduces some future sib's prospects by C, to which he or she has a degree of relatedness of $\frac{1}{2}$. Selection on offspring should therefore maximize $B-C/2$. This leads to a higher optimal PI from the offspring's point of view than from that of the parent (Figure 7.1).

We next mention a technical detail. It has been argued (e.g. Lazarus and Inglis 1986) that any parental benefits and costs should be $B/2$, $C/2$ and not B, C—because the parent itself has a degree of relatedness of $\frac{1}{2}$ to all offspring. So, it is argued, Trivers should (effectively) have doubled the benefits for the offspring, rather than halving the costs. It is quite clear that this makes no difference to the optimal solutions as shown in Figure 7.1. Maximizing either $W_p(I) = B(I) - C(I)$ or $W_p(I) = [B(I) - C(I)]/2$ both give the parent's optimal PI, I_p, as

$$\frac{dB(I_p)}{d(I)} = \frac{dC(I_p)}{d(I)}$$

while maximizing $W_o(I) = B(I) - \frac{1}{2}C(I)$ gives the offspring's optimal PI, I_o, as

$$\frac{dB(I_o)}{d(I)} = \frac{1}{2}\left[\frac{dC(I_o)}{d(I)}\right].$$

The difference in optima for the parent and offspring depends on the slopes of the $B(I)$ and $C(I)$ curves; offspring will be selected to take more PI than the parent is selected to give (Figure 7.1). If some of its offspring do take this higher amount of PI, then the reproductive success of a parent will be reduced, although the inclusive fitness of the 'conflictor' offspring is increased.

The idea of POC was swiftly contested by Alexander (1974), who argued that a gene causing the summed fitness of offspring produced by a parent to be reduced would be quickly eliminated by selection, in effect because offspring that conflict become parents that produce conflicting offspring. The parent should therefore always win. This does, of course, treat parental interests as paramount. Dawkins (1976) pointed out that phrasing the argument the other way round (allowing offspring interests to be paramount) leads to the conclusion that the offspring always wins. Alexander's paper had the useful effect of stimulating a series of explicit population genetic models (e.g. Blick 1977, Macnair and Parker 1978, 1979, Parker and Macnair 1978, 1979, Stamps *et al.* 1978, Metcalf *et al.* 1979, Feldman and Eshel 1982) that showed that a gene[1] causing an offspring to take more than the parental optimum could indeed spread, and the overwhelming view now is that Trivers' argument is essentially correct. Debate has subsequently moved away from the question of whether there *can be* genetic conflict (the 'battleground' exists), to considerations of how this potential for conflict (between genes acting in the parent and those acting in the offspring) is likely to be resolved in behavioural terms (Godfray and Parker 1992, Mock and Forbes 1992).

What are the parental resources—the effort we term PI? Obviously, they are often food resources. But conflict can concern other aspects of parental care in addition to feeding, e.g. vigilance, risk in defence, or energetic costs. Strictly, PI is not measured directly by the quantity of parental resource provided, but as the cost of that resource to a parent, measured in terms of other offspring (Trivers 1972), a point that must be remembered in the following discussion.

How can the conflict over PI arise? Consider a dominant, mutant 'conflictor' allele in a species where the parents stay together throughout their life, so that successive offspring are full sibs. The mutant allele causes an offspring bearing it to take more PI. Its personal fitness (= probability of surviving × its subsequent reproductive success), f_c, is therefore greater than that of an offspring homozygous for the wild-type allele (for 'non-conflicting' activities), which has a fitness of f_w. The conflictor gene spreads if there are more replicas of it in the next generation than there are replicas of a randomly chosen wild-type allele. It achieves this by taking a disproportionate share of the parental resources, at the expense of the sibs bearing only wild-type alleles. We can imagine parental investment as a fixed resource (the circles in Figure 7.2) that

---

[1] This is quite a good example of a situation where one needs to be able to switch from an individual-selection perspective to a genic-selection one (Stamps and Metcalf 1980).

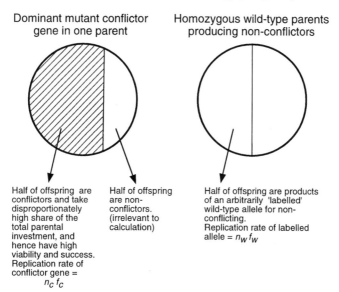

Dominant mutant conflictor
gene in one parent

Homozygous wild-type parents
producing non-conflictors

Half of offspring are
conflictors and take
disproportionately
high share of the
total parental
investment, and
hence have high
viability and success.
Replication rate of
conflictor gene =
$n_c f_c$

Half of offspring
are non-
conflictors.
(irrelevant to
calculation)

Half of offspring are products
of an arbitrarily 'labelled'
wild-type allele for non-
conflicting.
Replication rate of labelled
allele = $n_w f_w$

Fig. 7.2 Spread of a rare conflictor gene. Each circle represents a parent's total resources for investment in offspring (see text). (After Parker 1985.)

is gradually allocated to successive offspring; the parent dies when all its investment is used up. In a sexually reproducing species, if the conflictor allele is rare, parents bearing it will be heterozygous, and so on average half of the offspring will be homozygous wild-type and half will be heterozygous for the conflictor mutation. Suppose that some number, n_c, of homozygous (hence conflictor-phenotype) progeny are produced. They will take more than half of the parental resources, at the expense of their remaining n_c wild-type siblings (which are heterozygous for the allele, and thus non-conflictors). In contrast, an arbitrarily 'labelled' wild-type allele, drawn at random from the population, is likely to occur in a mating between two homozygous wild-type parents. Half of these offpring (n_w in number) will bear the labelled allele. Because none of these offspring take extra parental resources, there can be more of them, i.e. $n_w > n_c$. However, each non-conflictor will have lower fitness ($f_w < f_c$). The conflictor allele will spread if

$$n_c \cdot f_c > n_w \cdot f_w,\qquad(7.1)$$

which may be satisfied if the drop in numbers is offset by the increase in fitness. Note that the non-conflictor offspring in the brood containing conflictors are irrelevant to the calculation; they do not affect the mean fitness of non-conflictors in a large population.

We can develop the logic of conflict more rigorously using a form of marginal Hamilton's rule (see Section 3.3.4). Suppose that the relationship between offspring fitness, f, and the amount of parental resources it receives,

m, follows a rule of diminishing returns as discussed in Chapter 2 (see Figure 2.1). If m_w = the amount of PI taken by each non-conflictor, then its fitness is $f_w = f(m_w)$, while a conflictor sib's fitness is $f_c = f(m_w + \Delta)$, where Δ is the extra PI taken by each conflictor. Similarly, if each conflictor displaces a fraction, π, of a future offspring, having a probability r of carrying the conflictor allele, then the conflicting tendency will spread if the extra benefit to Self's fitness, f_c − f_w, exceeds the loss due to conflictor alleles displaced, $r\pi f_c$. That is, if:

$$f(m_w + \Delta) - f(m_w) > r\pi f(m_w + \Delta).$$

If Δ and π are very small, so that second-order terms (i.e. those containing Δ times π) can be ignored, then after Taylor expansion of this condition we get the marginal Hamilton's rule that

$$\Delta f'(m_w) > r \cdot \pi \cdot f(m_w),$$

i.e. the small increases in conflictor behaviour will spread so long as the marginal benefits to Self (=Δ times its rate of conversion into personal fitness around m_w) are greater than the costs to sibs bearing the conflictor allele (= r times the fraction of a sib displaced multiplied by its fitness). For reasons that will be apparent soon, it is convenient to express this inequality as

$$f'(m_w) = \frac{r\pi}{\Delta} \cdot f(m_w). \tag{7.2}$$

What is the ratio, π / Δ? Suppose that a female has a total investment of M units to allocate amongst successive offspring. Then if each offspring takes m_w units, she will produce a total of M/m_w. But if a fraction p of her offspring are conflictors, each taking $m_w + \Delta$, then she will produce a total of

$$n = \frac{M}{(1 - p)m_w + p(m_w + \Delta)} \tag{7.2a}$$

progeny. Since each conflictor displaces π of an offspring, then the total offspring displaced is

$$np\pi = \frac{M}{m_w} - n.$$

After a little manipulation, one can show that

$$\frac{\pi}{\Delta} = \frac{1}{m_w},$$

so that we can rewrite equation (7.2) as

$$f'(m_w) > \frac{r \cdot f(m_w)}{m_w}. \tag{7.3}$$

In terms of Hamilton's rule: small increases in extra PI will be favoured so long as an offspring's marginal rate of gain in personal fitness exceeds the average ratio of fitness/investment per offspring, devalued by the coefficient of relatedness.

Whether taking more PI is likely to spread depends on the current PI level, namely m_w. Consider two extremes of PI allocation (see Figure 7.3). In one population the amount of PI allocated, m_1, is low. The marginal rate of gain, the gradient of $f(m_1)$, is therefore high—much higher than the average fitness/investment per offspring, $f(m_1)/m_1$. Here it will pay the offspring to increase PI. In another population the PI allocation (m_2) is high, so the marginal rate of gain is less than r times the average fitness/investment. Here it pays the offspring to decrease the PI it takes. However, in both of these cases, the parent and the offspring *agree* about the direction of change in PI level; selection acts on both to increase or to reduce PI per offspring, depending on the current PI level. But between m_1 and m_2 there exists a range of m over which the parent will be selected to decrease PI, whilst the offspring will be selected to increase it. We now need to define the ranges of agreement and conflict about PI so that we can determine the optimum PI for each in the absence of any constraint by the other.

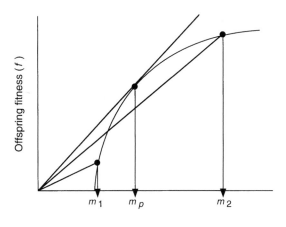

Parental investment (m)

Fig. 7.3 The fitness gain per unit investment in each offspring is given by the slopes to the curve, $f(m)$. The parental optimum is given by m_p, the m-value for which the tangent from the origin touches the curve $f(m)$. The best strategy for the offspring depends on what m strategy other offspring in the population are adopting. If the population is fixed at m_1, the marginal rate of gain (slope of the $f(m)$ curve) is much higher than the average fitness/investment per offspring, so offspring are selected to take more PI. In a second population (fixed at m_2), the marginal gain rate is less than r times the average fitness per investment per offspring, and it will pay the offspring to take less investment. The offspring ESS is always greater than m_p since $r < 1$.

7.2 What is the zone of conflict?

If there exists a range over which the best parental strategy is to decrease PI, but the best offspring strategy is to increase it, we have a genetic conflict of interest. In a sense this is a conflict between two genes or loci—one that determines the parental strategy and one that determines the offspring strategy.

7.2.1 *The parental optimum*

What amount of PI should the parent give to each offspring? Consider the typical case where parental care is restricted to female parents. It is generally assumed that the female parent will be selected to maximize her number of surviving progeny. Thus if she has a total of M units of PI, and allocates m to each offspring, her total numerical offspring production will be M/m. Each of these offspring will have a fitness of $f(m)$, and so the optimal parental strategy can be found by maximizing

$$W(m) = \left(\frac{M}{m}\right) \cdot f(m),$$

which gives the result that

$$f'(m_p) = \frac{f(m_p)}{m_p} \tag{7.4}$$

where m_p is the parental optimum. Rule (7.4) was first used in the context of optimal allocation of parental resources to progeny by Smith and Fretwell (1974)(see also Brockelman 1975). The parent should allocate extra resources to each offspring up to the point where the gradient of the $f(m)$ curve equals a tangent to $f(m)$ drawn from the origin at $m = 0$ (see Figure 7.3). This is, in fact, Charnov's (1976) familiar Marginal Value Theorem used in optimal foraging[2] (for the same rule applied to mate-searching see also Parker 1974, Parker and Stuart 1976). Rule (7.4) is valid, at least in the simple context of the present model, but may not apply if the prospects of each offspring vary (see Temme 1986).

What about biparental care? Parker (1985) argued that equation (7.4) will give the optimal allocation for each offspring only if there is 'true monogamy' (defined as a system wherein both individuals pair for life, and cannot pair again if their first mate predeceases them). With biparental care, the model makes no prediction as to how each parent should allocate PI amongst the various offspring. It would be equally satisfactory for each parent to give m_p

---

[2] Restating its application in the present context, the tangent line defined by equation (7.4) is the parent's optimum because it is the steepest fitness gain available to parents that is also realistic (i.e. by virtue of touching the curve). There are other lines that intersect the fitness curve (all non-tangents that touch it must actually do so twice), but these have visibly lower slopes and correspond with investing either 'too much' or 'too little' in each offspring.

exclusively to alternative halves of the total brood, or for each to feed $m_p/2$ to each offspring, or indeed for any allocation between offspring which amounts to m_p apiece.

If each parent has M units of PI, the biparental total is $2M$, so equation (7.4) might be interpreted as suggesting that twice as many surviving offspring can be produced. However, this assumes that the cost of each unit of resource supplied to an offspring is constant, so that as more units are supplied, the PI cost rises linearly. In a hypothetical bird, the cost (in terms of future offspring lost, i.e. PI units) of supplying offspring with 20 g of food per day may be much more than twice that of supplying 10 g daily. Hence two parents may produce *more* than twice the surviving offspring produced by one parent. Although equation (7.4) indicates the optimal PI, it tells us little directly about the optimal amount of resources to supply. We need to know how PI relates to parental input (Evans 1990).

Let $y =$ the food received by an offspring from a given parent. The cost, in PI, of giving y is $m(y)$ where m is an increasing function of y, and more specifically we expect that $m(y)$ will either be linear, or have an increasing or accelerating gradient (see Figure 7.4a). Offspring fitness is a function of the total food received. For uniparental care it is $f(y)$. We now deduce what amount of food should optimally be given per offspring, rather than the optimal amount of PI. For uniparental care, this is the maximum value for

$$W(y) = \frac{M}{m(y)} \cdot f(y),$$

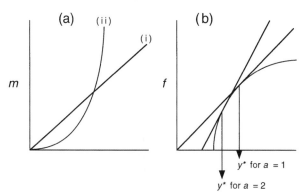

Food input by parent (y)

Fig. 7.4 **(a)** Parental investment, m, in relation to food input, y, supplied by parent. In curve (i), the costs of increased food input rise linearly; in curve (ii), the PI costs accelerate steeply with increasing food input. **(b)** Parental optimum for food input. This can be derived by the tangent from the origin if the costs of food input rise linearly ($a = 1$), but reduced food input occurs if PI costs accelerate with food input, as in the case shown where $a = 2$.

Differentiating W with respect to y and setting to zero gives the optimal food provisioning, y_p:

$$f'(y_p) = \frac{m'(y_p)}{m(y_p)} \cdot f(y_p) \qquad (7.5)$$

To interpret this equation, consider the form of $m(y)$. The simplest model is for $m(y) = by^a$, where a and b are positive constants with $a \geq 1.0$, which ensures that equation (7.5) is always a maximum. Then

$$f'(y_p) = \frac{a}{y_p} \cdot f(y_p) \qquad (7.6)$$

and so if $m(y)$ rises linearly ($a = 1.0$) from the origin, the rule for allocating food (equation 7.5) is the same as that for allocating PI (equation 7.4). Food should be allocated to each offspring in accordance with the tangent rule (Figure 7.4b). However, if there are *increasing PI costs* of each successive unit of food ($a > 1.0$), then the parent(s) should allocate **less** food per offspring than under the tangent rule.

For biparental care with 'true monogamy', the M is replaced with $2M$, and each parent provides $z = y/2$ of the food per offspring, and hence each sustains PI costs of $m(z)$. Assuming that the best strategy under true monogamy is for total co-operation[3], this is equivalent to maximizing the total production of the pair. This is achieved by maximizing

$$W(z) = \frac{2M}{2m(z)} f(2z).$$

Differentiating with respect to z and equating to zero, and remembering that $y = 2z$, we again obtain equations (7.5) and (7.6). Biparental care under true monogamy should not affect the total food that is given to the offspring, so that offspring fitness should remain unaltered. However, each parent will sustain costs equivalent to $m(y/2)$, in contrast to the $m(y)$ sustained by the single parent. If PI costs accelerate as y increases, then with biparental care more than twice the offspring should be produced than under uniparental care, since $2m(y/2) < m(y)$. One can go on to show that for the case where $m(y) = by^a$, the ratio of offspring (biparental/uniparental) will be 2^a. The notion that the main effect of biparental co-operation is to increase the number of offspring, and **not** the fitness (survival and reproductive success) of each one, is rather counterintuitive, and may repay investigation by the comparative method.

---

[3] This derives from the definition of 'true monogamy'. If the current partner is the only possible lifetime mate, then a selfish individual cannot get ahead by exploiting him or her. On the contrary, even a ruthlessly selfish adult does best by striving for both parents to have identical lifespans, and the partner reciprocates for the same reason. Accordingly, perfect co-operation is optimal.

But what about biparental care without 'true monogamy'? Strict examples of true monogamy are surely rare, if not non-existent. Often there appears to be little constraint on the future reproductive success of a given partner when a mate dies. Examples of biparental care without such constraint can range from cases where the parents remain together only for the production of a given brood and then separate (the next mating cycle is with another partner), to cases where the same two parents rear successive broods together until one partner dies and the other then 'remarries'.

Without the strictures of true monogamy, the appropriate optimization criterion is no longer to maximize the surviving young of a given pair. Parker (1985) analysed the case where both parents invest in a similar way in progeny (e.g. the same alleles control PI in the two sexes), so that a mutation affects PI strategy equally irrespective of the sex of the parent. This demands an ESS approach, because the best PI strategy depends on the partner's strategy (see also Chase 1980, Houston and Davies 1985, Winkler 1987). It can be shown (Parker 1985) that the ESS amount of PI for the parent has a Marginal Value Theorem solution:

$$f'(m_g) = \frac{2f(m_g)}{m_g} \tag{7.7}$$

where $m_g/2$ is the ESS amount of PI for each parent to give. Remarkably, equation (7.7) implies that the total PI provisioning *per offspring* should be **less** when two parents collaborate (without true monogamy) than under uniparental care (cf. equation (7.4) above), as shown in Figure 7.5. This paradoxical result arises because of the sexual conflict between the parents over how much PI each should give. That is, because each parent has an incentive to hold back some of its potential investment (for use with a different, future partner), the individual offspring necessarily receive less PI. There will therefore be more than double the uniparental number of offspring. This point is discussed further in Chapter 10.

How is this conclusion affected by looking at the optimal amount of food for parents to supply, rather than at their ESS PI levels? Following the same analysis as for biparental care with true monogamy, we assume that at the ESS, each parent provides $z_p = y_p/2$ of the food per offspring, and hence sustains PI costs of $m(z_p)$. A mutant parent that deviates unilaterally and supplies $z \neq z_p$ will produce M/z offspring instead of M/z_p, and the fitness of each one will be $f(z + z_p)$ instead of $f(2z_p)$. The fitness function for the mutant parent in the ESS population is

$$w(z,z_p) = \frac{M}{m(z)} \cdot f(z+z_p).$$

To find the ESS we set

$$\left[\frac{\partial W(z,z_p)}{\partial z}\right]_{z=z_p} = 0,$$

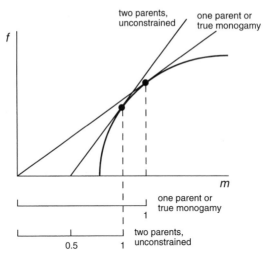

Fig. 7.5 The parental optimum for uniparental investment or true monogamy (tangent from the origin of the *f(m)* curve) and that for biparental investment (the other tangent) without true monogamy. For uniparental investment, the tangent to the *f(m)* curve is taken from the origin; for biparental investment, the tangent to the *f(m)* curve is taken from a distance to the right of the origin equal to half the optimal investment. Note that the optimal investment increases in the order: biparental < uniparental. (After Parker 1985.)

and ensure that the second derivative is negative (see Maynard Smith 1982a). This gives the result that

$$f'(y_p) = \frac{m'(z_p)}{m(z_p)} \cdot f(y_p), \tag{7.8}$$

and for the case where $m(y) = by^a$ (Figure 7.4),

$$f'(y_p) = \frac{2a}{y_p} \cdot f(y_p), \tag{7.9}$$

These equations show that the conclusion about reduced total PI under biparental care without true monogamy also extends to reduced total food; equation (7.9) relates to equation (7.6) in exactly the same way that equation (7.7) relates to equation (7.4). When they are unconstrained by the future survival of their partners, we expect biparental parents to supply less total food and to produce offspring of lower fitness than single parents! This prediction appears heretical and certainly remains to be tested, but as yet we can see no obvious flaw in this argument. For the present biparental model, we cannot directly compare the number of surviving offspring produced with uniparental care, because the total food provided is not the same. This can be done only if an explicit function is given for $f(y)$, but generally many more surviving offspring would be produced by two parents than by one, although fewer than under true monogamy.

7.2.2 *The offspring optimum*

From now on we shall consider conflict over the amount of PI directly, rather than over the amount of resources provided, since the conversion between resources and PI, although important for field studies, does not appear to affect conclusions qualitatively. Determination of the offspring's optimum again requires an ESS approach because the best strategy for the offspring depends on the current level of PI prevalent in the population. Earlier, we derived a condition (7.3) for the spread of a mutant allele causing an offspring to take marginally more PI than the current level, m_w (see also Figure 7.3). Small increases are favoured when the marginal rate of personal fitness gain exceeds the current fitness/investment ratio per offspring, devalued by coefficient of relationship, r. Suppose that this process of gradual increase in PI taken from parents were to continue. If the relationship between offspring fitness f and the parental investment m that it receives shows diminishing returns, there will come a point when the marginal rate of gain exactly balances the fitness/investment ratio devalued by r. This is the ESS level of m for the offspring to take. We shall call it m_o. From condition (7.3), m_o is

$$f'(m_o) = r \left[\frac{f(m_o)}{m_o} \right] \tag{7.10}$$

where $r = \frac{1}{2}$ for full sibs.

Parker and Macnair (1978) and Macnair and Parker (1978) derived these equations by the method of differentiating the fitness function, as we have just done for the sexual conflict game between parents. The levels of PI implied by these optima can be compared with the optima for a solo (uniparental) parent (Figure 7.6). Both offspring optima are for greater PI than is optimal for the parent, and where the displaced offspring are half sibs, the offspring will be selected to take more PI than when the displaced offspring are full sibs. Between the parent and offspring optima lies the zone of conflict in which parents will be selected to supply less PI, and offspring will be selected to take more. Note that for reasons of clarity, Figure 7.6 shows only the uniparental (or biparental, with true monogamy) optimum. As we have seen, the biparental optimum in the absence of true monogamy is less, so that the conflict zone is even greater under this more usual form of biparental care.

Extreme caution is needed when deciding which of the equations (full sibs or half sibs) is likely to apply to a given case (Parker 1985). The correct value for r cannot be deduced simply by calculating the average relatedness between sibs from a given parent; this is quite likely to give the wrong result (see Table 7.1). The best procedure is not to calculate relatedness between successive sibs at all, but to take a gene perspective and determine whether

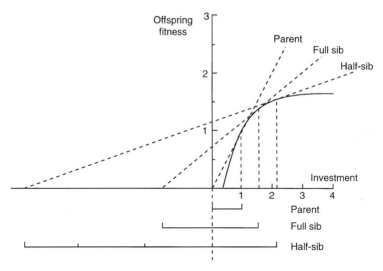

Fig. 7.6 The offspring optima for full sibs and for half-sibs. For full sibs, the tangent to the *f(m)* curve is taken from a distance to the left of the origin equal to the optimal investment. For half-sibs, the tangent to the *f(m)* curve is taken from a distance to the left of the origin equal to 3 times the optimal investment. Note that the optimal investment increases in the order: parent < offspring, full sib < offspring, half-sib. (After Godfray and Parker 1991.)

one parent or both are constrained by the effects of a mutant allele causing an offspring to take more PI (a conflictor gene). By 'constrained', we mean that some adverse effect is likely to be felt by the parent if its offspring takes more PI. This depends on the mating system.

For example, imagine the case where adults engage in promiscuity, and where only the female invests in the offspring. When carried in a male parent, a conflictor allele is *always* beneficial; each offspring of such a male achieves higher fitness at the expense of offspring to whom it is unrelated. The father is therefore unconstrained by the effects of conflict. When carried in a female parent, by contrast, the conflictor allele is constrained because all future sibs that are displaced will be full sibs. The averaged coefficient $r = \frac{1}{4}$ therefore comes about as the mean of zero and $\frac{1}{2}$, the respective values for the unconstrained and constrained parents. It would also be the r-value calculated from relatedness between successive broods of sibs, which are half sibs. However, $r = \frac{1}{4}$ also applies if both parents remain together until one dies, assuming that only one parent is sensitive to offspring demand signals (e.g. the female provides food and the male guards), and also assuming that the other parent can find a replacement mate if the 'sensitive' parent dies. Most of the offspring in successive broods here will be full sibs. As would be expected, true monogamy has $r = \frac{1}{2}$ (and hence a reduced offspring PI optimum). Less

Table 7.1 Effects of mating system on offspring's optimal investment. (After Parker 1985.)

One parent constrained			Both parents constrained		
Extremes of mating system	Relatedness of sibs in successive broods	ESS	Extremes of mating system	Relatedness of sibs in successive broods	ESS
Male and female stay together throughout life; only 1 parent invests directly; other can find a new mate if first mate dies.	Usually full siblings	$f'(m_o) = \dfrac{f(m_o)}{4m_o}$	Male and female may invest equally or unequally; stay together throughout life; no remating after being widowed	Full siblings	$f'(m_o) = \dfrac{f(m_o)}{2m_o}$
Promiscuity, no investment by male parent	Half siblings		Male and female invest equally but split up after each brood.	Half siblings	

intuitively, so does biparental care with two 'sensitive' parents, where the parents split up after each brood, and mate randomly for the next brood. Here the offspring in successive broods of a given parent are half sibs, but both parents are constrained by carrying the conflictor allele, unlike the case of promiscuous uniparental care where only one sex is constrained.

Table 7.1 provides a summary of the effects of the mating system on the offspring optimum. The greatest potential for genetic conflict (measured as disparity between parent and offspring optima) will occur with biparental care without true monogamy, which has the lower parental optimum and the higher offspring optimum. Before considering possible ways by which the conflict may be resolved, we need to outline two types of POC.

7.3 Types of parent–offspring conflict

We recognize two types of POC, depending on whether future or contemporary sibs are affected.

7.3.1 Inter-brood (Triversian) conflict

The parent alters its PI input to the brood (or single offspring if brood size is one) in relation to the demands of the brood. Thus if a given brood demands and receives more PI, that concession comes at the expense of *future* sibs. That is, the parent dies earlier so that sibs that would have been produced, had each conflictor offspring taken less PI, are now lost (Figure 7.7a). This is the type of conflct that we have been considering in the present chapter. It can be called Triversian conflict since this is explicitly the case that Trivers (1974) examined in his classic paper. The offspring optimum is defined in equation (7.10).

7.3.2 Intra-brood conflict

When brood size is greater than one, the parent allocation of PI *within* the brood can be disrupted by differential demands of individual siblings (without changing the total input of PI provided to this breeding cycle). This is a form of POC because the parent's optimal allocation of PI amongst the offspring can be affected against parental interests (Figure 7.7b). For this case, there is no offspring optimum equivalent to equation 7.10 (see Section 3.3.3).

It is difficult to claim that these are types of POC *rather than* types of sib competition because they are in fact both simultaneously. Without sib competition, there can be no potential for POC (see Figure 3.1). Inter-brood conflict is sib competition between present and future sibs; intra-brood conflict is sib competition between brood members. So, whether we refer to them as sib competition or POC depends on what we wish to calculate. In Chapters 2–4 we considered exclusively intra-brood sib competition, and were inter-

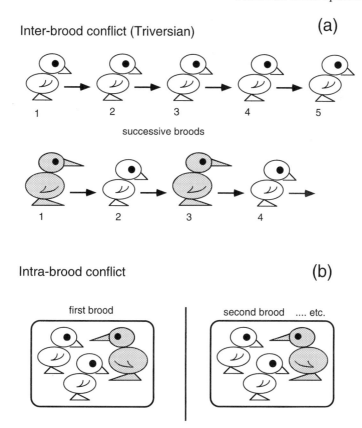

Fig. 7.7 **(a)** Inter-brood (Triversian) conflict. Increased begging in a given brood (here shown as a single offspring) increases the PI to that brood so that they gain extra resources (larger, shaded offspring). The extra PI taken displaces future sibs that could have been produced by the parent. **(b)** Intra-brood conflict. Increased begging in a given brood has no effect on future broods. The PI supply is not immediately affected by increased begging, although the allocation of PI to individuals in a brood depends on relative begging levels.

ested only in the immediate effects among rival siblings, rather than the effects on the parent. The distinction between inter- and intra-brood sib competition exerts an important influence on how POC will be resolved, as we shall see in the next chapter.

7.4 Variation in the two optima

In this chapter we have used a very general formulation of the way that parental investment, m, converts into offspring fitness, f, by using the function *f(m)*. This is convenient mathematically, but it obscures much of the rich biological diversity that surrounds the conversion of *m* into *f*. Much of this book is, in fact, devoted to detailing the reasons why the form of *f(m)* will vary according to the circumstances faced by individual offspring and individual parents. For example, offspring will vary in their condition or quality (Temme 1986) or their sex (Trivers and Willard 1973, Stamps 1990), which can generate reasons why parents should invest differentially, essentially because different offspring have different explicit forms for their *f(m)* relationship (see Section 10.2.3). Real battleground zones for POC may vary widely if we compare cases with radically different explicit *f(m)* forms. Finally, there are good reasons why parental age or condition may affect the optimal allocation of resources to each offspring, which add further to the variation in the two optima.

Summary

1. In most nursery taxa, parents deliver food and other materials from limited sources to their offspring. If the demand of the latter outstrips the supply, there may be disagreement as to how the available investment is best divided. Trivers' (1974) key insight was that of a genetic asymmetry in the perspectives of how parents and offspring view this problem (from Hamilton's rule), concluding that selection favours greater selfishness on the part of offspring and essentially an even distribution of lower PI by the parent.
2. An early dispute arose over whether different PI optima for parents and offspring could ever generate evolutionary change, which inspired a rush of supporting quantitative models. These established that the POC 'battleground' (a zone of evolutionary conflict across which plausible genetic mechanisms could produce shifts in gene frequencies) exists, but leave open the matter of how such conflicts are resolved (see Chapter 8).
3. One such 'battleground' model is presented, showing the conditions under which a selfish 'conflictor' allele can spread relative to a tagged wild-type allele at the same locus. This shows clearly that there is a zone of conflict between the best interests of parent vs. offspring, but also that there are substantial zones of agreement, where *both* parties favour increases or decreases to the focal offspring's PI share.
4. The zone of conflict is defined as the area of resource space between the parent's optimum for PI allocation to a given offspring and the offspring's own optimum. These optima can be found mathematically via modifications of the marginal value theorem.
5. When two parents participate jointly in providing the PI supply, an embedded 'game within a game' can arise, wherein *sexual conflict* between the mates may lead, paradoxically, to reduced quality of care delivered to the

offspring. Specifically, unless parents are inexorably tied to each other as lifelong partners without chance of replacement (a hypothetical condition termed 'true monogamy'), each can potentially benefit from expending somewhat less than its partner in the mutual tasks of parenting. This potential for partner exploitation leads to the counterintuitive prediction (as yet untested empirically) that biparental care should (without true monogamy) lead to larger numbers of poorer-quality offspring.

6. Equations describing the PI optimum for offspring differ, not surprisingly, for full vs. half siblings, but the application of the half-sibling designation in real systems is difficult. When one uses a standard value of ¼ for r in such cases, it is easy to forget that this is a *mean*, which is convenient for thinking about whole organisms, but not always the most precise definition of r. From a gene perspective, however, the appropriate r for half siblings is either 0 (allele from the non-investing male parent) or ½ (allele from the investing female parent).

7. This raises the issue of parental mating system, and specifically which parent delivered the hypothetical conflictor allele. If mating is promiscuous and the allele (assumed to be rare) arrived at the fertilization event in a sperm, then the chances are excellent that no future siblings from the same female will contain copies of it: $r = 0$. The male parent in such a case is said to be 'unconstrained' (i.e. its offspring's selfishness will not erode future sibling fitness). However, if that allele arrived for fertilization inside the egg, then the chances are higher for copies to exist in siblings, $r = \frac{1}{2}$, and the mother is 'constrained'.

8. Two types of POC are traditionally identified: (i) *Inter-brood conflict*, which affects siblings belonging to successive reproductive exercises (wherein current selfishness exacts a cost in parental ability to create successful siblings in the future) and (ii) *Intra-brood conflict*, which pertains to contemporaneous nursery-mates. The blurred distinction between POC and sibling competition is discussed.

8

Parent–offspring conflict II: models of resolution

The first half of our lives is ruined by our parents
and the second half by our children.

(Clarence Darrow)

Although the 'battleground' debate[1] over parent–offspring conflict has now subsided, it is far from clear how such disagreements over PI allocation should be resolved. In this chapter we present two modelling approaches to behavioural conflicts, addressing whether (i) offspring are likely to succeed in extracting PI from parents against parental interests (an outcome called 'offspring wins'), (ii) parents thwart such attempts ('parent wins'), or (iii) neither party wholly prevails (various levels of 'compromise'). An excellent review of the current models is given by Godfray (1995b).

8.1 Discrete strategy models

Much of the early analytical work following the controversy between Trivers (1974) and Alexander (1974) concerned population genetics approaches (e.g. Blick 1977, Macnair and Parker 1978, 1979, Parker and Macnair 1978, 1979, Stamps *et al.* 1978). A central aim of these papers was to demonstrate that parent–offspring conflict could occur between loci expressed in parents and those expressed in offspring. They were necessarily often restricted to **discrete strategies**[2] because of the difficulties of modelling continuous strategies under diploid genetics.

---

[1] Over whether genetic conflict *can* exist between parents and offspring over levels of parental investment

[2] This means that the alternative 'strategies' available to each player are not *continuous* (i.e. not continuously variable), and in this particular case there are only two options open to each player, an 'either-or' proposition. Once a strategy has been adopted it remains unchanged throughout that individual's life and we assess how such a policy affects fitness. Basically, the requirement of discreteness is a technique for making the model as mathematically simple as possible initially, whilst we gain a feel for how the system works.

Parker and Macnair (1979) analysed various discrete 2×2-strategy POC games in which the parent could either be sensitive or insensitive to offspring demands, while offspring could either be avaricious or not. Two alleles at a 'suppressor' locus (actively expressed only when the individual is a parent) determine whether the parent is sensitive to offspring begging (only *non-suppressors* tend to listen and respond) or, alternatively, simply ignores begging and allocates food according to the parent's own interests (*suppressors*). Similarly, two alleles at a 'conflictor' locus (active only when the individual is an offspring) determine whether an offspring demands an amount of food equivalent to the parental optimum (*non-conflictors* cause no trouble), or demands more (*conflictors* are selfish). The alleles at the suppressor locus are *S* (suppressor) vs. *s* (non-suppressor); those at the conflictor locus are *C* (conflictor) vs. *c* (non-conflictor). *S* is dominant over *s*, and *C* is dominant over *c*.

First, consider the case where there is just one *cost of insensitivity* to a parent. This cost arises because offspring carrying the *conflictor* allele continue to beg (thus wasting energy, attracting predators, etc.), because they fail to get the food they have demanded. The result is a reduction in both parental and offspring fitnesses, by whatever means, due to the continued begging. A brood's fitness is multiplied by 1.0 if no offspring show costly begging (i.e. if all are non-conflictors), and multiplied by $k < 1$ if some of the offspring carry the *C* allele and keep begging. In essence, k can be envisioned as a proportional reduction in survival prospects, a penalty or tariff levied by the presence in offspring of the *C* allele. If several offspring are produced, k is likely to increase with the frequency, p, of conflictor progeny in the brood. Because of this, it is possible for suppression (ignoring begging) to be advantageous to a parent when *some*, but not when *all*, brood members are conflictors.

Parker and Macnair (1979) examined a range of computer simulations for very large broods under the assumptions of 'true monogamy' and inter-brood conflict. *S* and *C* alleles begin as rare mutations in a *sscc* (phenotypically non-suppressor, non-conflictor) population. *Conflictors* are given the offspring optimum amount of PI, m_o, only if the parent is a *non-suppressor*; but they obtain only the parental optimum, m_p, if the parent they encounter is a *suppressor*. Non-conflictor offspring always obtain m_p. When conditions are favourable for the spread of the *S* allele (whatever the frequency of *C* within a brood), *C* invades very quickly at first and approaches fixation but then *S* begins to spread; the tables start to turn (see results of a computer simulation, shown in Figure 8.1a). Once the increasing *S* allele reaches a critical frequency, *C* suddenly crashes to extinction. Thereafter, the *S* allele becomes selectively neutral, and would be subject to drift (not included in these simulations). (One way that *S* could be maintained at high frequency by selection is through recurrent mutation of *c* to *C*.) Essentially, the parent wins this conflict.

By contrast, if conditions are such that *S* alleles are advantageous at low or intermediate frequencies of *C*, but disadvantageous when all brood members

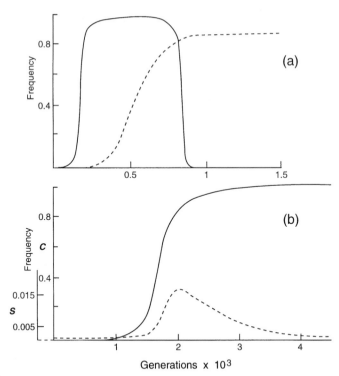

Fig. 8.1 Simulation of a parent–offspring game with two alleles at a conflictor locus (c demands a PI of m_p and C demands m_o) and two alleles at a suppressor locus (s gives each offspring its demand and S always gives m_p). The simulation is for inter-brood conflict with true monogamy. There are no costs of ignoring begging. C and S alleles start as rare mutants (frequency 0.0005) in generation 1. Solid line = frequency of C allele; broken line = frequency of S allele. **(a)** Begging costs $k(p)$ are always favourable for the spread of the S allele, whatever the frequency of conflictor offspring within a brood. **(b)** Begging costs $k(p)$ are favourable for the spread of the S allele at low or intermediate frequencies of conflictor offspring within a brood, but unfavourable at high frequencies. (After Parker and Macnair 1979.)

are conflictors (note the different scales for C and S frequencies in Figure 8.1b), the result is reversed. Now the offspring wins the conflict (a high frequency of C alleles is here maintained by selection). The C allele again invades more quickly than S, but this time it is the S allele that crashes early on: in short, S cannot achieve a high enough frequency to cause C to crash. For this particular game, in which both mutants start at low frequency, the critical determinant of the outcome appeared to be the relationship between the cost if all brood members beg at the conflictor level, $k(1)$, and the ratio $f(m)/m$ at the two optima (i.e. the marginal gain rate at the parent's optimum and that at the offspring's optimum). When $k(1) \cdot [f(m_p)/m_p] > [f(m_o)/m_o]$, a 'parent wins' solution was obtained, but reversing the inequality resulted in 'offspring wins' solutions. Thus, relatively low mortality costs of begging (= high k-values) are likely to generate 'parent wins' solutions; high mortality costs (= low k-values) yield 'offspring wins' solutions. When the forfeited future siblings are half-sibs, the probability of an 'offspring wins' solution becomes less likely[3], because $f(m_o)/m_o$ is lower with half-sibs, allowing $k(1)$ to be lower in order for S to spread.

However, both solutions can be precarious and may depend on the starting conditions. For example, consider the 'offspring wins' solution just shown. In the absence of C alleles, if S can somehow fixate in a non-conflictor (cc) population (e.g. by drift), then 'parent wins' can be an ESS even if $k(1) \cdot [f(m_p)/m_p] < [f(m_o)/m_o]$. Once S is fixed, C-bearing progeny always do worse than cc progeny because they have to pay the begging cost and never obtain any reward. The advantage of C is felt only with ss parents. Secondly, 'offspring wins' can often be invaded by a 'partial suppressor' parental strategy that removes just some of the begging costs by giving a little extra PI, in the range $m_p < m < m_o$ (Parker and Macnair 1979). However, both solutions become more robust if begging costs rise steeply as begging levels increase from zero; such high costs act to discourage any enhanced begging at all, thus tending to prevent the spread of any small changes (in parental or offspring strategies).

So far, these arguments relate to inter-brood (= Triversian; see Section 7.3) conflict, where increased PI to a given brood displaces only future sibs. For intra-brood conflicts, it is assumed that the parent is already functionally a *suppressor* in that it supplies only a fixed PI to each brood, at least in the short term: the family budget is flat. By demanding more, an offspring thus cannot displace future sibs; it can only disrupt the allocation of PI to one or more contemporaneous brood-mates.

Parker and Macnair (1979) examined a second 2×2 strategy game that differs from the first in just one respect: a second cost for being an insensitive parent is introduced. In addition to having brood fitness reduced by the usual $k(p)$, due to the fruitless begging of *conflictor* offspring, this additional insensitivity cost is felt regardless of progeny type in the brood. Parker and Macnair argued that imposing such a cost was biologically justifiable. For there to be a biological function for sensitivity (the policy of responding to offspring demands, rather than ignoring all such cues), it must logically have some selective advantage over insensitivity. A plausible reason for this would be that the begging cues must, at least on some occasions, be related to 'true need' (where both parent and offspring benefit if the offspring gains more PI) rather than always being in the zone of conflict (between the PI optima of

---

[3] Note that from equations (7.4), (7.7), and (7.10) we could rewrite the condition $k(1) \cdot [f(m_p)/m_p] > [f(m_o)/m_o]$ in terms of slopes $f'(m_p)$ and $f'(m_o)$, depending on the mating system (and how it affects r between siblings).

parent and offspring). The concept of true need is related to that of 'honest signalling', which we shall discuss in the next chapter. For present purposes, let us just assume that an offspring occasionally demands more because it has a true need (as just defined), perhaps because of illness or some other randomly occurring non-genetic reason. This modest additional tariff was modelled by multiplying the fitness of a suppressor parent's brood by a new constant, i (where again $0 > i > 1$). Thus an insensitive *suppressor* parent has a relative brood fitness of $k(p)i$, compared to a *non-suppressor*, which has a brood fitness of 1.0.

Introducing a small cost for being an insensitive parent was found to exert a radical effect on the outcome of the games. Suppose that both S and C start at low frequency in an *sscc* population and, as before, offspring demand either m_p or m_o units of PI, depending on whether they carry C, while parents give either m_p or m_o, depending on whether they carry S. We expect C to increase in frequency rapidly, and then for S to increase more slowly[4], until it reaches a threshold where C crashes (as it did before, as shown in Figure 8.1a). However, when C crashes this time, selection is no longer indifferent but starts to penalize S in favour of s (Figure 8.2a), because if the offspring are *non-conflictors*, it pays parents to be sensitive in order to escape the costs of insensitivity. Thus S crashes, and the cycle begins again. The cycles in Figure 8.2a show an increasing period and amplitude. Eventually one allele at one locus (here it happens to be S) is lost due to its frequency being too low to store in the computer (effectively, it is lost from the population). Regular limit cycles can be generated if alleles are maintained at low level by recurrent mutation (Figure 8.2b). Stamps *et al.* (1978) also argued that limit cycles are possible in games of POC (see also Maynard Smith 1982a for an account of similar cyclic games).

The reason for this cycling of strategies (alleles) is fairly simple. If we allow only one strategic change at each 'move' in a multi-move game, the cycling of strategies can be seen as follows: *non-suppressor–non-conflictor* loses to *non-suppressor–conflictor*, which loses to *suppressor–conflictor*, which then loses to *suppressor-non-conflictor*, which loses to *non-suppressor–non-conflictor*, and so on. As such, this resembles the familiar two-player 'Rock–Scissors–Paper' game (see Maynard Smith 1982a), where '*Paper*' beats '*Rock*' which beats '*Scissors*' which beats '*Paper*', and the only solution is to play a mixed strategy of one-third '*Paper*', one-third '*Rock*', and one-third '*Scissors*', with each move randomly chosen. Why doesn't the present parent–offspring game result in a balanced polymorphism for each of the four strategies? The difference is that in 'Rock–Scissors–Paper', each player must simultaneously chose a bid, **independently** of the other player. Were we to make 'Rock–Scissors–Paper' an indefinite and **sequential** game in which player 2 is able to see what player 1 took before committing his or her choice, and vice versa, the strategies of the two players would cycle continually, each choosing the best

---

[4] Because $k(1)i[f(m_p)/m_p] > [f(m_o)/m_o]$

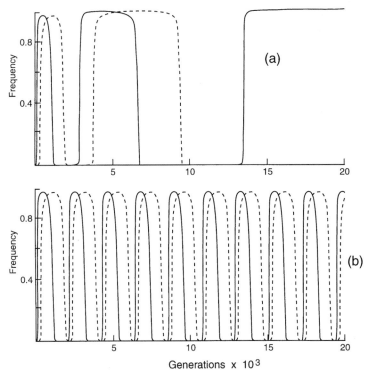

Fig. 8.2 Simulation identical to that shown in Figure 8.1a, except that there is also a small cost of being an insensitive parent and ignoring begging. **(a)** C and S have an initial frequency of 0.001, and there is no recurrent mutation. **(b)** C and S again begin at an initial frequency of 0.001, but alleles are maintained at low frequency by a constant low level of recurrent mutation. (After Parker and Macnair 1979.)

reply to the opponent's preceding bid. Our POC game is **'partially sequential'** in the sense that parent and offspring strategies are not independent—each generation plays against its own relatives from the previous generation, although of course some of the genotypes (strategies) arise from random mating under diploidy. It is therefore a type of mixture of simple 'Rock–Scissors–Paper', and sequential 'Rock–Scissors–Paper'. None the less it cycles, provided that (i) begging is costly, (ii) parental insensitivity is costly, and (iii) S can spread against s when the population is $ssCC$.

A criticism of discrete strategy models of POC is that they lack biological realism—there is no obvious reason why either parents or offspring should be constrained to adopt only one of two discrete strategies. The next model is perhaps more plausible. However, it must be remembered that mutations do occur as discrete events, so the supply of genetic variation at any given time need not be continuous. As a result, a certain amount of cycling may well occur.

8.2 Continuous strategy models

8.2.1 *Introduction*

Parker and Macnair (1979) and Parker (1985) analysed a POC game in which parent and offspring are allowed to vary strategies continuously. Since these models were presented, there have been various attempts to ascertain the resolution of POC in continuous strategy models (Eshel and Feldman 1991, Godfray 1991, 1995a, 1995b, Yamamura and Higashi 1992). We shall explain the Parker and Macnair (this chapter) and Godfray (see Chapter 9) models in detail. For an excellent comparative review of all four approaches, see Godfray (1995b). Yamamura and Higashi (1992) have proposed a very general model of conflict resolution between relatives, which has yet to be tailored specifically for POC. They assume that both parent and offspring can influence the resources obtained by offspring, and that any deviation from the ESS causes a conflict that has costs for both. The exact position of the ESS depends on the relative costs that the parent and offspring experience when attempting to determine the level of resources. In Eshel and Feldman's model, the parent monitors offspring behaviour and uses the information it obtains to determine its optimum allocation of resources (see also Godfray's model; Chapter 9). Effectively, they propose a mechanism in which the offspring alters the relationship between its fitness and the amount of resources it gains, such that it is optimal for the parent to provide more resources, leading to an increase in offspring fitness. This outcome requires a particular relationship between offspring behaviour and the value of obtaining extra resources (see Godfray 1995a, b). Costly begging might be favoured if it leads to a marked increase in the value of extra resources to the offspring, and a consequent increase in parental feeding.

The original models of Parker and Macnair have not proved easy to follow, so the following account is an attempt at clarification. We have also extended and generalized their analysis to identify general reasons for the effects noted. For simplicity, uniparental care and true monogamy (both parents constrained) are assumed, so that the progeny displaced are full sibs.

Strategies available to parent consist of a continuous choice of level of PI, m, to give to each offspring or brood.

Strategies available to offspring consist of a continuous choice of begging level, x.

Note that these are 'choices' only in terms of evolution: strategies can vary in evolutionary time only. For example, a mutation in parental strategy plays against whatever offspring strategy is currently fixed in the population, and vice versa. Simultaneous mutations in strategy do not occur. However, a deviation (mutation) in one strategy temporarily (in evolutionary time) brings about a change in the realized level of the opposing strategy, as we shall see below.

8.2.2 The ESS conditions

We are seeking an ESS pair of strategies, m^* for the parent and x^* for the offspring. To find this, we start by setting up some competing mutant strategies and testing whether they can invade against the ESS strategies. Let:

1. $W_p(m,x^*)$ = the pay-off of a mutant parent playing m (other than m^*) against offspring playing x^*; and

2. $W_o(x,m^*)$ = the pay-off to a set of mutant siblings playing x (other than x^*) against a parent playing m^*

For m^*,x^* to be an ESS, any unilateral deviation in either strategy must result in a lower fitness than the ESS itself. We can therefore derive the ESS by the usual technique of differentiating the fitness functions (see Maynard Smith 1982a):

$$\left[\frac{\partial W_p(m,x^*)}{\partial m}\right]_{m=m^*} = 0, \tag{8.1}$$

$$\left[\frac{\partial W_o(x,m^*)}{\partial x}\right]_{x=x^*} = 0, \tag{8.2}$$

and the second derivatives must be negative (for maxima). The first equation generates[5] m^*; and the second generates, x^*. It is also possible to find the ESS using versions of the 'marginal Hamilton's rule' as we have done elsewhere (see Section 3.3.4).

8.2.3 *Pay-offs and ESSs*

The Parker and Macnair model is analogous to a union-management wage bargaining game (e.g. see Maynard Smith 1982a), and for this reason the authors termed it the 'pro rata payment' model, the idea being that one party (the offspring) clamours for payment (of PI), whilst the other (the parent) allocates it grudgingly, but *in some proportion* to the demand.

Two distinct mechanisms shape the pay-offs of parent and offspring:

Mechanism 1 (the demand function).

PARENT: a strategic change in PI can alter observed begging.
Specifically, Parker and Macnair assumed that if a mutant parent unilaterally pays out more PI, then the offspring in its nest beg less (i.e. temporarily, in

---

[5] This is the standard modelling technique of differentiating the fitness function of a mutant in an ESS population, setting that equal to zero (the slope equals zero at the curve's highest point, as it turns downward), and solving for the variable of interest (see Parker and Maynard Smith 1990). When this is done for equation (8.1), that variable is m, so the resulting solution is, by definition, the value of m that confers the highest fitness, hence m^*. Repeating for x in equation (8.2) finds

evolutionary time, begging is reduced). Conversely, if a mutant parent uni-laterally pays out less, its offspring increase their begging. This mechanism, termed the 'demand function' by Hussell (1988), relates to the ethological concept of hunger. An offspring tunes its ESS begging level, x^*, to the current ESS input of resources, m^*, by the parent. If it receives *more* than it 'expects', it begs less, but if it receives *less*, it begs more. Although this is intuitively appealing, especially to human parents, it does require some adaptive justi-fication. Perhaps the best argument relates to the neurological mechanism by which the offspring monitors and acts on its own assessment of 'true need'. We have argued (see Section 8.1) that parental insensitivity could be dis-advantageous due to failure to accommodate the real needs of offspring, which will vary due to chance circumstances. Given that the parent should show some sensitivity (see also the honest signalling model in Chapter 9), there is good reason why the offspring should adjust its solicitation if there is a change in the 'expected' level of PI. Biological evidence for and against this mechanism is discussed in the next chapter. Its presence or absence, and the sign of the correlation between PI change and begging change (here negative), exert profound influences on the solution. Ideally, it should be modelled explicitly.

We first establish generally what effect *Mechanism 1* will have on the ESS allocation of PI. Following the notation of Chapter 3, we assume that an offspring's fitness is the product of $f(m)$, its gross gains f when it takes m units of PI (see Figure 3.2), and $S(X)$, its survivorship S when it begs to level X (see Figure 3.7).

The effect of *Mechanism 1* on the ESS, m^*, depends on whether a marginal increase in m about m^* causes the realized begging level, X, to increase or to decrease, i.e. it depends on the sign of $X'(m^*)$, as explained in Box 8.1. There are three possibilities.

Case 1. $X'(m^*) = 0$
Realized begging does not change at all. For this case, *Mechanism 1* must be null or inoperative; a change in the parental behaviour must be met with no change in the offspring's begging behaviour. If $X'(m^*) = 0$, then $U = 0$ in Box 8.1, and we would expect the parental ESS, m^*, to be equal to the parental optimum, m_p (as equation 8.3 converges to 7.4). In the absence of *Mechanism 1*, the parent plays its own optimum.

Case 2. $X'(m^*) < 0$
Here, increases in PI reduce begging, and decreases in PI increase it. This is the effect that Parker and Macnair anticipated, and used explicitly in their model. Then U in Box 8.1 is positive, so that $[1-U]$ is less than one, and the parent is pushed into providing *more* PI at the ESS than its optimum would be in the absence of conflict.

Box 8.1   Modelling *Mechanism 1*

For simplicity, consider uniparental care under true monogamy. The fitness of the parent is the product of offspring number and the fitness prospects of each offspring. If a parent has M total units of PI and allocates m to each offspring, it produces M/m progeny. Each one's reproductive success is the product of $f(m)$, its 'personal fitness' (future survival and reproductive success) and $S(X)$, its immediate survival prospects when its *realized* begging level is X.

What is meant by 'realized begging level'? If the parent deviates from m^*, X differs from the strategic choice of begging level, x^*. The offspring's *strategy* throughout is x^*, which is its strategic choice of 'baseline begging' at the ESS. Given the choice of strategy x^*, then if the parent shifts from the ESS, the realized begging level becomes X, which is not a *strategic* change on the offspring's part, but an *inevitable consequence* of *Mechanism 1*.

As usual, we assume that $S(X)$ is monotonic decreasing (see Figure 3.5), so that (plausibly) the greater the observed begging, the greater the reduction in survival prospects. The fitness pay-off of a mutant parent becomes

$$W_p(m,x^*) = \frac{M}{m} f(m) S(X),$$

and applying equation (8.1) to find the parent's ESS m^* we get

$$f'(m^*) = \frac{f(m^*)}{m} [1-U] \tag{8.3}$$

where

$$U = m^* \left[\frac{S'(X)}{S(X)} \right] X'(m^*).$$

$X'(m^*)$ is the rate of change in realized begging with m around m^*, and $S'(X)$ is the rate of change in survival with X around $X = x^*$. Note that $X'(m^*)$ is, in fact, *Mechanism 1*. It tells us how small deviations in parental strategy affect begging.

We can deduce the effect of *Mechanism 1* on the parental strategy by considering the sign of U. Figures 7.5 and 7.6 showed graphical solutions for equations (7.3), (7.4), (7.7), (7.8), (7.10) and (8.3), of the form such that $f'(m)$ equals $Z f(m)/m$. They demonstrate the following rule.

If $Z < 1.0$, the optimal m is greater than the parental optimum (m_p in equation 7.4). If $Z > 1.0$, the optimal m is less than the parental optimum (m_p in equation 7.4).

Hence if we can determine the sign of U, we can decide whether $Z = [1-U]$ is greater or less than 1.0, and thus determine whether the parent will give more or less than its optimum at the ESS.

Begging is assumed to be costly, so $S'(X)$ must be negative. Because both m^* and $S(X)$ have to be positive, the sign of U depends only on the sign of $X'(m^*)$, that is on whether realized begging rises or falls as the parent increases its provisioning marginally from its ESS, m^*.

Case 3. $X'(m^*) > 0$

This is the counterintuitive prospect that, if the parent increases its PI, the offspring increases its begging. Then U in Box 8.1 is negative, so that $[1-U]$ is greater than one and at the ESS the parent would be constrained to provide *less* PI than its optimum would be without conflict.

In their models, Parker and Macnair assumed that *Mechanism 1* followed a version of Case 2 in both inter-brood and intra-brood conflict (see Box 8.2). It was assumed that when the parent gives more PI, the solicitation level changes equally in both POC types. The ESS value of m^* exceeds the parental optimum. In contrast, in Godfray's model, *Mechanism I* follows Case 1 and m^* equals the parental optimum.

Mechanism 2 (the supply function)

OFFSPRING: a strategic change in begging level can alter observed PI allocation.

Parker and Macnair (1979) and Godfray (1991, 1995a, b) specifically assumed that if a mutant offspring begs more unilaterally, it extracts more PI from its parent(s); if a mutant offspring begs less, it receives less. Although this mechanism, named the 'supply function' by Hussell (1988), seems intuitively appealing, it requires justification. There are two interpretations. The first is that the parent has a fixed mechanism of allocating resources: it must pay out in direct response to begging, and hence an offspring that begs more obtains more. This was what Parker and Macnair had in mind when devising their model. The second interpretation relates to 'honest signalling'. If a parent has the policy of paying out more PI to offspring that signal more (because these

---

Box 8.2   Parker and Macnair's version of *Mechanism 1*

Parker and Macnair used the following explicit mechanism to deduce the parental strategy, m^*. They assumed that *Mechanism 1* would obey CASE 2, namely that $X'(m^*) < 0$. Specifically, they assumed that if a parent shifts its ESS level of PI (m^*) to a new level (m not equal to m^*), then the realized level of begging, X, changes from $X = x^*$ to:

$$X = x^* \left[\frac{m^*}{m} \right], \qquad (8.4)^{[1]}$$

which has the desired property that realized begging increases if $m < m^*$, and decreases if $m > m^*$. Then around the ESS, where $X = x^*$,

$$X'(m^*) = -\frac{x^*}{m^*} \qquad (8.5)$$

There exist, of course, many possible explicit ways of generating the qualitative requirements of *Mechanism 1* under CASE 2 (when increasing PI by the parent causes a decrease in begging). Provided that the offspring begging level varies continuously, and is hence differentiable, with the deviation in PI from the current level, then the qualitative predictions made by Parker and Macnair are often likely to be upheld.

---

[1] We note that equation (8.4) was stated incorrectly (as $X = x^*m/m^*$) in a review by Parker (1984).

indicate their true need by higher begging), then that parent is vulnerable to deceit: offspring also can get more PI by using a dishonestly high signal (Godfray 1991). We shall investigate honest signalling further in Chapter 9.

As before, the change by one party (here the offspring) is a **strategic** change, whereas the change by the other party (here the parent) is simply an inevitable **consequence** of the mechanism. We are now concerned with a strategic change by the offspring around the ESS (i.e. it shifts from x^* to a different level, x). The parent's strategy is m^* throughout (its 'baseline PI' at the ESS), but when the offspring shifts from its ESS begging rate, the 'realized PI allocation' to the offspring is μ. At the ESS, $\mu = m^*$, when $x = x^*$. Model details of the Parker and Macnair ESSs for inter-brood and intra-brood conflict are given in Box 8.3.

Box 8.3   Parker And Macnair's model of *Mechanism 2* for the two forms of conflict

For illustrative purposes, we again consider uniparental care under true monogamy. There are various ways in which begging costs can be felt by the offspring (see Section 2.5.4). For simplicity, assume that the cost of begging to a given level x is $S(x)$, which is felt by each offspring playing x (individual costs). The pay-off function depends on whether we are dealing with inter-brood conflict (taking extra PI displaces future sibs) or intra-brood conflict (PI to each brood is fixed, so taking extra PI harms contemporary sibs), as explained in Section 5.3 (see Figure 5.7). In each case we seek a function that expresses the sum of the fitnesses of the mutant offspring produced by a dominant mutant allele (for x not equal to x^*) in a parent. Thus half of that parent's progeny will bear the allele for x.
Inter-brood conflict—the pay-off (summed fitness) of mutant offspring is

$$W_o(x,m^*) = \left[\frac{\frac{1}{2}M}{\frac{1}{2}(m^* + \mu)}\right] f(\mu)S(x)$$

The value in the square brackets gives the number of mutant offspring (which is now no longer M/m^*). Half the offspring segregate as mutants playing x and receiving μ, and the others play the ESS x^* and receive m^*. Each mutant gains a personal fitness of $f(\mu)$ and has immediate survival prospects of $S(x)$.
Applying equation (8.2),

$$-S'(x^*) = \mu'(x^*)S(x^*)\left[\left[\frac{f'(\mu)}{f(\mu)}\right] - \left[\frac{1}{(m^* + \mu)}\right]\right] \tag{8.6}$$

where $\mu'(x^*)$ is *Mechanism 2*, namely the rate of change in PI with x around x^*, and $f'(\mu)$ is the rate of change in personal fitness with μ around $\mu = m^*$. Remember that $S'(x)$ is negative. To make biological sense, $\mu'(x^*)$ must be positive (increased begging must attract higher PI), so the value of (equation 8.6) lying within the widest square brackets must also be positive.
To deduce the offspring's best strategy, x^*, Parker and Macnair assumed that

$$\mu = m^*\left[\frac{x}{x^*}\right], \tag{8.7}$$

so that each mutant offspring gains μ and each non-mutant gains m^*. Remembering that at the ESS, $\mu = m^*$:

$$\mu'(x^*) = \frac{m^*}{x^*} \tag{8.8}$$

Intra-brood conflict—the parent fixes the PI supply per brood at its ESS level, so that an average offspring obtain m^*. Conflict concerns the allocation of M among the brood members. Thus a total of M/m^* offspring is produced within the parent's lifetime, whatever the mutant begging level. The pay-off of mutant offspring is

$$W_o(x,m^*) = \left[\frac{\frac{1}{2}M}{m^*}\right] f(\mu)S(x)$$

and applying equation (8.2) gives

$$-S'(x^*) = \mu'(x^*)S(x^*)\left[\frac{f'(\mu)}{f(\mu)}\right] \tag{8.9}$$

Parker and Macnair assumed for intra-brood conflict that parents employ the assessment method of 'mean-matching'. A mutant offspring receives a share of resources equal to its begging level relative to the mean level for the brood (see Section 3.3.2). Hence

$$\mu = m^*\left[\frac{x}{0.5(x + x^*)}\right] \tag{8.10}$$

$$\mu'(x^*) = \frac{m^*}{2x^*} \tag{8.11}$$

The Parker and Macnair ESS for POC resolution is derived in Box. 8.4. The useful feature of their result (equations 8.14a and 8.14b) is its remarkable simplicity. At the ESS, the parent's strategy is definable very simply in terms of the function $f(m)$ and constant α, while the offspring's strategy is equally simply defined in terms of $S(x)$ and constant β. Because α and β are constants (both depending on the mating system and type of conflict, etc.), we can easily represent the solution graphically (as in the ESS solutions shown in Figures 7.5 and 7.6). Following the reasoning developed earlier (in Chapter 7), when α is smaller than 1, we expect that at the ESS the parent will be giving more PI than the parental optimum for true monogamy. Yet the PI allocated at the ESS is never so great as the offspring's optimum for inter-brood POC; both α-values are greater than the equivalent for the offspring's optimum for full sibs, which can be seen to be $\frac{1}{2}$ from equation (7.8). Thus offspring will gain more PI than is best for parental interests, but less PI than is optimal for its own interests. The PI ESS for full sibs under true monogamy is shown graphically in Figure 8.3. The solution suggests a form of compromise in which offspring clamour for more PI even though they receive more than is ideal for their parents.

It is important at this point to stress a difference between the inter- and

Box 8.4   The structure of the Parker and Macnair ESS

Equation (8.3) gives the parent's ESS, and equations (8.6) and (8.9) give the offspring's ESS for inter- and intra-brood conflicts, respectively, expressed generally and without recourse to explicit formulations for *Mechanisms 1* and *2*. First consider the general equations for inter-brood conflict. Substitution of the value for $S'(x^*)/S(x^*)$ for the offspring into the parental equation gives the result that:

PARENT:

$$f'(m^*) = \frac{f(m^*)}{m^*}\left[\frac{1 - 0.5X'(m^*)\,\mu'(x^*)}{1 - X'(m^*)\,\mu'(x^*)}\right] \tag{8.12a}$$

and substituting the value of $f'(m^*)/f(m^*)$ for the parent into the offspring equation gives:

OFFSPRING:

$$-S'(x^*) = S(x^*)\left[\frac{\mu'(x^*)}{2m^*[1 - X'(m^*)\,\mu'(x^*)]}\right]. \tag{8.12b}$$

Note that the parental ESS depends on *Mechanisms 1* and *2* and on the form of $f(m)$. Similarly, the offspring ESS depends on *Mechanisms 1* and *2* and on the form of $S(x)$. The two are linked by the m^* term in the offspring equation; thus the parental ESS does also depend on the form of $S(x)$, and the offspring ESS on the form of $f(m)$.

Similar conclusions apply for intra-brood conflict:

PARENT:

$$f'(m^*) = \frac{f(m^*)}{m^*}\left[\frac{1}{1 - X'(m^*)\,\mu'(x^*)}\right] \tag{8.13a}$$

OFFSPRING:

$$-S'(x^*) = S(x^*)\left[\frac{\mu'(x^*)}{m^*[1 - X'(m^*)\,\mu'(x^*)]}\right]. \tag{8.13b}$$

Substituting equations (8.5), (8.8), and (8.11), which are Parker and Macnair's explicit formulations for the two mechanisms, $X'(m^*)$ and $\mu'(x^*)$, we get the following ESS:

WHEN PARENT, PLAY

$$f'(m^*) = \alpha\frac{f(m^*)}{m^*} \tag{8.14a}$$

WHEN OFFSPRING, PLAY

$$-S'(x^*) = \beta\frac{S(x^*)}{x^*} \tag{8.14b}$$

in which α and β are positive constants. For inter-brood conflict, $\alpha = 3/4$, $\beta = 1/4$; for intra-brood conflict, $\alpha = 2/3$, $\beta = 1/3$.

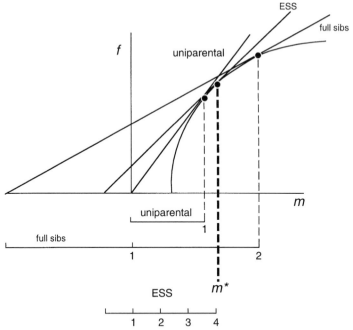

Fig. 8.3 The structure of the parental ESS, *m**. Under true monogamy, the optimum amount of parental investment, *m*, for the parent is found when α = 1 in equation (8.14a), and that for the offspring is found when α = ½. These optima are given by the tangent lines (see Figure 7.6). The ESS, *m**, for interbrood conflict (Box 8.4) has α = ¾. The tangent to the *f(m)* curve is taken from a distance one unit to the left of the origin, while the optimal investment occupies a distance three units to the right of the origin. Note that the optimal value of *m* increases as α decreases. The ESS is therefore intermediate between the parent and offspring optima.

intra-brood models. While inter-brood conflict clearly yields an offspring optimum in the absence of any begging costs (equation 7.10), this is not so in the present intra-brood model, where there is no definable offspring optimum in the absence of any begging costs, although offspring interests will always involve a higher PI than parental interests. (In Section 3.3.3 we showed that in the absence of any begging costs, a dominant mutant allele for escalated begging can spread that kills non-mutant siblings.) In these intra-brood models, an individual offspring can affect *only* contemporary sibs (by grabbing more of the available resource).

What about the offspring's strategy, the begging level, *x**? Figure 8.4a shows the graphical solution for the inter-brood case with true monogamy. At *x**, the tangent to the survivorship function, *S(x)*, must equal the slope of a line drawn from the point 4*x** on the abscissa to *S(x**) on the ordinate. The equivalent version for intra-brood conflict would have the line drawn from 3*x** on the abscissa (Figure 8.4b); the value of *x** is higher for this case. Note that

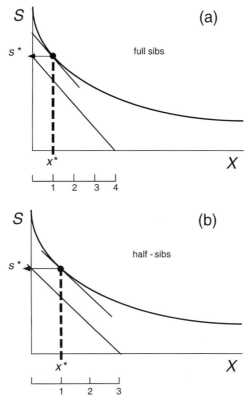

Fig. 8.4 The structure of the offspring ESS, *x**. The ESS amount of begging (*x*) for the offspring is found from equation (8.14b), and is determined by the constant β. **(a)** Inter-brood conflict under true monogamy. For this case, β = ¼. The tangent to the *S(x)* curve must be parallel to a line drawn from the abscissa to intersect the ordinate at *S**, which is equal to the *S*-value of the tangent. **(b)** Intra-brood conflict under true monogamy. For this case, β = ⅓. The tangent to the *S(x)* curve is again parallel to a line drawn from the abscissa through *S** on the ordinate. Its intersection on *x* is now three times the value of *x** given by the tangent. Note that the ESS value of *x* increases as β increases.

the higher the value of β, the higher the value of *x** (in contrast, the lower the value of α, the lower the value of *m**). Thus, in addition to featuring higher PI per offspring, the begging levels expected from the intra-brood model are higher than those for the inter-brood model.

8.2.4 *Inter-brood vs. intra-brood conflict*

If expected 'begging levels' are a fair measure of overt behavioural conflict, the intra-brood model appears to generate more conflict than the inter-brood model. It also results in higher PI. What causes these differences? Should we really expect intra-brood *genetic* conflict to result in higher observed *behavioural* conflict in natural systems with uniparental care?

There appear to be three possibilities.

1. They might arise from different forms of gene action. The present intra-brood model is one of 'unconditional gene action'. A mutant allele causes all of its bearers to beg at a deviant level and the action is not conditional upon role, since all sibs bearing the allele beg differently from those lacking it (see Section 3.3.3). If the mutation is for increased begging, extra resources are taken solely from the non-bearers of the mutant allele (see Figure 3.8). In the inter-brood model, the gene action is conditional upon the 'present' and 'future' roles, that is, an existing sib displaces a future sib, which may contain the same mutant gene (see Figure 3.9). However, the analysis in Section 3.3.3 shows that the same ESS obtains for intra-brood conflict whether gene action is conditional or unconditional, so the difference between the inter- and intra-brood models is unlikely to be related to conditionality.

2. They might arise from the different explicit forms, equations (8.7) and (8.10), used to model *Mechanism 2* under the two forms of POC (*Mechanism 1* operates similarly in both).

3. They might arise from the fact that there is no displacement of related sibs in intra-brood conflict: the fitness function contains no term for progeny number (an approach for distinguishing between possibilities 2 and 3 is explained in Box 8.5).

In summary, intra-brood conflict in uniparental care should act generally to increase the overt behavioural conflict, as measured by the observed level of escalated begging, provided that *Mechanism 1* either exerts no effect or obeys the rule that increased amounts of PI will decrease begging. In a general sense, the difference appears to relate to the fact that future sibs cannot be displaced in intra-brood conflict.

8.2.5 Implications: what effect do 'supply side conflicts' have on sibling rivalry?

Parker and Macnair (1979) and Parker (1985) carried out essentially the same procedure as outlined here, although their analysis was restricted to the explicit forms for *Mechanisms 1* and *2* in equations (8.7) and (8.10). However, they examined a wider range of mating systems and types of begging costs. The mathematical details of these alternatives will not be reiterated: suffice it to say that the same procedures were adopted to find the ESSs as those outlined above, using explicit equations (8.7) and (8.10). It is important to note that the sign of $X'(m*)$ is assumed to be negative: increased PI reduces begging. Parker and Macnair (1979) and Parker (1985) investigated both (i)

Box 8.5 Further details concerning possibilities 2 vs. 3

First consider the two explicit formulations used for *Mechanism 2*. In inter-brood conflict (equation 8.7), the parent is assumed to allocate PI in proportion to an offspring's begging *relative to ESS begging*. In intra-brood conflict, the parent allocates PI in proportion to an offspring's begging *relative to the mean begging for that brood*. Because of these different assumptions, which seem to be plausible ways of modelling the difference between the two cases, the value of a marginal increase in begging for inter-brood conflict (equation 8.8) is twice that for intra-brood conflict (equation 8.11).

What effect does this have? Let $\mu'(x^*) = jm^*/x^*$, where j is a positive constant that increases the value of marginal begging escalations; $X'(m^*)$ remains as $-x^*/m^*$. It can then be shown that increasing j *increases* the value of α and *decreases* the value of β in both sets (inter- and intra-brood) of general ESS equations. As might be expected, increasing j from zero to infinity (zero to maximum marginal values for begging) takes the PI from the parental optimum to the offspring's optimum. Since $j = 1$ for inter-brood conflict and $j = \frac{1}{2}$ for intra-brood conflict, then if all else is equal, we would expect *less* overt conflict when the biology follows the intra-brood schema. It follows that our results showing the intra-brood model to give higher overt conflict must relate to the difference between the general forms for the ESS equations themselves, not the difference between the explicit forms used by Parker and Macnair for the two mechanisms, because these would tend to act in the opposite direction.

This can be established by comparing the general ESS equations (8.13) and (8.14) directly, assuming that *Mechanisms 1* and *2* operate identically in inter- and intra-brood cases. This can be done by comparing the magnitude of the square-bracketed terms, which have similar effects to the α and β constants in the specific ESS equation (8.14). Provided that increased begging always results in more PI (i.e. that $\mu'(x^*)$ is always positive), then the level of overt conflict depends on *Mechanism 1*. Specifically, it hinges on the sign of $X'(m^*)$. If increased PI reduces begging, this sign is negative, and intra-brood conflict will always result in both higher ESS PI and higher ESS begging levels. If there is no effect of changed PI on begging levels, such that $X'(m^*) = 0$, then the PI level remains at the parental optimum for both cases. Nonetheless, the ESS begging level is higher for intra-brood conflict. If, paradoxically, increased PI increases begging, the sign of $X'(m^*)$ is positive and we should expect that inter-brood conflict will result in more PI at the ESS than intra-brood conflict, though (again, paradoxically) both will be lower than the parental optimum. Begging levels should now be lower with intra-brood conflict, if non-zero begging levels are possible. The full implications of positive $X'(m^*)$ have yet to be investigated.

'*Individual costs*', where the begging costs of deviating by playing x (not equal to x^*) are felt uniquely by the deviant individual, which has a resulting survival probability of $S(x)$; and (ii) '*Shared costs*', where the begging costs of deviating by playing x (not equal to x^*) are felt by the brood containing the mutant individuals, so that each offspring has a survival probability related to the mean begging level for the brood. This is $S((x+x^*)/2)$ for dominant mutations in large broods (see Section 3.3.5).

In all cases, the same form of ESS was found as that in equation (8.14)

(Box 8.4), with the only difference arising in the values for the positive constants, α and β, which indicate the expected level of overt conflict at the ESS. Remember that the higher the value of α, the lower the ESS PI (m^*), but the higher the value of β, the higher the ESS begging level (x^*). A summary of the results for uniparental care or true monogamy is given in Table 8.1a and of the results for biparental care without true monogamy in Table 8.1b. Under conditions of intra-brood conflict, brood-mates are usually assumed to be full sibs, although multiple paternity has proved rather common when sought (even in supposedly monogamous birds). A maximal degree of multiple paternity could be said to occur when each sib has the same mother but a unique father (for simplicity, in Table 8.1 this relationship is termed 'half sib',

Table 8.1a. Scale of increasing parent–offspring conflict for the case where only one parent invests in the offspring (or two parents invest under conditions of true monogamy). As explained in the text, α and β are positive constants: increasing values of α reduce the ESS for parental investment (m^*), while increasing values of β cause escalations in ESS begging levels (x^*). Thus, conflict increases going down the table. The parental optimum in the absence of parent–offspring conflict has α = 1. (After Parker 1985.)

α	β	Type of conflict	Breeding constraint	Type of costs
3/4	1/4	Inter-brood	Both parents constrained	Individual
2/3	1/3	Inter-brood Intra-brood	Both parents constrained Full siblings	Shared Individual
5/8	3/8	Inter-brood	One parent constrained	Individual
4/7	3/7	Intra-brood	Half siblings, maximum	Individual
1/2	1/2	Inter-brood Intra-brood	One parent constrained Full siblings	Shared Shared
1/4	3/4	Intra-brood	Half siblings, maximum	Shared

Table 8.1b Scale of increasing parent–offspring conflict for the case where only two parents invest equally in the offspring (and true monogamy does not hold). Both parents are affected by a mutant allele for investment level, and both can expend all their investment. As in Table 8.1a, the constants α and β affect m^* (negatively) and x^* (positively), so conflict increases going down the table. In the absence of parent–offspring conflict, the parental ESS has α = 2 (After Parker 1985.)

α	β	Type of conflict	Type of costs
1 1/3	2/3	Intra-brood	Individual
1 1/4	3/4	Inter-brood	Individual
1	1	Intra-brood Inter-brood	Shared

although we recognize that is best regarded as a hypothetical upper limit), in contrast to the lower limit where all the brood members have the same parents (the 'full sibs' case).

An interesting property of this ESS is that the sum of the two constants is always equal, whatever the mating system or form of costs. For uniparental care or true monogamy, $\alpha + \beta = 1$, and for biparental care without true monogamy, $\alpha + \beta = 2$. Thus for each case, as α decreases, β increases. A lower α-value implies higher PI, and a higher β-value implies an elevated begging level. Thus, at the ESS, greater shifts in PI from the parental optimum will be associated with a higher level of begging. The expected level of overt conflict increases down each of Tables 8.1a and 8.1b.

Calling α_p the value of α at the parental optimum (1 or 2, depending on whether the fitness of one or both parents is constrained by the conflict), and α_o the value of α at the inter-brood model offspring optimum ($\frac{1}{2}$ or $\frac{1}{4}$), all inter-brood solutions have the property that $\alpha_p > \alpha > \alpha_o$. The ESS PI level is a compromise between the two optima, representing parental and offspring interests. One intra-brood case (shared costs) has $\alpha = \frac{1}{4}$, which is equivalent to the inter-brood offspring optimum for cases where displaced offspring are half sibs (i.e. when one parent is constrained by the conflict). This does not mean that the offspring can push the parent into supplying it with its optimal PI—remember that in the intra-brood model, there is no offspring optimum that is equivalent to the offspring optimum for inter-brood conflict. The further away from the parental optimum that the ESS PI lies, the higher the escalation of begging, which is reflected by the increase in β. Interestingly, the highest levels of escalated begging should be found under biparental care without true monogamy, with shared costs. The sexual conflict over PI in this form of biparental care acts to decrease PI levels below the uniparental (or true monogamy) levels, and the sibling rivalry acts to push PI back up to the uniparental optimum again; the subversive effects of the two forms of conflict exactly cancel out.

Note that with biparental care and individual costs (Table 8.1b), intra-brood conflict should generate *less* costly begging than inter-brood conflict. The conclusion that the intra-brood model produces increases in overt conflict relates specifically to uniparental care or true monogamy. The reverse applies to biparental care without true monogamy.

8.2.6 'Offspring wins' and 'parent wins' solutions

The Parker and Macnair (1979) model suggests a resolution of POC by a compromise between parental and offspring interests, at which neither player wins outright, although there is nevertheless escalated begging by offspring. This solution has the attractive feature that it can explain the overt be-havioural conflict between parent and offspring, sometimes seen in nature (Trivers 1974; but see Chapter 11).

Parker and Macnair also found a special case where the 'pro rata payment' compromise ESS does not apply, in which either the parent or offspring can win the conflict outright, and that involves no costly escalation of begging (x^* is minimal). The PI is either at the parental optimum or (for the inter-brood case) at the offspring optimum. Such a solution can occur when the survival probability, $S(x)$, falls so steeply initially that either it cannot pay the offspring to escalate its begging from the minimal level ('parent wins'), or—assuming that increased PI reduces begging ($X'(m^*) < 0$)—it cannot pay the parent to precipitate escalated begging by reducing the PI ('offspring wins'). Thus the conditions for these two solutions depend critically on the slope of the survival function at the minimal begging level, x_{min}. If the slope $S'(x_{min})$ is zero, only the pro rata compromise ESS can apply, because there is no marginal cost to either player in having the begging level move away from x_{min}. Parker and Macnair argue that this is plausible if there is an optimal level of begging from the parental point of view; they regarded x as the number of times more investment that is 'demanded' than is received.

If we assume that $S(x_{min}) = 1$ (there are no extra survival costs to minimal begging), a condition can be found for the slope, $S'(x_{min})$, such that any given current level of PI, m, can be held stable. However, we are mainly interested in deducing when an 'offspring wins' or a 'parent wins' solution can be an ESS, so that the values for m that we need to consider are m_p and m_o. Theoretically, we need a condition for $S'(x_{min})$ for each player at PIs of m_p and m_o. However, it never pays the parent or the offspring to deviate from its own optimum. We therefore need only one condition for $S'(x_{min})$, not two. The necessary conditions for $S'(x_{min})$ to maintain 'offspring wins' or 'parent wins' ESSs are given in Table 8.2. Since there is no definable offspring optimum, m_o, under intra-brood conflict, we can include only the condition for the parental optimum for intra-brood cases. The values given in Table 8.2 are derived by assuming the explicit forms (8.4), (8.7) and (8.10) for *Mechanisms 1* and *2*. Given these assumptions, the conclusions are that (i) for comparable circumstances, it is more difficult to maintain either ESS ('parent wins' or 'offspring wins') under biparental than under uniparental investment, (ii) the same condition maintains each ESS when begging costs are of the individual costs type, and (iii) it is more difficult to maintain a 'parent wins' than an 'offspring wins' ESS when the begging costs are of the shared type, and when both parents are constrained by the conflict.

What factors might determine which ESS occurs? Can selection produce these ESSs? Neither of these questions is easy to answer (see Parker and Macnair 1979, Parker 1985). What happens depends on evolutionary history, i.e. on 'who plays what first', or the starting point of the game. Parker and Macnair concluded that in the absence of a 'pro rata' compromise, inter-brood conflict is more likely to result in an 'offspring wins' solution, and intra-brood conflict in a 'parent wins' solution. With intra-brood conflict, it is assumed that no future sibs are displaced: there is no built-in mechanism that causes parents

Table 8.2 Conditions necessary to maintain a 'parent wins' or 'offspring wins' solution.[1] (After Parker 1985.)

Model	Maximum starting slope for $S(x)$ to keep parent at offspring's optimum ('offspring wins')	Maximum starting slope for $S(x)$ to keep offspring at parent's optimum ('parent wins')
One-parent investment		
1. Inter-brood conflict		
(a) One parent constrained		
(1) Individual costs	$-\frac{3}{4}$	$-\frac{3}{4}$
(2) Shared costs	$-\frac{3}{4}$	$-1\frac{1}{2}$
(b) Two parents constrained		
(1) Individual costs	$-\frac{1}{2}$	$-\frac{1}{2}$
(2) Shared costs	$-\frac{1}{2}$	-1
2. Intra-brood conflict		
(a) Full siblings		
(1) Individual costs	NA[2]	$-\frac{1}{2}$
(2) Shared costs	NA	-1
(b) Half siblings; maximum		
(1) Individual costs	NA	$-\frac{3}{4}$
(2) Shared costs	NA	-3
Two-parent investment		
1. Inter-brood conflict (both constrained)		
(1) Individual costs	$-1\frac{1}{2}$	$-1\frac{1}{2}$
(2) Shared costs	$-1\frac{1}{2}$	-3
2. Intra-brood conflict (Full siblings)		
(1) Individual costs	NA	-1
(2) Shared costs	NA	-2

[1] See text for details. The solution is stable provided that the starting slope is less than the values indicated above. The lower the value, the less likely it will be that the solution is stable.
[2] Not applicable; see text.

to forage more if offspring beg more. If, therefore, the optimal parental strategy, m_p, has fixed before genes for escalated begging arise, then a steep starting slope, $S'(x_{min})$, will maintain m_p as an ESS. With inter-brood conflict, such a mechanism must exist in order to satisfy the requirement that future sibs can be displaced. If initially the parent is fully sensitive to offspring demands, then selection will proceed towards the offspring optimum, m_p. However, suppose that the offspring strategy is only an infinitesimal step away from m_o before the parent constrains the payment of PI. The conditions are then much more permissive for the parent to constrain PI than for the offspring to move to its ESS. If some constraint on PI payment then evolves, so that offspring must escalate begging in order to gain any further PI, then for the offspring to achieve m_o requires the implausible condition that $S'(x_{min})$ be positive (Parker 1985).

8.2.7 *Summary and comments on the pro rata payment type model*

We have dwelt upon the pro rata payment model of Parker and Macnair at some length because it represented a detailed attempt to evaluate the resolution of POC in relation to mating system and different forms of begging costs. In contrast to earlier versions (see Parker and Macnair 1979, Parker 1984, 1985), we have attempted here to present the model in a general form, stressing that solution depends critically on two mechanisms: (i) $X'(m^*)$, the effect that a marginal deviation in parental strategy at the ESS has on the observed begging level, and (ii) $\mu'(x^*)$, the effect that a marginal deviation in offspring's begging strategy at the ESS has on its PI gains. Whatever the exact nature of *Mechanisms 1* and *2*, we can see qualitatively what happens by looking at the sign of $X'(m^*)$ and $\mu'(x^*)$. We have assumed throughout that $\mu'(x^*)$ is positive, i.e. that increasing begging will increase PI. This means that at the ESS, offspring will beg in an escalated and costly fashion.

There is some evidence for positive $\mu'(x^*)$ and negative $X'(m^*)$ from ornithological field studies (see also Section 12.1.1), but some results indicate that the situation is not quite so simple. In classic experiments, von Haartman (1953) artificially enhanced the signal level of vocally begging pied flycatcher (*Ficedula hypoleuca*) nestlings by placing additional hungry chicks in a concealed chamber next to the nest box. This elicited greater food deliveries to the real brood, i.e. demonstrating a positive $\mu'(x^*)$. Several playback experiments, wherein taped nestling vocalizations are used to augment the brood's own signal, have replicated this result (e.g. Muller and Smith 1978, Bengtsson and Rydén 1981, 1983, Harris 1983). Similarly, von Haartman (1953) showed that temporarily food-deprived nestlings escalated their begging, i.e. negative $X'(m^*)$, which has also been frequently reported in other species (see Section 12.2.1).

Simple correlational studies have sometimes revealed positive across-nest relationships between offspring 'begging' levels and parental support, which are consistent with the above assumptions, but do not indicate which of the model components is/are responsible. For instance, the amount of food delivered to singleton cattle egret chicks correlates with the number of nestling scissor-grips of the parent's bill (an easily quantified component of the food-solicitation process: see Chapter 5). In full-sized broods of three, the C-chick's ingestion rate correlated with its scissoring rate, but no such relationship was found for either of the senior siblings (A. Fangmeier, unpublished data). Positive relationships between begging and provisions have also been reported for many other bird species as well (e.g. von Haartman 1953, Henderson 1975, Hussell 1988, Litovitch and Power 1992, Redondo and Castro 1992b).

Returning to $X'(m^*)$, on purely intuitive grounds Parker and Macnair expected that it should be negative, i.e. that increasing PI would reduce hunger and hence always decrease begging. If so, this could lead to PI levels being pushed above the parental optimum, since a parent can achieve a marginal gain (temporarily, in evolutionary time) by reducing begging costs by appeasing offspring and giving extra PI. A positive $X'(m^*)$, where increased PI increases begging, should lead to PI levels below the parental optimum. Here the parent can achieve a temporary appeasement of offspring by giving less PI. Once again, positive correlations between begging and provisions tell us too little. Do well-fed chicks beg more because they have more surplus energy (and their marginal cost of begging is low), or do parents respond more to chicks that are begging harder? Of course, if deviations in PI around the ESS have no effect on the begging level ($X'(m^*) = 0$), the PI should remain at the parental optimum. In summary, the biological evidence concerning *Mechanism 1* ($X'(m^*)$) is equivocal. There are a number of studies sugggesting that *Mechanism 1* is negative (that feeding causes an immediate decrease in solicitation effort by chicks that is due, at the proximate level, to reduced hunger). For instance, in penguins, food solicitation by vigorously chasing parents is reduced by decreased hunger level (e.g. Bustamente *et al.* 1992), as is the 'convulsion behaviour' of American white pelican chicks (Cash and Evans 1986b). However, Redondo and Castro (1992b) found that in natural broods of magpies, the effect was not immediate, but that begging was reduced by increased intake in the previous 1-h interval (again a negative *Mechanism 1*) and similar results apply in experimental conditions (Redondo 1991). However, nestling budgerigars appear to show a positive *Mechanism 1*, increasing their begging as their parents supply more food (Stamps *et al.* 1985), and a similarly positive effect may occur in starlings (A. Kacelnik, personal communication).

Finally, it must be stated that the Parker and Macnair model is essentially a heuristic device for making **qualitative** predictions about the nature of conflict resolution, and about the effect of mating systems and types of begging costs. For quantitative application, it has two problems: the simplifying assumptions, and the difficulty of measuring the relevant parameters in the field.

The simplifying assumptions are not a major limitation; all that is required is that further modelling technology be applied. For instance, Parker and Macnair assumed infinite sibships. It is possible to reformulate the model in terms of n sibs by using the marginal Hamilton's rule (Godfray and Parker 1992). As the brood size increases, the ESS begging level, x^*, increases for individual and shared costs, but decreases for summed costs (see Section 3.3.5 and Figure 3.10). The concept of PI as a fixed resource is also simplistic. PI is most likely to displace future sibs by reducing the survival prospects of a given parent. Models of this type can readily be constructed (see Charnov 1982, Forbes 1993), and behave in qualitatively similar ways to the 'fixed-PI' models; they share the common property that spending more parental effort now means reduced prospects of producing future offspring. The more serious difficulty relates to obtaining the field data necessary to establish the various relationships, even at a qualitative level (this is a general problem for all models of POC).

To validate the basic assumptions of the pro rata model and to generate quantitative predictions, measures of four relationships are required: (i) $f(m)$, how an offspring's reproductive value increases with PI; (ii) $S(x)$, the survival costs of begging; (iii) $X'(m^*)$ and (iv) $\mu'(x^*)$, i.e. the way that *Mechanisms 1* and *2* operate. A difficult parameter is $f(m)$; it is usually possible to measure an offspring's gain in weight in relation to food supply, but then we encounter the problems of converting weight into expected future reproductive success, and of converting food supplied by a parent into PI (but see Section 7.2.1). Also rather intractable is measurement of the impact of begging on survival, i.e. $S(x)$. If begging costs are mainly energetic (and hence of the individual costs type), an estimate may be provided by physiological studies using doubly-labelled water. (To date the only published attempt at such a measurement is that of Bryant and Tatner (1990), but the results of three seasons' work applying this technique to cattle egret siblicide are currently being analysed.) Finally, if begging costs are mainly related to predation risks (often then of the shared or summed costs type), the best prospects lie in field studies of nest predation. Apparent begging can be artificially increased using tape playbacks (the first such experiment is that of Haskell 1994). Alternatively, natural variation in begging levels could be correlated with predation episodes. Such data may prove difficult to obtain.

It may be easier to measure the operation of *Mechanisms 1* and *2* under field conditions. *Mechanism 1*, $X'(m^*)$, the effect of changes in PI on begging levels, could be ascertained by food supplementation experiments and *Mechanism 2*, $\mu'(x^*)$, the effect of changes in begging on PI gains, is readily accessible via playbacks.

Our discussion of models of resolution continues through the next chapter, which describes a substantially different approach that is useful for illuminating additional aspects of POC.

Summary

1. Given that parents and offspring can evidently disagree on optimal PI allocation, analytical models have concentrated increasingly on exploring how POC is likely to be resolved. When will parents-win, offspring-win, etc.? This chapter reviews and generalizes the POC models of Parker and Macnair.

2. A series of discrete-strategy models explores the relative spreading rates of a dominant mutant allele, C, that causes offspring to demand extra PI (*conflictor* phenotype) and another mutation, S, that causes parents to be insensitive to such requests (*suppressor* phenotype). For inter-brood

POC, either of these two alleles can reach fixation, depending on the starting conditions, offspring begging costs, and parental willingness to compromise. For intra-brood POC, the model assumes an additional begging cost (and increased 'honesty' re the signalling offspring's true needs), which has a major impact. The relative frequencies of S and C now cycle as the ebb and flow of phenotypes continually reshape the pay-offs (as occurs in '*Rock–Scissors–Paper*').

3. To add realism, a second set of models allows offspring and parents to develop continuously variable levels of begging and responsiveness (measured as amount of PI provided), respectively. Each party adopts a set level or strategy, the pay-off for which is affected by the strategy its 'opponent' uses. A hypothetical mutant allele leads to adoption of a different strategy by its player, which precipitates (temporarily, in the scale of evolutionary time) a change in the *realized* level of its opponent's response. According to a proposed *Mechanism 1*, a change in parental delivery (away from the ESS value, m^*) precipitates an immediate change in the offspring's *realized* begging response (symbolized as X). At the ESS, *Mechanism 1* can be defined as $X'(m^*)$. The new *realized* level is the result of the offspring's immediate response to the opponent's shift—it is not permanent in evolutionary time. Most reasonably, one imagines this mechanism to operate in a negative direction ($X'(m) < 0$): if a parental mutation causes higher food deliveries, the offspring's realized begging is expected to drop. Ideally, *Mechanism 1* would be allowed to evolve in models of POC. So far it has been assumed to have some explicit form.

4. According to the proposed *Mechanism 2*, an offspring mutation (causing a phenotypic shift in begging level away from ESS level, x^*) can deflect the parent's *realized* investment level to some new value, μ. This is the reverse analogue of *Mechanism 1*. At the ESS, *Mechanism 1* can be defined as $\mu'(x^*)$. Such a mechanism can be modelled either as some kind of automatic (and positive) function, wherein an offspring that begs more receives more (as modelled by Parker and Macnair), or as a complex function that incorporates consideration of the veracity of offspring begging (i.e. whether it reflects true needs; see Chapter 9). Either way, increased begging is assumed to result in an immediate increase in benefits ($\mu'(x^*) > 0$). Begging is also assumed to be costly to offspring fitness prospects, as described by a monotonically decreasing survivorship function, $S(X)$.

5. Solutions are derived for two kinds of POC, *inter-brood conflict* (where offspring selfishness reduces the production of *future* siblings) and *intra-brood conflict* (where the impact is felt by contemporary brood-mates). The parental ESS is stated formally but in general form as equations (8.12a) and (8.13a), and the offspring ESS as equations (8.12b) and (8.13b). These show that the parent's ESS depends simply on $f(m)$, the offspring fitness function, while that of the offspring depends on $S(X)$, the

begging cost function. This in turn leads to the conclusion of a compromise resolution: if $x'(m^*) < 0$, the ESS delivery of PI (m^*) is higher than the parental optimum and lower than the offspring optimum. Thus, offspring receive more than is best for parents, but not as much as they would like to receive.

6. These 'pro rata' models make some differing predictions for inter- vs. intra-brood POC. Under most conditions (and, especially, the reasonable assumption that the *Mechanism* 1 'demand function' be negative), offspring should beg more and receive more PI when competing against current brood-mates than when competing only against future siblings. POC should also be higher in the intra-brood context.

7. Various aspects of mating system impinge on POC through their effects on offspring relatedness. The models deal with this by considering which parties must pay the begging costs, which necessitates that such costs be sorted and classified. The personal energetic costs for escalated begging are obviously *individual* in nature, but other fitness-reducing effects (e.g. increased conspicuousness to predators) are clearly *shared* with other nest-mates, including those that do not carry the mutant allele. The type of costs can alter the model's predictions in terms of the extent of begging and the amount of PI given.

8. In addition to the usual compromise solution, extremely steep begging costs (where the survival probability, $S(X)$, declines sharply and immediately as begging starts) can produce one-sided ESS solutions. If the offspring cannot escalate above its minimum, a 'parent wins' outcome arises; alternatively, if decreased PI precipitates too great a cost of begging, the parent cannot reduce PI below the offspring optimum ('offspring wins'). The necessary conditions for these extremes depend largely on whether costs are individual or shared and on the evolutionary history of a given system (summarized in Table 8.2).

9. A very brief look at selected ornithological field studies indicates that at least some of the model assumptions are reasonable, but considerable variation is apparent.

10. Finally, we stress that the POC models developed here are intended primarily as heuristic devices that make qualitative predictions. To convert them for quantitative predictions would require some modification (especially with regard to size of sibship and the fixed nature of PI). Such an exercise would also require that data be obtained for four key parameters, notably $f(m)$ and $S(X)$ plus the two interactive mechanisms, $X'(m^*)$ and $\mu'(x^*)$, an empirical task that is likely to be substantial, but not impossible.

9

Parent–offspring conflict III: begging as an honest signal

There are times when parenthood seems nothing more
than feeding the hand that bites you.

(Peter De Vries)

We continue our review of resolutions models with a philosophically quite different explanation for escalated begging, which is related to the notion of 'honest advertising' (Godfray 1991). Here, instead of begging being the manifestation of a scramble competition between sibs, it represents an honest (but costly) signal of need for PI that the parent monitors and uses in allocating PI. Godfray (1995b) gives an excellent comparative analysis of approaches to POC resolution and fits his signalling model within the same framework as the models of Parker and Macnair (1978), Eshel and Feldman (1991) and Yamamura and Higashi (1992).

Godfray's analysis of begging is derived from a pioneering model by Grafen (1990a) of signalling between non-relatives. Grafen's model was couched in terms of mate choice, although it can be modified for more general applications. His argument can be summarized as follows. (i) Signallers differ in their quality (value to receivers), and give a signal (advertisement) that is less costly for high-quality individuals. (ii) Receivers cannot assess quality directly, but can monitor the extent of the signal so that they evolve an ESS rule for signaller assessment entirely in terms of the signal's level. (iii) At the ESS, signallers pick a level of advertisement that increases with their quality. It is this last connection that makes it an 'honest signal' (it reveals true quality) and as such was seen by Grafen as evidence for Zahavi's (1975, 1977, 1987) 'handicap principle', which stressed the notion that biological signals must be honest and costly.

Remarkably similar results were obtained by Andersson (1982) and Parker (1982) for sexual advertisement games in which the receiver is imagined to have a fixed rule for signaller selection (e.g. passive attraction to the strongest source of signal). The phenotypic optima for signalling are such that 'stronger' signallers will have higher levels of advertisement traded off against the cost constraint provided by increasing the signal. In Grafen's model, the signaller experiences benefits by being perceived as possessing high quality, and at the ESS, signalling level correlates with quality. In the Andersson/Parker models, the signaller experiences direct benefits by higher signalling levels. The fundamental difference between the two types of model is that in Grafen's model the receiver's response is allowed to evolve, whereas in the Andersson/Parker models it is fixed.

A similar kind of relationship applies between Godfray's and Parker and Macnair's models of the evolution of begging. Godfray allows the parental mechanism for allocation of food in relation to begging level to evolve (see also Harper 1986), whereas in Parker and Macnair's model it is assumed to be a fixed mechanism, following equations (8.7) or (8.10).

9.1 Godfray's model for inter-brood conflict

Godfray's model (1991; see also Godfray 1995a, b) can be summarized as follows. In common with Parker and Macnair, offspring fitness, f, shows diminishing returns with increased parental resources, m, and begging is expensive, so that costs rise monotonically with begging level. For simplicity of analysis, Godfray first analysed the case where offspring are produced singly, so that there is inter-brood (Triversian) conflict, but he later extended the analysis to intra-brood competition (Godfray 1995b). Similarly to Parker and Macnair, Godfray seeks the ESS PI level, m^*, and ESS begging level x^*, but the ESS he seeks is condition-dependent: x_c^* is a function of offspring physical condition, c, while m^*, in turn, is a function of x_c^*.

Godfray assumed that offspring vary in their need for parental resources, depending on their condition. The details of his model are explained in Box 9.1.

To determine the ESS, an assumption must be made about the offspring's physical condition. To solve the two equations for m^* and x_c^* in Box 9.1, an initial value is required mathematically to serve as a kind of logical anchor for the rest: this is provided by an offspring in maximum condition, c_{max}. Godfray reasoned that an offspring in maximum condition should not beg, i.e. $x^*_{cmax} = 0$. If the parent uses the offspring's begging level to recognize true condition, at the ESS an offspring in maximum condition cannot obtain any benefit by escalating above $x = 0$.

A numerical example (Godfray 1991) is shown in Figure 9.1 and mathematical details are given in the legend. If all offspring are equal in condition, there is no begging. If the offspring vary in condition, begging level increases with true need (i.e. as offspring condition decreases): the offspring gives an honest but costly signal (expressing its true need). The parent pays out its PI in relation to begging level, and the young in poorer condition beg and receive most. Interestingly, the greatest overall solicitation occurs when offspring needs are particularly *variable* (and there is none if there is no variation), as

Box 9.1   Godfray's signalling model for offspring begging

Let f be the personal fitness of an offspring, which increases with m, the amount of parental resources it receives, and decreases with x, its begging level. The *future* fitness of the parent is a second function g, which decreases with its expenditure, m, on the present offspring. Hence rg is the summed value of a given offspring's future siblings produced by the investing parent, devalued by the coefficient of relatedness r. The focal offspring's inclusive fitness is therefore $f + rg$, its current (or direct fitness) pay-off plus its future (or indirect fitness) pay-off. An offspring deviating from the norm by playing x_c ($\neq x_c^*$) gets a pay-off

$$W_o(x_c, m^*) = f[\mu(x_c)] + rg[\mu(x_c)]$$

where $\mu(x_c)$ is the PI attained by a mutant offspring of condition c deviating from the ESS by playing deviant strategy, x_c. Remember that the parental strategy is m^*, but a strategic change by the offspring (a mutation) causes the parent to give a new amount,[1] $\mu(x_c)$. In Godfray's model, if an offspring begs at new level $x_c \neq x_c^*$, it receives the amount of PI that would normally be given to an offspring signalling honestly at level x_c. (In other words, the parent's policy is to 'believe' the signal it receives and to deliver the amount that its own behavioural programme indicates. If an offspring truly needs x_c and 'says' so, the parent supplies an amount of food that is identical to the quantity it provides for a dishonest mutant offspring that 'says' x_c, but actually needs less.)   At the ESS, $\mu(x_c^*) = m^*(x_c^*)$ and applying equation (8.2) gives

$$\mu'(x_c^*)\{f_m[m^*(x_c^*)] + rg_m[m^*(x_c^*)]\} + f_x[m^*(x_c^*)] = 0, \qquad (9.1)$$

where, for example, $f_m[m^*(x_c^*)]$ is the derivative of $f[m(x_c)]$ with respect to m, evaluated at the ESS. The derivative $\mu'(x_c^*)$ is *Mechanism 2* of Chapter 8, the marginal resource gains due to begging around the ESS. In Godfray's approach, however, $m^*(x_c^*)$ is allowed to evolve, and so $\mu'(x_c^*)$ is the slope of m^* at x_c^*.

Godfray assumed parental fitness to be the sum of present and future reproductive success, that is, $f + g$. To calculate the parental optimum, $m^*(x_c^*)$, assuming that the parent can assess offspring condition correctly by monitoring begging performance, x_c^*, he used a version of the following expression for the pay-off to a parent deviating from the ESS by playing $m(x_c^*) \neq m^*(x_c^*)$:

$$W_p(m, x_c^*) = f[m(x_c^*)] + g[m(x_c^*)],$$

and from equation (8.1), at the ESS

$$f_m[m^*(x_c^*)] + g_m[m^*(x_c^*)] = 0, \qquad (9.2)$$

so that the parent must exactly balance the marginal gains to the present offspring against the marginal losses to future offspring. Note that any strategic deviation by the parent has no effect on the realized begging level of the offspring: the offspring begs at its ESS level regardless of what the parent does. In effect, therefore, Godfray's model assumes that *Mechanism 1* does not operate. More formally, in the notation of the preceding chapter, $X_c'(m^*) = 0$. Given this, the model in Chapter 8 would therefore predict the PI level to be at the parental optimum.

Note that, in equation (9.1), f_m must be positive (increasing PI amount enhances the present offspring's fitness) and g_m must be negative (increasing PI amount reduces the parent's future fitness). Because $f_m = -g_m$ (see equation. 9.2), and $0 < r < 1$, then $f_m + rg_m > 0$. The sign of $\mu'(x_c^*)$, therefore, depends on the sign of the begging cost function, f_x. This must be negative because begging is required to be costly, and so from equation (9.1) $\mu'(x_c^*)$ must be positive. At the ESS, the parent must give more PI to offspring that beg more.

---

[1]   In the same sense that Chapter 8's 'realized' values were responses and not evolved optima.

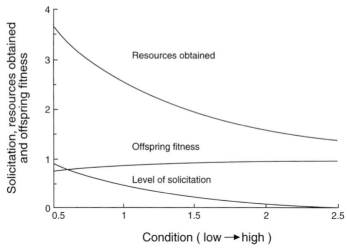

Fig. 9.1 Godfray's example of honest begging. The ESS level of begging is solved from equations (9.1) and (9.2) by assuming that $f(m,x_c)$ consists of two additive components, $u(m,c)-v(x)$, where benefits $u(m,c)$ are a function of the resources obtained and an offspring's condition, while costs $v(x)$ are a function only of begging level. Explicitly, $u(m,c) = U(1-\exp(-cm))$, where U is a positive constant defining the asymptotic maximum benefit, and the effect of lowered condition is to reduce the rate at which resources gained rise to this maximum. Parents are selected to invest more in offspring in poor condition. Begging costs are assumed to be linear: $v(x) = Vx$, where V is a positive constant. Provision of extra resources is assumed to cause a linear reduction in the parent's future reproductive success: $g(m) = G-\gamma m$, where G and γ are positive constants. Assuming that young in best condition have $c = 2.5$, this model gives $m^* = (1/c)[\ln(cU/\gamma)]$; $x^* = (m^*-m^*_{c=cmax})[\gamma(1-r)/V]$. In this example, $U = 1$, $\gamma = 0.08$, $V = 1$, and $r = 0.5$. (After Godfray 1991, © Macmillan Magazines Ltd, 1991.)

may happen towards the end of the parental care period when individuals are beginning to fend for themselves (Godfray 1991). Increases in costly begging around this time offer a novel interpretation for behavioural strife that Trivers had called 'weaning conflict,' supposedly a manifestation of offspring manipulating their parents. According to this classical view, the escalation arises when the parent wishes to terminate investment but the offspring continues to desire the easy lunch (but see Green *et al.* 1993 and Bateson 1994 for alternative interpretations). Godfray's model suggests that manipulation is not essential, because honest signalling would naturally lead to escalated signalling at this time.

Godfray's is a much 'freer' model than that of Parker and Macnair: it requires no explicit assumptions about the mechanism of food allocation. In fact, that mechanism itself evolves and at the eventual ESS the offspring signals its condition ('need') in an honest but costly fashion. The parent always wins in the sense that each offspring gets an amount of PI equal to the parental optimum (which depends on the offspring's needs, which are greater if it is in poor condition), but an offspring must show costly begging to signal its needs. In the Parker and Macnair model, the ESS is for an amount of PI that is intermediate between the two optima, and costly begging occurs even if there is no variance between offspring needs. However, the difference in PI allocation between the models is not due to the 'signal' vs. 'scramble' distinction, which is indeed the salient biological distinction. The two models are very similar in a mathematical sense, and the major formal difference between the two approaches relates to their very different actions of *Mechanism 1* (how a deviation in PI affects begging level). The ESS PI allocation depends critically on the action of *Mechanism 1*. In the Parker and Macnair model, *Mechanism 1* shows a negative relationship (increased PI temporarily reduces begging), whereas with Godfray's model *Mechanism 1* does not operate (changing PI has no immediate effect on begging). This is why Godfray's model generates a PI allocation at the parental optimum, and why Parker and Macnair's model generates a pro rata compromise between the parental and offspring optima. As we have noted, some species appear to show no change in begging if PI is increased, suggesting that Godfray's model is more appropriate for them. Others appear to show reduced begging (following Parker and Macnair, although Godfray could argue that condition has changed). The species that appear to show increased begging when PI is increased are not well served by either of the existing models.

Godfray concludes that, in one sense, Alexander's (1974) contention that parents will always win is vindicated. This can be so if *Mechanism 1* is non-functional, but not otherwise. Godfray points out that, in another sense, the potential for POC necessitates costly begging that reduces the fitness of both offspring and parent. Note that an indirect reduction in begging due to feeding is implicit in Godfray's model. If an offspring receives food, its condition improves so that it will beg less later. However, this effect is not explicit in the mathematics, and so there is no profit to the parent in feeding to reduce begging costs. Mechanism 1 could be incorporated into Godfray's model, should this be desirable. In effect, one uses the current Godfray model to solve the ESS at each feeding bout, with the simplifying assumption that each bout has an *independent* effect on fitness. Nevertheless, a parent that feeds more stands both to improve its own condition and to reduce total begging, and so must ultimately reduce begging costs as well as enhancing its personal condition.

In an interesting recent development, Johnstone (1996b) has extended Godfray's (1991) signal model of begging for a single offspring to consider the effect of parent–offspring interactions over a long period. In this analysis, parental allocation of resources to offspring precedes offspring solicitation as well as following it: the parent allocates an initial amount, I, independent of offspring condition (which is not yet signalled); the offspring then begs at level x, that can depend on its condition, and the initial allocation; and then the parent makes a final allocation that depends on the begging level, and hence possibly also on *c* and *I*. The model is otherwise rather similar to Godfray's. Selection always favours giving an amount of resources on the first step of the game which is superoptimal for the parent for at least some of the offspring (the ones already in the best condition). This cost to the parent is more than offset by the benefits of reducing offspring begging. Thus parents can indeed give more than is optimal in this more realistic signalling model, simply to reduce the subsequent level of costly solicitation by the offspring. This appears to be a method of adding *Mechanism 1* to Godfray's model.

A further interesting discovery is that Godfray's single-chick model has, in fact, two Nash equilibria (Rodríguez-Gironés *et al.* 1996), not just the costly signalling one just described. There is another equilibrium in which chicks do not beg at all, and the parental provisioning is optimal with respect to the expected distribution of conditions for the chicks. The expected fitness for both parents and offspring is higher at this non-signalling equilibrium than at the signalling ESS, and since non-signalling is likely to be unstable, there is the problem as to how the non-signalling ESS could evolve. Rodríguez-Gironés *et al.* suggest that in species which rear several young simultaneously, costly begging may initially have evolved through direct sibling fighting and scrambling, before the establishment of a parental response. The condition of a chick will be reflected in the intensity with which it scrambles against its nest-mates (Parker *et al.* 1989), allowing the evolution of selective feeding by parents in response to condition-dependent scrambling. Thus in many species begging scrambles could have preceded signalling.

Both signal and scramble models draw similar conclusions about costly begging, despite the fact that begging is a signal of condition (according to Godfray) vs. a result of a begging scramble (according to Parker and Macnair). Which view is more realistic? With signalling, the parent is assumed to be omnipotent. It measures begging levels precisely, and then allocates resources actively in exact accordance with the message prescribed by the signal. With scrambles, the parent is a passive vehicle, allocating each item of resource to the offspring currently presenting the greatest stimulus. Signalling assumes that the parent has control over the allocation, and scrambling assumes either that the offspring have control (by competition) or that the allocation mechanism is fixed. Because the parent is in the position of giving resources, and the offspring of receiving them, signalling at first seems more likely, especially when offspring are produced singly (the case considered explicitly in Godfray's 1991 paper). Godfray's model can apply, however, and with similar conclusions, to intra-brood conflicts (Godfray 1995b). Even so, parental control can sometimes be more easily usurped if offspring are produced in broods. The presence of nest-mates offers a target for various forms of selfishness that might be impractical if tried with a parent. Food can literally be snatched from a sib's mouth (or even the parent's!) by an escalating offspring, or sibs may scramble for teats without the mother being able to exert much (if any) control over who gets what. It is therefore tempting to suspect that many systems represent some form of compromise between the two philosophies of signalling and scrambling. Some evidence from experiments on European starlings has appeared recently (Kacelnik *et al.* 1995). Starling chicks were given more food if they begged more and also if they were in a position closer to the nest entrance. Whilst the former may be under parental control, the latter is not and depends on sib competition.

However, it is not easy to distinguish between the two models on the basis of the parental allocation alone. In both, an offspring that begs more (or expends more effort) receives more. Perhaps the best test would be to examine how parental food allocation correlates with offspring condition and, in particular, with relatively cryptic variance in offspring condition. More food should go to offspring in poorer condition. If the scramble model is extended so that offspring in better condition are able to beg more effectively at the same cost (see Section 3.3.6, Figure 3.11), then although sibs in better condition ('stronger sibs') are likely to have greater fat reserves and can thus probably survive on less food, it will be commoner for them to receive more food. If 'strong sibs' receive more food, this looks superficially like evidence for a scramble, although it must be established that the unequal allocation is not in parental interests. It is certainly most common for the strongest chicks in asynchronously hatched broods to gain most of the food input (see Chapters 5, 6, and 12). However, there is some evidence for judicious differential provisioning that may relate to true need (e.g. Stamps *et al.* 1985, Gottlander 1987, Redondo and Castro 1992b, Lyon *et al.* 1994), although of course this is not necessarily incompatible with the scramble model, because the effects of true need have not yet been incorporated into that model.

Thus, it will be very difficult to differentiate between the two models, because many of their predictions are similar. For example, in a detailed field study, Redondo and Castro (1992b) investigated the relationship between begging, chick nutritional state (condition), and parental distribution of food within natural magpie broods. As required for both models, offspring tended to receive more food if they begged more. There was no immediate reduction in begging effort with parental feeding (at first suggesting that $X'(m^*)$ may be zero, as assumed by Godfray), although there was indeed a reduction in begging in the ensuing 1-h interval (suggesting that $X'(m^*)$ may be negative, as assumed by Parker and Macnair). Smaller chicks begged more than larger ones, but despite this, the stronger large chicks obtained greater resource shares. Redondo and Castro (1992b) interpreted these results in terms of honest signalling, with a preference of parents for larger chicks, but they are perhaps even more compatible with the scramble model of competitive begging with unequal chicks (Section 3.3.6). In contrast to magpies, cuckoo chicks in the same nests showed no tendency to reduce begging after feeding (Redondo 1993). Cuckoos are, of course, unconstrained by kin selection, so both models would predict escalated begging.

A second difference in the models' predictions depends on the variance in condition among offspring. In scrambles, begging should be extensive even if all of the offspring are identical in condition. With signalling, a decrease in begging is predicted when variance in condition decreases (and without variance there should be no begging). One line of circumstantial evidence appears at first sight to argue against the signalling model. If hatching asynchrony is artificially reduced in species where it is typically present, begging and competitive interactions among chicks increase, rather than the reverse (see Section 10.3.3). However, note that the signalling model related to *cryptic* variation in offspring condition, rather than to conspicuous differences (e.g. those perceivable by the parent) such as may prevail in asynchronously hatching broods of birds. By synchronizing the brood, one could argue that the obvious variation is destroyed, and so the signalling models can predict more conflict in synchronous broods. It does, nevertheless, imply simplistically that conspicuously stronger sibs should be fed less (but see Temme 1986).

The relative importances of 'signal' and 'scramble' in determining how offspring can affect the flow of resources from their parents remains very much to be established. At present we appear to be on an exciting threshold, awaiting empirical evidence. At the moment an intuitive guess is all that can be achieved—ours would be that signalling may have a more important effect for inter-brood conflict, but that scramble effects could be very significant under conditions of intra-brood conflict.

9.2 Godfray's model for intra-brood POC

A second version of the signalling model addresses intra-brood (rather than inter-brood) conflict. Godfray (1995b) examined a situation involving two offspring in a brood that compete for a finite amount of PI provided by a single parent. Again, an ESS signalling system can apply, which is essentially identical to that of the previous section. Signals are costly but honest cues of true offspring condition, and parents allocate resources in an ESS relationship to the level of signal, with more resources going to more intense signals. The signalling ESS can operate because of the potential POC over shares of resources for the two offspring.

The two offspring are both aware of each other's begging levels, x_i and x_j. The parent can vary its PI levels to each offspring (m_i and m_j) but the total is fixed ($m_i + m_j = M$). The ESS prescribes what to do when the two offspring have any combination of conditions of true need, c_i, and c_j. (The explicit assumptions about condition and fitness of offspring i are as stated in the legend to Figure 9.2)

At the ESS, neither offspring can profit by changing its begging level: these are x_i^* and x_j^*, given that their conditions are c_i^* and c_j^*. The parent allocates resources as m_i^* and m_j^* (again given that the offspring conditions are c_i^* and

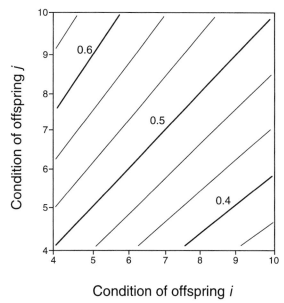

Fig. 9.2 Optimal share of resources, m_i, to be given to offspring i by the parent, in relation to offspring i's condition, c_i, and the condition of its brood-mate j, c_j. Fitnesses of i and j are defined as in Figure 9.1. The figure shows contours for a given resource share at intervals of 0.0333 with bold-labelled contours at intervals of 0.1. (After Godfray 1995a.)

c_j*), such that the marginal gains from allocating more food to each offspring become equal. Offspring can deviate by increasing begging, and so disrupt the allocation, but at the ESS allocation, the costs and benefits of marginal deviations exactly balance out for all three parties. Godfray first examined the case where the two offspring vary only in *cryptic condition* (they differ only in c_i and c_j not in U or V (see legend to Figure 9.1).

Figure 9.2 shows the contours of given resource shares for the two sibs, i and j, labelled from i's perspective. Offspring in identical condition receive exactly equal shares, as expected. As one offspring's condition decreases relative to that of the other, the parent supplies it with more food. Begging decreases as condition increases, as expected, but it also increases (weakly) with *declining* condition of a nest-mate. As the nest-mate's begging escalates through its increased need, Self's begging also increases, but much less strongly. This disparity in magnitudes may explain why sensitivity to the state of nestlings was not observed in their experiments on starlings by Cotton *et al.* (1996). As relatedness increases between the two sibs, lower (and less costly) signalling is predicted (Figure 9.3).

What happens if offspring vary in more obvious ways that can be perceived by the parent (such as may occur because of hatching asynchrony)? Suppose that the asymptotic fitnesses (U: see legend to Figure 9.1) of the two sibs

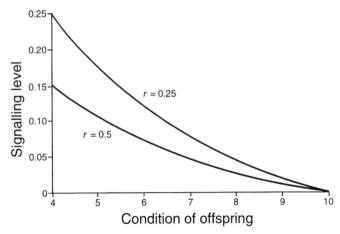

Fig. 9.3 Relationship between signalling level of offspring i and its condition, c_i, for full sibs ($r = \frac{1}{2}$) and half sibs ($r = \frac{1}{4}$). The fitnesses of i and j are defined as in Figure 9.1, with $c_j = 7$. (After Godfray 1995a.)

differ. The parent tends to bias resource shares towards the offspring having the lower asymptote (the marginal rate of gain in fitness per unit investment is usually steeper for this offspring). The other offspring begs more intensely when in poor condition.

So far, the model has concerned only individual costs of begging (costs of increased solicitation borne entirely by the offspring that performs the increase, e.g. costs are energy. What happens if the costs of an increase are shared by both sibs (e.g. costs are predation risk)? No ESS occurs if the costs are shared equally (exactly half to each). However, if the costs to chick i are apportioned as x_i and δx_j (where $0 \le \delta < 1$), then increasing δ reduces begging (Figure 9.4), rather as may occur in the scramble model (cf. Figure 3.11).

Godfray made a preliminary analysis of the case where there is both intra- and inter-brood conflict, which generated conclusions that are intuitively appealing from a biological viewpoint. One of these is that begging should increase with a relative decrease in relatedness to future, as compared to current, siblings.

9.3 Other approaches to honest signalling

Zahavi (1987) has argued that an offspring may reduce its own immediate prospects by taking on a costly handicap (e.g. costly begging) in order to create a situation such that the parent must invest more in the handicapped offspring (in order to increase its own fitness). Dawkins (1976) likened this to the offspring's crying 'Fox, fox, come and get me' (i.e. some risk is taken so as to force the parent to feed Self more than its 'fair share'). Some support for

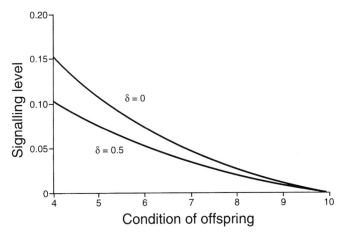

Fig. 9.4 Relationship between signalling level of offspring i and its condition, c_i, when costs of begging are felt by its brood-mate, j as well as by itself. If $\delta = 0$, all costs are felt by Self; when $\delta = 0.5$, j also experiences half of the costs suffered by Self. The fitnesses of i and j are defined as in Figure 9.1, with $c_j = 7$, except that the costs of begging are now $V_i(x_i + \delta x_j)$ instead of $V_i(x_i)$. (After Godfray 1995a.)

this form of handicap is claimed by Eshel and Feldman (1991), who used population genetic models, and we have already discussed the possibility that a parent might profit (under *Mechanism 1*) from providing extra food if so doing results in reducing the costs of begging. Perhaps the main question is whether an argument based on what may be a misleading human analogy is useful as a biological description of evolutionary adaptation: more neutrally, one may simply analyse the effects of the benefits of begging (increasing an offspring's food share) as balanced against the associated costs (e.g. effort and/or increasing predation risk).

Zahavi's notion of honest signals was rather less contentious, as we have seen. It is indeed possible for a begging signal to be an honest reflection of true offspring need. Although the idea of honest but costly signalling was embodied in Zahavi's 'handicap principle' (Zahavi 1975, 1977, 1987), Enquist (1985) was the first to show analytically that honest signalling could be evolutionarily stable in a model of conflict between non-relatives. Grafen's (1990a) and Godfray's (1991) signalling models are continuous games in that both the signal and the receiver's response can be regulated continuously and perceived perfectly (see also Grafen and Johnstone 1993).

Maynard Smith (1992) gives a much simpler analysis of a discrete strategy game, the 'Sir Philip Sidney Game'. Sir Philip—soldier, statesman and poet—was badly wounded in battle at Zutphen in 1586. After escaping from the melée and reaching the English camp, he called for a drink. A more severely-injured soldier was carried past as he raised the water-bottle to his lips. Rather than drinking, Sir Philip handed the bottle to the soldier, saying 'Thy need is greater than mine'. The basis of this game, then, is that one individual (X) may give some resource to another individual (Y) in response to a signal of 'true need' by Y (i.e. Y gives the signal only when its need is great). The resource has survival value to either party (depending on who consumes it), and each individual has some value to the other (e.g. they are related[1]). Maynard Smith's model gives essentially the same result concerning evolutionarily stable honest signalling as that of Grafen (1990a). It can be stable for Y to signal its need honestly, provided that the signal has a certain cost.

Johnstone and Grafen (1992a) give a continuous version of the Sir Philip Sidney Game, using essentially the same analytical techniques as the Grafen model. At equilibrium, X has a continuous response threshold for giving the reward to Y in response to Y's signal, and this threshold descends as X's own survival prospects increase. Individual Y gives a signal that increases in strength (and costs) with its true need—hence the signal is honest—and falls to zero if its survival prospects are fully restored (raised to probability 1.0). In a neat extension of the model, the authors consider the case where a third individual is involved. Now Y is the signaller, and Z is a 'messenger' that relays the signal (via a second signal) on to X. Remarkably, this model has a parallel in conflict during seed development in angiosperms, where X and Y are maternal and offspring sporophytes, respectively. These do not interact directly; offspring 'signals' must be relayed to the mother via the endosperm (see Chapter 16). This complexity tends to destabilize the signalling system and is unlikely to reduce costs.

A further extension is to allow communication to have some degree of error. Grafen's (1990a) and Godfray's (1991, 1995b) models assume communication to be error-free; at equilibrium, the signal gives perfect information about quality. Johnstone and Grafen (1992b) examine the case where the receiver has a probability distribution that a signaller is perceived as signalling at level p, when it actually signals at level a. Honest signalling remains robust provided that (plausibly) the higher the advertising level of the signaller, the more likely it is to be perceived as advertising strongly, and vice versa. Signals often have a 'typical intensity', that is they represent an 'all-or-nothing' message rather than a graded one. Johnstone (1994) has developed an ESS signalling model that, when perceptual error (on the part of the signal receiver) is incorporated, generates the evolution of discrete, 'all-or-nothing' signals. He has also shown that multiple signals can be stable, provided that signalling costs are strongly accelerating (Johnstone 1996a), something which may be important in the present context if parents inspect several aspects of begging (such as size of gape, strength of vocalization, brightness of pharynx, etc.).

Johnstone and Grafen (1993) show that biological signals need be honest only 'on average', and that the ideal of 'perfect honesty' will almost never be

---

[1] Presumably kinship did not apply in Sir Philip's case.

encountered. In a version of the discrete Sir Philip Sidney Game, they consider the case where there can be more than one class of signaller (the variance can be in degree of relatedness or in costs that the signaller must pay for the signal). A 'partially honest' signalling equilibrium is possible, in which the signal is costly for some individuals, which are constrained to honesty. For stability, the cost to the donor of being 'cheated' must be suitably low; honest signalling must predominate. Related to this effect, Godfray (1991) notes that his signal-begging model becomes more complex if the costs of solicitation vary between offspring having the same need. Offspring that are able to beg cheaply are expected to misrepresent their needs and to obtain more resources, whereas offspring for whom begging is expensive will solicit less and obtain fewer resources. Unequal costs for offspring in the same condition will lead to departures from the parental optimum at the ESS.

Finally, it must be noted that under certain parameter values in the discrete-strategy Sir Philip Sidney game, it can be stable for the signal to have no cost: honest signalling need not necessarily be costly (Maynard Smith 1992, 1994). This feature is lost in continuous-strategy versions of the game (Johnstone and Grafen 1992a). Maynard Smith (1994) has extended the game to include two-way signalling, so that the donor may signal that it is too costly for him or her to give resources even if the recipient is in need. This analysis has confirmed that signals between interactants need not necessarily be costly in order to be reliable (although they often must be so). If an interaction has several possible outcomes (1, 2, 3, . . .) and if the interactants place these in the same order of preference (measured as change in inclusive fitness), the cost-free signals can determine the outcome. In such circumstances, there is no conflict over the outcome. The degrees of preference need not be equal (e.g. one player may strongly prefer outcome 1 to outcome 2, while the other has only a weak preference for outcome 1). This conclusion applies if the outcomes are continuously distributed (e.g. the proportion of resource transferred). It may thus be possible for a parent to signal honestly to an offspring that there is no food available, but more analysis will be required to confirm this type of conclusion (Maynard Smith 1994). A very useful general classification of terms and models for animal signals is given by Maynard Smith and Harper (1995).

Summary

1. Honest advertising (and handicap principle) models concern the way in which signallers present themselves to prospective receivers. The basic idea is that signalling is costly, so—at the ESS—more capable individuals can afford to do more of it than weaker rivals. Accordingly, at the ESS, the receiver can use signal strength as a source of reliable, and hence 'honest', information about the signaller's condition.

2. Godfray (1991, 1995b) adapted this approach to the problem of offspring begging, seeking the ESS levels for both parental provisioning and offspring begging, the latter being *a true reflection of* an offspring's physical condition. Basically, offspring vary in their condition (and hence their need for PI) and direct costly signals to their parents in order to summon PI deliveries. In the model, offspring are allowed to beg at different levels and parents are responsive to them (even if the signals are untruthful); this follows *Mechanism 2* of the last chapter. However, strategic deviations in parental deliveries do not affect offspring begging. This differs sharply from the previous chapter's model in that no '*Mechanism 1*' is present here. Parental responsiveness is allowed to evolve, however, and at the ESS, parents must give more PI when offspring emit stronger (and more costly) signals.

3. An offspring's begging thus reflects its true needs. When it is weak, it begs more loudly and receives more PI; when it is strong, it begs less (saving the costs of begging). When in maximal condition, it does not beg.

4. When offspring condition tends to *vary* greatly, begging level should be highest. This feature has been offered as an alternative to Trivers' POC interpretation of 'weaning conflict.' Far from signifying a time of maximum conflict, the heightened signalling that seems to accompany the transition from parental investment to self-sufficiency may be a period of careful and co-operative monitoring of offspring needs. When it falls short of maintaining condition on a short-term basis, begging escalates.

5. A primary difference between this model and the rather similarly structured one in the previous chapter is not the 'signal' vs. 'scramble' distinction, but the built-in assumptions about how PI delivery immediately affects begging (*Mechanism 1*), which is inoperative here and negative in the other model. This alone can explain why Godfray's ESS lies at the parental optimum (offspring have no effective power), while Parker and Macnair's models derive ESS compromises.

6. Which of these two modelling approaches is the more suitable may depend ultimately on parental control. Where there is strong parental control, the signalling model is more appropriate, but where the parent has little control, the scramble model may be more suitable. The scramble approach may be significant where multiple siblings cohabit the nursery, although the signalling model can be applied to intra-brood conflict as well.

7. Empirical research is needed that takes changes in offspring condition more directly into account and in which parental ability to assess offspring condition is measured. The few data that exist at present are too fragmentary to evaluate.

8. Some theoretical attention has also been given to cases where two needy individuals must assess their own relative levels of need via signalling honestly (the 'Sir Phillip Sidney' Game), and so decide when to be selfish vs. altruistic. In one variant, a third party (the messenger) is involved as a

signal relay; adding this link to the chain reduces its evolutionary stability. Other promising leads incorporate elements of random error or differential costs in the signalling dynamics.

10

Parent–offspring conflict IV: clutch size and sexual conflicts

Thus for one seed to expand selfishly at the expense of its neighbours may or may not be advantageous to the inclusive fitness of its genotype, but it is almost certainly not in the interest of that of the parent plant.

(Hamilton, 1964b)

Up to this point, our attention has been directed toward the genotypic 'battle-ground' for POC and the resolutions one most expects for behavioural phenotypes. Trivers's original characterizations of POC were mostly behavioural (e.g. weaning conflict), yet there is no reason why other fitness-affecting components of phenotype should not be subject to the arguments, in which case some attention should be paid to life-history traits. This chapter addresses some of these issues and completes our treatment of POC theory with a consideration of how sexual conflicts may affect the 'supply side' in taxa with both biparental and uniparental care. The following chapter then deals with some of the interpretation problems that remain.

10.1 Disagreements over clutch size

We have seen (Chapter 4) that the number of sibs in a clutch exerts an important influence on the intensity of sibling competition. Parent and offspring interests need not coincide with regard to clutch size: the resolution of this conflict can therefore exert an effect on sibling rivalry. Indeed, POC over clutch size is evident in O'Connor's (1978) classic model that defined different thresholds (in terms of survival prospects of each offspring) under which it pays a parent and a dominant offspring to reduce clutch size by one (see Section 4.1.1). A conflict can occur when the survivorship difference between clutch sizes of n and $n-1$ lies within certain ranges. The threshold at which an offspring can profit by trimming brood size occurs at a lower survivorship than that for a parent (Figure 4.1), so a parent generally has a higher optimal clutch size than an offspring.

10.1.1 *Clutch-size POC in the hierarchy model*

O'Connor's model gives a dominant sibling its choice between two discrete strategies: to kill a subordinate sib or not to kill it. In the Parker *et al.* (1989; see Section 3.3) hierarchy model of sib competition, selfishness can be graduated continuously. A dominant sib can take as much or as little of the resources as it chooses, so its effect on a subordinate sibling can range from marginal reduction in personal fitness prospects to eventual death by starvation.

To recapitulate briefly, in the hierarchy model (see Figure 3.3) the most dominant chick (A) has first access to parental resources: it takes whatever amount will maximize its inclusive fitness, given that the next most dominant chick (B) makes a similar decision when A has finished feeding. The least dominant chick (C if there are just three brood-mates) has no decision to make; it simply uses up what is left for it. A specific case for this model for four chicks (A, B, C, and D) was shown in Figure 4.2, where the personal fitness of each chick was plotted against k—a measure of the abundance of food resources available for parents. This constant k defines the shape in the explicit form for fitness, $f(m)$, with exponentially diminishing returns (equation 3.2). When k is high, a chick's personal fitness increases quickly towards the asymptotic value (i.e. parental care is cheap because food is easy to supply), and when it is low the reverse applies. As k decreases (Figure 4.2), the personal fitnesses of each chick declines, but that of the most subordinate chick decreases fastest. The D-chick's personal fitness drops rapidly to its death at $k \leq 0.8$, at which point C's personal fitness plummets, but A and B gain. Similarly, C dies when k drops even further (to around 0.45), at which point A gains and B loses.

However, it is, of course, inclusive fitness that is maximized in this model. The inclusive fitness of each chick shows a much greater continuity than does personal fitness (Figure 10.1). In the computation used to generate Figure 10.1, the A-chick experiences a marginal gain as each successive sibling dies. This effect is related to the fact that, in this formulation, a minimal start-up amount of PI (m_{min}) is required in order that the personal fitness of a chick be non-zero (see equation 3.2). This minimal PI is re-allocated to the surviving chicks as each one dies. In reality, this cannot be an ESS; it would pay the A-chick to take more food at a slightly higher threshold of k than Figure 10.1 suggests, so as to cause reallocation of the m_{min}. This effect is lost if m_{min} is set to zero. However, there is a second effect. The inclusive fitness of the remaining next most subordinate chick shows a drop at the death of its sib. This consequence is not sensitive to assumptions about m_{min}, but relates to the fact that more dominant sibs alter their allocations at each death (which affects their personal fitnesses, but has much less impact on their inclusive fitnesses).

To allow a clear comparison with parental fitness, the latter can be depicted as half of the sum of the personal fitnesses of all of the surviving offspring (Figure 10.2). This 'realized' parental fitness is that resulting from the

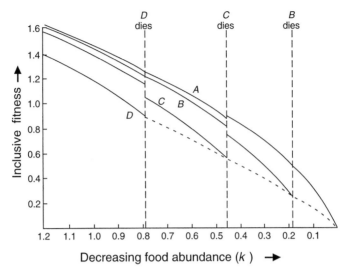

Fig. 10.1 Inclusive fitnesses of chicks in the hierarchy model with four full sibs. The explicit conditions are the same as in Figure 4.2 (see text).

hierarchy allocation with the deaths of chicks ordered (as in Figures 4.2 and 10.1). The 'ideal' parental fitness is that which the parent could maximally achieve were all of the chicks to take an equal share of the food input, without any brood reduction. Perhaps the most surprising feature of this logic (see Figure 10.2) is that the ideal parental fitness is not very much less than the realized fitness, despite the selfishness and offspring-controlled brood reduction in this model. If m_{min} is small, the realized parental fitness drops almost continuously, and it does not pay the parent to promote brood reduction. Thus, even if we assume that the parent has no control over the chick hierarchy, it should be unconcerned about the death of each successive chick. Indifference to brood reduction appears to be commonplace, but not universal, among parent birds (see Section 11.1).

Parents also often appear to be indifferent to overt aggression between siblings. One possibility is that they simply cannot prevent it, because offspring are often left unattended for long periods. There are also adaptive interpretations. The costs to the parent of suppressing the hierarchical food allocation to ensure equal food for all chicks may be significant. If these 'enforcement' expenditures outweigh the difference between ideal and realized parental fitness, parental indifference would be favoured by selection.

10.1.2 *Conflict over clutch size in begging scrambles*

Godfray and Parker (1992) examined the extent of POC over clutch size in begging scrambles. (The details of their model were given in Section 3.3.5,

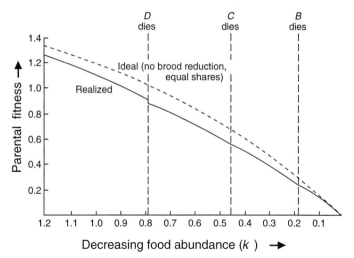

Fig. 10.2 Parental fitness in the hierarchy model with four full sibs. The explicit conditions are the same as in Figure 4.2 (see text).

which examined the ESS begging levels at different clutch sizes.) In this framework, all chicks are assumed to be equal, and food is allocated by 'mean-matching' (i.e. each offspring receives a share proportional to its begging costs relative to the mean begging costs for the entire brood, including the focal individual). The costs of begging can be 'summed', 'shared', or 'individual', as outlined in Section 3.3.4. ESS begging level varies with brood size (Figure 3.10) and, because begging is costly, this exerts an effect on the optimal clutch size for the parent to produce. The parent can retaliate by laying a different clutch size. Godfray and Parker point out that in species with no parental care, in which offspring are abandoned after oviposition, clutch adjustment may be the only possible form of parental retaliation for dealing with the high costs of sibling rivalry.

They first solved for an ideal optimal clutch size for the parent, under the assumption that offspring do not compete. This is the 'Lack clutch size' (Lack 1947), which produces the brood size that maximizes the productivity of the current reproductive venture. Parental fitness is simply the product of the personal fitness of each offspring, f, and the number of offspring in the clutch, n. So, if the parent has a total of M units of resource available for distribution to the brood, each offspring obtains $m = M/n$ units. Hence the Lack clutch size is the value of n that maximizes parental fitness

$$W_p = n \cdot f(m). \tag{10.1}$$

Differentiation with respect to n and equating to zero gives

$$f'(m_p) = f(m_p) \frac{n}{M}, \tag{10.2}$$

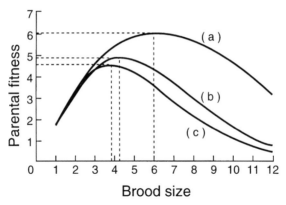

Fig. 10.3 Parental fitness (equation 5.30) as a function of brood size, with full sibs. $f(m)$ follows equation (2.7) with $k = 1.5$, and $S(x) = 1\text{-}bx$ with $b = 0.5$. The explicit model is explained further in Section 2.8.5 (see also Figure 2.13). Curve (a), no competition between offspring; the parent's fitness is scaled so as to be maximized at a clutch size of $n = 6$. Curve (b), sib competition with individual costs. Curve (c), sib competition with summed or shared costs (the same result applies for each). (After Godfray and Parker 1992.)

which (remembering that $m = M/n$) is the same as Equation (7.4). Thus the Lack clutch size is the number of eggs at which each offspring obtains an amount of resources that obeys the Smith–Fretwell rule (Figure 7.3). Note that we have assumed uniparental care. However, if the offspring *compete* for resources, the optimal clutch size for the parent changes. If $S(x)$ is the survivorship of an offspring begging to level x, parental fitness is now

$$W_p = n \cdot f(m) \cdot S(x^*).\qquad(10.3)$$

where x^* is the ESS begging level for the offspring at a given clutch size, calculated after Godfray and Parker (1992; see Section 3.3.5). This procedure assumes that the offspring play their ESS begging level *facultatively*[1] in relation to the size of the family in which they find themselves.

Some feel for how parental fitness can be affected by offspring selfishness (i.e. ESS begging levels) can be gained with a numerical example (Figure 10.3; from Godfray and Parker 1992). In the absence of sib competition (curve a), the ideal or Lack clutch size for the parent in this example (given the values substituted here) is six offspring. The other curves show parental fitness under the three types of begging cost (see below): curve b is for individual costs and curve c is for summed and shared costs.

As expected, sib conflict always reduces parental fitness, so the maximum in the absence of sib competition is always higher than the maxima where such is present. However, a second effect is also evident. Sibling rivalry favours the

---

[1] That is, after some assessment of current conditions (as opposed to *obligately*).

production of smaller clutches by the mother. Finally, the reduction in the parent's optimal clutch size is less when begging costs are felt just by the offspring begging to level x (individual costs) than when begging costs relate to the sum of all begging for the brood (summed costs) or to the mean begging level for the brood (shared costs). Summed and shared costs give the same optimum (in the specific example considered) because, although summed costs lead to lower begging levels (Figure 3.10a), the fitness penalties of competition are summed for the brood. The benefit and cost to the parent cancel exactly (in this specific example) to give the same parental fitness as when costs are shared (see Godfray and Parker 1992). The optimal clutch size reduction (from the 'ideal' maximum) in this example is of the order of 25–30%.

The reasons for this reduction in the parental optimum clutch size are interesting. Note that reducing clutch size means that each offspring gains more resources (logically, this should hold, although in real life there may be some short-term exceptions, e.g. see Section 6.3.1). As each offspring gains more, the value of each extra unit gained decreases (under the diminishing returns assumption). This means that the potential benefits of sib competition decrease, so the ESS begging level declines. The benefits of this reduction in begging costs to the parent exceed the loss due to the increased resources per offspring, so a reduction in clutch size is favourable to the parent.

This process is analogous to *Mechanism 1* in the pro rata payment model of Parker and Macnair (see Section 8.2.3) in which the brood size remains constant, but where the parent can increase the amount of resources through its own actions. In the present model, the parent cannot increase the total resources available per brood, but it can enlarge the per-caput shares by pruning clutch size. Both solutions result in the parent supplying each offspring with more resources (and hence more PI) than the ideal parental optimum.

Parental retaliation in the form of clutch-size cutbacks can have some interesting consequences, as shown in Figure 10.4. In all three graphs, the horizontal axis depicts the 'observed' clutch size (= the ESS clutch size *after parental adjustment*) and the numbers accompanying the points along each curve reveal the original (i.e. 'ideal', *before parental adjustment*) clutch sizes. First, the ESS begging levels (here referred to as 'Sibling Conflict') are plotted across different observed clutch sizes for our three types of cost (Figure 10.4a). This top panel can be compared directly with Figure 3.10a, which is the same plot except that its x-axis shows the ideal clutch size for the parent in the absence of begging costs, whereas this one shows the observed clutch size. Note that the qualitative changes in begging level with clutch size are very similar in these two graphs, which shows that adding the effects of parental retaliation does not greatly influence the general predictions. Conflict increases with clutch size for individual and shared costs, but at a decelerating rate. With summed costs, conflict declines for clutch sizes greater than two.

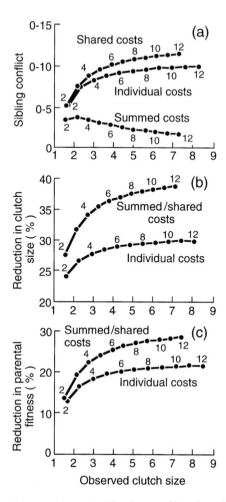

Fig. 10.4 Sib competition and parent–offspring conflict when the parent responds to sib competition by adjusting clutch size to its new optimum. The model has the same parameters as those shown in Figure 10.3 (see also Figure 3.10). The observed clutch size is the clutch size after adjustment, while the original clutch size is shown as the numbered points on the curves. **(a)** Predicted relationship between levels of begging (conflict) and clutch size. Thus, for example, a parent producing a brood of seven competition offspring would have been selected to produce a brood of about ten in the absence of sibling rivalry. **(b)** The percentage reduction in optimal clutch size caused by sib competition. A reduction from ten to seven is a drop of 30%. **(c)** The percentage reduction in parental fitness resulting from sib competition after clutch size adjustment. (After Godfray and Parker 1992.)

However, comparison of the two figures shows that parental retaliation via clutch size cutbacks markedly reduces the absolute magnitude of conflict. Godfray and Parker (1992) conclude that *the observed levels of sib competition may therefore greatly underestimate the importance of sibling rivalry to parental reproductive strategies.* Parental reproductive strategies may greatly modify the observed levels of sibling rivalry, because of previous adjustment of clutch size.

The optimal parental adjustment of clutch size can be substantial, particularly when the ideal clutch size is high, and costs are summed or shared (Figure 10.4b). For instance, for large broods the decrease in clutch size can approach 40% for summed/shared costs, and it can approach 30% for individual costs. Such effects do not appear to have been considered in empirical studies of optimal clutch size in birds. They are clearly able to exert a major influence in reducing the Lack clutch size prediction.

The decrease in parental fitness (Figure 10.4c) is less than the decrease in clutch size because of the savings gained via the lowered begging costs (which is the force promoting the reduction). This fitness decrease is, of course, much milder than the price that would have to be paid if the parents did not reduce family size. For very large clutches, the proportionate reduction in fitness falls from 0.33 to 0.23 for individual costs, and from 0.50 to 0.33 for shared and summed costs. This type of effect has implications for population dynamics, because intrinsic rates of increase are likely to be diminished considerably by these and other similar results of POC (Godfray and Parker 1991).

10.1.3 *Conflict over clutch size by siblicide in large broods*

The previous two sections outlined models designed mainly for small broods, where selfishness between sibs could be graduated. Parker and Mock (1987) analysed a model designed mainly for large broods in which offspring compete by siblicide. It is most relevant to invertebrates (see Chapter 15), in which brood size is typically large and where in many groups it is common for newly-hatched larvae to kill and also often to cannibalize some of their unhatched sibs (reviewed in Fox 1975, Polis 1981, Elgar and Crespi 1992).

Parker and Mock's model is a very simple and general one that does not include any costs of sib competition, such as begging costs. In this it resembles the hierarchy model. Similarly, it also assumes a strict hierarchy of power amongst the offspring. However, the mechanism of power use is very different from the hierarchy model. Certain senior sibs, e.g. the first to hatch within a clutch, can eliminate their weaker nursery-mates, without incurring any cost in terms of personal fitness. For simplicity, it is also assumed that no direct benefit is derived from eliminating siblings (e.g. no significant meal is obtained), so the incentive for killing involves only the liberation of more resources for surviving sibs (reduced competition within the nursery). An outline of this model is given in Box 10.1, which shows the 'battleground' for

Box 10.1   Brood reduction conflict for large broods

Offspring compete for a fixed larval resource (e.g. a host plant, host animal, dung ball, etc.), and the survival and reproductive success of each offspring, f, is a decreasing function of the number, n, of siblings competing for the same resource. For formal comparison with previous models, we might assert that $f(n) = f(m,n)$, where $m = M/n$ and M is the total resource available to the entire brood. However, to be more general, Parker and Mock (1987) did not assume any explicit re-lationship between f and m, and explored the consequences of $f(n)$ for the clutch size optima for parents and offspring (for a more complete approach to parental interests, see Parker and Begon 1986).

The parental optimum is assumed to be the Lack clutch size, i.e. the clutch size that maximizes the product of current offspring number and per capita fitness:

$$W_p = n \cdot f(n). \tag{10.4}$$

Differentiating with respect to n and equating to zero gives

$$-f(n_p) = \frac{f(n_p)}{n_p}, \tag{10.5}$$

in which n_p is the optimal clutch size for the parent. It is evident that these two equations relate to (10.1) and (10.2), but give a solution in terms of $f(n)$ rather than $f(m)$. While $f(m)$ has a positive slope (personal fitness increases with more resources; see Figure 7.3), we expect $f(n)$ to have a negative slope (personal fitness decreases with more sibs; see Figure 10.5). Both sides of eqn (10.5) are therefore positive. This equation represents the familiar 'Lack clutch size' (see Charnov and Skinner 1984; Parker and Courtney 1984; Parker and Begon 1986; Godfray 1987). This is in fact a maximum clutch size that can be optimal for a mother to produce. In invertebrates, this clutch size is approached only when the costs of searching to find an oviposition site are very high (Charnov and Skinner 1984, Parker and Courtney 1984).

Now suppose that the offspring perform some number, n_e, of 'eliminations' (siblicides) when a total of n_m eggs are laid initially by the mother. Because n = the number of offspring that remain to share the resource, $n = n_m - n_e$. We can rework the mother's optimal clutch size by writing equation (10.4) in terms of n_m, given that offspring eliminate n_e sibs. Solving for the optimal n_m gives

$$n_m = -\frac{f(n_p)}{f'(n_p)} + n_e, \tag{10.6}$$

which is the same as eqn (10.5). Note that both (10.5) and (10.6) give the optimal clutch size for the mother *assuming that eggs cost nothing* (we shall later include costs of eggs). On this assumption, the mother will increase n_m so as to end up with a final clutch size of $n_p = n_m - n_e$ eggs; i.e. it should account in full for all eliminations.

What clutch size should the offspring favour? Suppose that the offspring are all full sibs. Let us label the senior sibs 1, 2, 3, etc. We seek a number of junior sibs that each senior should eliminate, so that senior sib 1 should eliminate n_{e1}, and 2 should eliminate n_{e2}, etc. To find n_{e1}, we maximize the inclusive fitness of senior sib 1:

$$W_1 = f(n_m - n_{e1} - n_{e2}\ldots)[1 + \tfrac{1}{2}(n_m - n_{e1} - n_{e2}\ldots-1)],$$

by differentiating with respect to n_{e1} and equating to zero. To find n_{e2}, W_2 has the same value, but we differentiate with respect to n_{e2}, and so on. All these optimizations give the same result, namely that the total number of junior sibs to be killed is

$$n_e = n_{e1} + n_{e2} + n_{e3} \ldots = \frac{f(n_o)}{f'(n_o)} + n_m + 1, \qquad (10.7)$$

which simplifies to

$$-f'(n_o) = \frac{f(n_o)}{n_o + \phi} \qquad (10.8)$$

where n_o represents the optimal clutch size for senior sibs and ϕ (a constant affected by the relatedness of the sibs) is 1.0 (Godfray 1986). Equations (10.5) and (10.8) differ in the denominator terms n and $n + 1$, so that the parental and offspring optima are different. Substituting ¼ for ½ in the inclusive fitness function, and performing the same analysis gives a hypothetical limit for maximum multi-paternity, i.e. where all brood members have the same mother but different fathers. This again gives equation (10.8), except that $\phi = 3$. The result now differs from the parental optimum by replacing its denominator term, n, with $n + 3$.

POC over clutch size. The parental optimum is the Lack clutch size shown in equation 10.5, and the optimum for the senior sibs is equation 10.8, in which ϕ ranges from 1 (full sibs) to 3 (maximum multi-paternity, where all siblings are genetic half-sibs). The fitness f of each offspring is a declining function of n, the number of surviving progeny competing in the same nursery (as in Figure 10.5).

As we asked earlier *how much* the optimal PI should differ between parent and offspring (Section 7.2), we can now ask the same question for their preferred clutch sizes. If $f(n)$ is monotonic decreasing, it is easy to deduce that the optimal clutch size for parents is greater than that for full sibs, which is in turn greater than that for half sibs. More general conclusions are difficult to reach without some explicit form for $f(n)$. Parker and Mock (1987) used the form $f(n) = 1 - an^x$ (after Skinner 1985), where constant a is standardized as n_c^{-x}. The clutch size of n_c is the 'catastrophic' clutch size—at which the entire brood dies—and x is a severity measure for sibling numbers, that is it regulates the shape of $f(n)$ in its decline towards catastrophe (see Figure 10.5). The three clutch size optima (mother, full sibs and half-sibs) can be calculated in relation to increasing x (for details see Parker and Mock 1987) at any given catastrophic clutch size, n_c, for example, if we can contemplate either a modest-sized catastrophic clutch size ($n_c = 10$; Figure 10.6a) or a large one ($n_c = 100$; Figure 10.6b). In addition to increasing as the relatedness between sibs decreases, the relative difference between optima in this model also increases with decreasing x, and with reducing resources, i.e. low n_c.

Biologically, we would expect the greatest conflict in relative clutch size when (i) there is high multi-paternity in sibships, (ii) there is a sharp initial decrease in the personal fitness of each sib with increased clutch size and (iii)

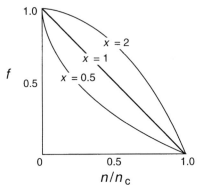

Fig. 10.5 Decline in personal fitness, *f*, of an offspring, in relation to the secondary clutch size (after siblicide), *n*. Clutch size is plotted as n/n_c, where n_c is the 'catastrophic clutch size' (see text). The shape of *f(n)* is determined by *x* (see text). (After Parker and Mock 1987.)

the Lack clutch size is small because few offspring can be supported by the available resources. O'Connor (1978) drew similar conclusions, noting that the loss of one chick has greater relative impact when brood size is small. However, Parker and Mock noted that the magnitude of conflict in clutch size (as measured in absolute numbers) is never large.

How may this conflict be resolved? Although there will be only a small difference in optimal clutch size between parent and offspring, its resolution may nevertheless involve extensive siblicide, particularly if the parental cost of laying extra eggs is small (in which case the mother can afford to over-produce prodigiously). Ultimately, the ESS extent of siblicide may depend entirely on the costs of laying extra eggs and on the costs and benefits of siblicide, not on the magnitude of difference between the clutch size optima.

As a heuristic demonstration, Parker and Mock (1987) gave the following example. The ESS is a Nash equilibrium[2] in which the mother lays $n_m{}^*$ eggs, of which $n_e{}^*$ are killed by the senior sibs. This can be an ESS only if the senior offspring are 'programmed' to kill a given number of sibs—so that they kill $n_e{}^*$ brood members even if the mother varies the number of eggs that she lays. Parker and Mock termed this 'victim-based' siblicide, the idea being that the senior siblings' strategy consists of choosing a set number of *victims*. This contrasts with 'survivor-based' siblicide, in which the senior offspring choose a set number of *survivors*. The distinction is important because a siblicidal ESS cannot exist in the latter case (see Box 10.2).

Some feel for this siblicidal ESS can be gained by plotting the proportion of eggs that are eliminated by senior sibs ($n_e{}^*/n_m{}^*$) against the catastrophic

---

[2] A 'Nash equilibrium' is simply a simultaneous solution for *n* players such that no one can profit by changing his or her strategy unilaterally.

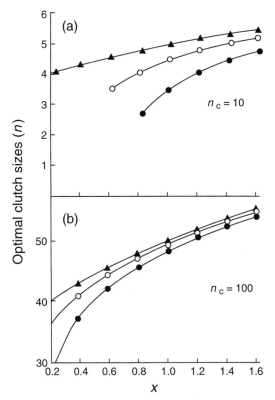

Fig. 10.6 Optimal clutch sizes for mother and offspring in relation to x (▲, mother; ○, full sib offspring; ●, half-sib offspring). When $x > 1$, sib competition in small broods is trivial; when $x = 1$, competition increases linearly with brood size; and when $x < 1$, it is intense, even for tiny broods. **(a)** Optimal clutch sizes for relatively small-brooded species (whose 'catastrophic clutch size', n_c, is only 10). **(b)** Optimal clutch sizes for larger-brooded species (whose 'catastrophic clutch size', n_c, is 100). (After Parker and Mock 1987.)

clutch size, n_c (Figure 10.7). Explicitly, we assumed that $f(n)$ is linear ($x = 1$, Figure 10.5), and that the cost to the mother of laying n_m eggs is $c(n_m) = bn_m^2$, where b is a positive constant that scales the cost of extra eggs (see Parker and Mock for mathematical details). As might be expected, the proportion of the clutch lost by siblicide increases as the degree of relatedness between sibs falls, and as the relative cost of each egg (decreasing with b) declines. More interestingly, the proportion of siblicides should decrease with n_c, concurring with the prediction that the relative conflict is greater in small clutches. Note that the proportion of the initial brood lost by siblicide can take any value between 0 and 1. Brood reduction by senior sibs can be massive or insignificant, depending on the parameters.

Box 10.2 The siblicidal ESS

Consider first the mother's ESS, $n_m{}^*$. Let the costs of laying n_m eggs be $c(n_m)$, which is the expected number of future surviving progeny lost by a female that lays n_m eggs now. Costs $c(n_m)$ are assumed to be monotonic increasing with n_m, so that $c'(n_m)$ will be positive. If the offspring play $n_e{}^*$, one can show that the female's fitness is maximized when

$$-f'(n_p) = \frac{f(n_p) - c'(n_m{}^*)}{n_m{}^* - n_e{}^*}, \tag{10.9}$$

which, as would be expected, yields a lower optimal value for n_p than if eggs are cost-free (cf. equation 10.6).

The ESS value, $n_e{}^*$, for the offspring can be calculated in a similar manner by expanding the inclusive fitness function for a given senior sib to include $rc(n_m{}^*)$, where r is the coefficient of relatedness between present and future siblings. We need not consider this term any more fully, because $rc(n_m{}^*)$ is undifferentiable in n_e. With victim-based siblicide, a change in n_e does not immediately affect n_m, and vice versa. Thus when we are solving the value of $n_e{}^*$ by differentiating the offspring's fitness function $rc(n_m{}^*)$ is lost and, as before, we obtain equation (10.8).

Now if this (equation 10.8) applies, the final clutch size is the offspring's optimum, n_o. The offspring must win. However, at the ESS, **the parent will nevertheless be laying more eggs than n_o** (for reasons to be explained shortly). How much siblicide will there be? The answer obviously depends on how many eggs *more* than n_o are laid by the female. Equations (10.8) and (10.9) must be satisfied simultaneously. Remembering that $n_o = n_m - n_e$, this gives:

$$-\phi f'(n_o) = c'(n_m{}^*). \tag{10.10}$$

Thus equation (10.8) tells us the value of n_o, and then equation (10.10) tells us the values of $n_m{}^*$ and $n_e{}^*$ (since n_e equals $n_m - n_o$). It follows that equation (10.10) can be satisfied only if one or both functions, $f(n)$ and $c(n_m)$, are non-linear. Furthermore, a siblicidal ESS demands that $c(n_m)$ has an increasing gradient, otherwise (implausibly) the mother lays no eggs or she lays all of her eggs at a single oviposition. If $c(n_m)$ is linear, $n_m{}^*$ and $n_e{}^*$ are undefined, even if $f(n)$ is non-linear and allows definition of n_o.

However, returning to the obvious, why should a mother lay more eggs than the offspring's optimum if these are doomed from the start? She is wasting n_e eggs at each oviposition, which have a cost both to herself and to her present progeny (albeit less so). The reason relates to the concept of victim-based siblicide, in which offspring are genetically 'programmed' to kill *a given number* of sibs. Remember that the female's optimal clutch size is greater than that of her offspring. At the ESS, it obviously cannot pay a mutant female to reduce her egg number, because her siblicidal offspring are programmed to reduce the clutch by $n_e{}^*$, regardless of her first move. Accordingly, a unilateral reduction by a mutant female would merely carry her eventual clutch size further from her optimum.

Suppose that the game begins with the mother laying a clutch size around the offspring's optimum. Would it be evolutionarily stable if the mother

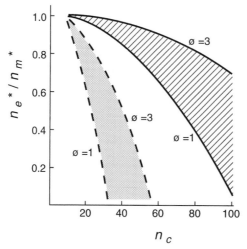

Fig. 10.7 ESS proportion of primary clutch size that will be lost to siblicide (n_e^*/n_m^*) in relation to the 'catastrophic clutch size' (n_c) at which all offspring die from sib competition (see Figure 10.6). These are expressed for single paternity ($\phi = 1$) and maximum multi-paternity ($\phi = 3$) conditions for two different maternal costs of laying extra eggs (b). The shading highlights the range of values between the boundary paternity conditions when eggs are either relatively cheap ($b = 10^{-4}$, solid line and cross-hatched area) or expensive ($b = 10^{-3}$, broken line and stippled area). (After Parker and Mock 1987.)

simply allowed the offspring to win, i.e. laying $n_m = n_o$ eggs? In such a population, there would be no siblicide, because the offspring's best strategy would then stabilize at $n_e = 0$. However, this cannot be an ESS because a mutant female laying more eggs would soon profit. Initially, there would be no siblicide, but this state would be ephemeral from the perspective of evolutionary time. A parental mutation for enlarging clutch size becomes profitable and the trait moves unobstructed toward the parental optimum, at least initially. As it does so, however, new mutations for small amounts of offspring siblicide only serve to push the clutch size higher; that is, parents respond to some egg losses by compensating. Thus, with victim-based genetic programming, large amounts of siblicide are possible (and such can be found in many insects; see Chapter 11), even though the 'evolutionary disagreement' over clutch size may be relatively trivial.

However, quite the reverse prediction is made for *survivor-based siblicide* (the strategy for choosing a given number of surviving sibs and eliminating any number in excess of this). Consider again the case where parents simply allow the offspring to win, i.e. females lay $n_m = n_o$ eggs. Although there is no evident siblicide, any mother that deviates by laying more eggs cannot profit from the act because her offspring are programmed to reduce the clutch back to n_o. Indeed, such maternal mutations are lost by selection because of the cost of the squandered eggs. Such a state is now an ESS, provided that the offspring's strategy remains constant as 'reduce clutch size to n_o'. This poses a technical problem. In the ESS population, there would be no siblicide, so the offspring's strategy is no longer moulded by selection; it can be subject to drift because a strategy for non-siblicide fares equally well. In a sense this is only a problem if we consider natural selection to be perfect. In a real population, deviations above n_o by mothers are likely to occur for a variety of reasons, so that the survivor-based siblicide strategy is likely to be challenged and, as a result, maintained. This example shows yet again how the underlying mechanism exerts a critical influence over resolution of an intra-familial conflict.

10.1.4 *Extreme siblicide under clutch-size conflict*

In some insects, such as parasitic wasps (see Chapter 15), and also in sharks (see Chapter 14), there is wholesale slaughter (without cannibalism) among larvae that are exploiting the same resource, so that only one survives. A very simple strategy appears to operate in such cases, which might be summarized as 'attempt to kill until all others have been destroyed'. Typically the deaths are siblicidal. Godfray (1987a) analysed the fate of a mutant 'fighting' allele that causes its bearer to kill all of its sibs. Siblicidal activity mows through the brood/clutch until only one larva remains; it is assumed that all bearers of the fighting allele have equal chances of being this sole survivor. Using a negative exponential form for the fitness function, $f(n)$, Godfray found that the fighting gene spreads in populations where females lay Lack clutch sizes of less than four. This fits the prediction in Section 10.1.3 that we expect the offspring clutch size optimum to be *most* reduced below that for the parent at low clutch sizes. Furthermore, the model in Section 10.1.3 indicates that, using a negative exponential $f(n)$, the disagreement between clutch size optima should be between 1 and 3, depending on the relatedness between sibs (Parker and Mock 1987). A fighting rule where offspring reduce clutch size to one can apply up to Lack clutch sizes of four.

As might be expected, Godfray showed that the gene spreads more easily if the mean relatedness between competing larvae is reduced. Hence super-parasitism (in which two unrelated females oviposit in the same host) increases the prospects for invasion. (Smith and Lessells (1985) also examined models for fighting between unrelated insect larvae competing within a food resource.) Once fixed, the population is characterized by solitary fighting larvae, which appears to be a locally absorbing state. The conditions for invasion of the fighting allele in a tolerant population are much easier to satisfy than the reverse—for invasion of a tolerant gene into a fighting population. In order for a dominant *tolerance allele* to invade a fighting population, it is necessary that the fitness of each larva increases, as clutch size shifts from one to two. Because per capita fitness must eventually decline with clutch size, the *tolerance allele* can invade only if there is a peak in fitness as a function of clutch size—i.e. it requires an 'Allee effect'[3] (Allee *et al.* 1949). However, for a dominant *fighting allele* to invade a population fixed for tolerance, an Allee effect is not necessary; the gene spreads whenever the clutch size is lower than a critical number, suggesting that small clutch sizes (e.g. two or three eggs) will be unstable. These conditions are slightly altered for invasion by recessive alleles, for unequal sex ratios, for incomplete penetrance, and when superparasitism is rare. The assumption of cost-free fighting also affects the analysis: there are plausible costs of the fighting armament, and in addition there is the probability that both combatants may be killed in the fight (e.g. Salt 1961). The spread of an allele for fighting is retarded when the parental optimum clutch size is below the 'Lack' solution (which maximizes clutch productivity). Once they have evolved, fighting mandibles are difficult to lose because any non-siblicidal mutant allele is at a great disadvantage.

Godfray's models very plausibly explain much of the mandible morphology and fighting behaviour in parasitic wasps. For example, a dichotomy rather than a continuum is found in parasitoid clutch sizes: species tend either to lay single eggs (which give rise to fighting larvae) or to produce larger clutches that develop as non-fighting larvae (Le Masurier 1987). A wasp species with fighting larvae typically rears one larva from a particular host, whereas a species of similar size but with non-fighting larvae may be able to rear up to 15 offspring from the same host.

Godfray's fighting gene rule, 'fight until all others have been destroyed', is a survivor-based siblicide rule for culling the clutch size down to one remaining survivor. Hence a female parent would profit by laying just one egg, and although this is common, some parasitoids with fighting larvae do indeed lay more than one egg. Why? One argument relates to superparasitism, which is not uncommon. Laying more than one egg increases the chance that one of Self's offspring will be the eventual survivor. This is a variant of the *insurance hypothesis* for parental over-production (see Chapter 1), in that extra progeny are made to maximize the chance that a smaller number survive various perils during a high-risk period. The basic insurance argument also applies, as there is always a finite chance that any given egg will fail due to disease, damage or externally caused developmental failures. (We note that protecting against infertility could not be an incentive for building 'marginal offspring' in parasitoids, all of which are hymenoptera, because unfertilized eggs become males in haplodiploids.) If developmental failures owing to genetic (intrinsic) factors are a problem, the *progeny choice* (= 'developmental selection') mechanism could also be important, and of course if sibling victims are subsequently consumed and if that nutrition is nontrivial, the *sibling facilitation* arguments (especially the '*icebox* hypothesis'; see Chapter 1) may

---

[3] An Allee effect occurs when per capita fitness first increases, then decreases with numbers of individuals (here clutch size).

also apply. Given the context, however, it seems that the most likely cause is superparasitism, and it would be interesting to know whether species with fighting larvae that lay more than one egg are more prone to superparasitism.

In several species of eagles and seabirds (see Chapter 12) the female lays two eggs, and the stronger hatchling almost invariably kills its weaker sib. Even if this were 'victim-based siblicide', the surviving offspring does not then kill itself, so if a mother were to lay just one egg she would save on egg costs. Infertility or other egg mishaps that imperil a given egg seem here to be the most plausible explanation for such modest over-production. This can apply only if the cost of additional eggs is low. Eagles which display obligate siblicide appear to have significantly smaller eggs relative to female mass than do eagles with facultative (i.e. dependent on conditions) siblicide or those that lay a clutch of just one egg (Edwards and Collopy 1983).

Finally, clutch-size POC can sometimes arise from (or exert an influence on) other forms of POC. Roitberg and Mangel (1993) analysed conflicts between a mother and her offspring over the issue of offspring emigration from the nursery. The models are devised mainly for insects in which a clutch is oviposited on a host plant and the ESS optimum for the mother involves greater offspring emigration than does the ESS optimum for the offspring themselves. They discuss how this disagreement relates to how many eggs should be laid in the first place.

10.1.5 *POC over clutch size when parents remain with the brood*

Rodríguez-Gironés (in press) has pointed out that where parents remain in close contact throughout the nesting cycle, as occurs with many birds, the use of siblicide to resolve POC over clutch size is complicated by two factors. First, whether a siblicidal strategy will be favoured depends on the parental response to reduced family size, and specifically on whether parental investment is readjusted. Secondly, optimal parental investment will depend on offspring reaction to the reduction. It turns out that the cost parents must pay to provide enough resources to prevent siblicide (e.g. in a hunger-sensitive system) may be smaller than the benefit they obtain from maintaining their preferred brood size. Rodríguez-Gironés (in press) sees siblicide in such circumstances as a form of evolutionary blackmail[4] because the risk that offspring will practise siblicide can be sufficient to cause parents to increase their provisioning levels. He argues that the level of sibling aggression may become a signal to which parents respond in terms of parental investment.

---

[4] The notion of blackmail forms of POC has been around, albeit informally, for at least 20 years, and Amotz Zahavi has been credited as the originator (Dawkins 1976:131).

10.2 Other theoretical issues in parent–offspring conflict

Before we go on to discuss how the imperfections of biparental care impinge on sibling rivalry and POC, we shall digress briefly on a few interesting points of theory that connect to empirical material reviewed in later chapters.

10.2.1 *Genomic imprinting and parent–offspring conflict*

The POC models outlined in this chapter have all been based on a classical assumption about inheritance, namely that a given gene has the same action whether it is inherited from a father or from a mother. Thus, in calculating the PI an offspring should extract from its mother, it is assumed that the optimal strategy would be the mean of the effect when carried by each sex (e.g. Parker and Macnair 1978). As an example, consider a species with inter-brood conflict, where only the mother cares for the offspring, and she mates randomly before each brood is produced. If a female carries a rare mutant allele, each of her offspring has a probability of ½ of carrying a copy of it. Thus for maternally transmitted genes, a selfish action by a given offspring has a probability of ½ of affecting the same gene in future offspring. If a male carries such a rare mutant allele, his mutant-bearing progeny cannot affect future copies because their mother is unlikely ever to mate again with a male bearing the same mutation. The relevant kin selection coefficient then becomes the mean of ½ and 0, i.e. ¼.

However, suppose that a gene were able to have its expression be conditional on the sex of the parent that transmits it. There is evidence that such phenomena do indeed occur; alleles that show differential gene expression depending on the parent of origin are said to show 'genomic imprinting' (Haig and Westoby 1989a; Haig 1992a, b). Whenever there are interactions between half-sibs (rather than full sibs), there is potential for conflict between maternal and paternal genes in the offspring. Thus, in the above example, in a zygote a mutant gene from a female parent would be selected to show restraint in its action because of kin selection, but no such restraint should operate on a gene from a male parent. Indeed, Haig and Westoby (1989a) proposed that the phenomenon of genomic imprinting has evolved in mammals for this very reason. An embryo's paternal genome is selected to take more resources from maternal tissues than is the offspring's maternal genome (see also Haig and Graham 1991, Moore and Haig 1991, Haig 1992a). Remarkably, in eutherian mammals the paternal genome appears to be particularly active in the development of the extra-embryonic membranes (for examples, see Sections 13.3.1 and 16.2.1).

Haig (1992a) has extended this idea, and points out that equation (7.10) with $r = $ ½ gives the offspring's optimum for a maternally derived gene with half-sibs (and, of course, the general ESS for true monogamy). In Haig's model, n males share the paternity of a set of offspring which compete over maternal PI. It could be likened to an inter-brood conflict in which the female mates with several males at the start of her reproductive life, and then produces offspring sequentially throughout her lifetime (e.g. as in honeybees). For a paternally derived gene, the offspring's optimum is

$$f'(m_o) = \frac{1}{2n}\left[\frac{f(m_o)}{m_o}\right], \qquad (10.11)$$

which converges to the offspring optimum (equation 7.10) if offspring are full sibs ($n = 1$), but if n is very large the optimum is to take a maximum amount of resource. Note that as n tends towards infinity, all interactions are between successive offspring, each of which has a different father.

In the absence of genomic imprinting, Haig's model gives the offspring optimum

$$f'(m_o) = \tfrac{1}{4}\left[\frac{n+1}{n}\right]\left[\frac{f(m_o)}{m_o}\right], \qquad (10.12)$$

which approaches Macnair and Parker's (1978) inter-brood conflict optimum for half-sibs (equation 7.10 with $r = \tfrac{1}{4}$) as n increases. (In the latter model, the female mates with a different male before each brood, so that offspring in successive broods *never* share the same paternal genes.) An interesting exception, which represents an extension of Charnov's (1982) approach, is a model analysed by Lundberg and Smith (1994). Applying a traditional life history approach in a population genetic model, they derived conclusions similar to those of other authors about the battleground for conflict. However, they do not see POC as a conflict, but rather as a trade-off between the benefits of acting 'assertively' while in the role of offspring vs. the costs of acting 'assertively' while in the role of parent. At the latter stage a focal allele (a rare dominant gene in a sexually-reproducing diploid population) faces half of its own offspring acting 'assertively' against Self.

Haig and Westoby's (1989a) idea that genomic imprinting has evolved because of POC is an attractive one. Alternative hypotheses for genomic imprinting consist of rather vague claims about greater flexibility of development and improving the accuracy of control of gene dosage. These theories do not explain why the same effects could not be achieved by some other means, without resorting to functional haploidy (but see Holliday 1990). Genomic imprinting has also been claimed to be a mechanism to prevent parthenogenesis. However, as Haig (1992a) points out, imprinting occurs in plants, but does not prevent apomixis.

10.2.2 The currency issue

Virtually all of the PI models in this book are of the 'cake' variety. A parent has a fixed amount of PI (the cake), which it must divide up into portions (the slices) for allocation to each offspring. Once its cake has been consumed, the parent dies. The cake principle has the required property that the larger the slice per offspring, the fewer the number of offspring that can be served. One version of this logic views the cake as a fixed reproductive effort budget (possibly even a fixed energetic expenditure) that a parent is able to spend. For semelparity, this interpretation seems a very good analogy, although it becomes less realistic for iteroparous species. Other rearrangements of the logic give the same result. For instance, the cake principle applies (directly, for a continuous breeder) if the slices are temporal periods allocated to each successive brood, and the cake is the expected adult lifespan. Alternatively, we can substitute 'cumulative energetic expenditure' for 'time'; because the total lifespan expectation is fixed, there is a total expected expenditure. As we have seen (in Section 7.2.1), the introduction of a conversion function between resources supplied to offspring and PI costs, as suggested by Evans (1990), need not have much effect on conclusions derived using the cake principle (see also the discussion of currencies in Forbes 1993).

The simplicity and mathematical tractability of this type of approach have made it very popular (e.g. Smith and Fretwell 1974, Brockelman 1975, Pianka 1976, Parker and Macnair 1978, 1979, Macnair and Parker 1978, 1979, Alexander 1982, Parker and Begon 1986, Temme 1986, Lloyd 1987), although more realistic extensions of the basic Smith–Fretwell model are now available that are set in a more conventional life-history theory framework (Winkler 1987, Winkler and Wallin 1987, Forbes 1993). Charnov (1982) has interpreted POC in this setting, as follows. Let P be the mother's yearly survival and S the survival of the offspring from conception to the end of the first year, when it becomes reproductive. There will be an inverse relationship between adult and offspring survival (Figure 10.8). For simplicity, assume that the primary sex ratio is 50:50 and that one offspring is produced per year after random

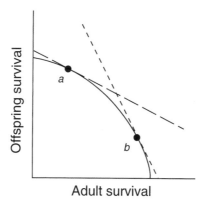

Fig. 10.8 Charnov's model of parent–offspring conflict over reproductive effort. A single offspring is born each year and its survivorship to breed is related to its mother's yearly survivorship. The mother's equilibrium is point *a*, the offspring's equilibrium is point *b* (see text). (After Charnov 1982.)

mating. Maternal fitness is maximized by maximizing $W_p = P + S/2$ (point *b* on Figure 10.8), at which the tangent line has a slope of -2. Offspring fitness is maximized by a point (*a* on Figure 10.8) at which the tangent line has a slope of $-\frac{1}{2}$. Charnov (1982) points out that life-history theory assumes that the maximization is always by the mother, something which need not be correct (but see Chapter 11 for difficulties in documenting these 'resolution' issues). An excellent review of parental investment, what it is and what its consequences are, is given by Clutton-Brock (1991).

10.2.3 Unequal PI to different offspring

The notion that a parent should invest equally in all offspring is correct only if all offspring are identical, and when other circumstances (e.g. resources available to parents) remain constant. Unequal investment by a parent can be favoured when these conditions are not fulfilled.

Temme (1986) considered the case in which offspring differ in quality, and argued that the optimal solution is then for the parent to adjust its investment so that the marginal gains from additional investment in each offspring are equal for all offspring. Furthermore, the marginal gain will be equal to the average return from all offspring (Haig 1990). Formally, when A, B, C, ... etc. are the offspring types, the optimal allocations, m_A^*, m_B^*, m_C^* ... etc. have

$$f'_A(m_A^*) = f'_B(m_B^*) = f'_C(m_C^*)... = G, \qquad (10.13)$$

where G is the average gain rate from PI. This is formally equivalent to the marginal value theorem in optimal foraging (Charnov 1976, see also Parker and Stuart 1976, Lloyd 1988). The marginal gains ($=$ slopes $=$ first derivatives) must all be equal. Haig (1990) extended this principle to consider selective brood reduction by a parent, as occurs in spontaneous abortion (or atrysia of oocytes) in both plants and animals.

Parents might commonly be expected to invest differently according to the sex of an offspring. This was first proposed by Wilson and Pianka (1963) and their prediction has subsequently been extended by various authors (summarized by Clutton-Brock 1991). Trivers and Willard (1973) predicted that mothers in good condition (and hence able to make greater investment at lower immediate cost) should produce and/or invest in the offspring sex with the greater variance in reproductive prospects. Imagine a polygynous ungulate, in which there is intense male–male competition for females, and where larger males tend to win. If a mother is in good condition and can invest heavily in her son, as an adult he is likely to win against other males and thus achieve a high level of reproductive success. His mother will therefore have profited from investing preferentially in him (as opposed to a typical daughter). Conversely, if the mother is in poor condition, her fitness will be better served by investing in a daughter (because a son raised under such circumstances would be unlikely to prevail in adult contests).

With similar fitness functions for the two sexes ($f_M(m)$ for male offspring = $f_F(m)$ for female offspring) the sex ratio at the end of parental care should be unity ('Fisher's principle', Fisher 1930) unless special conditions apply (for a review see Clutton-Brock 1991). However, if the fitness functions of offspring sexes differ ($f_M(m) \neq f_F(m)$), the ESS solution involves simultaneous adjustment of both the primary sex ratio and the investments, m_M* to male offspring and m_F* to female offspring (Macnair 1978). If the primary sex ratio is for some reason immutable (and there is evidence that for several species it is), the optimal strategy consists of a solution for just m_M* and m_F*, at which we should expect to find, following equation (10.13), that $f'_M(m_M*) = f'_F(m_F*)$. A formal treatment of this problem is given by Maynard Smith (1980), who constrained the sex ratio at 1:1 and showed that it is an ESS for a female to allocate more care to the sex that benefits most from higher investment under frequency-dependent selection of the type just discussed for our hypothetical ungulate. There is indeed evidence for this form of sex-biased PI pattern in mammals (reviewed by Clutton-Brock 1991; see also Chapter 13 and Stamps 1990) and this may have consequences for POC (Redondo *et al.* 1992; see Section 10.4).

The Trivers–Willard hypothesis is interesting in that the differential investment comes about because of variation in both offspring quality (gender) and parent quality (condition). Parental variation alone may suffice to generate unequal optimal investments across offspring. For example, Parker and Begon (1986) examined optimality models of egg size and clutch size from the mother's perspective, mainly devised for animals (e.g. many insects) that show no parental care. They analysed three components to an offspring's fitness. The first is the $f(m)$ effect, a measure of how an offspring's fitness is increased by the amount of maternal resources, m, it receives in its egg composition. The second is a clutch-size effect, a measure of how an individual offspring's fitness is decreased by the presence of other offspring in the same nursery (a density effect). Thirdly, there may be a 'hierarchy' effect, such that an offspring in the same nursery achieves a competitive advantage (e.g. due to its greater size). The first two effects can generate a complex interaction between female size and egg/clutch size for multi-clutch species. However, in single-clutch (semelparous) species, both egg and clutch size should increase with the amount of resources a mother has available for conversion into offspring (broadly related to her own size). The hierarchy effect increases overall egg size and can generate the intuitively paradoxical phenomenon of smaller females laying larger eggs than bigger females. Parker and Begon (1986) also considered the effects of various environmental conditions (the intensities of sib and non-sib competition in the nursery, the number of egg-laying females, and various aspects of seasonal development) and showed that these may all influence the mother's optimal investment strategy. Furthermore, Begon and Parker (1986) showed that for species with a fixed limit of non-renewable resources for conversion into offspring, the optimal investment per egg (egg size) should decrease with the mother's age, a trend shown in many insect species that do not feed as adults. The same effect also applies to clutch size (see also Parker and Courtney 1984).

In some instances, a parent may act to redress offspring asymmetries by differential feeding. Weaker offspring often beg more intensively than larger ones (see review of Stamps 1990), presumably at greater costs (see Sections 3.3.5 and 9.1), and selection may be expected to favour parental abilities to give special help to weaker progeny (see Section 9.1). In budgerigars (*Melopsittacus undulatus*) there is a pronounced size variation in nestlings due to hatching asynchrony. Remarkably, although the male feeds in relation to begging demand, the female feeds smaller offspring preferentially—several times more frequently than she feeds larger offspring (Stamps *et al.* 1985).

10.3 Sexual conflict

Sexual conflict is the third component in the intra-familial struggle (Figure 3.1). Much has been written about the implications of sexual conflict over mate desertion and PI, particularly in relation to the evolution of PI patterns and mating systems (Trivers 1972, Maynard Smith 1977, Grafen and Sibly 1978, Lazarus 1989). We are concerned here with the way in which sexual conflict can affect the total supply of resources to offspring, since this affects sibling rivalry (e.g. see Parker 1985, Queller 1994).

10.3.1 *Sexual conflict under shared investment*

We have seen that sexual conflict can occur between parents over how much PI each should give to the progeny and that resolution of this conflict can affect POC and sib competition (see Chapter 8). In addition, we outlined a simple PI conflict model in which the interests of each parent are essentially symmetrical and in which offspring fitness increases with diminishing returns with total PI (Parker 1985; see Section 7.2.1). To summarize the results of that model very briefly, *higher total PI* per offspring is expected if there is either a single parent investing with one mating per lifetime, or if there is 'true monogamy' (pairing for life with no prospects of 're-mating', such that each parent's fitness is constrained by the death of the other). With true monogamy and biparental care, the number of offspring is theoretically doubled. In contrast, *lower total PI* per offspring is expected if there is biparental care, but without true monogamy (e.g. pairing occurs for each brood, or monogamy lasts only until the death of one parent, or re-pairing is contingent on the partnership's previous success, etc.). Most biparental care in the real world is of this type (see reviews in Gowaty and Mock 1985, Mock and Fujioka 1990, Clutton-Brock 1991, Black *et al.* 1996).

There is therefore a paradox which is due to sexual conflict. Without true monogamy, simple models suggest that there is *less* PI per offspring if two parents supply PI simultaneously than if only one does. It never pays two parents to supply more PI than the optimum for the one-parent case. (With true monogamy, the optimum is the same, so there is no sexual conflict, but without true monogamy, a given parent can gain by unilateral reduction in PI.) The main effect of biparental care would seem to be to increase the number of progeny rather than the personal fitness of each one (see Section 7.2.1).

Rather few studies of relative parental investment actually take into account the above problem, and we know of no field tests of the predictions.

10.3.2 'Parental buffering' and unequal PI from each parent

The above model (Parker 1985) for the evolution of biparental care involves the simplistic, but critical, assumption of symmetry, i.e. that the cost/benefit implications of PI are similar for the two sexes (see also Winkler 1987). At the ESS, both parents adopt the same PI strategy. If one parent reduces its PI unilaterally, the deficit is not made good by the other parent, which continues to supply the ESS level.

Suppose that the deficit due to a mutant reduction in PI by one parent were to be met in full by the other parent, perhaps due to a compensatory increase in begging by the offspring (or via some alternative mechanism). There is complete 'parental buffering' in that the PI imbalance is fully redressed by the more solicitous parent. Obviously, biparental care could never be an ESS under such circumstances (unless there is true monogamy); a parent that defects without investment can always profit by 'using up' the PI of successive mates, without any loss of progeny fitness. For such an ESS to exist, there must be either no buffering or only partial buffering (Parker 1985). This conclusion does not depend on the assumption of symmetry; it also applies if pay-offs through PI are asymmetrical, and if PI strategies are conditional upon the sex of the parent. An experiment which demonstrates that buffering is only partial has been performed on European starlings (*Sturnus vulgaris*) by Wright and Cuthill (1989). By adding weights to the tail of one parent, its feeding rate could be reduced. It was found that the unweighted partner (male or female) compensated by increasing its feeding rate, although this buffering was only partial. In other studies, however, zero buffering (no change in investment in response to a mate's reduction) or near full compensation has been found (for reviews see Clutton-Brock 1991, Mock *et al.* 1996).

The models of Chase (1980) and Houston and Davies (1985) allow pay-offs and strategies to be independent for each parent (see also Grafen and Sibly 1978). These authors seek a Nash equilibrium pair of strategies for the PI or reproductive effort to be expended by each sex, E_m^* for the male and E_f^* for the female. The plot of the best strategy of 'initial parent,' i (its strategy of expenditure, E_i) against any given strategy (E_o) of the 'other parent,' o, is based on our expectation for these two strategies to have a negative

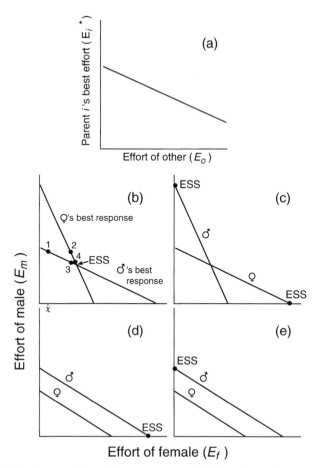

Fig. 10.9 ESSs in sexual conflict over parental effort from the 'reaction curve' models of Chase (1980) and Houston and Davies (1985). **(a)** The reaction curve: the ESS effort for parent i, E_i^*, in relation to the effort of the other parent, o. **(b)** A biparental ESS, given by the values: E_m^*, E_f^*, at the point of intersection of the two reaction curves. Note that the female's reaction curve has axes reversed so that the two curves can be plotted on the same graph. Now suppose that the female plays effort x. The male's best response is indicated by '1' on his reaction curve. The female's best response is then '2' on her curve, which leads to the male's '3', the female's '4', and so on, until both players converge at the intersection. **(c)** Reaction curves intersect, but there is no stable biparental ESS. Unlike **(b)**, the best responses diverge from the point of intersection and the ESS can be either all-male or all- female -care, depending on the starting conditions. **(d)** and **(e)**: Non-intersection reaction curves. The ESS is determined by the parent with the higher reaction curve. (After Houston and Davies 1985.)

relationship to each other, and for $E_i(E_o)$ to be decreasing (Figure 10.9a). This is because as *o* increases expenditure, it pays *i* to decrease expenditure. This plot has been termed the 'reaction curve' by Chase (1980). An ESS can occur in which both parents expend intermediate amounts (i.e. neither minimum nor maximum limits), provided that the slopes of both reaction curves are less than -1 at the region of intersection, which then defines the ESS, E_m^* and E_f^*. Under this condition (E_m' (E_f) < -1; $E_f'(E_m) < -1$), each parent responds to reductions in its mate's effort by *partial buffering* (i.e. by compensating to an incomplete degree). If one parent reduces PI, the other increases PI but by a *smaller* amount (as a result, if parents make alternating investment 'bids', strategies converge by ever decreasing amounts towards the point of intersection; Figure 10.9b).

With either complete compensation, or 'over-compensation' (when the slope of one or both reaction curves is equal to or greater than -1), the ESS can never involve biparental care. It must be uniparental care. If the reaction curves intersect (Figure 10.9c; note that the reaction curves for male and female parents are here reversed compared to Figure 10.9b), the intersection point is unstable. If one parent reduces expenditure, the response of the other is to increase its PI by a *larger* amount, and strategies diverge by ever increasing amounts from the point of intersection. There are two possible ESSs, either the male or the female parent may supply all of the PI; the ESS attained depends on the starting conditions. If the reaction curves do not intersect, the ESS is again for uniparental care (Figure 10.9d and e).

Houston and Davies (1985) extended this type of analysis to cover the case of three-way collaboration over feeding (two males assisting one female) of the type sometimes found in dunnocks (*Prunella modularis*). The total effort supplied sometimes increased and sometimes decreased as a result of the addition of the care by the second male.

The above discussion has focused on cases where an increase in PI by one partner should be met with a reduction in the PI of the other, i.e. where there is sexual conflict. This will apply if the fitness of the offspring shows diminishing returns with PI, as has been assumed throughout all our *f(m)* curves in this chapter and in Chapter 2. However, if *f(m)* were to show increasing returns with PI, co-operation could prevail because an increase in PI by one parent could then favour an increase in PI by the other. This seems biologically unlikely, however.

10.3.3 *Sexual conflict and hatching asynchrony*

Hatching asynchrony is one of the most pervasive causes of competitive differences and dominance hierarchies amongst nestling birds (see Chapter 11), and a potential sexual conflict over hatch timing has been proposed. Slagsvold and Lifjeld (1989) argued that asynchrony may be favoured by mothers when this would be against male interests. Because the female lays the eggs and performs all of the incubation in many species, she might be expected to have considerable control over the critical matter of when to start incubating. Specifically, if the female begins incubation before the clutch is complete, the offspring tend to hatch asynchronously (because the earlier-laid eggs commence embryogenesis before the others are deposited).

However, asynchrony often means that the mother must reduce her self-feeding effort (in order to incubate) at a time when egg metabolism makes high energetic demands. Hence the costs may be high to the female, and some positive advantage to her must be sought. One possibility is that she benefits because the male is manipulated into making a greater food provision effort. Slagsvold and Lifjeld (1989) proposed that the food-providing capacity of males can thus be extended over a greater period, causing the male to take on a larger share of the provisioning than would occur under conditions of synchronous hatching (Figure 10.10). In many species, the male also delivers food directly to the female during the brooding period, so that the female may be better able to control division of the food supply between herself and the

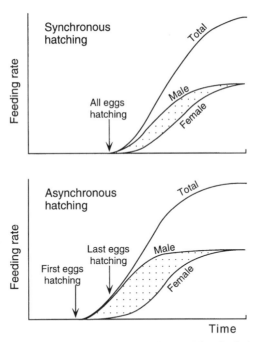

Fig. 10.10 Sexual conflict over hatching asynchrony. The shaded area represents the cumulative difference in food supplied by the two sexes. With asynchronous hatching, the male's share of the total feeding effort is much greater than that of the female. The female devotes most of her time to incubation and brooding, activities that require less energy expenditure than food collection. The female regulates the hatching asynchrony by starting to incubate during the egg-laying period. (After Slagsvold and Lifjeld 1989.)

brood, and perhaps to increase the male's effort thro\
Later on, during the nestling stage, males appear to ref
mate (see Slagsvold and Lifjeld 1989).

Slagsvold and Lifjeld also argued that advancing th
feeding by the male means that the male is constrained tc
to alternative reproductive activities, such as attracting a s
conflict with the female's interests (Figure 10.10). For exi
catchers, females mated to polygynous males receive less he
are mated monogamously.

Some evidence for the sexual conflict hypothesis for hatching asynchrony
comes from pied flycatcher data (Slagsvold and Lifjeld 1989; but cf. Hébert
and Sealy 1993, Nilsson and Svensson 1993). The total body weight of females
rearing asynchronous broods, measured at the end of the nestling period, was
higher than that of females rearing synchronous broods. The reverse situation
was not found for males, there being no significant dependence of male weight
on the degree of asynchrony. However, there may have been losses to males
due to the re-allocation of their efforts on extra-pair copulations towards PI.

An unexpected reversal of this type of conflict was recently documented in a
non-migratory population of blue tits (*Parus caeruleus*) (Slagsvold *et al.* 1994).
In this bird, all pairs are monogamous and share most parental tasks, except for
incubation, which is the exclusive province of the female, who usually effects a
fairly synchronous hatch (mean spread = 2.1 days). If this hatching spread is
manipulated experimentally (by swapping newly hatched young between
nests), such that parents provision either relatively synchronous (mean spread
= 0.6 days) or asynchronous (mean spread = 3.4 days) broods, parental
survival is altered. In contrast to unmanipulated nests, in which the two sexes
have very similar probabilities of surviving until the next breeding season (35%
of males vs. 37% of females), after rearing an experimentally synchronous
brood the sexes differed (25% of males vs. 47% of females) and after rearing an
experimentally asynchronous brood the pattern was neatly reversed (43% of
males vs. 29% of females). Slagsvold *et al.* interpret this result as follows: when
there are size disparities among offspring, females tend to feed smaller
offspring, and males larger offspring which are less energy-demanding to the
parent. Since only females incubate, they are able to win this investment
conflict by determining synchronous hatching.

10.4 Sex-biased parent–offspring conflict

As noted above (Section 10.2.3), offspring sex can affect optimal PI. For
example, in species in which males fight for sexual access to females *and* fight
outcomes depend on male size *and* additional parental investment contributes
to adult size, then mothers may prudently expend more PI on sons than on
daughters (Trivers and Willard 1973, Maynard Smith 1980). An interesting

of how such PI inequalities may affect POC is related by Redondo *et al.* (1992), who plausibly argue that when there is male-biased PI, sons may not only have a higher optimum for PI but also one that is more distant from that of their mother (even though she may favour some degree of bias toward sons). They thus predicted that sons should be more involved in POC than daughters, basing their argument on the pro rata compromise model (Parker and Macnair 1979). Evidence for this was found in the mother–lamb 'squabbles' of domestic sheep, in which sons solicited more often and were more frequently spurned by their mothers. In contrast, their data from rhesus macaques suggest a more complex story. Unlike sheep, macaques are matrilocal, with some daughters of high-ranking females inheriting the mother's lofty rank. There are no conspicuous differences in the way in which high-ranking mothers invest in sons vs. daughters (Gomendio 1990), but low-status mothers must pay more for daughters (Redondo *et al.* 1992). In particular, low-ranking mothers are less likely to reproduce in the next cycle after weaning a daughter. If this cost of reproduction is in any way imposed by the offspring (e.g. through extended oxytocin release caused by protracted suckling; see Section 11.4) and, most importantly, if it lowers the mother's lifetime reproductive success, this could represent a manifestation of POC.

10.5 POC theory epilogue

We have stressed that sibling rivalry must be considered within the broader context of intra-familial conflict (Figure 3.1). Parent–offspring and sexual conflicts influence the supply side—the amount of parental resources available to given offspring, which critically affects the level of sib-competition. Furthermore, POC can greatly affect the expression of sib competition by influencing the mechanisms whereby offspring receive pay-offs.

POC can clearly occur in theory, and there is even limited evidence that it also *does* occur in practise (see Chapter 11). The 'pro rata payment' or scramble model (Parker and Macnair 1979) offers a plausible explanation for much of the overt sibling competition observed in nature, especially under conditions of intra-brood conflict or other cases where parents may be able to exert relatively little control over the mechanism by which competing offspring acquire their respective shares of the resources. When the parent can control the allocation to each offspring very precisely, as seems likely if offspring are produced singly, Godfray's (1991) honest signalling model may offer a better explanation of the levels of sib competition. The two approaches are analytically very similar in structure, and actually form two limiting bounds: no parental control over the mechanism for allocation of shares to individual offspring (scramble), and total parental control (honest signalling).

As a result of these models, we also now know much more about how the details of mating system (whether one or both parents are constrained), type of conflict (inter-brood or intra-brood) and type of costs (summed, shared or individual) are expected to affect the observed levels of sib competition, and how the three dimensions of family conflict (Figure 3.1) interact with each other.

The models tantalizingly suggest that there are relatively few important parameters to be measured in order to make predictions from field data, e.g. both the pro rata and signalling models focus on just two functions, *f(m)* and *S(x)*. The bad news is that both of these are extremely difficult to measure in nature, and progress has evidently been hampered severely by this problem. For some of the models (e.g. the hierarchy model, asymmetrical begging scramble model and honest signalling model) we also need to know the condition or competitive ability of the competing siblings, which may be more easily accessed from field studies.

A further complication is that we generally need information about the underlying mechanisms. It will be especially difficult to determine the mechanism of gene action in begging scrambles, i.e. whether begging is conditional or unconditional upon role. Also of prime importance are the behavioural mechanisms whereby offspring compete and parents allocate food. In particular, the generalized version of signal and scramble models requires an estimate of how a change in parental resources affects begging (*Mechanism 1*; the 'demand function' of Hussell 1988) and of how a change in begging affects the parent's provision of resources (*Mechanism 2*; the 'supply function' of Hussell 1988). Mechanism 1 tells us the eventual level of provisioning, and Mechanism 2 tells us the eventual begging level (or other measure of the extent of sib competition). These may be relatively accessible empirically.

A further but related way in which POC affects sibling rivalry is that it determines the eventual clutch size, which is an important determinant of the level of sib competition. It is not surprising that parent and offspring can disagree over family size as well as more directly over the PI per offspring. Siblicidal brood reduction is simply a way in which a given offspring can acquire extra parental resources.

Hamilton and Trivers created the conceptual framework for our understanding of family relationships. In over three decades since Hamilton's (1964a,b) inclusive fitness papers and two decades since Trivers' (1974) insight into POC there has been an explosion of interest in this area of behavioural ecology, both theoretical and empirical. Yet much remains to be done, particularly with regard to the inter-meshing of empirical and theoretical approaches, and the understanding of mechanisms, some of which have not hitherto been considered important.

Summary

1. Although parent–offspring conflict theory was originally directed at explaining behavioural phenotypes, its applicability to various life-history

traits (e.g. offspring sex ratio) was apparent from the start. This chapter centres on strategic decisions about family size, and specifically how prospects for expensively wasteful sibling rivalry may constrain parents from creating their own optimal clutch size. The general idea is that offspring selfishness, such as siblicidal cannibalism, has the potential to deprive parents of their full array of possible clutch sizes. Some models are presented to address issues of control. In general, offspring favour smaller clutch sizes than parents.

2. In a *hierarchy model* (in which sibs are ranked and have despotic control of PI according to rank), stronger sibs may kill nursery-mates profitably. Parents may tolerate such siblicide for two very different reasons. Either they cannot enforce their preference for peaceful cohabitation at an acceptable cost (e.g. if the parents are absent at the critical time)—in which case true POC is present and the offspring are 'winning'—or the parents value their various progeny unequally (in which case POC may be be more subtle or may not apply at all).

3. In a *scramble model*, attention focuses on begging costs (energy and/or risk). As family size increases, each offspring's solicitations escalate and these expenditures consume more and more of the family's potential fitness dividend. It may pay parents to reduce clutch size so as to decrease offspring competition and begging. An ironic upshot is that, if parents respond to siblicidal selfishness in their offspring by cutting back on clutch size, we observe what appears at a behavioural level to be *less* sibling rivalry.

4. In very large clutches, as are common in cannibalistic invertebrates, POC over the degree to which family size should be trimmed can be modelled in terms of whether siblicides are 'victim-based' (i.e. each killer eliminates a set number of nursery-mates) or 'survivor-based' (i.e. executions continue until the offspring's optimal family size has been reached). Depending on a given system's parameters (relatedness, egg costs, etc.), siblicide levels are expected to range widely.

5. If siblicides are *victim-based*, a mother clearly cannot improve her fitness by producing fewer eggs. Less obviously, mothers may eventually be able to gain by laying more, until eventually they are constrained by the costs of eggs. If mothers initially play the offspring optimum, siblicide is clearly unfavourable for offspring. With all siblicidal tendencies removed from offspring by evolutionary change, a maternal mutation for increased laying would enjoy an advantage likely to remain unchecked until offspring mutations for some renewed siblicide happened to show up. Maternal over-production will ultimately be limited by escalating costs of egg production. Basically, victim-based siblicide is permissive enough to allow a parental rally and significant levels of sib destruction.

6. With *survivor-based* siblicide, however, such effects are prevented because the offspring trim the clutch size back to their own optimum: a

parental mutation for increase cannot spread. Siblicide is unlikely ever to disappear completely if there are recurring maternal mutations to start 'cheating' the clutch size upwards.

7. Extreme forms of siblicide occur without cannibalism, such that only one larva per nursery survives. This can lead to specialized 'fighting' morphology, which seems to be hard to shed once it has evolved (because of the risk of phenomena such as superparasitism). In general, parasitoid species with fighting larvae tend to have clutches of one (or mildly 'overproduced' variations in the vicinity of one), while taxa with non-fighting larval forms are characterized by much larger clutches.

8. 'Genomic imprinting' refers to phenomena in which gene expression differs according to whether a key allele was maternally or paternally derived. In taxa such as mammals and seed plants, considerable PI is extracted over time from maternal supplies, and monogamous mating is exceptional. A rare mutant allele selfishly directing its body to demand extra PI when paternally conferred would be most likely to compete with nursery-mates that do not bear copies of itself; the same mutant received from the mother, by contrast, would be surrounded by true sibs ($r = \frac{1}{2}$). Extreme selfishness is therefore favoured in paternally derived genes, but maternally derived genes are constrained by kin selection. It has been argued that genomic imprinting has evolved because of this dichotomy.

9. All offspring need not necessarily be equally valuable to their parents, in which case unequal investment is expected. In general, optimal PI allocation for parents should obey the 'marginal value theorem,' with each offspring receiving resources so that the marginal gains produced by increased investment are equal for each offspring. This point is readily appreciated in terms of offspring gender and a large literature on differential investments in sons vs. daughters has emerged over the past 60 years. Other disparities in offspring value should be subject to similar rules.

10. Where PI is contributed by both parents, most taxa should exhibit 'sexual conflict' over how much each contributes. ESS models of biparental care make the paradoxical prediction of less PI per offspring than under uniparental care (or an idealized 'true monogamy', wherein partners mate for life). In a biparental situation, if one mate unilaterally lowers its share of the load, its partner should compensate partially (thereby buffering the offspring from the full force of the unilateral deprivation), but not fully (i.e. not so as to make up for the total cut-back), unless there is true monogamy. The implications for sibling rivalry are clear. Selfish games between the care-givers can shrink the 'cake' on which the offspring depend, thereby accentuating their own incentives for selfishness. Some recent field studies of European songbirds provide some support for these and related predictions.

11

Tests of parent–offspring conflict vs. collaboration

The calf cannot fling its mother to the ground and suckle at will.

(Trivers, 1974)

Once tuned in, one is hard-pressed *not* to see conspicuous manifestations of parent–offspring 'conflict' everywhere. For even a casual POC connoisseur, five minutes in any supermarket is an observational feast: children ogle chocolate bars, chewing gum and brightly wrapped peppermints (thoughtfully displayed by store managers right where shoppers must linger in the check-out line) and pester their parents with farragos of whining pleas. Parents vary greatly in their responsiveness to such ploys, but one is drawn irresistibly to imagine that candy sales must be higher, possibly much higher, as a result of the begging. Next our connoisseur proceeds to a toy store . . .

Of course, this has nothing whatever to do with testing the generality of Trivers' POC hypothesis (if there are any *fitness* effects from investing in future tooth decay, they are unlikely to be positive). However, the above scenario may have a lot to do with why the concept has been warmly received in general. Anyone who has experienced parenthood has also endured successful small-time hustling, and everyone who can recall his or her own childhood knows that parents have inherent power advantages. The theory has enormous intuitive appeal.

It comes as a shock, then, to realize that more than 20 years have passed since Trivers first presented his theory, yet hardly any satisfying field tests have shown that it uniquely explains anything about behaviour. Before we defend that contentious position (the task of this chapter), let us briefly retrace our steps and set out the key elements that need empirical attention. As detailed in the preceding chapters, POC theory has the potential to offer much to our understanding of family structure and family size and, as befits a major concept, has been challenged in various ways (see Chapter 7). For reasons already discussed, we are convinced that the foundation is sound, and that a conflict between genes expressed at two stages of life can occur: parents and offspring can have evolutionary interests that are not congruent.

At a completely different level, though, we must consider the theory's value, namely its power to explain natural phenomena. Accepting that some genotypic conflict (the 'battleground') exists is not equivalent to crediting POC as a significant causal factor, a selective 'prime mover', that has shaped phenotypes of interest. One can use POC theory to look afresh at various *behavioural forms of 'conflict'* (which hereafter we shall call 'squabbles', reserving the word *conflict* for the specialized meaning it takes in a coevolutionary context; Mock and Forbes 1992). However, one must then test, as a separate exercise, whether those phenotypic traits arose as a result of the genetic asymmetries identified by Trivers. Alternative reasons for squabbling do exist. For example, it may be part of a complex signalling (negotiating) system that is inherently 'honest', although superficially messy (see Godfray 1991; summarized in Chapter 9), and no more tied directly to true evolutionary conflict than a tantrum over supermarket candy.

Many of the models we have discussed (Chapters 8–10) dealt with resolutions, and how significant levels of genotypic conflict might be resolved, and it appears that they can generate virtually any possible outcome ('parent wins', 'offspring wins', and many compromises), depending on the assumptions, constraints and starting-points of real-world systems. Such an array of modelling results may strike some empiricists as bewildering or numbing. It might even be viewed as an open invitation to publish virtually any data as 'consistent with' POC, since all possible outcomes seem to be predicted. As with any theory, presentation of data that are 'consistent with' some theoretical prediction is a mixed blessing. Supportive evidence enhances one's growing impression that the current explanation is correct, but Popperians would argue that it can give a false sense of security as well, and/or terminate the search through alternatives. Of the many empirical papers that present behavioural data as supporting POC theory, very few have set out a strong inference approach (Platt 1964) or specified criteria that could have led to its rejection. In short, we have a very popular theory that is widely believed to explain a great many traits, but almost no clear and convincing tests of it.

Having said that, we remain optimistic that the POC perspective has much to offer. Ironically, the uncritical reception it has received to date has probably not only impeded its being seriously tested, but also impaired its being applied usefully. This paradoxical lack of rigour was first pointed out when the POC concept had been discussed enthusiastically for 6 years (Stamps and Metcalf 1980), which struck those authors as rather a long honeymoon for such a hot idea. The ensuing dozen years added little, and a more recent review (Mock and Forbes 1992) found only one published behavioural study that provided data which were not also consistent with alternative explanations (see Section 11.4 below). There may be other such studies out there, but not many.

In this chapter, we shall review some problems that have thwarted empirical progress in this area. First, we describe the difficulties encountered during our own attempts to apply POC to egret siblicide, before going on to discuss general obstacles. Finally, we shall attempt to suggest steps that may improve the topic for future work.

11.1 POC and avian brood reduction

Avian brood reduction was introduced as possibly stemming from POC by O'Connor (1978). It seems to offer several attractive features. First, it involves concurrent siblings (as well as possible impact on sequential ones) in small broods. Secondly, it is a forum in which the physical power to influence sibling welfare is not overwhelmingly vested in either parents or nest-mates. Thirdly, in some species avian brood reduction involves observable behavioural phenomena, including extreme aggression. By contrast, mammalian *weaning conflict* is played out in the context of a stark physical mismatch between an omnipotent mother and frail juvenile (see chapter heading), forcing the offspring to rely on more nebulous 'psychological weapons' of deceit (for some neglected physiological weaponry, see also Gomendio 1993, Mock and Forbes 1993). At the other extreme, the balance ranges to 'worker–queen conflict' in social hymenopterans, where the workers may be physically at liberty to do whatever they want, including adjust the brood's sex ratio (Trivers and Hare 1976, Stamps and Metcalf 1980, Tepedino and Frohlich 1984). A nestling bird can promote itself in both dimensions of family structure (see Figure 3.1), begging at a parent and/or pummelling a sibling, in order to skew PI.

Recall that O'Connor's thresholds (Section 4.1.1) for family members in a brood reduction game focused on the matter of the last-hatched nestling's death, specifically on how that event affects the inclusive fitnesses of three classes of players: parent, surviving sib, and victim. From an elder sib's perspective, victim ought to die as soon as its continued existence would detract more from Self's direct fitness (e.g. via local resource depletion) than it could contribute to that senior's indirect fitness (via its potential for making Self's nephews and nieces). A similarly reasoned threshold for a parent has a higher value, due to the relatedness symmetry between parent–survivor and parent–victim. And even the victim itself has a threshold above which it should 'want' to die (when the indirect fitness thus gained by enriching its sibs' prospects exceeds the direct fitness lost; see also Alexander 1974). O'Connor added quickly that it is ordinarily a much better option for the victim if one of its sibs were to die instead (if such can be arranged), so 'adaptive suicides' are not generally to be expected. In general, then, all family members are predicted to do everything possible to survive.

One POC zone can thus be envisaged as the range of mortality rates over which parents and senior sibs 'disagree' about the victim's demise (Figure 4.1); parents and seniors might be expected to work at cross-purposes in this regard. That is, as competition sharpens just past the point where it pays the seniors for victim to die, there might be a period during which the parents prefer everyone to remain alive a little longer. Because seniors and parents are the parties with the most physical power to influence victim's fate, this is when behavioural manifestations of POC might be found (assuming that family members are able to recognize such moments when they arise).

O'Connor initially contended that sib fighting *per se* was clear evidence of such POC, i.e. that such is necessarily harmful to parental fitness: 'The fitness argument requires sibling aggression to be the outcome of conflicting sibling and adult interests' (O'Connor 1978: 89). This ignores the area above the parent's threshold where siblicide serves the parents' interests as well, some parameter space where parents should tolerate fighting (or, better yet, join in and complete the execution process cheaply). Between the thresholds of parent and senior sib, then, selection is predicted to favour parental counter-strategies, which might manifest themselves in such behavioural traits as inter-rupting sibling fights, foiling the influences of dominance on food distributions (e.g. by feeding the victim preferentially), and even punishing aggressors (*sensu* Clutton-Brock and Parker 1995). During our early egret field studies (through the mid-1980s), we kept these predictions firmly in mind, fully expecting parents to show clear opposition to siblicide.

11.1.1 *Great egret brood reduction and POC*

Before giving the behavioural details, we note that daily mortality rate values obtained from nest censuses suggested that great egrets might truly be ex-periencing just the kind of risks where overt conflict between parents and senior sibs might be found (from O'Connor's thresholds; see Mock 1987). Basically, we found no sign of overt behavioural conflict between parents and seniors. In almost 3000 recorded fights (at least one parent being present during 96.5% of these), the parents did nothing that could possibly be inter-preted as interrupting in over 99% of the fights for which they were present. (To be generous, we included all interruptive events, including accidents such as parents tripping while entering the nest and clumsily stepping on their embattled offspring.)

To determine whether something more subtle might be at work, we also scrutinized the final 10 meals for *C*-chicks that died, and compared them with age-matched *C*-chicks that went on to survive, to see if parents might have some hard-to-detect tricks for getting extra food to subordinate offspring in serious jeopardy. Once again we found no indication that parents were trying to sustain those lives. Specifically, parents did not duck or sidestep senior chicks so as to allow victims to gain extra scissor-grips on the parental bill, they did not produce food boluses on a higher proportion of victim scissorings and they did nothing that had the effect of increasing the *C*-chick's share. Furthermore, when *C*-chicks were ousted while still alive, the parents did not visit them on the ground (1–3 m away) or feed them (though feeding out of the nest has been observed when whole broods have been knocked down by storms). In short, egret parents appear to be completely blasé, leaving decisions about aggression and food allocation to the senior siblings.

Two observations run counter to this picture of totally non-existent parental influence in great egret broods. First, the few fights that occurred when parents were absent tended to be more severe-looking (more likely to involve the loser actively fleeing around the nest or over the rim; Mock 1987) than those in which a parent was on hand, suggesting that the mere physical presence of an adult may have some discreet (but passive) influence. Secondly, the broods that we hand-raised in captivity—where they were fed via puppets but did not have any kind of 'adult presence' on a continuous basis (see Section 6.1.4)—exhibited fighting that looked unusually vigorous (Mock *et al.* 1987a). Thus, we cannot rule out the possibility that parents have a tempering effect, perhaps posing some kind of tacit threat to the senior siblings that almost never requires more tangible enforcement. These caveats notwithstanding, the overwhelming impression given by these egrets, and paralleled by the other ardeids studied, is that the parents adopt a *laissez-faire* attitude toward the siblicidal activities of their elder chicks (Fujioka 1985a, Mock 1985, Ploger and Mock 1986, Mock 1987, Mock *et al.* 1987b). If parents and senior siblings in this population do have a non-trivial POC zone, we detect scant evidence of behavioural manifestations.

11.1.2 *POC and other avian brood reduction systems*

Greater detail on fatal sibling competition in various other bird taxa is provided in the next chapter, but it is appropriate to mention here that the anticipated role of parental interference has not been quantitatively documented for non-ardeids either (Drummond 1993). In the most closely observed siblicidal birds, namely kittiwakes (*Rissa tridactyla*; Braun and Hunt 1983), boobies (*Sula nebouxii, S. dactylatra*; Drummond *et al.* 1986, Drummond 1987, 1989, Anderson 1990b), pelicans (*Pelecanus erythrorhynchus, P. occidentalis*; Cash and Evans 1986a, Ploger in press, Pinson and Drummond 1993) and various raptors (e.g. Gargett 1978, Edwards and Collopy 1983, Simmons 1988), parents have been described explicitly as doing nothing to stop fights between nest-mates (for further review see Drummond 1993). The primary exceptions to date consist of anecdotal reports that South Polar skua (*Catharacta mccormickii*) parents may interrupt sibling aggression by settling on to chicks (Spellerberg 1971a ,b, G. Miller unpublished data) and by giving (apparently bogus) alarm calls that cause the fighting chicks to crouch immediately (Young 1963). Bald eagle parents may interrupt sib fights similarly by presenting food morsels to the combatants (Gerrard and Bortolotti 1988: 86). Skua parents, as well as some ground-nesting owls (Ingram 1959, 1962,

Parmelee *et al.* 1967) and cranes (Harvey *et al.* 1968), may separate their young chicks within the territory, which might be a parental counter-strategy against sibling aggression (although alternative explanations for segregating offspring are easily imagined; e.g. see Mock 1984b, Drummond 1993). Moorhen (*Gallinula chloropus*) parents usually feed whatever chick is closest—a position normally held by their larger (first-hatched) offspring. However, these adults may also seize individual chicks aggressively by the neck and shake them vigorously ('tousling'), which results in smaller siblings obtaining the next food item. Thus, one function of moorhen tousling may be to counter the effects of sibling competition, and specifically of larger sibs monopolizing parental food through positional manoeuvring (Leonard *et al.* 1988).

Clearly preferential parental treatment among brood members has been described with regard to which chicks are fed first in several avian taxa (e.g. Nuechterlein 1981, Braun and Hunt 1983, Bengsston and Rydén 1983, Horsfall 1984a, Boersma and Stokes 1995). In a detailed laboratory study, Stamps *et al.* (1985) reported that the two budgerigar parents use strikingly different feeding rules. Upon arriving at the nest, fathers give the food load to the first chick encountered (as do egret parents), but mothers actively seek and feed the smallest brood members. It is not clear whether this system helps or harms runts, and the fact that budgie senior sibs also frequently give food to their smaller nest-mates suggests that acute sibling competition is absent.

Finally, to underscore the illusory nature of behaviour that may resemble parent–offspring conflict, we close this section with an account of some recent work on 'terminal egg-neglect' in herring gulls (*Larus argentatus*). As with other gulls, parents hatch their three eggs asynchronously and the chicks vocalize from within their pipped eggs when they are chilled, thereby summoning renewed incubation. A problem seems to arise for the *C*-chick when it is the only one still in its egg, because the parents then spend more time standing up, tending to the needs of their liberated senior chicks, causing the *C*-chick's temperature to drop from the warm level experienced by the senior sibs (*c.* 36–38°C) down to levels below which embryo damage can result. Parents tending dummy (non-vocal) eggs at this stage gave them enough incubation for a live (metabolizing) embryo to experience a mean temperature of 33.9°C (Evans *et al.* 1995), roughly what *C*-chicks under field conditions achieve (33.4°C; Lee *et al.* 1993). Thus, this situation looked very much like a classic 'parent wins' POC over a limited and valuable parental commodity (time and attention). However, a technique for determining precisely the temperature that *C*-chicks actually want (pipped eggs were placed in an apparatus that supplies warmth in response to vocal signals, enabling the embryos to control their own temperature easily) revealed that they chose mean temperatures between 32.9 and 33.4°C (Evans *et al.* 1995). In short, although gull *C*-chicks receive less from their parents than their elder siblings do, they still obtain all they want (Lee *et al.* 1993).

11.1.3 *Parental withholding*

Set alongside this disinclination to referee the fights are indications that egret parents actually withhold resources that might be used to delay or even avoid brood reduction altogether. This was first suspected when we noted that parents often cease to expel boluses for their still-begging young while food was visibly present in the parents' throats (Mock 1985). This runs counter to an old report. Milstein *et al.* 1970) described a case in which grey heron parents were shot immediately after food presentations to their chicks and were found that their stomachs were completely empty. We did no shooting, but subsequent experiments on cattle egret hatching asynchrony showed clearly that parents are capable of delivering much more food than they normally do. Specifically, if egret nestlings are matched in age and size at hatching, the parents of these 'synchronous' (manipulated) broods provide 30% more food than the parents of unmanipulated control broods (Fujioka 1985a, Mock and Ploger 1987). Whatever the proximate cue(s) that trigger such a parental response, the finding clearly reveals that the adults routinely withhold food (or abbreviate their food-collecting efforts) while siblicidal starvation is ongoing. Recall also that, during temporary experimental reductions in brood size (Section 6.3.1), parent cattle egrets again adjusted the amount of food given to offspring dramatically and in both directions (Mock and Lamey 1991). (Brown pelicans also deliver less food to artificially reduced broods; Ploger, in press.)

Taken together, these results suggest that parents may have a notion of how much effort they are *willing* to expend on behalf of the current brood. That ceiling value may be influenced by whatever information the parents possess about both the economics of current food availability (prey abundance, weather, etc.) and the tougher calculus of life-history parameters (e.g. their own residual reproductive value). At present, we can only imagine the simple behavioural rules by which such decisions are made (although parental sensitivity to brood size *per se* suggests it is one of the major cues).

Of special relevance is the likelihood that parents are not necessarily aiming to sustain full brood size all the way to fledging (see also Section 12.3.1). Whereas O'Connor's models dealt only with the mortality risks to the current brood, it now seems clear that the parental threshold must be somewhat lower than his specification (i.e. our equation 4.3), as it must also be based on how current reproductive costs affect future fitness. Brood reduction can simply be thought of as a 'load-lightener', as that term is applied to the relief parents obtain from allo-parental contributions from helpers (Brown 1978) and which can also be demonstrated through experimental provisioning (e.g. Johnston 1993), although in the brood reduction context a burden is shed rather than shared. That benefit would be even more valuable if the relationship between parental effort and adult mortality were an accelerating function ('convex-up': Crick 1992). Thus, parents and siblicidal senior brood members may have rather more congruent opinions about the necessity and timing of brood reduction for this reason alone (Mock and Forbes 1992, 1995, Forbes 1993).

11.2 Reproductive value and unequal parental investments

It is also important to recognize a fallacy in one of the fundamental assumptions, namely that parents favour equal investments in all offspring. First, not all offspring have equal quality or prospects, so parents may do better by to make unequal PI allocations. Trivers' argument hinged on parents evaluating their offspring entirely by the equal coefficients of relatedness. This in turn contains a cryptic assumption, that parents know in advance how many offspring they can afford (Mock and Forbes 1992). However, if parents choose to gamble for any or all the potential benefits of over-production in a stochastic world (resource-tracking, replacement offspring and/or sib-facilitation), they add the risk of investing in expensive 'redundant' offspring (Forbes 1990, Mock and Forbes 1995). Such production of marginal kids is more affordable to the parents if the extras can be culled easily and inexpensively later on, as necessary. It follows that parents may have evolved ways of stacking the deck so as to protect their *core* investment, while deriving the benefits of initial over-production (Mock and Forbes 1995, Forbes and Lamey 1996).

In short, parents are likely to value certain offspring (their *core* brood) more than others (the *marginal* brood), despite the identical coefficients of relatedness (Temme 1986, Haig 1990; see Section 10.2.3). This is essentially another role assignment within a sibship, but as viewed by the parents. What matters is that the classic POC prediction, that parents ought to favour equal allocation of PI among all progeny, is incorrect: parents can and often do have favourites among their offspring. In making a similar case, Drummond *et al.* (1986) wrote of 'parent–offspring co-operation' during booby siblicides, depicting parents and senior nest-mates as collaborators in siblicide.

In egrets, for example, the initial hatching asynchrony is established mainly by parental incubation patterns. Even if one were to argue that such behaviour evolved solely to trim the costs of sibling rivalry or to gain some other benefit (e.g. to protect eggs from chilling, etc.), the fact remains that once hatching asynchrony exists the reproductive values of brood members are no longer equal. Senior siblings automatically represent an inherently more attractive vehicle for further parental investment. A similar argument may be applicable to particular cases of intra-clutch variation in egg size (e.g. Quinn and Morris 1986), egg composition (see review in Williams 1994a), egg additives (see Schwabl 1993), and so on. Extreme cases of the unequal-value principle exist in species where parents kill some of their eggs (by abandoning unhatched individuals or bludgeoning chicks outright; Section 12.2.2). In one sense, these could be held aloft as true POC (between parent and victim!), but such cases are lopsided 'parent wins' resolutions that simply do not require the particular (Trivers 1974) asymmetrical relatedness argument at issue here (Mock and Forbes 1992). They are more logically viewed as a life-history trait, with regard to optimal family size (Mock and Forbes 1995).

Clearly, then, one cannot safely infer POC from observations that some siblings obtain 'more than their share' of PI, because a degree of skew is probably the parental optimum as well (Mock and Forbes 1992, Forbes 1993). How the seniors' true optimum, the parents' true optimum and the realized distributions compare with each other is anybody's guess if we lack the ability to make independent measures of the first two—and we nearly always do lack that. In addition, with favouritism the already narrow 'POC zone' shrinks further and may even vanish altogether (Mock and Forbes 1992), defusing much of our puzzlement over why parents tolerate sibling fights and provide no succour for the suffering victim.

Even if asynchronous hatching is an adaptive part of the parental phenotype, there may still be scenarios in which parents might change their minds later and perhaps then fine-tune the competitive asymmetry. For example, because parents typically do all the food-gathering, they may be in a better position to monitor its availability as time passes. Through some process of Bayesian updating parents might detect that a 'good year' is unfolding (in which the full brood can survive), at which point they could profit from enforcing peace. Conversely, if parents are no more capable of such an assessment than the nestlings, they may be in no position to improve their pay-off by playing referee, and thus should let their offspring sort things out.

A final complication centres on what might be called the 'costs of enforcement' (Forbes 1993). In a true evolutionary conflict, each party gains fitness by 'winning' and surrenders fitness by 'losing,' but the possibility also exists that neither party can impose its preference on the other without going to unreasonable lengths. For example, even if egret siblicide *were* a net negative for parental fitness (which we clearly do not concede at this point), to what lengths might adults have to go to discourage it? On a fight-by-fight basis, especially during the first couple of weeks, little would be required. Simply settling on top of the chicks overwhelms the combatants with parental belly feathers, while lowering the bill (as if to regurgitate food) often has the effect of distracting the chicks' attention, etc. During this phase, refereeing looks easy. However, egret sibling hostilities persist for several weeks, well into the period when parents attend the nest sporadically. By stopping the earliest fights, parents could possibly make short-term gains in the energy conversion efficiencies of their offspring (more food being translated into growth, and less to aggression), but clear dominance relationships among the nest-mates would be less likely to be established. Several deleterious consequences might result, including the fact that fighting may simply be deferred until the chicks are larger, more powerful, and tougher. If so, the deferred costs of erecting a dominance hierarchy could be compounded, including the risk of damaging mutual injuries (e.g. hard blows to the eyes). By stopping fights parents might delay the resulting decline of the runt for an additional week or two, but perhaps no more. This means that the parents could well end up putting extra food into a doomed property, rather than cutting their losses. Worse still (under dire circumstances), the victim might not die at all, but sap resources and subsequent fitness from the core brood members (a 'redundant' offspring: *sensu* Forbes 1990). Recent models of 'punishment' (Clutton-Brock and Parker 1995) should be easily adapted to this problem.

11.3 Initial disparities among sibs

A cardinal property of any competition is the degree to which the contestants' abilities are equal. In a horse race with an even start, the lead may change hands, even see-sawing back and forth. By contrast, conferring sizeable advantages or handicaps on certain racers affects the symmetry and alters the situation fundamentally. Unlike horse races, contests for resources and growth are often self-perpetuating, in the sense that competitive disparities lead to ever improved physical condition, creating a positive feedback loop and widening the asymmetry. Just as money begets the power to make more money under capitalism, food begets the power to sequester more food in an avian brood. Lack (1947, 1954) identified the skewing of unpredictable food supply as a likely evolutionary incentive for parent birds to hatch eggs asynchronously, so as to fledge the largest number of robust young in any circumstance. It has since been argued that sibling asymmetries and brood reduction should sometimes be viewed as effects or even costs of hatching asynchrony, rather than its cause—a tariff parents must pay in order to gain various other benefits (Clark and Wilson 1981, Mead and Morton 1985, Amundsen and Stokland 1988, Magrath 1990, Beissinger and Stoleson 1991; reviewed in Stoleson and Beissinger 1995). Whatever the prime movers of the phenomenon, and they are likely to vary with the ecological and phylogenetic background anyway, there is no question that the style and intensity of the ensuing competition are affected by any 'unfair' parental manipulation.

These initial inequities among avian nest-mates can be divided into those of *time* (mainly hatching asynchrony) vs. *substance* (e.g. egg size polymorphisms), with the former category tending to have by far the more powerful effects in most birds (Slagsvold *et al.* 1984, Magrath 1992; cf. Simmons 1991). The greater impact of hatching asynchrony may be linked to avian growth speed; time is rapidly converted into size once the chick has escaped the confines of its shell. Even in the most extreme cases, namely the crested penguins (*Eudyptes* spp.), where one egg has on average a 20–80% greater volume than its lone clutch-mate (Warham 1974, Williams 1980a, b, Lamey 1990), chick survival correlates much more strongly with hatching interval than with egg size differences (Lamey 1992).

A third parental manipulation that has just come to light is hormonal. Captive canary (*Serinus canaria*) mothers deposit higher titres of testosterone in the yolks of last-laid eggs, boosting the aggressiveness of those chicks and making them better able to counter the temporal handicap that results from asynchronous hatching (Schwabl 1993, 1996). Because the generality of this phenomenon is not yet known (especially in species where brood reduction is a normal part of the chick-raising process), we mention this only as a new and provocative parental gambit whose functional significance remains to be explored. It is encouraging to note that the yolk androgen concentrations of cattle egrets are quite the opposite to this pattern in the canary, such that *A* and *B* eggs contain about twice as much testosterone and two related hormones as do *C* eggs (Schwabl *et al.* 1997).

In most species, the 'directions' of the first two advantage classes are aligned: eggs that are laid first also tend to enjoy any size enhancements that may exist. The cases in which the classes are in opposition provide the least ambiguous comparisons (e.g. *Eudyptes* penguins). It has been proposed that such cases, where the later-laid eggs tend to be larger, represent either (i) an adaptive fine-tuning of the intra-brood competitive balance (i.e. shaving off some of the crude handicap imposed by hatching asynchrony; Howe 1976) or (ii) a non-adaptive and largely inconsequential artefact of other forces impinging on egg size (e.g. fluctuating food levels available to the mother during egg formation; Clark and Wilson 1981). Whatever the nature of the interaction between these sources of sib asymmetry, our present concern is that parents probably control both to a large degree (see also Section 12.3.1). Egg size is set primarily by maternal metabolism (sometimes indirectly affected by paternal courtship feeding: Royama 1966, Nisbet 1973, Salzer and Larkin 1990) and/or by physical changes to the oviduct (i.e. if it shrinks, later eggs may be constricted: Williams 1994a). The species-typical hatching intervals correspond rather well with the onset of parental incubation (e.g. Inoue 1985, Magrath 1992; but see Bortolotti and Wiebe 1993), so it seems reasonable to assume that parents could either diminish the asymmetries (e.g. by deferring incubation till somewhat later in the laying sequence) or increase them (e.g. by extending the interval between eggs).

In some precocial species, the prehatched chicks themselves may influence the synchrony of hatching by acoustic monitoring of sibling noises (calls and respiratory clicks) and adjusting development rate to this information (e.g. Vince 1969). It is not known whether species whose chicks engage in vital post-hatching competition, including siblicide, also actively tamper with brood asynchrony, and there is some question about whether this 'talking eggs' effect really occurs under field conditions (Schwagmeyer *et al.* 1991). At present, we can only assume that parents have a much greater say in the initial intra-brood balance of power than do the chicks.

To the extent that parents control gross initial asymmetries among nest-mates, natural selection might be expected to shape their decisions so as to maximize parental fitness effects. Lack's brood reduction hypothesis was founded on the then radical assertion that some chick mortality can be in the parents' best interests, but a more general view suggests that parents should strive for an optimal degree of sibling competitive asymmetry and the pay-offs need not necessarily take the form of partial brood losses. This idea was first proposed by Hamilton (1964b), who envisioned a parentally favoured level of sibling rivalry under which unpredictable resource shortages could be resolved decisively and efficiently, without undue levels of 'wasted energy' being shunted to protracted battle between evenly matched siblings.

The logic of asymmetrical contests (Maynard Smith and Parker 1976) is applicable to conflicts involving potentially high costs wherein rivals are mismatched in terms of how much each values winning and/or how each assesses its chance of victory, and the contestants' simultaneous solutions determine the expected contest costs at the ESS. There are essentially two approaches: (i) the Asymmetric War Of Attrition (Hammerstein and Parker 1981), in which before the trial begins opponents pick a time for which they are willing to persist in the contest, and (ii) the Sequential Assessment Game (Enquist and Leimar 1983, 1990), in which contestants continue to refine their estimates of relative fighting ability during the contest itself. Both generate the prediction that contests should be more protracted (and hence more costly) if opponents are more evenly matched (in fighting abilities and re-source values). Therefore anything that makes the critical assessment of relative fighting abilities more clear-cut should tend to reduce the frequency of fighting, with conflicts being settled more parsimoniously via some appro-priate convention, such as 'biggest rival eats first.' Conversely, evenly matched combatants should have frequent, lengthy, and/or especially vicious fights. Because these contest models were designed specifically for non-relatives, the theory must be modified for kin. In general, relatedness between opponents is likely to reduce contests costs, although the problem is complex (Hines and Maynard Smith 1979).

11.3.1 *The sibling rivalry reduction hypothesis*

In the present context, parents may seek to minimize the sib fighting needed to get critical differences settled, in the hope that nobody will have to lose; alternatively, they may prefer a quick and efficient verdict. Hahn (1981) tailored Hamilton's concern about 'wasted energy' to the problem of hatching asynchrony in a brood-reducing bird. In such species, parental interests may be best served by creating an optimal degree of competitive asymmetry, and balancing the facilitation of brood abridgement if and when a crunch occurs against the risk of premature or unnecessary losses. A corollary of this is that parental policy should promote the highest *parental efficiency*, defined as the ratio of fledged young per unit parental effort (Mock and Ploger 1987).

Hahn (1981) tested the basic Sibling Rivalry Reduction Hypothesis in a

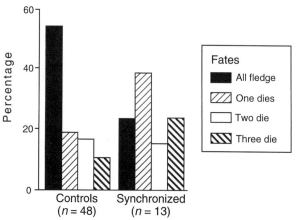

Fig. 11.1 Asynchronous hatching produces dramatically higher reproductive success for laughing gull parents. Hahn swapped eggs to create 13 synchronized broods (in which all three chicks hatched on the same day) and compared their fledging success (survival to day 30) with 48 naturally asynchronous control broods. Controls produced significantly more fledgelings (mean = 2.13 vs. 1.62), mainly because of broods that escaped mortality altogether (filled bars). (From Hahn 1981.)

laughing gull (*Larus atricilla*) colony by exchanging hatchlings between nests so as to create experimental broods that were artificially matched in size. As predicted by the hypothesis, these 'synchronous' nests had a lower level of fledging success than did controls (Figure 11.1). However, Hahn did not provide details about what disputes arose, how they were settled, or which parties exerted the greatest control over limited resources during the nestling period.

11.3.2 *Testing egret hatching asynchrony*

The hatching intervals of herons and egrets match those of laying rather well (Inoue 1985), although the *A–B* gap tends to be somewhat shorter than that for *B–C*, *C–D*, etc. (Werschkul 1979, Mock 1985), suggesting that effective incubation commences sometime between the laying of the *A* and *B* eggs. Unlike the situation in some raptors (Bortolotti and Wiebe 1993), no reversals of laying order/hatching order have been found in egrets (D. Mock, unpublished data). Junior eggs are also slightly smaller (Custer and Frederick 1990), but the consequences of this for sibling rivalry have not been studied. The first experimental manipulation of ardeid hatching intervals, which predated the sibling rivalry reduction hypothesis, was performed with little blue herons (*Egretta caerulea*) in order to test whether asynchrony influences the evenness of food apportioning among nest-mates (Werschkul 1979). Chicks were swapped between nests to produce five synchronous three-chick broods plus 10 synchronous four-chick broods. The resulting weight gain patterns of these *synchronous experimental* (hereafter abbreviated to 'SE') broods served as the indirect diagnostic for food distribution. Compared with controls of normal hatching asynchrony, Werschkul's SE broods provided two intriguing results (Figure 11.2). (i) They showed less within-clutch variation in growth rates (as predicted) and (ii) they grew much more rapidly than 'average' chicks in the control nests. In fact, all SE sibs tended to grow at the high rates of *A*-chicks in normal broods! Because the broods were not kept under close observation (i.e. they were measured during brief visits every 2–7 days), no information on sibling aggression or other behavioural forms of sib competition was available. Werschkul noted that, while the accelerated SE growth rates might have stemmed from increased parental food deliveries (for which he had no data), he thought that parents must normally be 'unable to distribute food evenly among nestlings'. He concluded that the typical bias in consumption (caused by the asynchrony in normal broods), rather than the total amount of food delivered, was the primary factor accounting for the stunted growth of junior sibs in normal broods. In terms of POC theory, his interpretation would be an 'offspring wins' resolution—larger chicks use

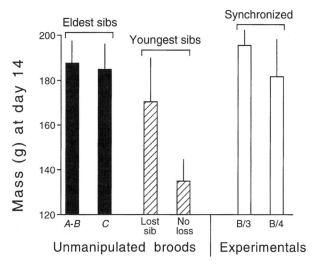

Fig. 11.2 Little blue herons normally produce a hierarchical sibship through asynchronous hatching, such that by the 14th day, the three eldest siblings in broods of four or five (the two filled bars = means + 95% confidence intervals) are substantially heavier than the one or two youngest sibs (hatched bars). These, in turn, are much heavier when someone in the brood has predeceased them ('lost sib' column) than when no such losses ('no loss' column) have occurred. Interestingly, the young of artificially synchronized broods (open bars) of three (B/3) or four (B/4) chicks grow as rapidly as the most privileged members in hierarchical sibships, suggesting either that the normal situation is remarkably wasteful or that parents of experimentally synchronized broods deliver far more food. (From Werschkul 1979.)

their size advantages to deprive parents of the parentally preferred even distribution.

The first experimental studies to incorporate the detailed observations of brood activities were begun independently (in Japan and Texas) and simultaneously (literally, within 2 weeks of each other), both using cattle egrets. Fujioka (1985a) created 12 SE broods (the number of chicks in each matching the original complement for that nest (i.e. two to five), which he compared with 38 control broods (also of varying original sizes) that had normal hatching asynchrony. Seven of his controls were composed of swapped non-sibs, to allow any kin-recognition effects to be detected (none were apparent). The behaviour of the nestlings was recorded during all-day observations of five SE and six control nests once every 5 days until the chicks were about 10 weeks old.

The design of Mock and Ploger (1987) was quite similar, except that it was restricted to natural broods of three hatchlings and included a second manipulated treatment, the *asynchronous experimentals* ('AE'), in which the mean successive inter-hatching interval was doubled from approximately 1.5 days to 3 days. This study included 42 SE nests, 38 AE nests, and 54 control nests (nine of which were composed of non-sibs, an exercise that once again failed to detect any kin-recognition effects). By using a rotating team of observers, the Texas study provided detailed full-day activity records for 10 SE, 7 AE, and 12 control broods every second day till broods were at least 25 days old.

The Japanese study included growth measurements that corroborated Werschkul's growth phenomenon: SE chicks gained weight as rapidly as *A*-chicks in control broods and showed no signs of the starvation ordinarily suffered by junior sibs. Without further details, this result could have suggested that hatching asynchrony imposes a net cost to chick growth. However, Fujioka also provided missing pieces of the puzzle. His SE broods begged more often (especially during the first 30 days) and somehow inspired their parents to work harder! SE parents were less idle and provided more food for their chicks during that first month. In Texas as well, SE parents were solicited more and delivered about 30% more food than parents of control broods (Mock and Ploger 1987). Extra food notwithstanding (see Chapter 6), fighting was much more frequent in SE broods, where rather stable linear dominance hierarchies emerged and the attacks gradually came to be concentrated on the weakest sib (Fujioka 1985a). That is, after a brief round-robin tournament involving all dyads, one habitual loser descended into the victim role (ordinarily pre-set by hatching order). In the Texas study, where the single brood size simplified statistical comparisons, the average SE fighting rate was three times that of controls and five times greater than that of AE chicks (Mock and Ploger 1987).

By reducing sib fighting, hatching asynchrony probably does reduce energy wastage (a point currently under study), but the fact that parents are sensitive to the higher begging rates of SE broods and seem to be 'tricked' into delivering more food somewhat complicates our understanding of just what hatching asynchrony accomplishes. At the very least, it suggests that parents are not bringing as much food as possible. Once one realizes that parents do 'slack off' on potential effort, the contribution made by asynchronous hatching for any one breeding cycle needs to be evaluated in terms of its effects on parental efficiency (which presumably enhances future success).

A simple index was devised to express units of parental efficiency as *nestling success* (the number of chicks surviving the 25-day period) divided by units of *parental effort* (food delivered to the nest). On that basis, Texas cattle egret parents achieved significantly better results with their normal hatching interval, which outperformed the AE treatment by 26% and the SE treatment by 45% (Figure 11.3). Doubling the asynchrony may also impose further costs, such as prolonged exposure to egg and small chick predators (Clark and Wilson 1981).

In a repeat of this protocol with blue-footed boobies, control broods (with a mean hatching interval of 4.1 days) once again produced higher parental efficiency scores than either synchronized (interval = 0.2 days) or exaggeratedly asynchronous broods (interval = 7.5 days; Osorno and Drummond, 1995). As with cattle egrets, synchronized broods of boobies produced similar numbers of fledgelings as controls, despite receiving 50% more food. Without initial size differentials, the siblings fought more (60% more blows) in establishing dominance relationships, which contributed to parents being about 19% less efficient.

Of course, in times of unusually high food availability, the parental efficiency argument could favour tighter synchrony in hatching (i.e. if delivering all of that extra food is truly affordable; Fujioka 1985a). One might

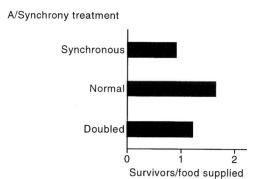

Fig. 11.3 Parental efficiency defined as the number of chicks surviving through the first month, divided by the amount of parentally delivered food) was significantly higher in cattle egret broods that had normal (mean = 1.5 days) hatching intervals than in those for which the intervals had been eliminated (synchronous) or enlarged (doubled). (From Mock and Ploger 1987.)

speculate that this underlies the uncommon habit of switching the asynchrony patterns between broods within a single season. A few birds, such as great tits and American coots, hatch their early-season/first broods synchronously, when food availability is believed to be more predictable, and then change to asynchronous hatching.

Finally, parent American kestrels (*Falco sparverius*) actively manipulate the degree of hatching asynchrony to match current local food conditions (Wiebe and Bortolotti 1994a). Correlational data across four field seasons in northern Saskatchewan showed that hatching tends to be more asynchronous during years of low food availability (as measured by small-mammal trapping) than in years of high food availability. Within-season variations similarly showed that parents on low-food territories and/or with poor female body condition at the time of laying tended to hatch their eggs with greater asynchrony than parents in better circumstances. To separate the usual covariation problems, a field experiment was performed over three summers in which some randomly chosen pairs were provisioned (two dead mice were left on top of the focal nestboxes each day) for about 2 weeks (halting when the first egg appeared in two summers and continuing through incubation in the third), while time-matched control pairs were sham-visited but not fed. Both groups produced clutches of the same size (five eggs), but food-supplemented mothers laid slightly larger eggs (6–7% by volume) that hatched over consistently shorter spans (i.e. were more synchronous), resulting in less within-brood size variation.

The functional significance of kestrel hatching asynchrony was explored in a separate experiment (using different pairs), following the general chick-swapping protocol used with gulls, egrets and boobies. Once again, parents caring for artificially synchronized broods made more provisioning trips (up to 31% more), but produced lighter nestlings. From a modified index of parental efficiency (brood mass at fledging divided by parental delivery rate when the oldest chick was 25 days old), parents of asynchronous broods were shown to be 26–56% more efficient than those rearing synchronous broods (Wiebe and Bortolotti 1994b). This result is highly consistent with the fledging success data from their provisioning experiment. When parents were 'tricked' into facultatively increasing the degree of hatching synchrony, their mean fledging successes dropped by about 20% (Wiebe and Bortolotti 1984a).

11.3.3 *Hatching asynchrony experiments and life history concerns*

The optimal degree of hatching asynchrony for a given bird is likely to depend on multiple factors, including the relative frequencies of benign vs. harsh resource levels (Pijanowski 1992) and the long-term impact of parental effort on lifetime reproductive success (Mock and Forbes 1994). We are certainly in no position at this time to argue that the hatching intervals produced by parental behaviour are truly optimal for cattle egrets or any other species, but some published reviews of the experimental results seem inclined to bury that possibility—prematurely in our opinion. Part of the confusion may derive from the fact that Hahn's (1981) original paper presented such dramatic results that field-workers may simply have come to expect immediate short-term differences in current brood survival, and thus to regard anything less definitive as running *counter* to the general prediction that observed levels of asynchrony should enhance parental fitness. If so, this is an extremely onerous criterion. Seldom are phenotypic traits *required* to produce effects that can be measured in terms of significant and speedy differences in death rates. (For example, there would be, no explorations of 'optimal foraging' in such a scientific climate.)

Of 30 hatching-asynchrony field experiments that have now been conducted on 25 different species (Amundsen and Slagsvold 1991), most show that synchronized broods produce more fledgelings. These differences were not always significant, but 22 studies showed higher means for synchronized broods, and only six showed higher means for asynchronous broods, a pattern that is highly significant (signed-ranks test) regardless of whether the unit of analysis was the individual study, species, genus or family (Amundsen and Slagsvold 1991). On the other hand, chicks from asynchronous broods tended to be heavier, although this trend was less strong than that for fledging success. These results certainly call into question Lack's original suggestion that asynchrony exists in order to maximize single-year nestling *production*, and they are leading some to suspect that hatching asynchrony is unavoidable and deleterious.

While it is wonderful to have so many studies to think about, it seems reasonable for today's focus to be less on Lack's views of 1947 and 1954 and more on the life-history aspects of parental strategy, which have been prominent for three decades (Williams 1966b). While the effects on lifetime success are vastly more difficult to measure, they are of disproportionate importance. When all of these experimental studies began 18 years ago, the possibility that parents would respond differently to experimentally syn-chronized broods was not anticipated. Either Hahn's gull parents were not inspired to deliver more food (parental effort was not quantified at those nests; D.C. Hahn, personal communication) or their efforts were overridden by other factors. It appears that in only four of the 30 experimental studies reviewed by Amundsen and Slagsvold (1991) were data collected on parental effort. All four of these (Fujioka 1985a, Hébert and Barclay 1986, Mock and Ploger 1987, Gibbons 1987) plus another four completed since (Osorno and Drummond 1992, Hébert and Sealy 1994, Wiebe and Bortolotti 1994b, Machmer and Ydenberg submitted) found evidence that parents provide considerably more food (30–50% more food delivered) to synchronous broods and thus respond in ways that would tend to mask asynchrony's advantages. While we can only guess as to *why* parents do this (perhaps synchronized broods generate super-normal begging stimuli and parents respond strongly to unusually robust broods), the exercise of caution seems to be in order.

In general, we have very little information about the lifetime consequences for parents of working above their normal level, although that literature is growing. A simple quantitative model for the life history consequences of brood reduction (Mock and Forbes 1994) shows, at least in long-lived species, that even a slight rise in parental survival (e.g. for raising an easily trimmed asynchronous brood) may suffice to compensate for loss of a victim. Indeed such survival benefits can absorb surprisingly high short-term losses.

The main implication of this for POC studies is, once again, that we are handicapped by the complexity of identifying the parental PI optimum in most cases. Once the possibility of adaptive favouritism has been acknowledged, it quickly becomes apparent that very sophisticated experimental designs will be essential for exploring POC empirically, and the key question remains the morbid one—when offspring are killing their sibs, are the parents approving?

11.4 Other empirical tests of behavioural POC

A very different kind of avian social system provides another case of behaviour that is apparently illustrative of POC, but which crumbles under close scrutiny. In co-operatively breeding white-fronted bee-eaters (*Merops bullockoides*), a young male's attempts to breed are often actively and effectively opposed by his own father, which results in the son returning to serve further as a helper at his parents' nest. At first glance the POC interpretation is obvious. Surely the son is being dragged home, sullen and mutinous, to be exploited. However, when the options are evaluated with data, quite a different picture emerges. The son achieves virtually the same fitness reward from helping (0.47 'offspring equivalents') as he would have obtained by breeding (0.51), suggesting that the interaction that superficially seemed so overbearing was, upon reflection, a toss-up to the offspring (Emlen and Wrege 1992). We may see a 'squabble' (paternal interference), but no discrepancy in genetic interests.

In mammals, the most popular POC cases involve 'weaning conflict' over when parental largess (milk) should be discontinued. Of the species for which this has been proposed, the case for red deer (*Cervus elaphus*) rests on perhaps the most convincing field data. The fitness of a yearling calf (in terms of growth and over-wintering survival) is enhanced if it is allowed to continue suckling, but that of the mother is maximized (in terms of lifetime production of surviving offspring) if she is impregnated each year (Clutton-Brock *et al.* 1982a, Clutton-Brock 1991). These two events are mutually incompatible, so if there are any steps the calf can take to obstruct its mother's oestrus and/or fertilization, it should probably do so. One ploy explored by Gomendio (1991,

1993) involves the well-documented link between vigorous suckling and the prevention of ovulation (e.g. Louden *et al.* 1983), although it is hard to imagine why the mother's physical superiority would not be used in defence of her own best interests (Mock and Forbes 1993). In most (perhaps all) such cases, maternal toleration of contraceptive levels of suckling probably occurs when conception is not optimal for her, although further study of this mechanism may prove quite interesting. In cases where it seems superficially that the mother is imposing her wish for an earlier termination of nursing than the offspring might prefer, it is important to remember that the offspring has to make a number of physical changes in gut physiology, and thus must antici-pate the dietary change in advance and start to make adjustments (Bateson 1994). For example, it has been proposed that a very few genes (only one in humans) that control the production of lactase in the small intestine may cease functioning at an optimal time and that this change may drive weaning for reasons that are beneficial to the offspring (Lieberman and Lieberman 1978).

The equivalent transition to (or toward) independence in birds has received rather little attention, probably because newly fledged offspring cease to be sessile and are soon far more mobile than earth-bound ornithologists. There has been some discussion, none the less, of which party (parents or chicks?) controls the timing of fledging. Davies (1976, 1978) postulated that parents may reduce the frequency of deliveries to the nest as a means of lowering the relative pay-off associated with passively waiting for easy food. Because young birds of most taxa are notoriously inept at their own early foraging efforts (e.g. Orians 1969, Desrochers and Ankney 1986, Weathers and Sullivan 1989), they presumably would 'prefer' to defer that responsibility. In a few species, parents may even use direct physical abuse to encourage greater independ-ence in older chicks (Leonard *et al.* 1988). However, Nilsson and Svensson (1993) have suggested that fledging may be more clearly interpreted as a direct manifestation of sibling competition, at least in altricial species. They see the race to acquire flight (but see also Bustamente *et al.* 1992 for an account of comparable hustling by non-volant penguin young) in terms of how that skill enables chicks to intercept parental deliveries away from the nest. It may be that *brood division* (wherein certain siblings move around with each parent; see Harper 1987) follows as the direct result of such pursuits.

In general, the most compelling demonstrations of POC are likely to be of the 'offspring wins' variety, for the simple reason that 'parent wins' is the default (Mock and Forbes 1992). We have always known that adults can often manipulate their own young offspring willy-nilly (e.g. brooding male fish may consume some of their eggs in order to sustain their brooding of the others; Sargent 1992, FitzGerald 1992). Such traits are not counterintuitive discover-ies that require Trivers' insight about relatedness asymmetries for an explana-tion. The potential elegance of POC is that it suggests reasons why offspring might rise, and effectively so, against the tyranny which they often face. Not surprisingly, then, the clearest cases for POC are likely to arise from systems

"Gee whiz . . . you mean I get a THIRD wish too?"

Fig. 11.4 'Man bites dog' is news. 'Parent controls offspring' is not. In this cartoon Gary Larson captures the priorities of a magically empowered offspring. Now *this* would constitute a compelling demonstration of POC!

where the offspring have some power (Figure 11.4). Accordingly, the best case published to date of behavioural traits that are likely to have been shaped by POC (Mock and Forbes 1992) is probably that of brood sex-ratio conflict in primitively social bees.

The idea that hymenopteran brood sex ratios should evolve according to the predictions of POC theory originated with Trivers (1974) (see also Trivers and Hare 1976). In many social ants, bees and wasps the mother abdicates brood care to senior offspring (workers), thereby empowering them. At the same time, the genetics of haplodiploidy automatically creates strikingly different investment optima for the mother and her worker daughters over the future brood. In particular, the mother's coefficient of relatedness is 0.50 to all her offspring, while a worker can have r-values as high as 0.75 to her future sister (if they share a common father), but is always related to a future brother by only 0.25 (because he is haploid). For this reason, depending on the mating system (true r between sisters) and other factors, workers might prefer the brood sex ratio to be substantially skewed toward female siblings, while the mother favours sex-ratio parity. This argument has been tossed back and forth vigorously (e.g. Alexander and Sherman 1977), but has recently passed what appears to be a very strong twofold test. First, when the original foundress (the mother) dies in a nest of sweat bees (*Halictus rubicundus*) she is replaced by one of her daughters. At such a moment, all of the non-reproductive workers' relatedness to the next brood is channelled through their sister, the new 'queen'. For that reason, they no longer experience the asymmetrical relatedness to the new brood (now composed of their nieces and nephews) that they did to previous broods (their brothers and sisters). The theory predicts that such workers should no longer favour female larvae, as before, and indeed males increase at this time (Yanega 1988, 1989). Secondly, experimental removal of the original foundress in a related bee, *Augochlorella striata*, reducing the problem of possible covarying factors, produced the same result: when worker asymmetry-of-relatedness disappears, male frequency rises (Mueller 1991). Thus, it appears that the offspring workers ordinarily skew brood sex ratio toward the production of females (to their advantage and against that of the mother), and then switch to produce more male larvae as soon as their relatedness to offspring eliminates the haplodiploid effect (see reviews by Boomsma 1991, Seger 1991).

11.5 Empirical tests of life-history POC

So far, we have restricted our discussion to behavioural manifestations of POC, largely because that was the class of phenotype for which the theory was originally introduced, and because of the ubiquity of squabbling. The case of brood sex-ratio manipulation may sit astride the border between behaviour and life-history traits, but it clearly does involve behaviour (the actions by which the life-history trait is altered). In the preceding chapter (Section 10.1.4), Le Masurier's (1987) review of parasitoid clutch sizes provides a strong candidate for an 'offspring wins' POC. In species where larvae have highly evolved warrior morphology, only one egg is oviposited in a given host; species without special weaponry lay many more eggs. It is quite plausible that mothers have conceded the clutch-size issue to their despotic progeny.

11.6 Measurement problems for POC

As with behaviours that look like altruism, the important conceptual distinction must be made between phenomena that resemble POC and those that really constitute an *evolutionary* conflict. Our working definition of POC as 'a disparity in the evolutionary optima for PI between parents and offspring' makes this point explicitly. The empiricists' nightmare stems from Trivers' definition of *parental investment*, which, strictly speaking, requires that both the benefit (to offspring survival) and cost (to parent's future success) be *demonstrated* before any parental effort can qualify as PI.

The following argument is adapted from Thornhill's (1980) discussion of forced copulation ('rape'), where he stated that, to qualify as true evolutionary sexual conflict, one must show that the act simultaneously increases the perpetrating male's fitness and decreases that of the female. Otherwise, it might be either maladaptive for the male or a cryptic benefit for the female, which leads to endless, if familiar-sounding, debate over whether the female is a victim or an active solicitor. In its essence, then, evolutionary conflict requires demonstration that the two parties not only *appear* to have a conflict of fitness interests—behaving as if in opposition to one another (the phenotypic conflict)—but that they be enhancing their own fitness while depressing that of their opponent (the genotypic conflict). In POC, this means that successful parental efforts must depress the fitness of certain offspring *while enhancing parental fitness*, and that the successful offspring's behaviour must reduce parental fitness *while enhancing* the individual's own inclusive fitness (Stamps and Metcalf 1980, Mock 1987, Mock and Forbes 1992).

Pragmatically, there are two ways in which this could be achieved. Genotypic conflict requires either a large amount of descriptive (correlational) data on everybody's production of grandchildren, or some clear-cut experimental manipulations that expose the inclusive fitness consequences of different outcomes. Ideally, POC studies should include *reciprocal 'winning' criteria* and evidence that (i) parents and offspring act *as if* at cross-purposes (documentation of phenotypic conflict[1]), i.e. that parents are trying to do X while offspring are trying to do Y, (ii) when X occurs ('parent wins'), net offspring fitness suffers measurably, and (iii) conversely, when Y occurs ('offspring wins'), net parental fitness suffers measurably. These are onerous criteria but, in our opinion, constitute a worthy goal.

Summary

1. A distinction is made between true evolutionary (genotypic) conflicts-of-interest between generations and the phenotypic (especially behavioural) 'conflicts' we can readily observe. For clarity, we label the latter 'squabbles'. Conceding that the battleground for parent–offspring conflict (POC) exists, we focus on whether a given behaviour is likely to have evolved *because of* the specific underlying genetic asymmetries indicated.

---

[1] We note that there could be genetic POC that produces no visible 'squabbles' if one party wins outright and the other concedes, rather than putting up a losing struggle.

2. Although a great many squabbles have been interpreted as evolved manifestations of POC, few convincing tests support such claims.

3. To illustrate the problems, we focus on avian siblicide, which has been proposed as a clear example of POC, the idea being that fatal sibling aggression must necessarily reduce parental fitness while enhancing that of the victor. Parents were predicted to interrupt fights and/or to offer special succour to a victim that was being socially excluded from food. However, field observations of egrets and other siblicidal birds have found no such thing: with poorly-documented exceptions, parents are reported to stand by passively during lethal attacks. Furthermore, considerable evidence shows that parents are capable of delivering much more food to their current offspring than they actually do, even though starvation underlies most brood reduction.

4. In considering why parents act in a *laissez-faire* manner during siblicides, attention has shifted increasingly toward long-term parental strategies that incorporate short-term compliance.

5. In this stochastic world parents can derive several benefits from creating an optimistic clutch size, the net value of which can be enhanced by features that facilitate brood reduction later on, when needed. If one takes the view that parents create a modest *core* brood (basically the number of offspring that they realistically expect to carry through the whole dependency period), and then add one or more *marginal* offspring, it may be very much in their interests for such marginal individuals to be maintained in a state that does not threaten the core. This perspective accounts for many levels of parental favouritism, and quickly draws one away from the assumption that parents value all offspring equally. Although equally genetically related to all offspring, they may confer handicaps/benefits unevenly across the brood as part of an overall strategy for dealing with environmental unpredictability.

6. In many brood-reducing birds, the handicap of choice appears to be a temporal one. By incubating prior to the completion of laying, parents give the first-produced embryos a head-start that leads to hatching asynchrony and a consequent size hierarchy among brood-mates. Because it appears that parents can control the hatching span, its effects on their fitness (as measured by the growth and fledging rates of offspring) have been used to evaluate whether the observed levels of hatching asynchrony (and perhaps hormone allocation) are beneficial. This has been pursued primarily through experimental exchanges of chicks between nests, designed to reduce or eliminate the sibling size hierarchy.

7. Dozens of such studies produced a most puzzling pattern. The asynchrony normally produced by parents often seems to have either no effect or deleterious ones. However, a few studies in which the swap protocol was accompanied by sampling of parental effort uncovered a complicating artefact. For some reason parents supply considerably more food to

artificially synchronized broods than to controls. This not only shows that such parents routinely withhold available food from their families, but it also seriously complicates the interpretation of studies in which parental effort was not monitored because accelerated food deliveries are likely to inflate nestling growth and survival. All the (still few) studies where degree of hatching asynchrony has been evaluated in terms of its impact on 'parental efficiency' have shown that parents fare better with asynchronous hatching.

8. American kestrel parents on poor-food territories (or experiencing poor-food years) tend to hatch their eggs at longer intervals than those in better situations. Furthermore, if food is artificially supplemented to some pairs but not others, the provisioned broods hatch more synchronously than controls. Thus it appears that parents of this species, at least, adjust the hatching span facultatively in relation to cues that may normally augur well for conditions of high food availability. When brood reduction is likely to be necessary, the handicap is enlarged.

9. Life-history models show that only a very small enhancement of parental survival can compensate it for a lost offspring. Thus, in long-lived taxa, parents may be better off taking quite a substantial short-term loss so as to come out ahead in the long run. This may help to explain why parents of such species often seem lazy, selfish and indifferent to suffering offspring.

10. The illusion of POC can also be seen in the context of co-operatively breeding bee-eaters, where attempts by sexually mature sons to breed independently are often thwarted by harassment by their fathers. Paternal pestering apparently coerces the son into returning to his parents' nest and resuming duties as a helper. Although this type of squabble looks as if POC is involved, estimates of the son's pay-offs from the two options (breeding vs. helping) show them to be equal.

11. In mammals, weaning conflict in red deer is a likely candidate for POC. The yearling would almost certainly gain by being allowed to continue nursing (yearlings that do continue enjoy higher rates of overwintering survival), but maternal fitness is better served by weaning the yearling and becoming impregnated again. As a philosophical point, cases of probable POC in which the offspring has no real power are of little interest because they explain nothing especially novel. Cases abound where parents impose their will on their smaller and weaker progeny (consider filial cannibalism). For this reason, cases like the red deer contribute little to our understanding of parent–offspring conflict. The calf simply seems unable to realize its preference that the mother skip mating.

12. It follows that the most compelling cases of behavioural POC are found where offspring have some real power, namely in primitively social insects where both reproductives and their worker daughters play active roles in matters subject to dispute. The relatedness asymmetries of haplodiploidy lead to the prediction of worker–queen conflict over the sex ratio of

reproductive broods: the queen is expected to favour parity; the workers are predicted to favour a female skew. Of special interest is the time when the old queen dies and a worker takes her place. In the ensuing cycle, workers are predicted to change their preference and to favour increased numbers of males (their choice is then between nephews and nieces, not brothers and sisters). Some field studies, including an experimental demonstration, have supported these predictions.

13. It is hoped that future research on POC will include experimental manipulations of both parties. When offspring are allowed, artificially, to 'win,' the fitness of the parents should be depressed. Conversely, when parental wishes are enforced, offspring fitness should drop. These two criteria, together with evidence that offspring have some capacity for getting their own way, should elevate this concept above the 'interesting theory' stage.

12

Sibling rivalry in birds

My reason for suggesting such an improbable idea as the fratricide hypothesis here is that I want to make a general point. This is that the ruthless behaviour of a baby cuckoo is only an extreme case of what must go on in any family . . . Even if we cannot believe that outright fratricide could evolve, there must be numerous lesser examples of selfishness . . .

(Dawkins, 1976)

In the remaining chapters, we present a preliminary review of the sprawling and uneven empirical and natural history literature pertaining to sibling rivalry in various taxa. The diversity of traits that relate to, and may have resulted from, sibling competition is far richer than can be imagined from the study of one group alone. We shall start by surveying the ornithological work, the largest component and the one most clearly connected to the egret studies detailed earlier, but now approaching the material more broadly. Nestling birds compete with nursery-mates in different ways, including growth rate, unilateral effort (begging), and a full range of aggression (from the occasional peck to pushing from cliff ledges or ripping sibs open with hooked beaks).

As a quick reminder (see Chapter 1), we reserve the term *brood reduction* for those cases where partial brood loss results from competitive sibling interactions *per se* (usually as revealed by unequal growth patterns), treating siblicide as an aggressive subset (Mock 1994). Sometimes, however, published studies are less explicit about mortality causes than one might wish, which introduces some problems in summarizing across eras, taxa, and investigators. Here we commence with relatively mild 'begging scramble' rivalry (which can be fatal, if not bloody), and then move along the spectrum towards increasingly rigid brood reduction processes.

12.1 Non-lethal sibling rivalry in birds

Most of the examples given in this book are severe, with losers expiring quickly and pathetically, or succumbing after fiery struggles. This is an incomplete picture of the phenomena. In most bird species for which patterns of partial brood loss have been documented (primarily via censuses), there are no epic battles, no visible wounds. In many cases, we know little about cause of death and infer much after noting that eggs and hatchlings are regularly more numerous than fledgelings. Werschkul and Jackson (1979) also proposed that resource shortages per se may lead selection to favour 'races' in the growth of nestmates, an idea that was initially questioned (Ricklefs 1982) but more recently resurrected for further examination (Ricklefs 1993): certainly the very rapid growth of certain brood parasites points to this as a promising direction for future studies within competitive sibships also.

The behavioural mechanics of begging scrambles have been studied in several species. Not surprisingly, larger siblings often use their superior size to skew food deliveries toward themselves, although in practise it is difficult to separate the food-controlling actions of the nestlings (e.g. the jockeying by heavier sibs to position themselves favourably with respect to the parents' usual direction of approach; Bengtsson and Rydén 1981, Gottlander 1987, Khayutin *et al.* 1988, Stamps *et al.* 1989, McRae *et al.* 1993) from those of parents (e.g. if there is a tendency to favour the largest nestlings; Lack 1956, Ricklefs 1965, 1968, Hussell 1972, Rydén and Bengtsson 1980, Bühler 1981, Boersma and Stokes 1995). Obviously, the relative efforts made by all family members toward investment skews are of interest (Brockelman 1975, O'Connor 1978). For example, in six pied flycatcher nests, videotapes showed that two aspects of chick action, begging vigour and position in the nestbox (relative to the parent), significantly affected food allocation, with position being the more important predictor of success (Gottlander 1987). When broods were deprived briefly in order to intensify sibling competition, heavier individuals moved to the more favourable region near the parents. Even without artificial food stress, heavier chicks enjoyed a 'position effect' (i.e. obtained a higher number of feeds when near the parents) than lighter sibs. Thus, even in birds with relatively mild brood size hierarchies and no overt fighting, subtle advantages often accrue to the slightly larger nestlings. The frequency with which these skews result in fatalities (brood reduction) in this species varies considerably according to habitat (Lundberg *et al.* 1981, Lundberg and Alatalo 1992), whether the male assists in feeding the offspring (Alatalo *et al.* 1982), and so on. Such competition is also likely to be affected by the degree to which the parents' direction of arrival can be predicted by their offspring. In cavity- or burrow-nesters there is a definite 'best place' to be, but in cup-nesting species many points of entry can be equiprobable.

The term 'scramble competition' becomes a *double entendre* for some colonial birds, such as chinstrap penguins (*Pygoscelis antarctica*), because mobile chicks actively pursue returning parents and the sibling that wins the race gets the food. In this species, parents caring for only a single offspring delivered the same amount of food per visit as parents with two offspring, and the latter were far more likely to be chased—before, during, and after the meal—and their chases lasted eight times longer (Bustamente *et al.* 1992).

Similar chases in jackass penguins (*P. demersus*) also serve to skew food distributions (van Heezik and Seddon 1996). Many volant fledgelings also pursue their parents and the ability to intercept arriving parents has been suggested as an interpretation for the process of fledging itself (Nilsson and Svensson 1993).

12.1.1 *Passerine begging behaviour—nonviolent scrambling*

In some competitive avian sibships (including some with regular starvation), the need for sibling aggression may be obviated by certain parental traits. Specifically, if parental acts create *core brood* advantages (which may be rather meagre initially) or participate in a positive feedback loop that extends the *core's* lead, there may be little reason for chicks to incur the additional costs of fighting. Even if the parents are totally impartial, such that each meal is a totally 'fair' contest, modest pre-existing asymmetries among nest-mates may suffice without escalation. In great tit broods, for example, most feeding visits involve single prey items, with success for the smallest members hinging on position and timing of their begging. Energy spent on random and spontaneous begging would presumably increase the likelihood that such a chick might already be begging when the parent chanced to arrive, which might in turn improve its chance of receiving the item. However, parents pre-announce their arrival with a 'feeding call' that elicits the onset of such begging outbursts. The larger chicks tend to occupy the best positions, if such exist, and need only respond quickly to any cues of parental arrival in order to maintain their competitive edge (Bengtsson and Rydén 1981). Speed of response has been shown to be crucial by itself (Price and Ydenberg 1995, Leonard and Horn 1996). If parents give a signal (the feeding call), they probably devalue spontaneous begging, with its associated individual and shared/summed costs (e.g. attracting predators: Haskell 1994), and may simplify the seniors' food control. Other taxa beg in response to other cues (e.g. the tremble caused by a landing parent).

Field workers have begun to address the game-theoretic nature of begging scrambles explicitly, by testing how a given chick's solicitation and rewards obtained are affected by its nest-mates' activities. The begging behaviour of an individual American robin (*Turdus migratorius*) chick does influence its chance of receiving a given food item (Smith and Montgomerie 1991). When a focal chick is successful, it has usually been more vigorous (quicker to beg, more persistent) and better positioned than its sibs. Moreover, within-individual analysis shows that its successes occur when it begs above its own mean begging level; failures occur when it is relatively lazy. Chicks deprived of food for 1–3 h beg harder and increase their likelihood of being fed next. Most interestingly, the *nest-mates* of deprived chicks also increase their own begging (Figure 12.1). A similar result was obtained with European starlings when a focal chick was moved temporarily to nests where it experienced competition

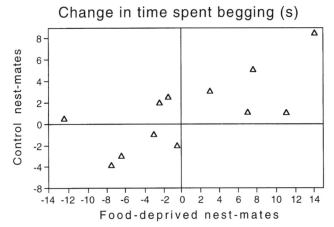

Fig. 12.1 American robin nestlings match the begging levels of their nest-mates. One chick was removed from each of 12 nests, when broods were *c.* 10 days old, and held for 1 to 3 h. These 'food-deprived' individuals begged more than before, and their un-deprived (control) nest-mates tended to match their begging intensity. Data plotted are mean values for each nest. (From Smith and Montgomerie 1991.)

that was either heightened (oversized brood), lessened (reduced brood) or normal. Upon return to its home nest, its begging intensity and probability of being fed were logically related to the degree of competition it had just faced (Kacelnik *et al.* 1995).

Of course, parents may also truly *prefer* to feed the largest chicks in gaping competitions (Lack 1956, Ricklefs 1965, Hussell 1972, Rydén and Bengtsson 1980), a predilection that backfires disastrously when the fastest growing individual happens to be a cuckoo. Although the proximate decision rules for parental allocations are receiving increasingly close attention (reviewed by Stamps 1993), clear-cut cases of subtle parent–senior collaborations will often be difficult to tease apart. In other cases, parents may take relatively direct action against individual nestlings, including selective infanticide (Section 12.2.2). Finally, the task of identifying which parties control nestling food allocation has become somewhat more complicated by the discovery that begging efficiency becomes increasingly temperature-dependent as the young develop endothermy. One-day-old red-winged blackbirds (*Agelaius phoeniceus*) stretch to nearly the same maximum height when cool (body temperature 25°C) as when warm (40°C), but 1 week later do not even respond at the lower temperature (Choi and Bakken 1990). Whether temperature affects the competitive performance levels of contemporaneous nest-mates that differ in age by only a day or so (passerines) or up to 2 weeks (parrotlets and barn owls) has yet to be explored. Another enticing layer of complexity is added by observations that barn owl *A*-chicks may control prey that they do not eat, selectively delivering it to their smallest siblings (Marti 1989).

12.1.2 *Long-term effects of non-lethal sibling rivalries*

Both growth rate and survivorship typically decline with increasing brood size (reviewed by O'Connor 1984) and with lower hatching rank (reviewed by Spear and Nur 1994, Stoleson and Beissinger 1995). These trends can sometimes be muted by artificial food supplementations (Figure 12.2; see also Graves *et al.* 1984, Simons and Martin 1990, Verhulst 1994). By focusing initially on competitions that retard growth of some or all nestlings, the issue of possible delayed mortality arises. A chick may obtain less food than it apparently 'wants' (i.e. it strives for more), which may or may not affect its fitness (it may not know its true needs or its prospects regarding future meals, etc.) after all other factors have been considered. Still, one must start somewhere, and some very reliable data address the relationship between fledging size and probability of future survival (recruitment into the breeding population).

Some species exhibit positive correlations between mass at fledging (or at the end of parental care) and subsequent survival. Not surprisingly, these data are especially convincing for the long-studied, highly sedentary population of great tits (*Parus major*) at Wytham Wood (Perrins 1965, Garnett 1981), for which the relationship between chick fledging weight and probability of future recapture is quite linear (a 2.5% increase in survival for each additional gram of body mass; Perrins and Moss 1975[1]). Body weight was negatively correlated with brood size in each of 15 years (Perrins 1965). The shape of that regression varied dramatically between years, reflecting relatively good and bad times for fledgelings of intermediate weight (Klomp 1970). In this species, linear body dimensions of fledgelings also faithfully predict both adult size (e.g. for tarsus length, $r = 0.917$) and post-fledging success in dominance contests over food (Garnett 1981).

Certain philopatric colonial seabirds have also shown positive survival effects of greater fledging weight. Manx shearwater (*Puffinus puffinus*) chicks from England overwinter in Brazil, and then return to their natal colony to breed. Heavier chicks survive at significantly higher rates (Figure 12.3), a pattern also found in cape gannets (*Sula capensis*: Jarvis 1974). On the other hand, significant relationships between fledging weight and survival have been sought but not detected for various other species, including common guillemots (*Uria aalge*: Hedgren 1981), puffins (*Fratercula arctica*; Harris and Rothery 1985) and European sparrowhawks (*Accipiter nisus*: Newton and Moss 1986).

The exact mechanism by which being underfed as a nestling (fledging at relatively low body weight) translates into survivorship is seldom known precisely. Some laboratory and field data for carrion crows (*Corvus corone*) show that experimentally underfed nestlings achieved lower asymptotic body

---

[1] A similarly linear relationship has also been shown for this species in The Netherlands, where the causality has been demonstrated by experimental brood-size manipulations (Tinbergen and Boerlijst 1990).

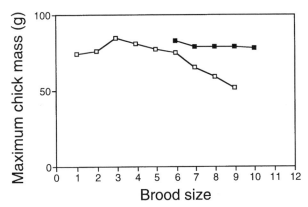

Fig. 12.2 European starling nestlings relying on parental food deliveries achieve greatest maximum size (mass) when raised in broods of up to about six chicks, while those in larger broods are stunted (open squares). The fact that this is apparently due to food shortages is revealed when the largest broods are experimentally provisioned (filled squares). (From Crossner 1977.)

mass than controls fed *ad libitum*, although their linear measurements were not stunted. Nonetheless, these undersized individuals achieved lower social status in the adult flock hierarchy (Richener *et al.* 1989).

Probably the best demonstration of long-term survival effects of *relative* competitive abilities of siblings in a brood-reducing bird rests on recent analyses of western gulls (*Larus occidentalis*) nesting on South-east Farallon Island, California. These long-lived birds are so philopatric to the natal colony that three cohorts of chicks banded as hatchlings in 1978–1980 could be followed over many years. Spear and Nur (1994) divided the survivorship records for over 1000 chicks from one-, two-, and three-chick broods into four temporal stages and examined how hatching date, intra-brood hatching order, and initial brood size accounted for variation in survival. Of central interest to us are the results for the 352 three-chick broods in the sample. As expected, the two senior ranks (*A* and *B*) had much greater success through the 'natal' stage, when all are dependent on parentally delivered food (Figure 12.4a). Proportionately more seniors than *C*-chicks also survived the following 9 months (juvenile stage). Beyond that point, however, the proportions of *A*-, *B*-, and *C*-chicks that survived through 2 years of sub-adult life and became reproductively active were quite equal, so that the eventual skew in recruits to the breeding population (Figure 12.4b) is attributable to rank differences during the nesting and post-fledging periods. Spear and Nur (1994) speculated that the residual rank effect during the juvenile stage is related to the young gulls' reliance on localized and defendable food sources (e.g. garbage dumps) during that phase. Perhaps the dominance/subordinance acquired during intra-brood skirmishes extends into the later context as well. However, the gulls' hatching date provides an even stronger determinant of both early

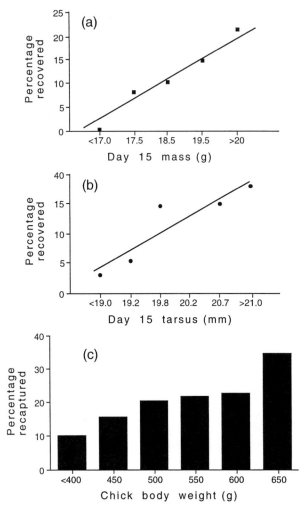

Fig. 12.3 Larger fledgelings often realize higher survivorship, as shown in these recovery data for **(a)** great tit mass at fledging; **(b)** great tit tarsus length at fledging (both from Garnett 1981); and **(c)** manx shearwater mass at fledging (From Perrins *et al.* 1972.)

survival and eventual recruitment in this population (Figure 12.4c). Thus, these sibling rivalry dramas are played out against very different backdrops, which could in turn be due to at least two mutually compatible causes. Early-nesting broods may receive more food and achieve greater success because of declining food availability over time, or because higher quality parents tend to breed earlier, or both (if better parents are the only ones capable of breeding during the food-rich early weeks). Clearly, different broods may experience very different resource levels despite being neighbours.

12.2 Facultative brood reduction in birds

We follow the useful, if arbitrary, convention of dichotomizing fatal sibling rivalry according to its frequency across nests. *Facultative* brood reduction systems are those in which even parentally handicapped siblings have a reasonable chance of surviving (e.g. egrets), while in *obligate* brood-reducers

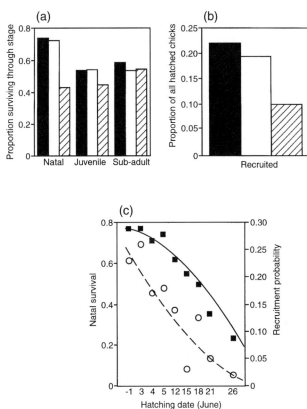

Fig. 12.4 Survival probabilities of western gull *A*-chicks (solid bars), *B*-chicks (open bars), and *C*-chicks (hatched bars) were calculated for 1,056 hatchlings from 352 three-chick broods. (**a**) It can be seen that only about 25% of the senior sibs but >50% of the *C*-chicks died during the 'natal' stage (until independence at age 3 months). Of the 673 chicks known to have passed that test, significantly more senior sibs than *C*-chicks were still alive at the end of the 'juvenile' stage (age 12 months). Of the 304 chicks followed thereafter, only about 50% survived through the entire 'sub-adult' stage to the minimum age of first breeding (33 months), but the relative losses were no longer correlated with hatching order. (**b**) The 'recruited' group (the 194 individuals passing through all of these stages and reaching the point of having eggs to tend) are, therefore, heavily skewed toward senior siblings. (**c**) Hatching date within the season exerted an even stronger influence on survival through the natal stage (■) and eventually recruitment into the breeding population (○). From Spear and Nur (1994).

the death of one sib is virtually guaranteed (e.g. in 90% of nests: Simmons 1988). While obligate brood-reducers are surely just the tail-end of a continuous distribution (Simmons 1991), the dual categories focus attention on different system features and questions, and thus are heuristically useful.

As with the egrets, immediate causes of nestling death vary considerably, and fatality often involves multiple contributing factors. In many taxa, the victim simply withers and starves (e.g. Howe 1978). Even in raptors, the most common (and in some species the exclusive) form of brood reduction has been called 'selective starvation' (e.g. Meyburg 1974, Newton 1977). In other taxa (e.g. blue-footed boobies, ospreys, owls), sib-inflicted injuries are common and may constitute the immediate cause of death (e.g. from unchecked bleeding, decapitation, etc.) or contribute to food deprivation.

Brood reduction may also be effected through physical eviction of nest-mates. It is suspected that this may occur even before (asynchronous) hatching is complete, essentially a 'cuckoo-method' of reducing food competition by elder nestlings (Alvarez *et al.* 1976). In cliff-nesting kittiwakes, siblicide sometimes consists of the *A*-chick driving its lone nest-mate off the home ledge and letting gravity do the rest (Braun and Hunt 1983, Dickins and Clark 1987). Similar patterns have been noted for cactus wrens (*Campylorhynchus brunneicapillus*; Simons and Martin 1990), where the fallen victim may be impaled on cactus spines, and for laughing doves (*Streptopelia senegalensis*; Walsh 1980). Most brood reduction in Mexican blue-footed boobies results from *A*-chicks driving *B*-chicks from the nest scrape, and thus out of the safety zone; once at large, most *B*-chicks are quickly killed by adult neighbours (Drummond *et al.* 1986).

12.2.1 Fatal starvation—intra-brood hierarchies

Like egrets, most brood-reducing birds hatch their eggs asynchronously to some degree, and the resulting initial size/age disparities strongly affect the competitive dynamics when demand exceeds supply. A second factor, egg-size polymorphism, can contribute to offspring fitness (Parsons 1970), but its effects on sibling rivalry are usually swamped by asynchronous hatching (e.g. Howe 1978, Lundberg and Vaisanen 1979, Poole 1982, Slagsvold *et al.* 1984, Lamey 1990, 1992, Magrath 1992). For example, in blue-eyed shags (*Phalacrocorax atriceps*) egg volumes differ only slightly and the mass of a freshly hatched *C*-chick is on average just 4 g less (*c.* 10%) than the masses of its senior siblings *when they were newly hatched*. However, the temporal disadvantage is such that the two seniors are already 70% and 110% heavier by the time *C* emerges from its shell (Shaw 1985). That harsh reality only gets worse; 5 days later, the seniors on average weigh 102% and 183% more (Figure 12.5). The situation can be far more extreme in other species. In one bald eagle brood, for example, the three chicks hatched at comparable weights of 85–90g (albeit on different days), but showed such sharp subsequent divergence that by the

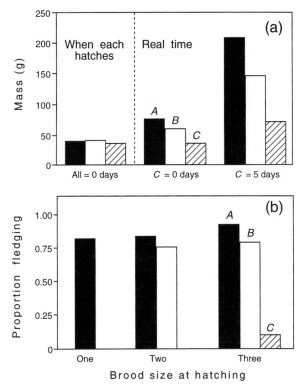

Fig. 12.5 Blue-eyed shags. **(a)** Initial size differences due to hatching asynchrony. The youngest (*C*-chick) is nearly as large at hatching as its more senior nest-mates were when they hatched (left panel), but its temporal disadvantage means that the seniors are substantially larger in real time when *C* hatches (*C* = 0 days old) and this difference typically expands (*C* = 5 days). **(b)** Survival consequences. The *A*-chick in each brood size (solid bars) enjoys very high success, as does the middle chick, *B* (open bars), but the youngest, *C* (hatched bars) seldom even fledges. (From Shaw 1985.)

C-chick's sixth day it had actually lost weight (down to 80 g), while its senior siblings weighed 477 g and 260 g (Gerrard and Bortolotti 1988).

The magnitude of initial handicaps can dramatically influence whether the last-hatching sibling survives to fledge (e.g. Shaw 1985, Skagen 1987). Gibbons (1987) showed that the fate of the youngest jackdaw (*Corvus monedula*) chick is strongly predicted by its weight relative to that of its nest-mates (Figure 12.6). Nearly 90% of the mortalities involved the brood's youngest chick, most of which had been visibly emaciated just before dying. Although the total food requirements of avian broods are typically well within the parents' limits during at least the first few days (Bryant 1975, Perrins and Moss 1975), some young may still go underfed even then. If the hierarchy were generally disadvantageous to parents (Bryant 1978a, b, Skagen 1988), this early period

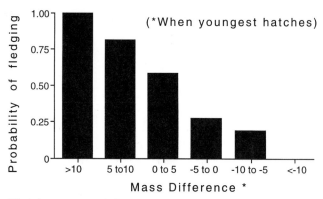

Fig. 12.6 Fledging success of the youngest member of jackdaw broods varies as a function of its nestmates' relative sizes on the day it hatched, expressed here as its own mass minus the brood's mean in grams. (From Gibbons 1987.)

would seem an opportune moment for them to nullify the sibling size disparities. That is, a correction should often be possible, e.g. by feeding the smallest young selectively, or by presenting food in bursts so as to swamp the monopolization capacity of senior chicks. Such favouritism towards the smallest nestling has been reported in a few taxa (e.g. Stamps *et al.* 1985, Gottlander 1987, Lyon *et al.* 1994, Leonard and Horn 1996), but seems to be quite exceptional. Moreover, the possible influence of yolk hormone depositions in shaping brood hierarchies (see Chapter 11) is just beginning to be explored. However, it appears that, in many species, the absolute magnitude of the initial size hierarchy among nest-mates does not narrow (O'Connor 1975, Bengtsson and Rydén 1981, Ohlsson and Smith 1994), and may indeed widen. This probably depends on whether food is limiting. An early headstart is likely to be amplified if the superior rivals consume virtually everything provided (bouncing them quickly into the steep stretch of an S-shaped growth curve), while lesser siblings struggle to hang on. Conversely, if food is not limiting, the gap should close eventually (e.g. when the younger nestlings hit their growth spurt and the seniors' gain rate levels).

Parental behaviour during the critical early brood period is often consistent with the view that the hierarchy benefits them overall. For example, when three-chick herring gull (*Larus argentatus*) broods received very modest food supplements (a single can of cat food per day) during just the first 5 days after hatching of the *A*-chick, all three offspring grew more rapidly and more *C*-chicks fledged, compared to unprovisioned control broods (Graves *et al.* 1984). This finding is particularly impressive because these provisioned *C*-chicks had access to the cat food for only 2 or 3 days. It would seem that parents could easily supply that much additional food, but for some reason they do not, which suggests that they are simply aiming for a smaller *core brood*: see Section 1.2) than the one they have hatched (see also Nisbet and Drury 1972, Parsons 1975, Graves *et al.* 1984).

Considering the prominence of Lack's 'resource-tracking' hypothesis, it is surprising that the fundamental interaction between hatching spread and realized food levels has seldom been demonstrated experimentally. In a fine series of field studies of European blackbirds (*Turdus merula*), Magrath (1992) showed that the intra-brood size hierarchy is, in fact, controlled by the early onset of maternal incubation (which accounted for almost all of the subsequent size variation among siblings, far outstripping egg size differences and other candidate factors). Mothers incubated intermittently prior to the completion of laying and did not necessarily commence with the penultimate egg, suggesting that they, like American kestrels, may fine-tune the overall hatching span facultatively. The blackbird sibling hierarchy thus created produced more and better (heavier) young under conditions of food shortages than did artificially synchronized broods, while achieving lower success when food was more plentiful (Magrath 1989). Importantly, the disadvantage to synchronous broods could be nullified via provisioning with extra food during the shortage, which supported Lack's hypothesis that the hierarchy advantage results from an interaction between asynchrony and food levels (Figure 12.7). Magrath (1991) also showed that heavier nestlings are subsequently recruited into the local population significantly more often than lighter ones (when analysed across broods), documenting a long-term advantage to parents of raising robust offspring. Similar results from other avian taxa were summarized in Section 11.3.2. Some especially compelling parallel evidence now exists for American kestrels (Bortolotti and Wiebe 1993, Wiebe and Bortolotti 1994a, 1994b).

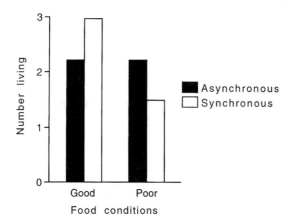

Fig. 12.7 European blackbird broods that were artificially synchronized (open bars) achieved significantly greater fledging success when food conditions were favourable, but broods that were naturally asynchronous (filled bars) prevailed when conditions were poor. (From Magrath 1989.)

Once it has occurred, the loss of a brood member may impinge on the personal fitnesses of surviving family members in different ways (O'Connor 1978), not all of which are positive for the remaining sibs. The death certainly eliminates one rival, but it may also affect the brood's overall thermal properties, alter the likelihood (and consequences) of predation, and influence subsequent parental effort and commitment in complex ways (for an example, see Figure 12.8). Considering the last point in isolation, POC models based on the 'cake' assumption (that parents fix the amount of PI that they will provide; see Section 10.2.2) automatically assume that killing a rival will enlarge each survivor's slice of the cake (e.g. Husby 1986). However, as we have seen for cattle egrets (Section 6.3.1), parents may adjust their delivery rates to match brood size. If parents do this in a precise 1:1 manner, the survivors can actually take a cut, since they were probably consuming all or part of the victim's share prior to brood reduction (Mock and Lamey 1991). In these cases, at least, brood reduction appears to increase the parents' allocation of food to themselves (see similar discussion in Martins and Wright 1993a, b). Conversely, if the cake is not affected by brood reduction, loss of a victim frees up whatever resources it had been consuming. In three-chick herring gull broods, for example, brood reduction can significantly increase *B*'s food intake and mass (Graves *et al.* 1984). In these same broods, *A*'s mass was not affected, suggesting that the top-ranking sib truly takes its cut 'off the top' (as required by the hierarchy models: Chapters 3 and 8) and is relatively buffered from lesser siblings. Survival rates reflect this same privilege of rank and parental behaviour may also be re-shaped in the aftermath (Figure 12.9). As with great egrets, the continued existence of the youngest chick in blue-eyed shag broods has no effect on *A*- or *B*-chick survival, although *C*'s own chances soar if either senior fails to hatch.

Keeping our focus within the confines of a sibship for a moment longer, it is not necessary for the fitness of surviving siblings to be enhanced in any dramatic or immediate way when a victim sibling dies. This point has confused some workers, so it is worth pondering the simple fact that such kin are likely

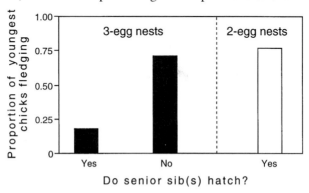

Fig. 12.8 The fate of the youngest blue-eyed shag chick usually hinges on whether a senior sibling fails to hatch in three-chick nests. (From Shaw 1985.)

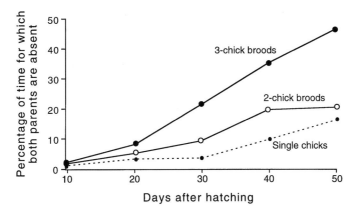

Fig. 12.9 Herring gull parents spend more time away from the nest (and presumably foraging) when their three-chick broods are intact than when these broods have reduced to two or one. (From Graves *et al.* 1984.)

to have dubious future prospects from the very start and to embody rather little investment at the time of their demise. In addition, the key consideration usually centres on how much future investment will be required (Dawkins and Carlisle 1976) and how likely that sufficient resources will be available (Figures 12.10 and 12.11).

To understand the functional significance of brood reduction fully, of course, one must also take the parents' interests into account (Williams 1966b). Because parents are often highly sensitive to the magnitude of the task before them and are known to adjust their effort level accordingly (e.g. Figure 12.12), the degree to which current effort may affect their future reproduction has become an active research topic. Experimental tests of parental effort are notoriously difficult to perform under field conditions and have produced mixed results (see reviews by Nur 1988, Clutton-Brock 1991, Hochachka 1992), but the logic and some of the strongest evidence (e.g. Figure 12.13) make a compelling case. It seems likely that many aspects of parental behavior, at least in long-lived species will prove to be best explained in terms of long-term pay-offs. And even in short-lived species, brood reduction effects on parents may prove to be surprisingly clear (Hõrak 1995), once they are sought.

12.2.2 *Filial infanticide*

The reported examples of selective parental infanticide (i.e. excluding whole-brood abandonments) were few and rather anecdotal (see Mock 1984a) until fairly recently. At least 18 species have now been observed practising filial cannibalism (Stanback and Koenig 1992), and the number of parents that kill their own offspring without ingesting them may be much higher (e.g. Aguilera 1990). In addition, many parental acts that contribute to offspring mortality

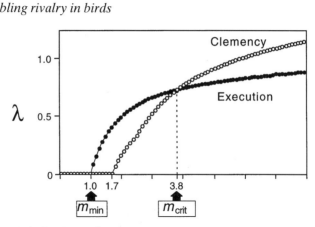

Fig. 12.10 Inclusive fitness, λ, of a senior sibling as a function of differing amounts of parental provisioning, M, with respect to the two behavioural options of 'execution' (●) and 'clemency' (○). At some relatively low level of provisioning, resources sufficient for one offspring to survive are reached, m_{min}, scaled here to 1.0. Above that level it initially pays the senior to execute its sib, at first because the junior sib is of no value until its own minimal start-up costs have been reached, and for a considerable range of resource values beyond that because the senior's direct fitness is not devalued by coefficient of relatedness. Eventually, however, a point is reached where resources are sufficient for both to do reasonably well; if M will exceed m_{crit}, the senior's fitness is better served through clemency. (From Forbes and Ydenberg 1992.)

are coming to light, ranging from subtle cases of favouritism (e.g. Magellanic penguin parents feed one chick preferentially and give only 'excess' food to its marginal sib: Boersma 1991, Boersma and Stokes 1995, Blanco *et al.* 1996) to outright executions. Hooded grebe (*Podiceps gallardoi*) parents may desert unhatched *B*-chicks after *A* has hatched successfully; the *core* chick literally rides off on a parent's back (Nuechterlein and Johnson 1981). Similarly, African white pelican parents (*Pelecanus onocrotalus*) may discontinue incubation after *A* hatches, such that soon thereafter '… forsaken eggs littered the breeding ground' (Vesey-Fitzgerald 1957: 128), while *A*-chicks congregated in crèches. Desertion during hatching has also been reported for wattled cranes (*Grus carunculata*; Arbrey 1990), and permanent neglect of unhatched second eggs has been reported in two eagle species (Rettig 1978, Gerrard and Bortolotti 1988). American and European coot (*Fulica americana* and *F. atra*) parents may also lead their partially hatched brood away, deserting junior eggs (Gullion 1954, Horsfall 1984a, b). In addition, parents of the latter species kill some of their new chicks by 'tousling' (seizing them by the head and worrying them violently; Horsfall 1984a).

Filial infanticide is a fine example of something that is rare, and hence difficult to study quantitatively, but potentially important in terms of fitness pay-offs. Thus data showing that filial infanticide may be a routine part of brood-reducing processes are especially interesting. The clutch size of

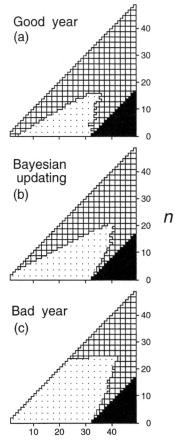

Fig. 12.11 A stochastic dynamic model shows how the senior sibling's optimal decision (concerning execution vs. clemency) is likely to change as the breeding season's resource quality is realized. As time *t* passes (horizontal axis, expressed in units of 'moves'), the senior sibling must assess and reassess its alternatives with regard to the daily number of food deliveries, *n*, the parents make. **(a)** In a good year, this value will tend to be high, and the better behavioural strategy will usually be clemency (cross-hatched zone), except where sustained periods of 'bad luck' have meant too few deliveries, in which case execution is favoured (stippled zone). There is always the chance that deliveries will be so consistently infrequent that even the senior dies (black zone), in which case decision-making is moot anyway. **(c)** By contrast, in a bad year the senior is much more likely to find itself in a situation where clemency is simply not affordable. **(b)** Interestingly, it may be possible for the senior sibling to use its *memory* of how the season has been going to a given point (Bayesian updating) to predict whether the current year is good or bad. The behavioural effect of such improved information is that the potential fitness gains offered via clemency can remain open as an option for longer. (From Forbes and Ydenberg 1992.)

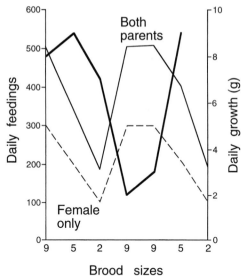

Fig. 12.12 Repeated adjustment of parental effort with brood size manipulations at one experimental nest of European starlings. The original brood size (five nestlings) was altered repeatedly from high values of nine chicks to low values of two chicks across seven consecutive days (ordinate). Parental effort was adjusted every time, with both the total daily deliveries from both parents (thin solid line) and female-only deliveries (thin broken line), spanning a twofold range. These parental adjustments failed to compensate fully, however, as per-capita food (bold line) was lowest when the brood was largest and vice versa. (From Drent and Daan 1980.) See also Westerterp *et al.* (1982).

Heerman's gull (*Larus heermanni*) in western Mexico is two or three, and mean fledging success is low. In one study of three-chick broods, the youngest sib died in 13 out of 14 cases, and in seven of these the victim was observed being pecked by its own parent while still alive (Urrutía and Drummond 1990). Victims were not killed outright by these pecks, but the chicks were 'probably inhibited' from begging and eventually died from starvation or thermal stress, since parents provide not only food, but also essential shelter from tropical solar stress. In two other cases, the victim was left unbrooded and exposed. The authors speculated that the third egg's primary function was as insurance against hatching failure (see Section 12.3).

Parents were also observed killing the youngest brood member in nine of 63 white stork (*Ciconia ciconia*) nests during the first 2 weeks of nestling life (Tortosa and Redondo 1992). These killings occurred in relatively large four-chick broods, and this report substantiates earlier descriptions of 'Kronism'[2]

---

[2] An old term for filial infanticide, coined from the mythical Titan Kronos, who ate his own children.

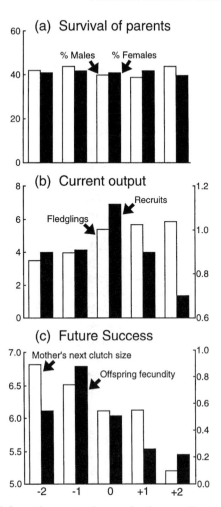

Fig. 12.13 Collared flycatcher cost-of-reproduction results. A total of 320 clutches were manipulated, either through additions (+1 or +2 eggs), subtractions (−1 or −2 eggs), or reciprocal exchanges (change of 0). There were no effects on adult survival rates **(a)** for either sex. Larger families did produce more fledgelings **(b**, open bars) but control nests recruited the most offspring (filled bars). **(c)** In the following year, both maternal fecundity (measured as number of eggs laid; open bars) and offspring fecundity (measured as number of recruits produced by surviving daughters; filled bars) declined significantly with experimental family size. (Redrawn from Gustafsson and Sutherland 1988.)

in the species (Schüz 1957). There is also reason to suspect filial infanticide in acorn woodpeckers (*Melanerpes formicivorus*), an altricial species in which brood reduction occurs very early (usually during the runt's first day post-hatching, when its yolk reserves are still present) in most nests. Some victims bear adult peck marks (Stanback and Koenig 1992). And in royal penguins, mothers actively evict the lesser of their two eggs provocatively soon after it has been laid (Miskelly and Carey 1990, St Clair *et al.* 1995).

Parental destructiveness (see also Section 12.2.2) raises the intriguing and unresolved question of why parent birds do *not* provide the executioner's muscle during brood reduction as a routine matter of course. Superficially, parents seem eminently qualified, by virtue of their power and experience, to handle a job that is otherwise slow, difficult and probably costly for gawky nest-mates. Inasmuch as parents of many species are willing to abandon whole broods when breeding conditions deteriorate to the point of making continued investment futile (e.g. Kahl 1964, Pomeroy 1978, Mock 1984a), why do they not make similar partial brood decisions? One possibility is that there are incentives for dragging out the process, and for keeping the victim alive and in limbo for as long as possible within the constraint of minimal investment. So long as the victim is alive and capable of recovering from short-term deprivation (e.g. after being underfed for several days or weeks), it carries some insurance value (see Section 12.3). Ironically, adults may typically be 'too efficient' at killing. Along the same lines, by leaving the subjugation process to one or more senior offspring, parents may obtain valuable information, a test of sorts, about the mettle of *core brood* members. An older nestling that cannot protect its privileged rank from junior sib challenges may be less worthy of future investment than the obvious alternative (a variant of the Progeny Choice Hypothesis).

Finally, mention must be made of the 'guano ring' hypothesis for brood reduction in blue-footed boobies. In the Galápagos, S.J. Gould (1982) noticed that his approach to a booby nest elicited no alarm response from a brooding parent until his toe actually crossed the white ring of dried excreta surrounding the nest, at which point the adult became highly defensive. Noting also that some chicks were dying in interstitial zones between nests, he reasoned (probably correctly) that they had been driven out initially by an aggressive senior sibling. However, his conclusion, which surely counts as an outstanding 'Just-so Story' (*sensu* Gould and Lewontin 1979), was that sibling aggression evicts the junior sib to a point outside the parent's defensive perimeter, and that an extraordinarily rigid (but simple) behavioural rule of the parent (not to allow anything except its mate to cross the guano ring) serves to render that eviction fatal. In essence, the parent was depicted as a very active collaborator in a senior sibling–parent infanticide pact against the victim. What a story! Sadly, there is no evidence that it is true, as neither Gould nor anyone else has actually observed the parental rejection on which the argument rests. On the contrary, a simple experiment showed that it is highly unlikely. At five booby nests, Dave Anderson tied monofilament fishing line to the junior sib's leg or wing and retired to a concealed spot 20–30 m away. After 10 min he gently eased the chick out from under its parent and well beyond the guano ring before relaxing the tension on the line. In every case, the chick returned home without eliciting parental discord, although the parent typically watched it do so, and one senior sibling gave it a 54-peck reception (Anderson 1991).

12.2.3 *Aggression in facultatively siblicidal birds*

In avian broods aggression subserves the general function of buffering senior siblings from competition and, especially, from role reversals by upwardly mobile junior nest-mates (Stinson 1979: 1221). Fighting is often fiercest between the penultimate and ultimate chicks in the brood hierarchies and can often be regarded fruitfully as a battle for next-to-last place, which is often the last *core* slot. (This is because parents in most avian taxa produce only one *marginal* egg, so only one junior sib typically must die for the supply–demand balance to be reached.) Thus, it seems likely that most sib fights should be initiated by the penultimate chick, and directed toward its younger nest-mate. In two-chick kittiwake broods, for example, senior sibs started 118 of 120 fights (Braun and Hunt 1983), while in three-chick cattle egret broods, nearly 75% of all attacks were launched by *B* toward *C* (Mock and Lamey 1991).

Across taxa, the distribution of overt fighting by avian nest-mates is poorly understood. Except when fatal, aggression is likely to be overlooked and under-reported, perhaps because it is both relatively uncommon and brief: most field studies do not involve long periods of observation at nests. This is not intended to suggest that closer study will eventually reveal all avian nestlings to be aggressive. Here, we draw attention to three general patterns of fighting behaviour in facultatively brood-reducing birds: (i) *none*; (ii) *facultative aggression*; and (iii) *obligate aggression*. Perhaps the most surprising members of the first group are some penguins and motmots that have extremely high rates of brood reduction without overt aggression (see Section 12.3), plus a few others that are taxonomically close to highly aggressive species. For example, several species of small raptors appear to have all of the necessary prerequisites for adaptive sib fighting (including resource shortages, monopolizable prey items offered by parents, weaponry, crowded nest conditions and asynchronous hatching), yet show few signs of using combat (Balfour 1957, Newton 1977, 1979). In such taxa, critical information is lacking as to whether such siblings (i) are truly engaged in a real competition (i.e. resources are limiting) or (ii) have more cost-effective alternative means of resolving competitions.

Facultative aggression is the ability to adjust fighting rate and vigour to external conditions. The only clear, experimental demonstrations of this ability have been for blue-footed boobies and ospreys (which regulate their fighting to hunger or nutritional state; Section 6.2.2), cattle egrets (which respond to brood size; Section 6.3.1), and great blue herons (which respond to prey size variations; Section 6.1.5). However, other species have been reported to show changes in fighting rates that are probably correlated with environmental predictors of future resource shortages. These include kittiwakes (Braun and Hunt 1983), South Polar skuas (Procter 1975) and various medium-sized raptors (*Buteo* and *Accipiter*: Newton 1977). Furthermore, Newton (1977, 1979) pointed out that sibling aggression in raptors is affected positively by the amount of time that chicks are left unbrooded, which may be linked indirectly to the food supply. When food is scarce, the female must hunt more, thereby leaving the chicks freer (unbrooded) to fight.

In a literal sense, many facultatively siblicidal birds (e.g. great egrets, cattle egrets, great blue herons, and Mexican blue-footed boobies) show *obligate aggression* in that one observes at least low levels of fighting in every brood, regardless of extrinsic conditions. This contrasts with ospreys, where combat seems to be entirely lacking in some high food areas (Poole 1982) or seasons (Stinson 1977, McLean and Byrd 1991), but fatal when conditions are poor. The distinction between brood reduction systems that always feature what might be called 'low-grade' fighting and those that do not might be related to the overall predictability of resource levels. Investing modestly in a latent dominance hierarchy could reflect (i) the likelihood that brood reduction will prove necessary eventually, plus (ii) the higher costs of creating fully functional hierarchies *de novo* if and when a deferred crunch arises (see review in Mock *et al.* 1990).

An especially interesting context for sibling aggression arises in the case of cavity-nesting birds whose nest compartments occur at the ends of narrow burrows. Not only is the parent's direction of arrival entirely predictable (as with any cavity nest), but the hallway itself can simply be occupied at a bully's convenience. This means that the temporal guesswork of choosing the perfect moment to beg is also eliminated: whoever can arrange to hold the position nearest to the burrow's mouth is guaranteed to make first contact with a returning parent. Thus, birds such as motmots (Scott and Martin 1986), kingfishers and bee-eaters seem to be logical candidates for effective and low-cost use of sibling aggression. In fact, at least two species of bee-eaters are known to practise brood reduction (Lessells and Avery 1989, Bryant and Tatner 1990), although the behavioural details of the way in which size and aggression may be used to control the key position are not known. Hatching asynchrony spans a relatively large span (up to 9 days for the whole brood) in European bee-eaters (*Merops apiaster*; Lessells and Avery 1989). In blue-throated bee-eaters (*M. viridis*), nestlings possess hooked mandibles that are used in siblicidal fights, and which may in fact be weaponry specialized for such fights (Bryant and Tatner 1990). If so, this parallels some cases in mammals, amphibians, social insects and plants (see Chapters 13–16), where morphology and physiology are sometimes adapted for the destruction of sibs. It also represents a neat case of convergent evolution with the nestlings of honeyguides (Piciformes, Indicatoridae), which also have mandibular hooks with which they kill nest-mates. The telling difference is that honeyguides are interspecific brood parasites, so their victims are of the host species and their selfishness is unconstrained by relatedness (see this chapter's header).

12.2.4 *Cannibalism after brood reduction*

It was long assumed that chicks dying in the nest were routinely eaten by other family members, especially siblings, and that the food value thus gained, rather than the competitive release, was the primary pay-off. Siblicide was suspected of being '. . . in all probability invariably followed by cannibalism' (Ingram 1959: 225). The assumed connection with food shortages made raptor siblicidal aggression seem reasonable, but was hard to reconcile with the usual clutter of prey present during the killings. Ingram (1959) also reported that short-eared owl (*Asio flammeus*) parents created a 'larder' of prey near the nest in order to control the senior sibs' destructive instincts. It was subsequently proposed that parents might create an extra offspring during periods of relative abundance early in the season (at laying) so that its tissues could be consumed by stronger siblings later, when food became scarce, an idea sometimes referred to as the 'icebox hypothesis'.

Although there have been occasional observations of sibling birds consuming nest-mates and/or discoveries of half-eaten sib carcasses (Ingram 1959, 1962, Heintzelman 1966, Beecham and Kochert 1975, Pilz 1976, Steyn 1980, Baker-Gabb 1982, Bechard 1983, Kojima 1987, Bortolotti *et al.* 1991), the frequency and nutritive importance of such events are very much open to doubt (Stinson 1979, Mock 1984a, Stanback and Koenig 1992). For example, in one northern goshawk (*Accipiter gentilis*) brood whose female parent had vanished for reasons unknown 3 days earlier, the *A*-chick attacked, killed, and immediately began to consume pieces of its *C*-sib, pausing only to chase *B* out on to a precarious perch from which it eventually fell to its death (Boal and Bacorn 1994). In many of the most-studied siblicidal species (e.g. kittiwakes, boobies, egrets, pelicans), by contrast, the victim's body is typically not eaten, but left to rot on the nest floor. Furthermore, the victims are generally very young and commonly starving, suggesting that they offer little food value. Finally, even where they are eaten, the question of whether victims are killed *for that purpose* (as opposed to being scavenged after death; McNicholl 1977) is hard to resolve. In the only extensive data set from a species that does regularly consume brood-reduction victims, Bortolotti *et al.* (1991) found that such corpses were eaten in 20–63% of American kestrel nests that experienced the death of one or more nestlings. This ingestion was associated with times of low availability natural prey and on territories with relatively scarce food (i.e. starvation was most likely), but contrary to the logic of the icebox hypothesis, smaller victims were *more* likely to be consumed than larger chicks that died. In this, as in the overwhelming majority of avian brood reduction cases, then, the *principal* benefit gained from the death of a nest-mate probably relates mainly to decreased competition (Bortolotti *et al.* 1991).

12.3 Obligate brood reduction in birds

In the great majority of obligate brood-reducing birds, the parents produce two eggs, and then fledge only a single chick. In most of these cases, the victim's death is preceded and apparently caused by early and tenacious assault by the first-hatched chick. There are a few exceptions, including brown pelicans (three or four eggs, one fledgeling; Ploger 1992), brown boobies (*Sula leucogaster*) (occasionally three eggs, one fledgeling: Nelson 1978, Cohen Fernandez 1988) and the six species of crested penguins (*Eudyptes*) (only two eggs but no overt aggression en route to a single fledgeling: e.g. Lamey 1990), that offer some relief from this pattern. However, even the simplest two-egg siblicide system poses general questions related to the parents' vs. the bully's perspectives. The obvious first question concerns why parents bother to produce an egg/offspring that is doomed. (By this point, readers will be unsurprised by our orientation toward insurance logic here, but the matter is fundamental and ours is by no means a universally shared emphasis.) The second question concerns what the senior sib gains by inexorably sacrificing such a large fraction of its inclusive fitness?

Obligate brood reduction has been reported for approximately three dozen avian species and has been examined in varying detail for a handful of them. Overall, the data are rather sparse, and various reviews (e.g. Bortolotti 1986a, Simmons 1988, Lamey 1990, Anderson 1990b) have disagreed as to whether whether certain species 'qualify' as truly *obligate*. The practise itself appears to be most concentrated in four orders (penguins, pelecaniforms, cranes, and large raptors), the first two of which are colonial (and thus relatively convenient for study). Exceedingly little is known about crane brood reduction (see Novakowski 1966, Miller 1973, Walkinshaw 1973), even though several of the species are endangered with extinction and at least one major population restoration measure (the rearing of whooping crane chicks by sandhill crane foster parents) is presumably based on the fact that the *B*-chick is marginal and typically hatches into a near-hopeless situation. On a purely pragmatic level, measures that rescue such chicks might foster population expansion.

12.3.1 *Three obligate brood-reducing species*

Brief sketches of three relatively well-studied birds—the black eagle, American white pelican and rockhopper penguin—will illustrate some of the behavioural and ecological diversity found in obligate brood reduction and also reveal the underlying issues more fully.

Black eagle

Of 158 nest records for 2-egg clutches (the typical number laid in this African species, occurring at more than 75% of all nests; Gargett 1977), there is only one case (0.6% of the total) in which both chicks apparently survived to fledge (Brown *et al.* 1977). Various attempts have been made to increase this yield artificially (e.g. Gargett 1967, 1970a; see also Meyburg 1974 and Arbrey 1990 for other species), but even herculean efforts have so far failed to protect the victim from its relentless nest-mate. Similarly, in several other obligately siblicidal eagles the successful production of both young is sufficiently news-worthy to merit public proclamation (e.g. Gargett 1968, Steyn 1982, Hustler and Howells 1986).

The first full account of any avian siblicide may be Rowe's (1947) description for one black eagle brood; the most detailed is undoubtably Gargett's (1978) for the same species. In the latter case (Figure 12.14), the *A*-chick was 4 days old, healthy and feeding well on morsels of meat offered by its mother when *B* hatched. *A* even pecked twice at the pip hole in *B*'s shell, but *B* enjoyed 16 h of peace after hatching before the first full attack came. When *B* tried to feed for the first time, *A* began to peck its face, briefly but hard, and *B* withdrew. Gargett noted that there were 5.7 kg of mammalian prey (rock hyrax) carcasses on the nest at that time. An hour later, the first really sustained assault began, with *A* pecking *B* more than 30 times over a 3-min period. By the middle of *B*'s first day its face had been bloodied.

As the offensive resumed the next day, *B* grew increasingly timorous. Whenever it retreated, *A* followed it and persisted in pecking. The third day differed only in that *B* was too weakened by then to flee, and the mother ceased even trying to offer it food. It died that day. In all, *A* pecked *B* at least 1569 times during 38 separate attacks. The *A*-chick had gained 50 g during the coexistence period, while *B* lost 18 g. At the time of *B*'s death, there were still six hyrax carcasses on the nest, leading Gargett (1978: 62) to conclude that '... battle confers no advantage to the survivor.'

Fighting does not occupy an especially large fraction of the time prior to the

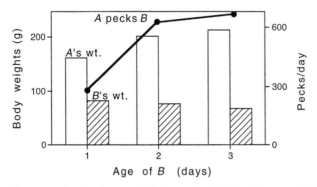

Fig. 12.14 The most closely observed black eagle siblicide (Gargett 1978) showed a pattern soon to be repeated in species that are easier to study. Aggression commenced immediately (line) and allowed the *A*-chick to eat well and gain weight (open bars), while the *B*-chick lost weight steadily (hatched bars).

brood reduction event. In Gargett's nest, attacks represented 9.2% of the total daylight hours. Rowe's (1947) observations were generally similar, except that his A-chick also had the habit of squatting on its nest-mate's back or head while receiving food. In this nest (where the killing took 7 days), 3.5% of the 39 observed hours were spent in fighting, and A spent another 4.6% resting on top of B.

The singular feature of black eagle aggression is its repetitive nature. Hoping to ease the younger sibling past what might conceivably have been a temporary period of viciousness on the part of the A-chick, Gargett (1970a) tried various hand-raising programmes, including alternating weeks where A was left in the nest and B was hand-fed in captivity, and vice versa. During change-overs, the sibs were allowed some time together on the nest, serving as brief tests to determine whether peaceful cohabitation were yet possible. However, 'the aggression reaction between black eagle siblings is as marked at 6, 7, and 8 weeks as it is in the first few days' (Gargett 1970a: 35). For many additional details on black eagles, see Gargett (1990).

American white pelican

In this species, obligate siblicide takes a different twist—although sib attacks are common, death seldom results directly from the wounds inflicted. Instead, the victim faces socially enforced starvation (Cash and Evans 1986a, b), differing from that of great egrets mainly in its higher mortality rate, or occasional sibling ejections that lead to predation (Evans 1984). The pelican A-chick consistently hatches from a slightly larger egg (c. 4% difference, by volume; O'Malley and Evans 1980) and earlier than its nest-mate by an average of 2–2.5 days (Cash and Evans 1986a, b, Evans 1996). Thereafter, the senior chick occupies the more forward brooding position for 97% of the time. Parents hold regurgitate in the gular pouch, offering it to the more anterior chick 89% of the time. The A-chick actively defends this profitable area, attacking B almost every time (94%) a brooding parent rises (e.g. to preen), and desisting only when B moves rearward. Attacks last from 34 s to 300 s and quickly produce visible superficial wounds on B's back and head. Parents display no favouritism toward the posterior chick, even when it begs loudly, nor do they interfere with the assaults. Most brood reductions occur in the first few days (Evans and Macmahon 1987). Chicks that survive for about 3 weeks move into large crèches where they are sustained via parental visits until fledging. Mortality during the crèche stage is very low (Knopf 1979, Evans 1984).

The evidence that starvation is the primary cause of death is slightly piece-meal, but consistent. First, during over 200 h of direct colony observations (pre-crèche phase), victims were not beaten comatose, but merely excluded from food. Only a few fell victim to predators and attack by neighbouring adults, although sibling harassment probably renders them more vulnerable to both (Evans and Macmahon 1987). In the most closely observed nests, *B*-chicks were seen to feed on only two occasions, both of which occurred when *A* had not attacked (Cash and Evans 1986a).

Secondly, *B*-chicks that eventually survived (usually after *A* had vanished) were significantly less plump (as measured by 'condition index,' the ratio of weight to culmen length) than surviving *A*-chicks of the same age (Evans and Macmahon 1987). Thirdly, there are indications—but no direct data in support of them—that parents withhold food during the brood's earliest days, in which case they may miss opportunities to get food to *B* after temporarily 'stuffing' *A*. At this point, the chicks are so tiny that even small amounts of food satiate them. Indeed these same parents soon deliver boluses (to their crèche-age offspring) that are larger than the tiny hatchlings' whole bodies (Evans and Macmahon 1987). Thus, because parents appear to have little difficulty in meeting at least one chick's extended needs (as reflected in the stable growth rates and high survivorship of one chick per brood), it is hard to imagine that they could not support both chicks temporarily, if doing so were in their best interests. It is more likely that parents are 'playing for one' all along (see Graves *et al.* 1984), effectively collaborating with *A*-chicks to facilitate *B*'s early death. The singleton eventually raised is *B* about 18–20% of the time (Cash and Evans 1986a, Evans and Macmahon 1987), when *A* dies young (Figure 12.15). The smaller size of the *B*-egg does not seem to reduce its ability to produce a viable chick, once the bully has died (or been experimentally removed: Evans 1996, 1997).

Rockhopper penguin

Finally, not all obligate brood reduction is siblicidal; some is free of fisticuffs, where the victim is eliminated via starvation without the extra burden of physical injuries. Parents of the six crested penguins (*Eudyptes* spp.) have a unique system of conferring size asymmetries of such magnitude that overt aggression is apparently rendered unnecessary (Lamey and Mock 1991). The most peculiar *Eudyptes* feature, the significance of which is not yet fully understood (Johnson et al. 1987), is that the first-laid egg is the sharply disadvantaged one. It is 15–42% smaller by volume (Warham 1975), by far the greatest egg-size dimorphism in Class Aves (Slagsvold *et al.* 1984), and is not incubated during the 4 to 5 days before the large *B* egg is laid (Gwynn 1953, Warham 1975, Lamey 1990) (Figure 12.16a). Interestingly, a similar pattern of obligate brood reduction via starvation of the lesser chick also occurs in Magellanic penguins (*Spheniscus magellanicus*), which lack the marked egg dimorphism. Here the allocation decision seems to be entirely parental (Boersma 1991, Boersma and Stokes 1995).

Once the clutch is complete, *Eudyptes* parents keep the *B*-egg in the more rearward position under the linear brood patch (Lamey 1992, St Clair 1992), which may or may not provide it with greater warmth while rendering it less

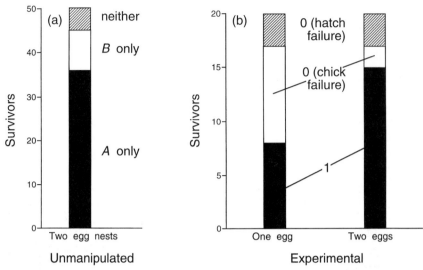

Fig. 12.15 The insurance value of American white pelican second-laid eggs can be seen in two ways. (**a**) In 20% of unmanipulated nests where eggs were marked at laying, the *B*-egg produced the sole surviving chick in 20% of all nests that did not fail completely. (**b**) Furthermore, nests where the back-up egg was experimentally removed experienced a much lower success rate than nests allowed to keep their second egg. (From Cash and Evans 1986a.)

conspicuous to skuas and other predators. In most (60%) rockhopper nests both chicks hatch, but *B* almost always hatches on the same day or even a day earlier than *A* (Lamey 1993, St Clair 1996, St Clair and St Clair 1996). Thus the second-laid *B*-egg hatches into a situation embracing both temporal and physical advantages.

In Tim Lamey's field study of *E. chrysocome* in the Falkland Islands, 76% of all second-hatching chicks died within the first 10 days and 85% of 168 two-chick broods had lost at least one nestling by the crèche formation stage, 2 to 3 weeks post-hatching (Lamey 1992), by which time they are still only 20–30% of the way through their dependency period (Warham 1975: 228-9). Fighting between siblings was negligible. The *B*-chick simply used its superior bulk to reach the parent's mouth, from which it fed first. The small *A*-chick commonly found itself sandwiched between *B*'s large body on one side and the parent's even larger body on the other, and could not manoeuvre effectively to compete for food. This pattern is quite common in the three *Eudyptes* spp. that regularly hatch both chicks (Figure 12.16b). As a result, the *A*-chick often fails to receive any food whatsoever and quickly starves.

In approximately one out of seven two-chick rockhopper broods, however, the survivor was the 'marginal' *A*-chick, not *B*. The factors that enable the lesser chick to turn the tables and outsurvive its nest-mate are not clear, but presumably include various density-independent mishaps that can strike individuals at random. Because single losses also occur regularly during incubation (Lamey 1992), *A*-chicks may have a non-trivial chance of fledging.

12.3.2 *Shared features in obligate brood-reduction taxa*

Several traits are common to the three species profiled above and virtually all other birds that practise obligate brood reduction (Edwards and Collopy 1983, Simmons 1988, Mock *et al.* 1990). First, clutch size is minimal, so the relative fitness and resource stakes are maximal. (Interestingly, this is not the case for

Fig. 12.16 Crested penguins. **(a)** The most extreme egg-size dimorphism in Class Aves is evident from this clutch of royal penguin on Macquarie Island. **(b)** In species where both eggs hatch, the small chick usually starves, sometimes as a result of being sandwiched between parent and nest-mate, as seen here for a Fiordlands crested penguin. (From photographs by Colleen St. Clair.)

some non-avian practitioners of obligate brood reduction; see Chapters 13–16.) With just two offspring, loss of the victim cancels half of the brood's potential Reproductive Value and doubles the share of food per capita in one fell swoop. Secondly, these species tend to have slow-growing nidicolous[3] offspring, which require comprehensive and protracted parental care. Thirdly, they tend to be very long-lived, suggesting that the production of fewer de luxe-quality offspring may pay eventual dividends on investment. Obligately siblicidal raptors acquire their definitive adult plumages at a mean age of 4.8 years, substantially later than the 3.6-year average for facultatively siblicidal raptors (Simmons 1988). Finally, these species exhibit the most exaggerated competitive asymmetries among siblings, both in terms of hatching asynchrony (Edwards and Collopy 1983, Bortolotti 1986a, Anderson 1989) and egg dimorphism (Slagsvold *et al.* 1984, Lamey 1990).

The small size of the brood is perhaps the most conspicuous of these features. When viewed by conservationists keen to replenish vanishing stocks of the largest raptors, siblicide must have seemed as frustrating as it appeared nonsensical. Ingram (1959: 22) argued that siblicide in times of food abundance must be maladaptive because '. . . it would then unnecessarily curtail a natural increase in the birds' population.' Wynne-Edwards (1962) cited raptor 'Cainism' and white stork 'Kronism' as examples of reproductive restraint that helped to keep populations comfortably below carrying capacity, which he interpreted in terms of his group-selection framework. Indeed, 'for-the-good-of-the-species' arguments have been invoked repeatedly with respect to these phenomena. Gargett (1978: 62) suggested that selection may be powerless to oppose black eagle siblicide until such time as the species is poised on the brink of extinction, and she also thought the insurance value of the second egg too low to make a 'serious contribution to the survival of the species' (Gargett 1991). Taking another tack, Brown *et al.* (1977: 70–71) simply concluded that 'the notorious Cain and Abel battle . . . is still an inexplicable example of apparent biological waste.' While we remain far from having all the answers, it now seems clear that explanations based on individual (genic) selection should have no difficulty in accounting for these phenomena.

12.4 · Brood sex ratios and sibling competition

Lack (1954) pointed out that gender differences in growth rate could affect the competitive dynamics of avian siblings, pointing specifically to raptors (where females are generally the larger sex). Discussion of this point was renewed by Edwards and Collopy (1983), who expressed concern about how certain gender/rank combinations might affect nest success. As a class, birds

---

[3] Literally 'nest-bound', but generally referring to relatively underdeveloped, hence helpless, hatchlings.

exhibit provocative variation in sexual size dimorphisms, with males being much larger in some altricial taxa (e.g. rooks and New World blackbirds) and females in others (e.g. many raptors and owls). If one assumes (i) that eventual adult size dimorphism correlates with unequal growth rates as nestlings (and not just duration of the growth period), (ii) that the changing relative sizes of such nest-mates affect their competitive dynamics and (iii) that a younger sib survives the brood reduction process long enough to overtake a senior nest-mate of the smaller sex—then the potential for some very complex and tantalizing sibling (and parent–offspring) dynamics exists. However, these assumed conditions are by no means universal, so a given species that is both size-dimorphic and brood-reducing may or may not show interesting patterns of this type. Even so, in the light of increasing evidence that offspring gender is not always randomly related to hatching order (see reviews in Clutton-Brock *et al.* 1985, Clutton-Brock 1986, Gowaty 1991, 1993), such effects having been reported for gulls (Ryder 1983), snow geese (*Chen caerulescens*) (Ankney 1982, but see Cooke and Harmsen 1983), boobies (Drummond *et al.* 1991) and various raptors (e.g. Bortolotti 1986b, Dijkstra *et al.* 1990, Olsen and Cockburn 1991, Anderson *et al.* 1993), secondary effects on competitive sibling dynamics are now under scrutiny. At present, it is probably fair to say that many interesting and controversial results, but no clear patterns, are emerging from the published work. We here present a brief sampler.

A fundamental premise of Fisherian sex ratio arguments is that offspring of the larger sex are more costly for parents to raise, a point that has been addressed in a number of size-dimorphic birds and generally supported (e.g. Cronmiller and Thompson 1980, Bancroft 1984, Teather and Weatherhead 1988). Applying this to problems of sibling rivalry, Howe (1976) showed that male common grackle nestlings (*Quiscalus quiscula*) suffer more than their (smaller) sisters when sibling rivalry is intensified via experimentally in-creased hatching synchrony. Røskaft and Slagsvold (1985) predicted and found that experimentally produced all-male broods of rooks (*Corvus monedula*) show more brood reduction than all-female broods. However, in addition to gender differences in food needs, sexual size differences can affect competitive abilities as well. Male red-winged blackbird nestlings (*Agelaius phoeniceus*) use their height advantage to intercept more parentally delivered food over their sisters' heads (Teather 1992).

Other predictions relate sexual size dimorphism to aggressiveness in siblicidal species. For raptors, Edwards and Collopy (1983) reasoned that if the younger sib belongs to the sex that grows more rapidly while sharing a nest with a sibling of the smaller sex (i.e. eagle broods in which *B* is female and *A* is male, typically abbreviated to M–F), the senior's supremacy may be threatened early and the cost:benefit ratios for each player's actions reshaped accordingly. For example, such an *A*-chick might be under stronger selection pressure to kill its sibling immediately, lest it face demotion. From the parents' point of view, growth reversals like this might well increase the overall costs of sibling rivalry. Conversely, these authors supposed that the reverse order, F–M, might generate a harmfully wide sibling size disparity (i.e. premature brood reduction), although the need for aggressive enforcement of social rank would probably recede. These arguments have led to very mixed empirical results.

From field data on bald eagles (*Haliaeetus leucocephalus*), where adult females are 25% heavier than males, Bortolotti (1986b) reported a statistically significant paucity of M–F broods, although all other dyadic combinations (M–M, F–F, and F–M) existed at their expected frequencies. This dearth of the M–F combination was interpreted as an adaptive parental manipulation for avoiding the pattern with accentuated rivalry, exactly the reverse of Edwards and Collopy's (1983) prediction. Bortolotti (1986b) justified his argument on the grounds that male bald eagle nestlings grow very rapidly immediately after hatching, and thus enjoy about a week of exceptional size supremacy (after which their advantage has decreased to that of the other dyadic combinations, with younger sisters likely to exceed the male *A*-chick in mass after 1 month). This is certainly a provocative result, although the sample sizes were not large (as one might expect, considering the difficulty of censusing eagle nests). The picture for golden eagles (*Aquila chrysaetos*) is different, perhaps because their growth pattern takes a different form to that of bald eagles, but is consistent with the Edwards–Collopy prediction: partial brood losses are apparently more frequent when *A*-chicks are female (Edwards *et al.* 1988).

In the Harris's hawk (*Parabuteo uncinatus*), Bednarz and Hayden (1991) found a dissimilar pattern. *A*-chicks tended to be male (in 69% of 95 broods), and such broods fledged significantly more young than those in which females hatched first. Here the dimorphism may be qualitatively different because females achieve their substantially (42%) greater adult size by taking more time than males, not by growing more rapidly. As in other brood-reducing taxa, junior sibs are significantly (but not by much) more likely to die as nestlings; survival was very high for all ranks in this communally breeding species. By the age of independence, the sex ratio is only slightly male-biased, perhaps in keeping with that sex's lower total production costs (Fisher 1930).

The most detailed study of gender effects in a siblicidal bird involved blue-footed boobies (Drummond *et al.* 1991), a species in which adult females are about 27% heavier than males, and younger sisters in M–F broods do overtake their brothers (despite a 4-day hatching interval) by their 37th day, on average. All the same, gender plays little or no role in the siblicidal process. Despite inevitable size reversals in these broods, no *dominance* reversals accompanied the *size* reversals in five of six closely monitored M–F nests. Furthermore, brood sex composition did not seem to affect growth rates or survival of the siblings. In particular, males in F–M broods fared as well as those in M–M broods. The early and persistent domination of booby junior sibs may endure because there is no moment at which it pays the younger sib to escalate. This in turn may be a function of fighting qualities other than size (e.g. speed and co-ordination) that improve with maturation (and thus are promoted by the wide hatching interval and may even be faster in the smaller sex, in any case) and are reinforced by early social experience (Drummond and Osorno 1992). It may be that parents have evolved the great hatching asynchrony as a means of putting such reversals effectively out of reach (Slagsvold 1990).

If broods of American kestrel are divided into 'early' vs. 'late' broods, according to the median hatching date, a significant tendency for the youngest sibs to be female (62% in 1990) in the early nests, but male (61% in the same year) in later nests occurs in some years, but not in others (Wiebe and Bortolotti 1992). This may be due to parental condition, fathers that were able to maintain their own condition through the pre-laying period raising female-biased broods, while males that lost condition did the opposite. Similarly, the mothers of female-biased broods had better condition scores than those of male-biased broods.

Meanwhile, Anderson *et al.* (1993) showed that inside American kestrel nests female nestlings use their greater size to out-scramble their brothers for food, without using aggression, provided that prey items are monopolizably small (see Section 6.1). In many cases, a declining food base forces a shift from more profitable vertebrate prey to large insects (orthoptera), which are small enough to be swallowed immediately. As a result, the sexual size advantage may come into play only when it is most needed for brood reduction. These authors suggested that a male bias in secondary sex ratio (i.e. at hatching) may exist in order to compensate for the subsequent higher female survival.

A marked within-season sex ratio skew occurs as a function of intra-brood rank in the European kestrel, which appears to be related both to deferred aspects of recruitment into the breeding population and to immediate intra-brood rivalries (Dijkstra *et al.* 1990). Early-hatching broods tend to have male senior sibs (*A* and *B*), a mix of middle sibs (*C* and *D*), and female juniors (*E*, *F*, and *G*), while later-hatching broods show exactly the opposite pattern (Figure 12.17). This is related to the differing prospects for sons vs. daughters in establishing themselves quickly (breeding successfully as yearlings, when nearly one-third of lifetime egg production is realized). Sons from the very earliest broods (hatching by mid-April) have a greatly enhanced probability of breeding in their first year, while those hatching later seldom acquire the necessary territory. By contrast, the prospects for daughters do not change with season. This skew in male pay-offs is believed to create an adaptive incentive for parents that produce robust sons early on (laying those eggs early in the clutch) and for parents that favour daughters thereafter (Dijkstra *et al.* 1990).

Finally, progress in understanding the possible interactions between offspring gender, hatching order, and avian sibling rivalry has been hampered

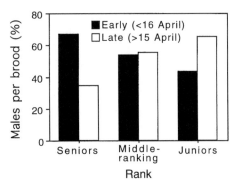

Fig. 12.17   European kestrel rank-specific sex ratios change as a function of season, with males as senior sibs and females as juniors in the early broods, but the reverse situation in later broods. These differences are highly significant, except for the middle-ranking (*C* and *D* chicks), which are nearly sex-balanced. (From Dijkstra *et al.* 1990.)

by frequent difficulties in identifying nestling sex immediately after hatching (the moment when such rivalry begins, and before early mortality has had an opportunity to introduce bias through differential losses). Most studies have relied on marking very young individuals and subsequently assigning sex on the basis of plumage/size dimorphism or laparoscopic examination. Happily, a recent proliferation of new techniques has made this task much simpler (in the field, if not in the laboratory), enabling one to determine sex from very small blood samples collected at the time of hatching and assessed via flow cytometry (Nakamura *et al.* 1990, Tiersch *et al.* 1991, Tiersch and Mumme 1993) or various DNA probes (e.g. Rabenold *et al.* 1990, Longmire *et al.* 1993, May *et al.* 1993).

12.5 Avian sibling rivalry after fledging

Although the great majority of ornithological sibling rivalry studies have focused on begging, brood reducing and siblicidal fighting during the nestling period *per se*, some families remain together long after the 'nest' has been abandoned. We have mentioned special cases such as grebes (where the parents become floating pseudo-nests) and some highly precocial species such as cranes where the chicks can follow their parents but are unable to feed themselves (a very similar situation occurs in oystercatchers, which may also be siblicidally aggressive in competition for food items that parents must process; Safriel 1981, Groves 1984). With the invocation of such mobile family units, our usage of the **nursery** concept may seem a little stretched (as it no longer represents a finite spatial entity), but its essence remains—young crane and oystercatcher chicks are so dependent on shared parental care that they cannot afford to leave. In short, the critical element of forced proximity to each other persists, and this may be limiting to eventual reproduction (e.g. Berglund *et al.* 1993). Where the 'benefits of philopatry' (Stacey and Ligon 1987) defer dispersal for varying periods, sibling rivalry can persist as part of the social landscape. We shall illustrate this first with two cases of merely slow dispersal before moving on to species in which siblings may remain together for life.

Barnacle goose (*Branta leucopsis*) families, containing up to six goslings, typically stay together for nearly a year and feature a dominance hierarchy that helps to determine which siblings are allowed to remain longest. Retention appears to be beneficial to individuals, as they are attacked less, feed more, and grow fatter when associating with the protective family (Black and Owen 1989a, b). Although all hatching occurs within 24–36 h, an initial rank order is established, often in accordance with hatching order, because the first chick starts to peck the others as they hatch. After about 40 days, however, adult size has been reached and the (now larger) males become dominant over their sisters, such that a new, renegotiated rank order is in place by approximately day 50 (Black and Owen 1987). This affects the order in which the offspring disperse 6 to 9 months later; the first to leave are the subordinates, which are threatened and assaulted by parents and dominant siblings alike. To support itself, a barnacle goose must feed almost constantly throughout the daylight hours (Owen *et al.* 1992); the subordinate moves off when intra-family harassment exceeds the level that would be experienced among non-family flock members (Black and Owen 1984, 1989b). By this process, the largest gosling remains with its parents the longest (Black *et al.* 1996). In other geese, parental guarding itself has been argued to be a limiting commodity, such that loss of brood members eases the parental load and may frequently be beneficial (Friedl 1993, Williams *et al.* 1994).

Dominant grey jay (*Perisoreus canadensis*) fledgelings expel their lesser sibs from the natal territory at the age of 55–65 days, usually forcing them into a dangerous solitary existence (although a few catch on as 'third birds' with pairs that had been unsuccessful in rearing their own young). These evictions occur in June, when food supplies are peaking, but have been explained as a mechanism for trimming group size to that supportable on the territory during the harsh boreal winter (note 'pending competition' redux). Furthermore, these birds cache summer foods that are important for winter survival (Strickland 1991), and food that has been hidden by a soon-to-be-banished sibling may be unrecoverable, and hence of little value, since the crucial site information presumably departs with the disperser. On the other hand, if it lands a position on a new territory, such a bird still has the whole summer in which to lay up winter provisions, and thus to make itself useful.

The opposite pattern is found in many other co-operative/communal birds, where multiple siblings remain on the home territory, usually serving as 'helpers' while waiting for an appropriate (or even high-quality) reproductive slot on a nearby territory (many reviews; e.g. Emlen 1991). Such viscous family groups commonly have sibling dominance hierarchies, and more than one same-sex sibling may be a candidate for a given opening. One especially vivid example of sibling conflict in a communal species has been reported for acorn woodpeckers, where adult sisters have been observed actively sabotaging each others' nests (Mumme *et al.* 1983). One female steals her own sister's egg, breaks it open and may then even share the resulting meal with the egg's mother. Such behaviour may be a residuum from much earlier nestling dominance relations; from the time of banding, larger siblings (wing measurements are assumed to indicate slight age differences resulting from asynchronous hatching) tend to dominate smaller brood-mates. While it seems likely that relative size/age confers some increased likelihood of surviving to breeding age (Stanback 1994)—more a function of avoiding accipiters than of finding food in this acorn-caching species—the habit of sharing sexual partners with same-sex siblings offers a great opportunity for complex and fitness-affecting relationships later on (as are found in social mammals and insects). Similar carry-over effects from juvenile dominance to adult co-operative breeding have been shown for Arabian babblers (Zahavi 1990).

Finally, a fascinating reversal of perspective occurs in the development of osprey nest-mates. Although they are facultatively siblicidal when young (reviewed in Mock *et al.* 1990), after fledging the siblings benefit from remaining in association and learning to fish together. In addition to the usual tasks associated with learning aerial pursuit of prey, plunge-diving birds face additional problems with glare, refraction, etc., that may make the transition to independence especially challenging (Orians 1969, Carl 1987). Ospreys that fledge with a nest-mate, however, learn more rapidly how to cope with wind and how to exploit the greater opportunities which result from diving from a height than do birds which forage alone. As a result, individuals belonging to sibling pairs tend to capture more fish during a substantial portion of their post-fledging adjustment (Edwards 1989), apparently via social learning (e.g. from their partner's mistakes).

Summary

1. The great majority of avian brood reduction systems probably results from non-aggressive scramble competition for parentally delivered food, where initially superior nestlings use their size and strength advantages to out-position rivals, thereby setting up a positive feedback loop that maintains some degree of size hierarchy. Some field experiments have tested game-theory predictions about how individual chick signalling should be predicated on the relative begging levels of nest-mates. If one American robin chick in a brood is artificially deprived of food, so that it escalates its begging when it is returned to the nest, its siblings increase their efforts, too.

2. Many altricial birds appear to be food-limited with regard to how many young can be raised at once, as reflected by a widespread negative relationship between brood size and offspring growth or survival. Sometimes this can be muted by experimental provisioning. The long-term effects of having been underfed as a nestling are more difficult to measure, but some studies do show that heavier fledgelings enjoy higher recruitment rates. In many cases, there may be a threshold value for fledging condition, above which the value of extra size is swamped by stochastic factors. Recent data on western gulls show clearly that hatching order affects survival during the 9 months after fledging (over and above its influence during the nestling stage).

3. As with egrets, size hierarchies resulting from asynchronous hatching are quite common in brood-reducing taxa. The youngest chicks frequently grow more slowly and may perish at higher rates even if they survive the nursery period to reach independence, presumably in part because of lower fat reserves. In one study of gulls, it seemed likely that parents could have enhanced the youngest chick's chances substantially by providing just a little more food after hatching.

4. Lack's 'resource-tracking' hypothesis for hatching asynchrony has been demonstrated most clearly for European blackbirds. When food conditions were poor, the youngest chick died early and its senior sibs developed well, whereas experimentally synchronized broods experienced greater mortality and weighed less. Under better food conditions, synchronous and asynchronous broods fared equally well. The overall disadvantage experienced by synchronous broods under poor food conditions could be obliterated by food supplementation, indicating that low food supply *per se* was the cause.

5. In densely colonial birds that also practise brood reduction, victim sibs may encounter such brutality and deprivation at home that it pays for them to move about, seeking meals or a new family to join. Neighbouring adults and young typically resist intrusions by such wanderers, but several studies have reported evidence of successful insinuations that amount to brood parasitism by the propagule itself.

6. Filial infanticide (i.e. that perpetrated by the victim's own parent) in the form of deserting still unhatched eggs (e.g. grebes, pelicans, cranes) and physically assaulting downy chicks (e.g. gulls, storks, woodpeckers, coots, moorhens) has been reported somewhat anecdotally. This raises the question of why it is not more common. Some reasons for allowing the feeble offspring to sort things out slowly are discussed.

7. The distribution of overt aggression across brood-reducing avian taxa remains poorly understood. Small brood size appears to be a necessary, but not sufficient, precondition. There are many species that appear to have all the necessary elements for effective use of intra-brood aggression, but remain enigmatically peaceable. At least one species (osprey) may

vary geographically in this regard, but many other species routinely establish some semblance of a sibling dominance hierarchy, which may or may not play a crucial role in resource control.

8. Despite initial assumptions that the primary ecological significance of avian siblicide was related to the subsequent ingestion of the victim's tissues (the 'icebox hypothesis'), relatively little consumption occurs in most birds. Where it does occur the cannibalism aspect is probably less important for promoting the killing than for release from acute competitions. The value of that release to the surviving siblings and parents has been shown in studies where brood size and/or resource levels were manipulated and where brood reduction was disallowed.

9. In nearly all obligate brood-reducing birds, the parents lay two eggs and raise just one chick to independence. Three case studies are presented. Black eagle victims are killed quickly as a result of repeated attacks, American white pelican victims are beaten enough for them to become feeble and then starve, and rockhopper penguin victims, which hatch from the first-laid egg, are not struck but seldom obtain any food, and thus quickly starve.

10. The primary riddle, namely why parents in obligate species bother to create an apparently-doomed marginal offspring, is best understood in terms of its insurance value. In most studies where eggs are marked individually at laying, a sizeable proportion of lesser siblings end up as the sole survivor. That is, the extra egg is far from being automatically doomed. The marginal egg's insurance value has also been demonstrated once through experimentation. When second eggs were removed from two-egg white pelican nests, parental success dropped sharply.

11. In brood-reducing species where sons and daughters represent different investment costs to parents, the possibility of adaptive manipulation of gender with laying order has been raised repeatedly. Where sexual size dimorphism also arises during the nestling period and affects the fighting abilities of nestlings, additional predictions have been proposed and partially tested in eagles, kestrels, hawks, and boobies.

12. Finally, extended rivalries occur in some bird species where siblings maintain extended contact, either by the retention of family structure well past fledging, or by not dispersing at all. In barnacle geese, families remain together for months and higher-ranking goslings sequentially drive their sibs off into an independent state. In highly philopatric acorn woodpeckers, which show many co-operative dimensions in their habit of communal nesting, females actively destroy their sisters' nests and receive the victim's servitude in return.

13

Mammalian sibling rivalry

A piglet's most precious possession
Is the teat that he fattens his flesh on.
He fights for his teat with tenacity
Against any sibling's audacity.
The piglet, to arm for this mission,
Is born with a warlike dentition
Of eight tiny tusks, sharp as sabres,
Which help in impressing the neighbours . . .

(Fraser and Thompson, 1991)

13.1 Features of mammalian family structure

Mammals differ from birds in several important features that influence both the expression of sibling competition and our ability to study it. As with birds, parental care is extensive, but unlike the situation in most birds the burdens of parenthood normally fall exclusively on the mother. When either the mother's capacity for investment is exceeded or non-maternal resources prove inadequate for all young to reach independence, brood (litter) reduction follows. These squeezes take many forms, although comparatively little concerted research effort has been directed toward their study. Whereas avian embryos subsist on spatially segregated food supplies (yolks within shells), eutherian development begins with a lengthy gestation during which embryonic litter-mates share a common food source, namely the mother's circulatory system. Mammalian embryos are in close physical apposition with their siblings and therefore have the potential for prenatal interactions. Furthermore, litter-mates are born almost simultaneously (typically within minutes or hours of one another), so the chief means by which avian parents establish a size hierarchy among nest-mates is lacking (Godfray and Harper 1990).

Finally, the routing of the postnatal diet through multiple outlets (separate mammary glands) diminishes the degree to which food can be monopolized. The family of secretions we call 'milk' (with their various mixtures of water, amino acids, vitamins, minerals, fats, antibodies, etc.) comes from histologically similar mammary gland tissues and almost always reaches the neonate through a mouth-fitting port called a *teat* (if it contains a cistern in which substantial milk accumulates between nursing episodes) or a *nipple* (if the storage volume is smaller). In some groups the storage capacity is truly prodigious (e.g. the gland cistern of ruminants), while in others milk flows through a system of galactophore vessels that may contain modest storage chambers (e.g. the milk sinuses of humans) or that run straight from numerous tiny alveoli (e.g. cats, rats, pigs, and rabbits) where the milk itself is metabolized from various plasma substrates (Wakerly *et al.* 1988). In this last case, the mother's milk supply at a given moment is roughly proportional to the number of (undrained) alveoli subserving all her nipples.

Shortly before birth, oxytocin receptors appear on the myoepithelial cells surrounding these mammary gland alveoli, setting the final link in a neuroendocrine feedback loop regulating milk delivery. These receptors literally vanish after weaning (Wakerly *et al.* 1988). Stimulation of the nipple by offspring sucking causes oxytocin to be released from the mother's posterior pituitary, triggering reflex contractions in the myoepithelial cells, which effect milk ejection (also known as 'milk let-down'). Thus '... the milk ejection is an intermittent response to an effectively constant stimulus' (Drewett 1983). The milk flows to the storage organ or outlet, depending on each species' mammary architecture. In taxa with cisterns, that milk ejection replenishes the stored supply (which can be removed at any time, independently of ejection); in species that essentially lack storage, each ejection provides immediate access to a typically modest volume of milk (e.g. 0.05–0.5 ml per nipple in rats; Cramer and Blass 1983). As a result, there is substantial variation in the temporal nature of milk deliveries. Mother rabbits accumulate their whole day's supply (150–200 g of milk) before making themselves available to their young for a mere 4–5 min. By contrast, rat pups receive milk meals consisting of a dozen or so tiny doses (Cramer and Blass 1983), with the constituent letdowns spaced by 5–20 min (Grosvenor and Mena 1974) and the meals themselves being provided at 1- to 2-h intervals throughout the day (Wakerly *et al.* 1988).

It is probably safe to characterize the maternal milk allocation of mammals as much less directly under parental control than is the comparable food-item allocation of parentally delivered avian food. Uneven distributions of PI among nursery-mates is more likely to result from scramble competitions than from contests, which in turn are more to be expected than active parental favouritism (although both predictions have some interesting exceptions; see below). However, to the degree that scrambles prevail, the models of Parker and Macnair (Chapter 8) may fit the mammalian landscape better than the 'honest signalling' approach (Chapter 9).

The importance of kinship in shaping sibling selfishness and competitiveness has been emphasized throughout this book. In mammals, coefficients of relatedness among litter-mates are seldom clearly known, at least not to researchers. In species where there are frequent multiple mating by females (i.e. mixed paternity within litters), the degree of relatedness between mammalian litter-mates may be lower than in their avian equivalents (although DNA tests keep showing the situation in birds to be far more complicated than previously thought). Mammals are certainly far less *socially* monogamous than birds (Kleiman 1977, Dunbar 1984), with male reproductive strategies often involving little or no mate-guarding (e.g. Schwagmeyer and Parker 1987, 1990).

Most mammalian sibships are inherently more difficult to study than their avian counterparts. Many species are nocturnal and/or give birth in isolated and frequently subterranean dens. Nevertheless, an emerging literature shows both pre- and postnatal rivalries for mammals, plus some competition between sequential (non-contemporaneous) siblings, that parallel and extend the diversity found among birds. For example, nursing per se influences the production of future sibs, as shown in savanna baboons (*Papio cyanocephalus*), where 'young infants have the effect of a perfect contraceptive, older ones an effect more like a faulty contraceptive' (Altmann *et al.* 1978: 1029). Similarly, large rat litters receive greater amounts of milk and reduce the probability of maternal ovulation more strongly than small litters (Drewett 1983). Although this may provide the mother's system with valuable information on which to base birth-spacing, one can also imagine circumstances under which such postponement could compromise the reproductive value of future siblings, even to the detriment of maternal interests (but see Gomendio 1991, 1993; Mock and Forbes 1993).

In general, mammals have proved more convenient than birds for studying certain key life-history relationships, such as how offspring body condition at independence influences future survival, eventual body size and lifetime reproductive success (e.g. Clutton-Brock *et al.* 1982a). Whereas birds often migrate or disperse over great distances, mammals (especially females) frequently remain conveniently close to their birth sites, facilitating the documentation of individual case histories. The non-dispersive nature of social mammals also has the interesting property of extending the potential period for sibling rivalry far into adulthood. Mammalian sibs may therefore be more likely to contend over actual reproductive opportunities (Greenwood 1980, Waser and Jones 1983, Waser 1985, 1988; but see McGuire *et al.* 1993), a pattern likely to exist in only the most viscous of avian populations (especially as a correlate of co-operative breeding; see Emlen 1991 and other reviews). In short, mammalian sibling rivalry is likely to begin earlier (as embryos) and last longer (until adulthood) than the avian versions, although its overall frequency, intensity and importance are usually far from clear.

Mammalian sibling relations may also be based on subtle kin-recognition abilities that have not yet been demonstrated for most other vertebrates (Blaustein *et al.* 1987a, b). In some ground squirrel (*Spermophilus*) species, for example, where multiply sired litters have been documented under field conditions (e.g. Hanken and Sherman 1981, reviewed by Schwagmeyer 1988), offspring have demonstrated the ability to discriminate between full and half siblings in laboratory choice experiments (Holmes 1986). In fact, ground squirrel litter-mates can even distinguish unfamiliar paternal half siblings with which they have had no previous contact whatsoever from unfamiliar non-relatives (Holmes and Sherman 1983). Furthermore, in the field, siblings of *S. beldingi* associate preferentially with one another (Holmes and Sherman 1982).

As in most other mammals (Greenwood 1980, Waser and Jones 1983), juvenile dispersal in ground squirrels is male-biased, with young females moving little if at all from their natal home ranges (Vestal and McCarley 1984). It seems likely that full and half sisters probably compete with one another, as well as with unrelated females, for limited high-quality breeding space. Finally, the reproductive physiology of a few domesticated species (rabbits, mice, rats, pigs, and various primates) has been studied in enormous detail for biomedical and psychological purposes. This literature offers far more material regarding issues of prenatal contact, nutrient allocation, and maternal limitations in mammals than can be explored in our brief review.

In this chapter, our goals are as follows: (i) to draw attention to a few systems that offer clear cases of sibling rivalry, plus a few that look especially promising in that respect; (ii) to illustrate the variations in sibling rivalry most likely to typify mammals generally; and (iii) to consider which mammalian families have the reproductive features most conducive to the evolution of such behaviour. We conclude that mammals may well prove to manifest sibling rivalry as richly as birds.

13.2 Litter size

In some mammalian species, life-history trade-offs have been found between offspring quantity and quality, such that mothers often incur higher personal costs for larger-than-average litters. According to studies of laboratory rodents (Mendl 1988), compared to offspring from small litters, those from large litters tend (i) to be lighter at birth (due to crowding *in utero* and/or sharing of maternal nutrients), (ii) to grow and develop more slowly (Figure 13.1), (iii) to suffer higher postnatal mortality and (iv) in the case of daughters, to be delayed in their own onset of sexual maturity (or to show reduced fertility). Furthermore, compared to mothers of small litters, those of large litters typically (i) consume more food and produce more milk, (ii) delay subsequent reproduction, (iii) are more likely to kill one or more nestling(s) and (iv) suffer higher mortality themselves. Similar patterns of trade-offs have also been reported from comparisons of single and twin litters in beef cattle (Gregory *et al.* 1990). Some field data on Columbian ground squirrels (*S. columbianus*) show suggestive correlations, with maternal survivorship and offspring recruitment both declining in families larger than the median only

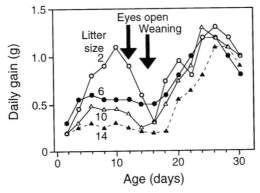

Fig. 13.1 The effects of mouse litter size on nursing period weight gains. Litters of laboratory mice were adjusted at parturition from intial litters of 7 to 11 pups. Individuals in the smallest litters (two pups; ○) grew much more rapidly, more than doubling their weight, compared to those in the largest litters (14 pups; ▲) through day 12, when the eyes opened and mobility (exploration) began. Growth in all litter sizes was minimal during the transition from milk to solid food, after which all grew pups at similar rates. At the end of 1 month, the early nursing advantage to small litters remained. Relative to L/14 pups, L/10 individuals averaged about 22% heavier, L/6 about 43%, and L/2 about 60% heavier. (Adapted from Fuchs 1982.)

during food-lean seasons (Festa-Bianchet and King 1991). There are indications that individual mothers may adjust litter size, either prenatally (Mendl 1988) or postnatally (Day and Galef 1977), so as to match more closely either the current ecological situation or parental capacity (e.g. see Bronson 1989). If so, then in studies of the effects of *x* on *y*, experimental manipulations of litter size need to be performed relative to an *individual mother's initial choice* of litter size, rather than to the population mean litter size (Mendl 1988; for avian parallels, see Högstedt 1980, 1981, Nur 1986, Gustafsson and Sutherland 1988).

Considerable behavioural data also demonstrate differential investment in male vs. female offspring, especially in highly dimorphic taxa such as many pinnipeds (Trillmich 1986), ungulates (Clutton-Brock *et al.* 1981) and coypus (*Myocastor coypu*; Gosling *et al.* 1984, Gosling 1986). In some cases, raising the more expensive sex has been related to maternal fitness; for example, red deer mothers are less likely to survive the winter after raising a male calf (Clutton-Brock *et al.* 1982a), which parallels the situation in antechinuses (Cockburn 1994). Similarly, in Mongolian gerbils (*Meriones unguiculatus*) mothers rearing experimentally arranged all-male litters nurse more, produce fewer independent young from the ensuing (post-partum) oestrus, are slower to show vaginal opening (and hence to copulate) and exhibit longer intervals between litters than those mothers raising all-female litters (Clark *et al.* 1990).

One logistical difficulty encountered when studying early sibling rivalry in viviparous animals such as mammals is that 'initial' litter sizes (i.e. those at fertilization, uterine implantation or placentation) often remains unknown under field conditions until much (and frequently all) of the potential competitive period has elapsed. Sometimes data drawn from multiple sources reveal diminishing numbers of litter-mates with the passage of time, suggesting sequential competitive litter reductions. For example, studies of *Spermophilus richardsonii* showed average *embryonic* litter size to range from 5.2 (lowest mean value) to 9.3 (highest mean), while litter size at parturition ranged from 4.9 to 8.3, and at first emergence from the burrow ranged from 3.8 to 7.2 pups (Michener and Koeppl 1985). In marsupials, of course, counting the young becomes much easier once they inhabit the pouch: in the honey possum (*Tarsipes rostratus*), for example, litter size has been shown to drop from a mean of 3.6 when the pouch-mates are very tiny (2 mm trunk length) to a mean of 2.5 when they are ready for pouch-exit (24 mm long). In this species, larger mothers tend to retain their young more than smaller mothers (Wooller and Richardson 1992). While many factors other than fatal sibling competition can account for such losses, the pattern's similarity to avian brood reduction invites further study. To put it another way, if black eagle nests were censused *only* at fledging, then initial clutch size would be totally unknown (or falsely assumed to be one) and obligate siblicide in this eagle would have been missed.

A very large number of mammals do, of course, give birth to single offspring. Many of these are not burrow-dwellers, which means that their monotocous pattern has been confirmed by observations both in the field and in zoos. In such taxa, postnatal sibling rivalry is mainly sequential (e.g. Trivers' 'weaning conflict'), wherein the current infant is envisioned as attempting to extract a disproportionate share of its mother's milk (or other form of parental investment) at the expense of future siblings. As mentioned above, there is evidence for such reproductive costs in mammals (but see Galef 1983). However, before we turn our attention to polytocous species (i.e. those with multiple-offspring litters) and this chapter's primary emphasis on contemporaneous siblings, we note again that a single birth does not necessarily mean that only one zygote was produced.

Two general types of very early partial brood loss have been documented for mammals, both of which may lead to significantly fewer offspring surviving long enough to be counted: (i) *polyovulation*—more ova are produced and fertilized than implant and develop (Birney and Baird 1985), so the rest of the zygotes die; and (ii) *litter pruning*, which typically involves some type of filial infanticide (often maternal cannibalism), by which parents pare offspring numbers (Day and Galef 1977, Gandelman and Simon 1978, Lee and Cockburn 1985, König 1989, Elwood 1992), and can also include fatal sibling rivalries, before or after parturition. Clearly, a considerable need exists for repeated-measures data on litter size at *various points* in the reproductive cycle, including counts of corpora lutea (e.g. Polge *et al.* 1966, O'Gara 1969), placental scars[1], and embryos (e.g. *in-vivo* censuses via x-ray or ultrasound or after collections of pregnant mothers; e.g. O'Gara 1969, Fujimaki 1981, Michener and Koeppl 1985, Gosling 1986) to supplement traditional counts at parturition and at independence (e.g. at post-weaning emergence from the nursery). Some early estimates for voles and deer mice showed prenatal losses ranging from 9 to 22.5% (Beer *et al.* 1957).

13.3 Resources for which mammalian siblings compete

The search for competition among concurrent mammalian siblings begins with a closer look at those taxa whose reported (postnatal) litter size consists of two or more young, an exercise of limited value by itself until the resource base (i.e. the commodity over which siblings are likely to compete) is considered. We shall consider four resource categories in turn.

13.3.1 *Uterine space and nutrients*

Not surprisingly, the earliest rivalries between mammalian siblings centre on survival (prenatal food and space), and most adult sibling rivalries focus on sex. By the time the mammalian proto-offspring has become a blastocyst, its first order of business is attaching to the uterine wall. In species that ovulate only one egg (or a very few) per cycle, success or failure probably depends overwhelmingly on interaction between the individual offspring and the uterine lining, with both offspring and mother usually co-operating toward successful implantation. By contrast, polyovulation introduces potential for conflict among sibling blastocysts and/or between parent and offspring over the value of attachment. Paralleling the broader issue of why parents of many taxa commonly over-produce, several explanations have been offered for the adaptive significance of mammalian polyovulation *per se*. In polyovulating taxa with relatively large litters, such as domestic swine and tenrec shrews, the supernumeraries have been proposed to serve as insurance against various sources of failure (e.g. infertility, early developmental irregularities, etc.), to ensure full-sized litters at birth (Wimsatt 1975, Dziuk 1977, Birney and Baird 1985). It has also been proposed that competition among siblings in poly-ovulant species allows investment to be restricted to only the most viable individuals (Cohen 1969), a progeny-choice argument. For swine, a species in which 30% of the potential litter is lost during the first month after mating, asynchronous fertilization and unequal development of blastocysts have been interpreted as maternal strategies for producing full-sized litters when the timing of the uterus's peak condition (for implantation) cannot be known by

---

[1] *Mammalian Species*, a journal published by the American Society of Mammalogists, provides these data for many of their 400+ species summarized to date. See also Lindström 1994.

the mother (Dziuk 1987). In four polyovulating groups (five elephant shrews, one rodent, two bats and the pronghorn), litter size is both small and invariant, typically two (one in each uterine horn), yet an average of five ova are shed in the elephant shrew (*Elephantulus intufi*) and at least 845 ova have been found in the plains viscacha (*Lagostomus maximus*: Weir 1971a, 1971b)! These polyovulating taxa are not closely related, so the pattern is apparently convergent.

Birney and Baird (1985) proposed how natural selection might have shaped female polyovulation strategies. Polyovulators share several interesting features: (i) large, precocial young; (ii) litter size limited to one individual per uterine horn and (iii) highly mobile mothers. Birney and Baird envisioned the production of two expensive offspring as the initial life-history state, but pointed out that ovarian control (the release of a single egg per ovary) is not an efficient method for limiting litter size to two, given that some failures are bound to occur. If a litter size of two is optimal, deferring litter size determination until a moment well past ovulation (namely to an intra-uterine culling mechanism) makes sense, especially if temporary sustenance for extra zygotes is relatively cheap. The final (birth) litter size of two may be determined by the mother's need for manoeuverability in evading predators, while the over-production of ova may be an insurance gambit to ensure an adequate supply of blastocysts. While this account of polyovulation may seem stretched by the extravagant numbers of ova produced by plains viscacha or one elephant shrew (*Elephantulus myurus*) (average of 49.0 ova; Tripp 1971), we suspect that such a judgement rests on a tacit overestimation of egg production costs. If the true metabolic or nutritive value of each discarded egg is extremely small, the cost of 'prodigal polyovulation' (Birney and Baird 1985) may be trivial, and consequently unchecked by natural selection.

Whatever the ultimate reason(s) for polyovulation, its existence creates an automatic scramble competition among zygotes/blastocysts for the limited number of attachment sites. In some cases more offspring manage to 'implant' than will be carried to term (e.g. big brown bat, *Eptesicus fuscus* and eastern pipistrelle, *Pipistrellus subflavus* implant 3 to 7 offspring, but only one survives in each uterine horn; Wimsatt 1945), so further partial brood losses must be occurring *in utero*. Parallels can also be found in humans (*Homo sapiens*); during the first trimester sonagrams show two to four times more dizygotic twin 'litters' than are carried to term, with the smaller one typically disappearing in the interim (Landy *et al.* 1982, Schneider *et al.* 1979, summarized by Anderson 1990c). It is also worth pondering why monozygotic twinning is not more common in humans. A theoretical model (Gleeson *et al.* 1994) presents this as a possible parent–offspring conflict, a unilateral and asexual attempt by the offspring *to duplicate itself in utero*. On the other hand, multiple-birth infants often have special problems. In a sample of 335 twin deliveries in a Finnish hospital, about 70% of the twins had complications: one-third were premature, one-third showed 'intra-uterine growth retardation' and the rest suffered from respiratory difficulties (Moilanen 1987). Other Finnish data, gleaned from the 18th and 19th centuries (Haukioja *et al.* 1989), indicate that double birthing was selectively neutral for that population, in which case the spread of a self-replication trait could be opposed by the costs associated with bearing triplets, quadruplets, etc. (Anderson 1990c). In any case, twin birthing is relatively uncommon in humans (<5%), and most cases are dizygotic.

How this prenatal reduction problem is resolved in mammals (and by whom—mother or offspring?) has been partially elucidated for pronghorn, *Antilocapra americana,* a North American ungulate in which the losses occur at two distinct periods (O'Gara 1969). In the early days after fertilization, three to seven spherical blastocysts typically float freely in the uterine cavity and reshape themselves into greatly elongated tubes (up to 12 cm in length). During this so-called 'thread stage' the blastocysts often become entangled with one another, sometimes forming tight overhand or granny knots. During the thread stage there is a 30% decrease in embryo numbers (15 maternal tracts dissected just after the thread stage contained an average of 4.47 corpora lutea, but only 3.07 embryos: O'Gara 1969), with no more than two embryos implanting in each uterine horn. It is not known which blastocysts survive beyond the thread stage or why. One might speculate, for example (i) that individual blastocysts somehow managing to escape entanglement, either by 'skill' or by good fortune, win by default; (ii) that O'Gara's 'entanglement' is actually a snapshot of results from some kind of direct combat (although this seems unlikely since it is apparently suicidal); or (iii) that additional (unknown) factors, which perhaps have nothing to do with the knots, are responsible for the embryo losses at this time.

There are two implantation sites in each uterine horn, one of which is closer to the common birth canal than the other. For convenience, we shall call these the 'front' and 'rear' sites, respectively. The front site has the richer maternal vascularization (O'Gara 1969) and this embryo grows a long extension of chorioallantois (Figure 13.2), the 'necrotic tip', that pierces and kills its sibling. O'Gara (1969: 219) described removing the victim's membranes from a dissected necrotic tip as 'like [rolling] a stocking from a leg.' Poetry aside, the upshot is that no more than two young—one per horn—survive. Although data are lacking, we suspect that an occasionally defective front embryo fails to kill its 'horn-mate' and is eventually supplanted. If so, the rear embryo may represent adaptive maternal insurance.

More generally, in many polytocous mammals, embryos seem to wander about as blastocysts before attaching within one or two species-specific areas. Interestingly, embryos normally end up with a fairly uniform, minimally crowded, spacing (Mossman 1987). Embryonic over-dispersion could result from (i) maternally defined attachment sites, (ii) some form of 'territorial defence' by individual embryos (i.e. the newly attached individual creating a refractory zone around itself; Mossman 1937) or (iii) some combination of these processes. In swine, crowding of embryos has no *immediate* effect on

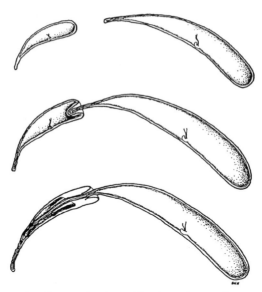

Fig. 13.2 Intra-uterine litter reduction of implanted pronghorn embryos. The front embryo's chorioallantois ('necrotic tip') penetrates that of its distal sibling entirely. (From O'Gara 1969.)

survival, but leads to significantly increased fetal mortality later in pregnancy (Dziuk 1987, Wu *et al.* 1989); the fact that space *per se* is something for which swine fetuses compete has been demonstrated by experimentally manipulating embryo crowding (Webel and Dziuk 1974, Dziuk 1985).

After placenta formation, mammalian embryos can continue to affect one another development directly via hormone secretion (Gandelman et al. 1977, vom Saal and Bronson 1980, Meisel and Ward 1981, vom Saal 1984, 1989, Clark and Galef 1988, Clark *et al.* 1990, 1992, 1993). Much attention has been paid, for example, to the heightened 'masculinization' of female rodent fetuses when positioned between two brothers *in utero*. Unborn mammals can also suppress the growth of litter-mates. In sheep, male siblings apparently do this more effectively than females (Burfening 1972). We know of no data, however, which show that individual embryos/fetuses can either shunt nutrients toward themselves or extract nutrients more rapidly than their siblings so as to sequester extra investment. However, because mammalian neonates commonly show intra-litter size variation, which is in some cases quite marked (e.g. runts), it may be worth exploring the mechanisms responsible for size variation and determining the degree to which its control rests with maternal or offspring physiology. Because sibling size variation has dramatic fitness consequences in hatchling birds and also in a few mammals (e.g. pigs; Section 13.3.3), such variation merits close attention.

Another aspect of intra-uterine sibling competition that is receiving increased attention for mammals involves 'genomic imprinting' (Section 10.2.1): 'the process by which a gene comes to be expressed differently in an individual, depending on whether the gene is derived from the individual's mother or father' (Haig and Westoby 1991). Mammals are excellent candidates for paternal 'rip-offs' of this sort because (i) parental investment comes almost entirely from the mother and (ii) prenatal siblings often have different fathers. From the paternally conferred gene's point of view, a maternal half sib is no kin at all. In mice there is evidence that one gene responsible for the production of *insulin-like growth factor* ('*IGF-II*') makes this polypeptide only when it arrived in the zygote via a sperm. When present, *IGF-II* causes its bearer to extract extra nutrients from the mother during embryonic and fetal periods. Furthermore, it appears that an oppositely imprinted gene may have evolved to nullify the indirect sexual conflict. The *mannose-6-phosphate* receptor offers a dummy binding site, where *IGF-II* molecules are captured and degraded before they can link up with the primary binding site (Haig and Graham 1991).

13.3.2 *The marsupium as nursery*

In marsupials, the over-production initiated by polyovulation can be sustained during the abbreviated gestation and corrected after parturition. The following examples are extracted largely from two excellent reviews of marsupial reproductive physiology (see Tyndale-Biscoe and Renfree 1987, Cockburn 1989). In *Dasyurus viverrinus*, a pouchless marsupial, the neonates remain continuously attached to the teats for about 2 months. More young are born than the six teats can accommodate, so partial brood loss is automatic, occurring immediately after parturition. Direct observations of *Dasyuroides byrnei* and *Antechinus swainsonii* during birth make the point vividly: '. . . in both species the female stood on all four legs with the hips raised. The young travelled downwards to the teats very rapidly. Young in excess of the teats were discarded on the ground' (Tyndale-Biscoe and Renfree 1987: 37). If further young are lost before attachment, the surviving sibs grow at faster rates, indicating that the growth of individual offspring is normally food-limited. Similar over-production is found in other dasyurids.

The basic pattern of maternal over-production and ensuing partial brood loss can be found across a diverse range of marsupial families. Tyndale-Biscoe and Renfree (1987) have summarized the relevant data for several well-studied taxa, including *Isoodon macrourus* (Peramelidae), *Pseudocheirus peregrinus* (Petauridae), *Dasyurus viverrinus* (Dasyuridae), *Gymnobelideus leadbeateri* and *Cercartetus concinnus* (both of Burramyidae). In each case, the number of corpora lutea exceeds the number of uterine young, which exceeds the number of teats, which exceeds the number of pouch young.

Many other marsupials can overlap their single-offspring 'litters' temporally, such that siblings from successive mating cycles coexist while they are still dependent on maternal investment. In most macropodids (wallabies and kangaroos), ovulation is not blocked by lactation and occurs a few days after birth. Development of the new proto-offspring halts before the blastocyst has 100 cells ('embryonic diapause') and ovarian cycling also stops, so most of the previous joey's nursing period elapses with a sib essentially 'on hold'. At times of severe ecological stress to mothers, the accessibility of pouch young for maternal eviction may provide a general advantage over eutherian modes of investment (Stoddart and Braithwaite 1979, Morton *et al.* 1982).

The pouches of the large *Macropus* kangaroos contain four teats, but usually harbour only one 'pouch young' at a time. Rare instances of simultaneous (so-called 'twin') pouch occupations in captive *M. eugenii* (plus a few experimental attempts to foster additional young to vacant teats) have led to mortality in nearly every case. Normally, a diapausal embryo resumes growth and is born (moves to the pouch) immediately or very soon after its sibling has vacated. The new pouch occupant attaches to a different teat (its tiny mouth fits on to a small bud at the end of an unused nipple) and the elder non-pouch young continues to reach in to nurse sporadically on its own teat (Figure 13.3). These two teats respond to vastly different levels of stimulation and even secrete milk of dissimilar composition. In wallabies, the individual mammary gland changes in sensitivity to circulating oxytocin levels, yielding steady and rhythmic contractions to even basal (i.e. minimal) concentrations of oxytocin during early lactation. This provides a small trickle of milk to the continuously attached small joey. Later, the same gland requires much larger oxytocin pulses as the joey grows stronger and feeds less regularly (Wakerly *et al.* 1988).

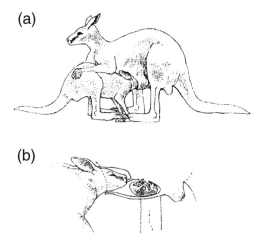

Fig. 13.3 Nursing by an older and much larger kangaroo offspring when a new pouch-living sibling is present, **(a)** as viewed from the outside and **(b)** as viewed in cutaway. (From Tyndale-Biscoe and Renfree 1987, originally published by Sharman and Calaby 1964.)

Clearly, two dependent and mismatched joeys compete only when their mother's overall ability to provide is surpassed (e.g. during severe droughts), as each draws only on the whole parent's nutritional economy.

13.3.3 *The milk supply*

In considering possible eutherian nursing competition, attention focuses first on species that produce more young than they have nipples/teats to support (e.g. Hill 1910, Hartman 1929, Petrides 1949). Although young in some such species participate in nipple 'time-sharing,' individuals thus fed may suffer compared to those with free access to suckling sites, experiencing either fatal starvation (Egoscue 1962, Cameron 1973, Schadler and Butterstein 1979, Fuchs 1982) or slower growth (see Figure 13.1). For example, in two out of 10 captive litters of 13-lined ground squirrels (*Spermophilus tridecemlineatus*) from Oklahoma, the number of young exceeded the number of functional nipples, although the mean pup-to-nipple ratio was 0.73. Wild litters showed a very strong negative relationship between the number of pups and mean mass at weaning ($n = 14$, $r = -0.86$, $P < 0.001$; P.L. Schwagmeyer, unpublished data). Although it is unknown whether retarded early growth affects eventual survival and reproductive success in this species, the observed pattern suggests milk limitations and sibling competition.

In a survey of 266 rodent species from 26 families, Gilbert (1986) found that litter size averaged about 46% of the number of functional nipples—a relationship he dubbed the 'one-half rule'. The significance of this seeming conservatism may involve several contributing factors, including (i) the possibility that natural selection has matched mammary number with the highest number of young typically produced (i.e. to accommodate extremes), (ii) the possibility that each offspring needs an average of two teats (e.g. because of refractory requirement for milk replenishment and/or because the lower row of teats may be inaccessible when the mother is lying on her side) and (iii) the possibility that the frequency of non-functional ('dry') nipples is higher than generally appreciated, perhaps reflecting maternal condition (Gilbert 1986).

In at least some mammals there are productivity gradients across nipples, some providing milk more generously than others. The patterns vary taxonomically. In ungulates (especially pigs), perhaps some marsupials (Lee and Cockburn 1985), and a few others the better nursing sites tend to be toward the mother's anterior (teat number 2 of 4 for coypu; Gosling *et al.* 1984). By contrast, the rear mammary glands are more productive in felids (Rosenblatt *et al.* 1961, McVittie 1978), binturongs (*Arctictis binturong*; Schonecht 1984) and common opossums (Cutts *et al.* 1978). In several species, variation in nursing site quality corresponds with site preferences by nurslings, extending to 'teat fidelity/constancy' and, more especially, 'teat order,' wherein individual offspring tend to use one pair of nipples faithfully, and may even actively defend their nipples from siblings (e.g. Ewer 1961, Schonecht 1984, Fraser and Thompson 1990). In rats, newborn pups initially locate the mother's nipples by olfactory cues provided by their mother's saliva (demonstrated by experimentally washing nipples and making them harder to find) and subsequently 'home' to their own nipple by seeking the scent of their own saliva (Teicher and Blass 1976, 1977). Active defence of inguinal teats has been reported briefly for two species of free-living hyraxes (Hoeck 1977) and for captive binturong (Schonecht 1984), but has been elucidated most carefully for domestic swine in an elegant series of experiments by David Fraser and his colleagues.

Barnyard piglets are born in unusually large litters (for ungulates), and are armed from birth with precocially erupted deciduous eye-teeth (canines and canine-like incisors) that angle out from the jaw (forward and slightly sideways), making them useful weapons in the defence of nursing position. With fast lateral jabs, piglets use their eye-teeth to slash the face of a litter-mate encroaching from the side (Figure 13.4). Within a few hours after birth, siblings establish a stable teat order—the slightly larger individuals tend to occupy the more productive anterior positions (Scheel *et al.* 1977, Fraser 1990). It is not known whether the superficial wounds produced by such slashing would lead to secondary infections under 'more natural' conditions, but even in modern research barns the smallest piglets spend more time trying to nurse than do their better situated larger siblings. Nursing is a risky business, and the small individuals suffer significantly higher mortality than their larger siblings as a result of accidental crushing by the ponderous sow (Fraser 1990). Erupted eye-teeth and fighting behaviour by suckling young have been found in both suids (e.g. wild boars) and tayassuids (peccaries) (reviewed by Fraser and Thompson 1990). Although these teeth are not oriented sideways in the latter group, large litters of collared peccaries (*Tayassu tajacu*), i.e. those containing three young, display considerable squabbling over the two nursing positions, and one piglet may starve to death within 48 h of birth (J. Packard, personal communication). What little information exists for peccary field biology (e.g. Byers 1983) does not address this situation directly.

The effectiveness of the domestic swine eye-teeth as sibling weapons has been demonstrated experimentally (Fraser and Thompson 1990). It is routine agricultural practise for farmers to clip those eye-teeth at birth (largely out of concern for the mother's comfort), but if certain piglets in large litters are allowed to retain their eye-teeth while the eye-teeth of size-matched siblings are removed, the 'armed' individuals achieve an average pre-weaning weight gain 11% higher than that of their clipped rivals. In smaller families, where sibling competition is presumably relaxed, the effect is lost. Finally, if *only* the lightest piglets are allowed to retain their eye-teeth, they are able to overcome their usual disadvantage and match the weight gain of larger siblings (Figure 13.5).

(a)

(b)

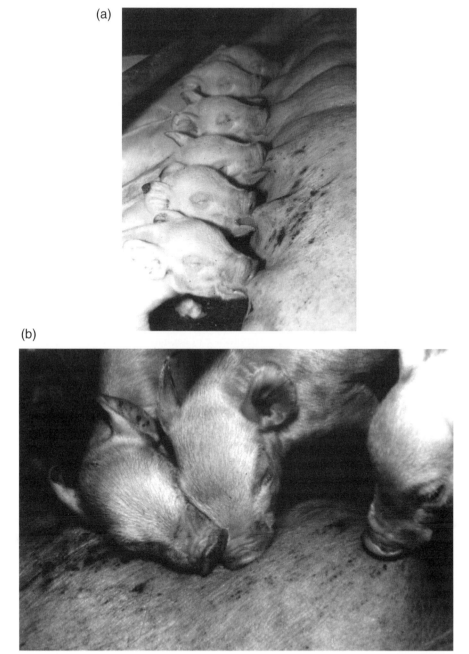

Fig. 13.4 **(a)** Piglets in very large littlers jostle for position to obtain the preferred anterior. **(b)** Sideways face-slashing motions put the precocial eye teeth to work in the usurpation or defense of higher-quality teats. (Photographs by D. Fraser.)

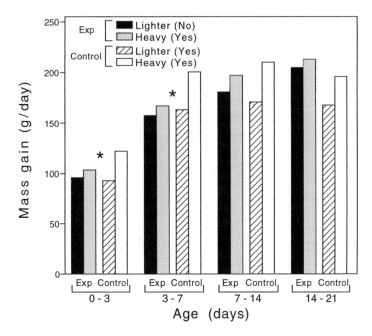

Fig. 13.5 To explore the functional significance of neonatal weaponry, experimental litters of domestic swine were divided according to birth weights and some smaller piglets were allowed to retain their eyeteeth. Those teeth were clipped on the heavier siblings (solid grey bars: 'Heavy [yes]'), but not on the lighter sibs (filled black bars: 'Lighter [No]'). In control broods, the teeth were clipped for both the heavier (open bars 'Heavy [yes]') and lighter (hatched bars 'Lighter [yes]') sibs. Thus armed, the lighter piglets did not gain weight more slowly than their larger rivals, as occurred in the control litters (especially during the first weeks: * indicates $P < 0.05$). (From Fraser and Thompson 1990.)

A more subtle form of scramble competition, called 'nipple-switching' (or 'nipple-shifting'), is likely to be widespread in mammals that lack cisterns for milk storage. The idea here is that, after draining the mammary gland associated with one nipple, a pup moves quickly to another nipple and collects what it produced in the most recent let-down (Cramer and Blass 1983). During their first 2 weeks, neonatal rats apparently have no control over their appetites and, if maternal regulation is rendered inoperative (experimentally), will actually consume so much milk that it backs up in their throats and complicates respiration. Older pups show greater control and often practise nipple-switching, especially when experimentally deprived of food (Cramer and Blass 1983).

Milk is metabolically expensive for mothers to produce, such that lactation generally constitutes a greater energy drain than gestation (e.g. Hanwell and Peaker 1977, Millar 1979, Galef 1981, 1983, König *et al.* 1988, Creel and Creel

1990, Kenagy *et al.* 1990, Sikes 1995a), even when seasonal changes in food availability have been taken into account (Clutton-Brock *et al.* 1989). The amount of milk available to nurslings is critically linked to the balance between the mother's food base and her personal maintenance expenditures. If she cannot afford full milk production, family trimming is often more viable than either complete abandonment or a self-sacrificing attempt to nurse everyone. Brood reduction has been documented under field conditions for bandicoots (*Isoodon macrourus*, Gemmell 1982), desert-dwelling *Sminthopsis* spp. (Morton 1978), antechinuses (Cockburn 1994), and various other small marsupials (reviewed by Lee and Cockburn 1985: 56). Maternal desertion of partial litters has also been reported for bears (Tait 1980), where it has been interpreted as infanticidal litter abridgement for the purposes of paring down current maternal effort. Underfed mothers may also employ direct infanticide and cannibalism (e.g. Day and Galef 1977, König 1989) or selective neglect (McClure 1981) for litter pruning. In some single-offspring species, such as sea otters (*Enhydra lutris*), sick mothers are known to abandon their only pup and sometimes recover to breed again (Garshelis and Garshelis 1987), essentially sacrificing one offspring on behalf of future sibs. As in birds, mammalian litter reduction tends to be associated with the production of relatively altricial young (Eisenberg 1981; see also Figure 3.6 in Lee and Cockburn 1985).

The central proximate role of low food supply in maternal brood reduction decisions has been implicated experimentally in the laboratory, where mother woodrats (*Neotoma floridana*) were provided either with a diet sufficient to sustain lactation or one that was inadequate (McClure 1981). In the food-stressed condition, partial brood loss occurred. Specifically, some male pups were found cold and well apart from the rest of the litter. Underfed mothers were assumed to have rejected these pups (McClure 1981), thereby alleviating the competition for milk. These results were interpreted mainly in the light of the effects on litter sex ratio (Trivers and Willard 1973, but see also Marsteller and Lynch 1983). However, in the absence of close behavioural observations, the case for maternal control of this brood reduction is essentially *de facto*: the sibs seem to be too feeble to have forced the victims' isolation. The meaning of these results are complicated further by a follow-up study of the same species using the same experimental protocol that did not find the same bias in mortality (Sikes 1995b), and some evidence that the surviving offspring of food-stressed mothers, while smaller at weaning, grew to the same average adult size as those from well-fed mothers (Sikes 1996), which erodes a key premise of the Trivers–Willard argument. In a similar study of bushy-tailed woodrats, *N. cinerea*, pup mortality was apparently not due to maternal rejection. Frequent observations of the nursing litters showed that pups died despite remaining attached to nipples at the same high frequencies (*c.* 86% of records) as litter-mates that survived (Moses *et al.* submitted). The greater vulnerability of male sibs *per se* during competition for limited milk is proposed to be related simply to that gender's greater metabolic needs in this size-dimorphic rodent.

A different kind of milk-related selectivity is found in African lions (*Panthera leo*), a communally-nursing species. Closely related adult females, including many siblings, typically allow each others' cubs to suckle, but this tolerance depends in part on each mother's current litter size. Specifically, mothers with a single cub are significantly more likely to provide milk for non-offspring than mothers with two of their own cubs (Pusey and Packer 1994).

In *Antechinus stuartii*, an Australian marsupial 'mouse' with incomparably bizarre life-history features, litter sex ratios are apparently adjusted by select-ive maternal cannibalism, sometimes favouring sons, while at other times favouring daughters (Cockburn 1994). Discrimination against daughters, which are completely philopatric and thus likely to remain in competition with one another as adults, is believed to be driven by local resource competition (*sensu* Clark 1978), while discrimination against sons appears to be more related to whether the male offspring are likely to be sufficiently robust to achieve breeding status (Trivers and Willard 1973). Sons are also more expensive to raise. All mothers usually wean at least one daughter (culling others, as needed); those mothers with high personal life expectancies raise all male offspring, but those with no future breeding ahead of them generally raise no sons at all (Cockburn 1994).

Finally, an interesting bias has been detected in cheetahs (*Acinonyx jubatus*), where mothers with two or more male cubs tend to provision their offspring more lavishly (and accept lower personal condition) than mothers caring for only single sons (Caro 1990). This pattern has been interpreted in terms of the males' future need for sibling assistance (coalitions are usually composed of brothers) in order to breed. To state this the other way round, singleton sons are more likely to lack co-operative support as adults, and hence may be poorer prospects for creating grandchildren; in any case, such males are physically smaller as adults. No similar patterns were found among daughters.

13.3.4 *Post-weaning food competitions*

Weaning refers generally to the dietary changeover from milk to solid food. It is a gradual process which requires substantial and non-reversible changes in gut anatomy and physiology that often impose a temporary slowing of growth during the transition (Bateson 1994). Much discussion has centred on whether this transition is also a crucible for parent–offspring conflict (e.g. Section 11.4), but our interest here is on how the diet shift affects the probability and form of intra-litter rivalries. Clearly, after weaning, the young of many mammals remain highly dependent on support from parents (and alloparents in some taxa), and thus remain subject to possible fluctuations in resources that are still being shared. As with birds, this dependency is typically protracted in predatory species, where the offspring's future livelihood requires mastery of complex hunting skills. In red foxes (*Vulpes vulpes*), for example, by the end of their first month siblings establish what appears to be a strict dominance hierarchy that dictates subsequent access to solid food. From popular accounts (Henry 1985a, b), the mortality patterns in foxes are highly suggestive of a facultative siblicide system. During their second week, the pups' eyes open (at age 10–12 days), growth accelerates and the milk teeth erupt ('canine' fangs first). By day 25, their brief periods above ground include short but serious fights with each other (Figure 13.6) that are sometimes, but rarely, fatal. A social hierarchy begins to form, after which overt fighting tends to diminish (and is replaced by threat). It is not clear what, if anything, is gained by combat during the nursing period *per se* (e.g. we know of no reports on suckling positions). However, fighting precedes the transition to solid food, which spans weeks 5 to 8. Meat is apparently always delivered to the first pup that intercepts a parent returning from hunting, and the scramble competition for priority leads to hungry pups remaining above ground and watchful (a pattern akin to that seen in penguins and some other birds; see Section 12.1). Dominant pups also take food from weaker siblings by force. One might speculate that a subordinate's chances of receiving food depend both on whether the morsel being delivered can be swallowed very rapidly (hence minimizing its exposure to pirates) and on how well its dominant litter-mates have been fed. If food is insufficient, the runt dies first, then the second-most subordinate pup, and so on (Henry 1985a, b). Early fighting and establishment of dominance hierarchies also seem to be the rule in other canids (e.g. Wandrey 1975, Zimen 1975, Bekoff 1981). Similarly, northern grasshopper mice (*Onychomys leucogaster*) switch from milk to a diet of maternally delivered prey, which may be limiting to the litter-mates. It is unclear at present whether siblings in captive litters fight amongst themselves (Ruffer 1965, J.D. Moodie, unpublished data).

Fig. 13.6 The scuffling of red fox pups looks much like the harmless play of domestic dogs, but can apparently escalate to fatal levels. (Redrawn from Henry 1985b by Coral McAllister.)

13.3.5 *Reproductive opportunity*

In species with low dispersal, same-sex siblings may remain in sufficiently close contact to vie directly for reproductive opportunity when they are adult. Because the general mammalian pattern is for males to disperse further than females (Greenwood 1980, Waser and Jones 1983, Roff 1992, Stearns 1992), it seems likely that mammalian sisters engage in adult sibling rivalry more often than brothers. Adult sibling encounters presumably peak in the highly social taxa (e.g. lions, wolves, dwarf mongooses, prairie dogs, etc.), where there are often complex elements of both co-operation and conflict among group members. It has also been pointed out that the mother's rank in social taxa can affect the relative costs and benefits of investing in sons vs. daughters (Gomendio *et al.* 1990).

One dramatic mammalian version of adult sibling rivalry takes the form of *reproductive suppression*, wherein only one (or a few) dominant individual(s) of each sex imposes some kind of endocrine/behavioural control on the breeding of same-sex group members, which often include siblings. Reproductive suppression often leaves subordinates with no better option than investing (as a 'helper') in the offspring of the dominant individual. In some cases, attempts at pre-emptive control are not entirely successful—that is, subordinates breed anyway—and subsequent corrective measures, including infanticide, may be added to the repertoire of reproductive controls (Wasser and Barash 1983). Such behavioural and physiological reproductive controls are often interrelated. In dwarf mongooses (*Helogale parvula*: Figure 13.7), for example, dominance position within the female hierarchy (which includes close genetic relatives, including many sisters, with a measured mean r of 0.33: Creel and Waser 1991, Keane *et al.* 1994) may be related both to the likelihood of suspected postnatal infanticide (Rood 1980) and to conception itself (Creel and Waser 1991). Risk of infanticide has been interpreted as a tool for enforcing reproductive suppression in many communal taxa, whereby despotic breeders reserve exclusive (or nearly so) breeding rights for themselves, forcing other group members into servitude (Packer and Pusey 1984). For these mongooses, however, mechanisms of endocrinological suppression appear to pre-empt the costly 'gestation-and-destruction' process of infanticide, while still providing despotic control. Social status firmly influences baseline oestrogen levels such that subordinates are less attractive to males and less receptive than are dominants. In a high-ranking female, greater mating frequency feeds back positively on oestrogen levels until the release of luteinizing hormone is eventually triggered, precipitating ovulation (so-called 'reflex' or 'induced' ovulation). Because subordinate females start with very low baseline oestrogen levels, mating seldom occurs at rates sufficient to stimulate ovulation, thus sparing subordinates the considerable expenses of gestation and lactation that would, in any case, be lost via infanticide (Creel *et al.* 1992). In older and higher-ranking subordinates, however, there is a higher incidence of

Fig. 13.7 Dwarf mongooses compete with their siblings for reproductive opportunities throughout life. These three females remained in their natal pack. In the centre is the pack's new α-female, the 6-year-old daughter of the previous α-female and old α-male. On her left stands the current β-female, a year younger, and α's full sibling; on her right is the 3-year-old γ-female, half-sister to α (sired by an α-male that replaced α's father). Neither of the subordinates produced any young when this photograph was taken in 1990. (Photograph provided by Scott Creel.)

pregnancy than in younger, lower-ranking females. Any young produced by the former animals may be combined with the (numerically greater) alpha female's litter. However, it has now been shown by means of DNA fingerprinting that subordinate females do produce some offspring (15% in one sample; Keane *et al.* 1994). Thus, there may well be a sliding scale in the degree of reproductive suppression among females, which could compensate subordinates just enough to retain their services as helpers. Intriguingly, most subordinate reproduction by female dwarf mongooses seems to be achieved by relatively high-ranking individuals that are less closely related to the group's alpha female (Keane *et al.* 1994). This feature was predicted by Creel and Waser's (1991) modification of Vehrencamp's (1983) 'power-sharing' model. The counterintuitive point is that the dominant female should be less generous with her siblings than with non-kin simply because the related helpers are receiving some of their compensation via indirect fitness. (The same argument can be cast in terms of a 'weak queen' (Strassmann 1993). The dominant female's despotic control slips as an increasingly potent rival arises, the challenge being more strongly pressed when that individual is not a genetic relative.)

Some parallels exist among dwarf mongoose males, except that the mechanism of control is more behavioural and less effective. Females tend to mate with more than one male (even the alpha female does), and subordinate males obtain a considerable share of copulations. In contrast with subordinate females, low-ranking males pay only a modest cost for such activity (mainly the likelihood of taking a beating), which they often seem willing to accept. Creel and Waser (1991) have also suggested that the larger testes of dominant males may also play a key role in the ensuing sperm competition, but a direct link between higher sperm competition and paternity has yet to be shown.

Reproductive suppression via infanticide by the dominant female (where the victims are progeny of subordinate females) has also been reported in wild dogs, foxes, and wolves (reviewed in Packer and Pusey 1984). Where such acts involve sibling adults, as seems plausible for some canids, the reproductive sabotage parallels that reported for communally breeding birds (see Section 12.5) and for social insects (see Section 15.5.3).

The fine line between sibling co-operation and conflict is apparently tracked in other ways by social carnivores. African wild dog (*Lycaon pictus*) yearlings routinely serve as 'helpers', providing food to young kin when sufficient food is available, but actually stealing it from the same pups when food is scarce (Malcolm and Marten 1982). The dual kinship structure of African lion prides is particularly interesting in this respect, as both the matriline (containing some adult sisters) and male coalition (containing some brothers) behave co-operatively much of the time. Females generate food for the whole pride by hunting in co-ordinated joint manoeuvres and by nursing communally; males likewise work as a unit to achieve and retain possession of a pride. Just as female co-operation erodes when key resources become limiting, male–male rivalries within the pride reach their azimuth when contesting access to oestrous females. Overt within-coalition combat for oestrous females is not uncommon, and is probably constrained more by the economics of male aggression (high costs and low value per mating) than by relatedness *per se* (Packer and Pusey 1982, Packer *et al.* 1991). Similarly, black-tailed prairie dogs (*Cynomys ludovicianus*) exhibit communal nursing (Hoogland 1983, Hoogland *et al.* 1989) and respond far more 'amicably' to siblings and other same-sex kin than to non-kin (Figure 13.8). Hostile behaviours directed at non-kin (which include chasing and fighting) look as if they may be expensive in terms of time, effort and risk of injury, so the ability to 'switch off' such actions may be advantageous. Interestingly, the discrimination between kin and non-kin conspecifics vanishes for each sex at the points in the annual cycle when intra-sex rivalries are strongest (Hoogland 1986). For example, females become dramatically more pugnacious during pregnancy and nursing, when female competition for burrow systems is acute. Competition for burrows may underlie infanticide by lactating females even

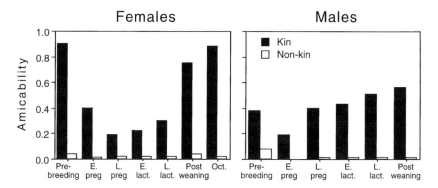

Fig. 13.8 Both male and female black-tailed prairie dogs are much more 'amicable' (non-hostile) to kin (solid bars) than they are to non-kin (open bars), with females being generally more amicable than males. Interestingly, however, female sociability drops dramatically when personal reproductive stakes are high, that is from conception through to weaning, when the female's pups are vulnerable to infanticidal attacks from kin and non-kin alike. (From Hoogland 1986.)

of members of their home coterie (Hoogland 1985). Data from 19 females suggest that individuals show one extreme or the other in the treatment of related babies (i.e. either communal nursing or infanticide), but not both (Hoogland *et al.* 1989). On the male side, adult sibling competition apparently centres on mating opportunities, such that 'amicability' drops almost to non-kin levels just when females enter oestrous (Figure 13.8).

Finally, it should be pointed out that sibling rivalries are likely to be resolved so far in advance of sexual maturity that the remote issue of sexual opportunity can be hard to identify clearly as the goal of early squabbling. (This is confusing for the same reason that obligate siblicide in well-stocked black eagle nests has struck some observers as unrelated to food shortages—it can be a 'pending competition'—but here the events may be many months or even many years ahead of the pay-off.) Nevertheless, eventual breeding opportunity must be considered as a likely bone of contention when same-sex altercations are severe enough to force subordinate individuals to disperse while dominant siblings inherit the farm (so to speak). Conflict over breeding opportunity, as well as some parallel arguments derived from this (Bekoff 1977), provide a plausible explanation for the initial dispersal in various taxa. For example, female wild dogs (*Lycaon pictus*) depart from their natal pack as soon as the most dominant sister of a given sibship starts to breed (Frame and Frame 1976, Frame *et al.* 1979). The exodus presumably indicates concession that the one local opportunity has been seized. Similar suggestions have been made for status influences on dispersal in coyotes (e.g. Bekoff 1977, 1981), wolves (Mech 1970) and marmosets (Kleiman 1979, 1981). In its milder forms, sexual competition may underlie the ebb and flow of mutual toleration between same-sex sibling neighbours (e.g. in marmots; Frase and Armitage 1984). Moreover, local resource competition has also been invoked to explain why red deer mothers invest more in sons than in daughters during nursing. The latter remain near their mother throughout life, depressing forage for each other and for her (Clutton-Brock *et al.* 1982b; see also Cockburn *et al.* 1985). At the other extreme, truly ferocious competition for limited breeding slots may partly underlie the evolution of sex-biased obligate siblicide (see discussion of spotted hyenas, below).

13.4 Fatal sibling aggression

The existing literature on postnatal mammalian partial brood loss in the wild is generally lacking in behavioural and ecological details, with most documentation simply showing a decrease in family size. Various other kinds of indirect field data suggest a key role for sibling aggression. For example, meadow vole (*Microtus pennsylvanicus*) litters in rural Pennsylvania show (i) a negative correlation between pup mass and litter size (indicating possible food limitations), (ii) frequent partial losses in otherwise successful litters, (iii) decreasing litter size over time (averaging about one fewer pup per week) and (iv) highest mortality for the lightest nest-mates. Before they die, the lightest vole pups sometimes have cuts and bruises that could have been inflicted either by an adult (perhaps the mother, actively culling her litter) or by siblings (whose teeth have just erupted) (McShea and Madison 1987). In many such cases, overt aggression by siblings can be excluded on logical grounds (e.g. if death occurs when nest-mates lack strength and weaponry and/or if the deceased bear no marks of physical abuse). Thus siblicide *per se* is easier to disprove than to document in species that are difficult to observe.

True siblicide (i.e. involving overt aggression) has been documented in only a very few wild mammals, owing at least in part to logistics. It seems likely, for example, that wild swine and peccaries use their precocial dentition and motor skills as do domestic pigs, but this will not be easy to investigate. To date, the best examples involve canids, spotted hyenas, and Galápagos fur seals.

Arctic fox (*Alopex lagopus*) and red fox pups are known to practise siblicide, although most of the evidence is indirect. Macpherson (1969) found that the number of arctic fox placental scars (mean of 10.6 per parous vixen, $n = 118$) far exceeds post-weaning litter size (mean $= 6.7, n = 27$), and noted that the number of scars does not vary significantly across seasons (while litter size does). From this he concluded that partial brood loss is commonplace. The role of sibling attacks in causing pup mortality was inferred from one excavated burrow that contained six dead and three live pups. All of the deaths were apparently due to fractures at the base of the cranium plus severe internal haemorrhaging. Two captive pups of similar age were actually seen fighting—once over a prey item and the second time fatally. In the latter instance, the stronger sibling seized its victim by the upper jaw and literally shook it to death, producing fractures and haemorrhaging identical to those found on victim corpses in the wild. In all, 20.3% of 330 arctic fox specimens bore facial injuries (puncture wounds on the face or palate) like those made during the one observed death-grip (Macpherson 1969). As described earlier, the behavioural components of facultative siblicide have been observed in red foxes (Henry 1985a, b), but have not been presented in any detail.

Coyotes (*Canis latrans*) show some strong behavioural parallels to foxes, engaging in serious fights when 3 to 6 weeks of age. The significance of this aggression remains unclear. Bekoff (1981: 325) asserted that 'high-ranking individuals do not have greater access to food and simply seem to be avoided more than other litter-mates.' His preliminary functional interpretation is that subordinates may disperse because the perilous step of dispersing may be preferable to that of remaining in an area where they must avoid dominants.

Spotted hyenas (*Crocuta crocuta*) have been reported to perform siblicide that is singularly contingent on litter sex composition. The account that follows is something of an amalgam of field census data and laboratory behavioural observations, collected by Lawrence Frank and his colleagues at Berkeley and Heribert Hofer and Marion East of the Max Planck Institute. The mother hyena gives birth to two, or rarely three, cubs at the mouth of a vacant burrow excavated by some smaller animal (typically an aardvark or wart hog), which is too small for an adult hyena to enter. For the first 2 to 6 weeks, the infants live in the burrow, coming to its mouth during maternal visits. When the young are then relocated to a communal den, litter size has often already dropped to one.

Although field observations of within-burrow behaviour have not been made, several lines of indirect evidence make a strong case for siblicide as the cause of litter reduction. First, like piglets, neonatal spotted hyenas are armed with fully erupted incisors and canines (Figure 13.9)—about 3 weeks ahead of the average for other carnivores, including the related striped hyena (*H.*

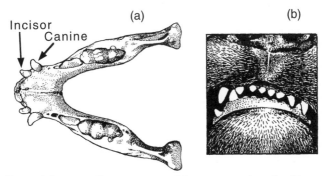

Fig. 13.9 Precocial mammalian weaponry. **(a)** pre-erupted teeth of barnyard piglets. (From Figure 1, Fraser and Thompson 1990). **(b)** Erupted incisors of neonatal spotted hyenas (Frank *et al.* 1991). (Redrawn from photographs by Coral McAllister.)

hyena)—and they also exhibit the species-typical 'bite-shake' killing behaviour within 40 min of birth. Secondly, the two cubs are born about an hour apart and, in captivity, the older cub typically attacks its sibling within min. Over the first few days they struggle until a clear dominance relationship has been established, and they struggle less frequently thereafter. Thirdly, litters containing two members of the same sex are significantly under-represented 2 to 3 weeks post-partum (only one of 15 two-cub litters). Finally, in two litters of very young same-sex infants taken into captivity (one containing two sons, and the other two daughters), one member of each litter was extremely aggressive and excluded its subordinate sibling from food (Frank *et al.* 1991).

The sex-biased siblicide of spotted hyenas is unique (so far) among vertebrates, but has striking parallels with certain hymenoptera, including fig wasps and honey-bees (see Chapter 15), where siblicide is obligate, practised only on same-sex victims, and occurs early in life. The standing explanation for siblicide in those insects is *local mate competition* (Hamilton 1967), i.e. that pre-emptive killing of brood-mates eliminates rivals in a closed system of sexual opportunity. On the basis of current information, a similar argument may account best for siblicide in female–female sibling pairs of spotted hyenas (L.G. Frank, personal communication), if less well for male–male pairs. Certainly, the major hypotheses for obligate siblicide in birds are not sufficient to explain these hyena data well. Although hyena singletons are known to grow more rapidly during the nursing period than members of two-cub litters (Hofer and East 1993), sibling aggression for the purposes of relaxing competition over limited food makes no obvious predictions about gender-biased siblicide.

However, spotted hyenas sibling pairs may remain in their original burrow (and in potential competition for food) for several months if food conditions are so poor as to retard normal development. Observations on sib dyads living in such conditions seldom include gender identification (presumably they are heterosexual after the first few days), but the dominant member of a sibling pair is known to be highly aggressive at the infrequent meals, routinely denying its sibling access to the burrow mouth (and hence to the mother). Eventually the subordinate dies of starvation (H. Hofer and M. East, personal communication). Thus, spotted hyenas may fight with opposite-sex rivals for limited food late in the lactation period.

Alternatively, both the aggressiveness and precocial weaponry that produce hyena siblicide can be interpreted, at least partly, as a side-effect of selection for elevated levels of circulating androgens (Frank *et al.* 1991). According to that argument, adult female dominance is so important as to favour the pre-natal androgen exposure that underlies an extensive masculinization syndrome (which produces extremely aggressive females, a remarkably enlarged and erectile clitoris, a pseudo 'scrotum,' etc.). The resulting early sibling aggression creates a hostile social environment, spawning a secondary arms race, and producing behavioural and dental adaptations useful for winning. It is not clear how this hypothesis for the evolution of siblicide in hyenas can account for the observed gender bias.

Our final example of mammalian siblicide concerns offspring that are a full year apart in age. Milk available to a suckling Galápagos fur seal (*Arcto-cephalus galapagoensis*) appears to be limited by maternal condition, which varies from year to year amidst dramatically variable environmental circumstances (e.g. 'El Niño'). As an equatorial pinniped totally reliant on patchy cold upwellings for food, the pup's nursing period is both protracted and variable (1–3 years) in comparison to that of pinnipeds from higher latitudes and this duration, in turn, leads to overlap between successive breeding cycles. Although litter size is always one, and despite the fact that continuously nursing mothers (those with yearling or 2-year-old offspring) are only 50% as likely to give birth as females without young (F. Trillmich unpublished data), one often finds concurrent but grossly mismatched siblings (Figure 13.10). In such cases, the neonate gains little or no weight. Despite nursing in longer bouts than infants without older siblings, 80% of such younger pups die of starvation within 1 month of birth (Trillmich 1986). This mortality is probably related to the mother's milk delivery system, which—as in other mammals—

Fig. 13.10 A mother Galápagos fur seal threatens her yearling offspring as its newborn sibling nurses. Continued nursing by the yearling may lead to the neonate's starvation, or the yearling may abscond with its sibling, killing it in the process. (Photograph by Fritz Trillmich.)

requires increasing stimulation for oxytocin release and milk let-down as offspring grow. If so, the essential mechanism that sacrifices the neonate on behalf of its nearly grown sib resides in the *mother's* physiology. Furthermore, some older pups attack and kill their newborn siblings, either by stealing them from their mother's side and tossing them repeatedly into the air, or by seizing one end as the mother grabs the other, thus creating a fatal 'tug-of-war' with the infant (F. Trillmich, personal communication).

Summary

1. Sibling rivalry has been less well studied in mammals than in birds for various reasons, including the more cryptic nature of some cases (e.g. occurring prenatally and/or in burrows) and the more likely extension of competition into adulthood.
2. Maternal adjustment of litter size involves quantity–quality trade-offs that clearly implicate competitive interactions. When stressed, mothers of many taxa have physiological and behavioural ways of culling family size (embryo resorption and pup cannibalism, respectively). Because the selective value of such traits lies in avoiding pending competition, they are analogous to pre-emptive obligate siblicide in predatory birds, although here the killing is effected by a parent instead of a sib.
3. In many 'polyovulating' species, females create surplus offspring—sometimes hundreds of them—that must then compete for a limited uterine wall space. Obviously, many siblings fail at this point.
4. In pronghorn, two embryos commonly develop in each uterine horn, but one grows rearwards to skewer its horn-mate. This initial over-production is provisionally regarded as an insurance system.
5. With truncated gestation followed by lengthy lactation, marsupials have the option of retaining their marginal offspring until birth, and then pruning the litter to match pouch capacity and food availability. In some species, many extra young are lost once the few teats have become occupied.
6. In placental rodents, the number of functional nipples generally exceeds mean litter size by about twofold, but in a few other taxa there can be more offspring initially than there are feeding sites. Teats may also vary in value, with some providing richer milk flow, and siblings compete for such sites. Domestic piglets even possess precocious eye-teeth jutting out from the jaw that are used as effective weapons during fights over the best teats.
7. In mammals that breed co-operatively, siblings often remain together throughout their lives and compete as adults for reproductive opportunities. The most dominant group member controls the other same-sex members by various (behavioural and/or chemical) signals that are backed up by the option of infanticide for any offspring produced by subordinates.

Despotic reproductive suppression in dwarf mongooses may sometimes be relaxed partially as a means of discouraging emigration by subordinate siblings. Similar conflicts are evident in African lions, black-tailed prairie dogs and various canids, as well as in many social insects and birds.

8. Detailed accounts of siblicidal behaviour are relatively (and understandably) scarce for mammals, but indirect evidence exists for voles, foxes, and coyotes. The best data for siblicide in non-domesticated mammals are for arctic foxes, spotted hyenas, and Galápagos fur seals. Like domestic swine, these hyenas are born with precocial teeth, which they can use lethally on their single litter-mate. For reasons that are not yet clear, hyena siblicide seems to be restricted to same-sex litters. Unpublished data on Galápagos fur seals show that newborn pups are routinely killed when competing with a 1- or 2-year-old sibling, if food conditions are poor. The victim may die of starvation (failing to suckle successfully from a mother whose delivery system is calibrated to the stronger nursling) or it may be snatched by the sibling and killed by direct physical abuse that sometimes even involves a grisly 'tug-of-war' with the mother.

14

Sibling rivalry in ectothermic vertebrates

Examination of the embryos began in a startling way. When I first put my hand through a slit in the oviduct I received the impression that I had been bitten. What I had encountered was an exceedingly active embryo which dashed about open-mouthed inside the oviduct.

(Springer, 1948)

14.1 Parental investment by vertebrate ectotherms

In most fishes, reptiles and amphibians, offspring receive no post-zygotic parental investment, shareable or otherwise, and have little to contest with siblings they seldom if ever contact. Because most species are oviparous (hatching from eggs outside the mother's body), sibling embryos are physically separated by shell membranes and thus remain isolated from one another until hatching, after which they are on their own (Shine 1988, Clutton-Brock 1991). Additionally, in contrast to birds and mammals, parental feeding of offspring is extremely rare. In some anurans, however, parents place larval offspring in small aquatic **nurseries** (restricted bodies of water suitable for early development, including rain-filled puddles, tree-holes, leaf axils, bromeliad 'tanks,' ponds, and other ephemeral catchments). Many of these are prone to shrinkage and/or resource depletion from various consumers. Alternatively, offspring may be affixed to parental body parts. Either way, key resources may become limiting and highly developed sibling rivalry sometimes arises.

Among live-bearers, the term 'viviparous' embraces an impressive diversity of reproductive modes. At one end of this spectrum are reptilian species that create and retain fully shelled eggs within the oviduct (or analogous structure) until after they have hatched. Such siblings are presumably buffered from one another. However, in other viviparous groups, the ovulated eggs are not shelled (the shell glands themselves have vanished secondarily), but develop in the oviduct, where they either rely solely on yolk stores or receive varying amounts of maternal resources. Such dependence on internally transferred nutrients, called *matrotrophy*, reaches its reptilian zenith in some neotropical skinks (*Mabuya* spp.), which produce by far the smallest reptilian eggs (volume = 0.5 mm³, containing little or no yolk) and then deliver more than 99% of the embryos' nutrients via a complex chorio-allantoic placenta during a protracted 10–12-month gestation (Vitt and Blackburn 1983, Blackburn *et al.* 1984, Blackburn and Vitt 1992), during which offspring increase in mass by 38 400%. Although this is the most extreme reptilian convergence with the familiar eutherian investment pattern, other taxa also have complex placentae, and some sharks, rays and pouch-brooding fish have a 'placenta-like system' for passing nutrients from mother to offspring. In most cases it is not known whether brood-mates sharing the same maternal nutrients ever compete (e.g. when maternal condition declines during periods of ecological stress), much less what forms such prenatal rivalries might take.

Some fishes and amphibians produce truly vast numbers of zygotes, with clutch sizes commonly two or more orders of magnitude greater than those of birds, mammals and even reptiles (Dominey and Blumer 1984). In such taxa, fecundity varies widely as a negative function of parental care. Amphibian clutch size ranges from 40000-plus zygotes in some large anurans (e.g. bullfrog, *Rana catesbeiana*, and Great Plains toad, *Bufo cognatus*; e.g. Krupa 1988a, b, 1993) to very few in others (e.g. *Dendrobates*; Wells 1978, Weygoldt 1980) and just one in *Sminthillus limbatus* (Duellman and Trueb 1986). Fish clutch sizes also vary, from many millions in large pelagic species (e.g. 300 million for ocean sunfish, *Mola mola*: Hart 1973) to a small handful in nursery species with relatively lavish parental care. In general, the second group tends to consist of species that inhabit environments with relatively low prey densities (Winemiller and Rose 1993).

Offspring in nursery taxa are more likely to impinge on the fitness of siblings than those in dispersing taxa for two general reasons. Most obviously, by remaining in proximity they have greater temporal and demographic opportunities to interact with siblings (both positively and negatively). Secondly, because most nursery taxa have relatively small families, each offspring represents a greater proportion of parental investment. Because the potential indirect fitness contribution that a given clutch-mate represents is a product of its expected reproductive value (Fisher 1930) and its coefficient of relatedness to Self, even a full sibling can be virtually worthless (e.g. if it is doomed anyway; Eickwort 1973, Milinski 1978, Charlesworth and Charnov 1980).

Of course, if the nursery-mate is a half-sibling or non-sibling, it is even more expendable, so the matter of genetic relatedness remains important. External fertilization is typical in most fishes and anurans (cf. caecilians and salamanders; Duellman and Trueb 1986). The impact of this on within-brood relatedness is not yet clear, although it has been postulated to increase a given male's control over fertilization (Trivers 1972), which might increase mean relatedness among brood-mates. On the other hand, some externally fertilizing fish almost certainly face sperm competition from satellite males ejaculating at or near the spawn site (reviewed by Taborsky 1994), which could mix paternity and decrease *r* among brood-mates (e.g. Ridley 1978). Similar effects on relatedness may arise for anurans that oviposit communally (Howard 1979, Gross and Shine 1981, Halliday and Verrell 1984, Caldwell 1986).

Ecological factors can also devalue a sibling's potential for delivering indirect fitness and potentially enhance Self's incentives for depriving or killing it. As in other taxa, non-cannibalistic sibling rivalry is most likely to occur when larvae depend on severely constrained resources. In ectothermic vertebrates, the two most obvious situations involve a few species with (i) expensive and non-shareable parental investment (either *in utero*[1] or later) and/or (ii) the combination of meagre or ephemeral resources at the oviposition site plus 'indirect development' (the larval form is dissimilar to the adult), circumstances that essentially establish a race to metamorphose before pending catastrophes materialize.

14.2 Sibling competition in viviparous and egg-retaining ectotherms

In taxa where parents carry zygotes within their bodies for all or part of the embryonic period, providing them with physical protection and in some cases with nutrients and thermal benefits (e.g. heliothermic reptiles; Shine 1985), the vulnerable time of early development can be abridged. Among fishes, most sharks, skates and rays have internal fertilization, and half or more of these are viviparous to varying degrees (Wourms 1981, Wourms *et al.* 1988). In the most primitive forms, the embryo is metabolically autonomous from the mother[2], but in other groups a great diversity of more complex, matrotrophic relationships has also evolved. The size increases possible for fishes are even greater than for skinks, including dry weight increases of up to 842 900% for the teleost *Anableps* and 1 200 000% for sand tiger sharks (*Eugomophodus (= Odontapsis) taurus*)(Wourms *et al.* 1988)! Although livebearers generally produce fewer young than their oviparous counterparts, their offspring are believed to have much higher levels of survivorship (Wourms 1977, Baylis 1981, Wourms *et al.* 1988). In bony fishes, although livebirthing is rare, occurring in only *c.* 3% of species, these taxa exhibit as much diversity of form as their oviparous counterparts (Wourms 1981, Wourms *et al.* 1988).

Parental feeding is an exceptional form of parental investment in fishes, and its rarity automatically drives the manifestations of sibling rivalry in circumscribed directions. As Dominey and Blumer (1984: 62) state, '. . . fishes do not

---

[1] 'intra-oviductal' is technically preferable, but we use the more familiar mammalian term for consistency and simplicity.

[2] Called '*lecithotrophy*', meaning strict reliance on egg-borne nutrients, in technical literature.

feed their offspring [so] fratricide of the type practiced by birds as resource competition is generally not expected.' Accordingly, we focus on other forms of non-shareable parental investment when seeking sibling rivalry in these taxa, especially space, and this leads to a consideration of mouth- and pouch-brooding forms.

The possibility that brood space becomes increasingly inadequate as fry grow has been considered briefly for mouth-brooding cichlids (Yanagisawa and Sato 1990), but not explicitly with regard to how it affects sibling competition. Indeed, the tendency for offspring numbers to decrease with the passage of time (Welcomme 1967) has been attributed to filial cannibalism, a reasonable inference given that larvae are sometimes found in the brooding parent's stomach (Dominey and Blumer 1984: 55). However, the habit of swallowing some offspring when the brood's space requirements exceed buccal cavity volume could easily be a secondary parental trait (as opposed to outright ejection of supernumerary fry, it has the obvious merit of nutrient recapture) that might well be incidental to the sibling competition. That is, ingestion may be effect, not cause. Moreover, some swallowing may simply be accidental. Brood size might easily be limited by cavity dimensions, and the usual suite of incentives for parental over-production (see Chapter 1) could still favour the parental practise of aiming high when setting clutch size. The protective nature of the oral cavity is suggested by the behaviour of fry (which hasten back inside when threatened). The value of that space is underscored by the existence of at least one interspecific brood parasite, a mochokid catfish, that bothers to steal it. As with avian cuckoos, the catfish mother slips into the spawning nest and oviposits before the cichlid mother takes all the eggs into her mouth. The catfish fry hatch sooner and eat the eggs and hatchlings of the host (Sato 1986).

In other mouth-brooders (e.g. *Cyphotilapia frontosa*), larvae eat food particles inhaled by the mothers, which appear to seek food-rich micro-habitats actively (e.g. they move to introduced clouds of brine shrimp in aquaria), as if for this purpose (Balon 1991). The parallel with parental feeding by birds and mammals is evident. These cichlids also tend to have much larger eggs and longer associations with their less numerous young, relative to neighbouring species that provide scant amounts of fully shareable parental investment (e.g. nest-guarding only; Balon 1991). In *C. frontosa* and two similar cichlids (*Tropheus duboisi* and *T. moorii*), the benefits of this maternal feeding have been demonstrated under field conditions: young triple in size (wet mass) while still inside the mouth (Yanagisawa and Sato 1990, Yanagisawa and Ochi 1991). However, possible density-dependent effects on offspring growth have apparently not been addressed, except for noting that these species have small broods and the overall importance of sib competition in mouth-breeders remains unclear.

Sea-horses and pipefishes retain eggs inside specialized and highly variable male brood pouches (reviewed by Vincent *et al.* 1992) and recent studies have focused more directly on possible sibling rivalry consequences, especially in captive pipefishes. Male pipefish (*Syngnathus typhle*) have closed brood pouches (which are transparent, allowing the counting of eggs within) into which as many as three or four females oviposit until a pouchful of 40 to approximately 150 eggs is on board (depending on male size). Eggs spend a month in the pouch, benefiting from predator protection and an osmo-regulated environment (Quast and Howe 1980), while also obtaining both oxygen (Berglund *et al.* 1986) and nutrients (documented by tracking the passage of labelled amino acids from male to offspring; Haresign and Schumway 1981) via a placenta-like structure laced with paternal blood vessels. Under these conditions, eggs do not lose mass (as fish eggs normally do, from respiration), but are sustained by continuous paternal provision. Even after hatching, the young remain inside the male's pouch for a few more days, with all parental care ending at parturition.

Pipefish clutchmates apparently do *compete* for limited oxygen and/or nutrients while in their father's pouch, as has been shown by a negative relationship between numbers and weights of newborn (Ahnesjö 1992a, b, 1996). This in turn may be affected by a twofold variation in egg size, related to maternal body size, because large eggs generally respire more rapidly than smaller ones. As a result, small eggs from small mothers may fare less well at intra-pouch scramble competitions for resources. Furthermore, at least in a captive experiment, neonates emerging with larger body size tend to grow more rapidly as juveniles, and are more likely to escape early predation (Ahnesjö 1992b). It therefore appears that sibling rivalries during the pouch phase are likely to have substantial fitness consequences in this species.

In the closely related sea-horses, which also show reversal of sex roles (with males being effectively pregnant; see Vincent *et al.* 1992) but are apparently monogamous, a similar competition among full siblings may occur (however, accounting for all eggs through the opaque sea-horse brood patch is con-siderably more difficult than for pipefish). In laboratory studies of *Hippo-campus fuscus*, brood size at emergence from the male's pouch was about half the clutch size that went in, although the fates of the missing eggs/young remains unknown. Interestingly, the missing fraction of offspring may be smaller for parents that had been together longer prior to mating (Vincent 1990). Inside the pouch, sea-horse embryos obtain paternally conferred benefits rather similar to those enjoyed by pipefish (namely protection, aeration, osmotic regulation, some nutrients via a 'placenta', etc.), and these also terminate when juveniles leave the pouch. It is unknown whether comparable trade-offs exist between offspring numbers and size at the terminus of parental care, although the pouch is often crammed full at the time of oviposition. At emergence, however, brood-mates may vary fourfold in dry mass! If larger initial size confers higher survival probabilities, sibling differences in rates of consumption or control over room for personal growth may be adaptive in this context.

Internal partial brood losses can also be inferred for *Dermophis mexicanus*, a viviparous caecilian, whose oviducts sometimes contain desiccated, partially developed embryos. The mean *ovarian* clutch size, 19.6, is more than double the mean *oviducal* clutch size of 7.0 (Wake 1980). Whether such prenatal mortalities are due to ontogenetic failures, maternal manipulation (e.g. selective abortions à *la* progeny-choice), sibling competition, or combinations thereof is not known.

An alternative form of amphibian egg retention involves transport on or in a parent's body (see Section 14.3.1). Internal retention of amphibian embryos to the point of live birth is believed to occur in relatively harsh environments, such as alpine zones (Salthe and Mecham 1974), where extra protection for the developing young presumably carries high selective value (but see Dopazo and Alberch 1994). Among salamanders, giving birth to live young is especially rare, with only a few montane populations of normally oviparous species practising egg retention (producing either larvae or fully developed young) and two or three other species being truly viviparous (Duelmann and Trueb 1986, Dopazo and Alberch 1994). By contrast, viviparity is found in about 75% of all caecilians (Wake 1977, 1992), about which little else is known.

In the caecilian *D. mexicanus* the young finish their meagre yolk supplies early and then feed on nutritive secretions from the oviduct lining (Wake 1982). These young become very large (the brood may total 65% of the mother's post-parturitional mass: cited in Stearns 1992), but we know of no studies of whether siblings in any way constrain each other's prenatal growth. Something similar occurs in *Salamandra atra*, in which 20 to 30 eggs are produced in each oviduct, but only one or two of them are fertilized. After the viable offspring have consumed their own yolk reserves, they eat these eggs (Duellman and Trueb 1986), which have not been resorbed in the mean time by the mother.

More complex forms of prenatal cannibalism have been discovered in *S. salamandra*, a highly variable species in which populations differ greatly with respect to reproductive mode (Dopazo and Alberch 1994). In one population females give birth to small aquatic larva, in a second population they give birth to fully metamorphosed and large young, and in a third to multiple development patterns that differ between—and even within!—individual females (Figure 14.1, upper panel). In a single reproductive tract, Dopazo and Alberch (1994) found hatched larvae, aborted embryos (at varying stages of development) and unfertilized eggs. The stomachs of all larvae contained material from unfertilized eggs and smaller siblings, which the authors interpreted as evidence of siblicidal cannibalism. It would certainly be helpful to know whether the ingested larvae died as a result of maternal actions (i.e. *filial infanticide*—perhaps to trim brood size and/or to provide food for core brood members) or from sibling attack *per se*. The existence of unfertilized eggs could be seen as implicating some level of maternal collaboration—an

(a)

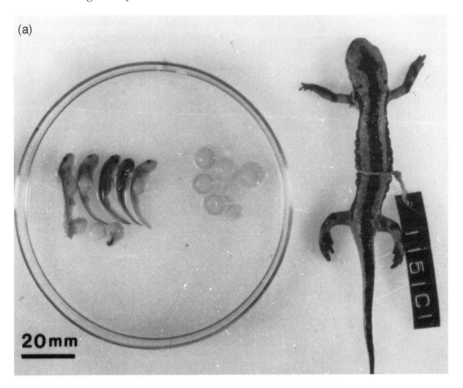

20 mm

(b)

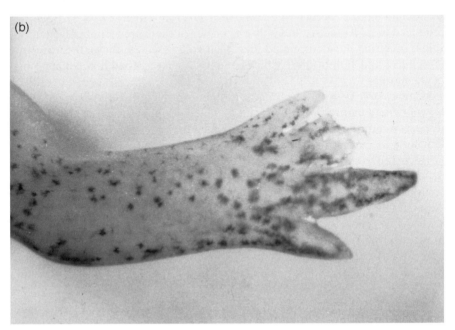

Fig. 14.1 Brood polymorphism in *Salamandra salamandra*. **(a)** Adult female with the brood of five embryos plus seven unfertilized (presumably 'trophic') eggs she was carrying when collected in northern Spain. **(b)** A larval forelimb in which digits II and III apparently have been bitten off by a sibling *in utero*. (Photographs supplied courtesy of Hernán Dopazo and Per Alberch, Museo Nacional Ciencias Naturales, Madrid, Spain.)

effort to fortify the core brood without use of a more direct nutrient transfer system between mother and embryo, such as a placenta. However, trophic eggs are sometimes found in one oviduct while all viable consumers are in the other, suggesting that matrotrophy may not be a highly evolved system. Indeed, the observed oophagy may be a mere side-effect of incomplete fertilization. Nor is it clear whether the 'cannibalism' is anything other than opportunistic scavenging of embryos that died for other reasons (i.e. unrelated to either maternal manipulation or siblicidal attack), although recent scrutiny has uncovered evidence of what might be sib wounding (Figure 14.1, lower panel). On balance, it seems most likely that the sibs do, in fact, kill and consume otherwise viable womb-mates, although it is also possible that ingesting siblings in such cases may be incidental to reducing sibling competition for space and/or nutrients (Dominey and Blumer 1984: 63). In short, cannibalism may exist primarily because of its undoubted efficacy in curbing other forms of sibling rivalry.

Prenatal cannibalism has also been reported for several live-bearing fishes (e.g. Meffe and Vriejenhoek 1981), the most celebrated being the sand tiger shark. As in Springer's (1948) persuasive description of the original behavioural evidence (the heading for this chapter), it appears that only one embryo completes development in each oviduct, that it is not sheathed inside a shell membrane, and that it consumes both eggs (oophagy) and smaller embryonic siblings[3] (Figure 14.2), truly an extreme form of matrotrophy (Springer 1948, Wourms 1977, 1981, Gilmore *et al.* 1983, Wourms *et al.* 1988). The development of this shark is notably precocious in two areas: first, an extensive lateral line appears when the embryo is only 30 mm long, with dentition arising by the 40- to 45-mm stage. Thus, functional dentition is in place when only 4% of the term embryo length has been achieved (cf. 60–100% in non-oophagous sharks) and the embryo is just about to hatch from its own egg. These findings have been interpreted as true adaptations for sibling consumption (the lateral line senses presumably assist in locating prey). Consumption of eggs begins when the offspring is about 50 mm in length, and consumption of hatched sibings when it has reached 80–100 mm (Wourms 1977, 1981). Numerically, the consumption of siblings is prodigious in this species, a typical fetus being estimated to eat approximately 17 000 eggs, most

[3] The term 'adelphophagy' is commonly used in the ichthyological literature for this practice (Wourms 1977, 1981, Wourms *et al.* 1988).

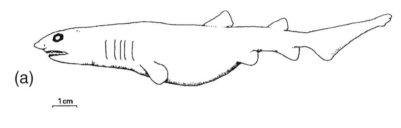

(a)

(b)

Fig. 14.2 Cannibalistic sand tiger shark embryos, drawn to the same scale. Note **(a)** the cannibal's distended gut and **(b)** the residual external yolk sac on the victim's belly. (From Gilmore *et al.* (1983), redrawn by Coral McAllister.)

of them fertile, with egg cases being supplied continually throughout gestation (Wourms *et al.* 1988). The whole package of traits is regarded as a suite of strategies for attaining neonatal gigantism; after 9–12 months of gestation, the surviving pup is over 1 m long and can weigh more than 10 kg. At least 14 species of mackerel sharks (Lamniformes) are known to engage in similar practises (Wourms *et al.* 1988, Wourms and Demski 1993), and fossil evidence for shark oophagy exists as well (Lund 1980).

There is truly an impressive diversity of maternal–embryonic relationships across the fishes generally (Wourms *et al.* 1988), ranging from straight egg-laying (oviparity), through reliance on yolk-sac nutrients while being retained inside the mother's body (known as 'lecithotrophic' viviparity), to three broad forms of matrotrophic viviparity that seem intrinsically suited to prenatal sibling rivalry. These include the uptake of maternal nutrients via the embryo's skin/gut epithelium ('trophodermy'), the familiar placental exchanges ('placentotrophy'), and sib cannibalism. The first two suggest potential for scramble competitions among siblings, while the latter is more direct. For example, in at least 11 viviparous but non-placental sharks in the order Lamniformes, mothers periodically shed eggs that the embryos then eat (Table 1 in Gilmore 1993). In coelacanth (*Latimeria chalumnae*), the ancient teleost, viviparity is also complicated, with pup development relying on multiple nutritional sources. First, each egg is roughly cabbage-sized (19 cm in diameter and 334 g in mass) with large yolk stores that support the embryo initially until a placenta develops that supplies additional materials. This is supplemented further by the ingestion of supernumerary eggs; the mother initially produces 19 to 30 of these enormous eggs, but can accommodate substantially fewer than half that number of growing embryos within her single functional oviduct

(Wourms *et al.* 1991). In short, there is a crowded nursery, some access to circulating maternal nutrients, and the ingestion of proto-siblings. Assuming that maternal over-production of enormous eggs is costly, the extras are likely to represent an adaptive parental strategy for setting nutrients aside as 'icebox' offspring, although the same eggs (if fertilized) might be simultaneously available as insurance offspring (Mock and Forbes 1995). In brotulas, *Ogilbia cayorum* and *Lucifuga subterraneus*, late-term embryos have been observed with bulb-like evaginations of maternal tissues in their mouths. These bulbs, which are fluid-filled and capillary-rich, may serve as 'buccal placentae', although it is also possible that they are simply harvested and swallowed (Wourms *et al.* 1988). Whether bulbs are ever in short supply, and hence the focus of prenatal competition, is not known.

Embryo crowding also seems likely in viviparous teleosts, which are unique among vertebrates in using the ovary as the site of both egg production and gestation. In many of these taxa, embryos are supported by matrotrophy that can lead to massive size increases (up to 24 000%: Wourms *et al.* 1988).

An interesting puzzle is suggested by indications that female adders (*Vipera berus*) that mate with multiple males carry a higher proportion of offspring to term than do females which mate with just one male (Madsen *et al.* 1992). If these broods actually have mixed paternity, then one might speculate either that mothers value mixed sibships more (and hence invest more in them) or that offspring selfishness is less constrained by intra-brood relatedness such that half-sib embryos take more—at the expense of future maternal production (i.e. parent–offspring conflict). The latter possibility might arise as a result of paternal alleles expressing themselves through increased consumption of maternal PI ('genomic imprinting'; see Sections 10.2.1, 13.3.1 and 16.2.1).

14.3 Sibling rivalry in ovipositing taxa

In ectothermic vertebrates that lay eggs rather than retaining them internally, most subsequent parental care is shareable[4] (e.g. building nests, guarding eggs, fanning eggs, and removing fungus-infected eggs; Keenleyside 1978, 1980, Blumer 1979), and thus offers few incentives for sibling-sibling conflict (Townsend *et al.* 1984). For example, there are frogs in which fathers moisten the eggs by transferring water from the male's ventral surface (Taigen *et al.* 1984) or by emptying their bladders over the clutch (Myers and Daly 1983). Egg-guarding is especially widespread in salamanders (65% of species; Salthe and Mecham 1974; see also Nussbaum 1985), but also common in anurans (10–20% of species, from 14 of the 20 families; McDiarmid 1978, Wells 1981)

---

[4] Indeed, most of the shared parental care in these taxa is unlikely to be limited, and thus should be uncontested.

and fishes (Blumer 1979). Among reptiles, such guarding is very irregular, occurring in all crocodilians, 3% of oviparous snakes and just 1% of oviparous lizards (Shine 1988). Some salamander parents concentrate their own egg cannibalism on diseased, fungus-infected eggs (Tilley 1972, Forester 1979), the removal of which enhances survival of the remaining eggs (Crump 1992). A few reptiles provide additional services. For example, Indian pythons (*Python molurus*) and related species (Boidae) are reported to perform maternal incubation (coiling around the clutch and twitching their muscles continuously in order to elevate their body temperature by a few degrees; Hutchison *et al.* 1966, Vinegar *et al.* 1970). Furthermore, crocodilian mothers not only tend the nest, but also respond to the hatchlings' calls, carry them to the water, and accompany them for a year (Shine 1988). Nevertheless, there are other taxa in this group where parental care is of a less shareable nature, taking the familiar currencies of space and food.

14.3.1 *Offspring transporting*

As with mouth-brooding fish, parental space may constitute a limited commodity, and hence fodder for conflict, in frogs that actively transport their tadpoles, even if just for a single trip from nest to water. Among oviparous reptiles and amphibians, it is the male parent which retains the eggs in only one anuran genus, namely the midwife toads (*Alytes* spp.) (Boulenger 1912). Here the father carries several string-like clutches simultaneously entwined around his legs. The similar habit of back-packing tadpoles, by contrast, occurs in all Dendrobatidae plus at least six other species. In *Dendrobates auratus*, for example, hatchlings from the half dozen eggs squirm through the nest's jelly, over the father's legs (which are flattened and extended backwards to make a ramp) and up his sloping back. Those that try to wriggle straight up his side usually fall off; indeed, Wells (1978) reported that 'many tadpoles fell repeatedly before securing a firm hold'. Because the male typically carries only one tadpole of several on a given trip, and because the availability of water repositories is likely to be limiting to male reproductive success (demonstrated experimentally in *D. pumilio*; Donnelly 1989), getting aboard appears to be simultaneously very important and accomplished at some cost to sibling rivals, whose transport is both delayed and increasingly improbable. Two back-riding species that carry multiple tadpoles also seem particularly well suited to sibling rivalry because the entire process of larval development is completed there (i.e. the ride is of considerable duration; Inger 1966, Duellman and Trueb 1986). In this context, it would be interesting to know how often individual tadpoles fall off during translocations, whether dorsal sites vary in their attributes for safe carriage, how sites are claimed, and whether individual larvae actively supplant each other. Females of the species *Colostethus inguinalis*, for example, carry very large numbers (*c.* 35!) of tadpoles that remain with them for up to 9 days while adding nearly 40% to their body length (Wells 1980a, b). Which tadpoles are released first, and with what consequences, is unknown.

Across taxa there are various methods for adhering to, or being embedded in, the parent's dorsal integument. In hylid 'marsupial frogs' (e.g. 39 species of *Gastrotheca*[5]), transportation involves a special dorsal pouch on the female, from which newly hatched tadpoles are released into ponds (Duellman and Maness 1980). In the two 'gastric-brooding frogs' (*Rheobatrachus silus* and *R. vitellinus*), the mother actually swallows the fertilized eggs, and then fasts while the eggs incubate (and later when brooding the tadpoles) inside her stomach. The mothers appears to be extremely full during the later stages (Corben *et al.* 1974), suggesting the potential for space shortages. Eventually, the young froglets emerge from the mother's mouth (Tyler and Carter 1981) and receive no further care. Space limitations during development may also arise for sibships of *Assa darlingtoni*, which develop into froglets within paternal inguinal pouches (Ingram *et al.* 1975, Heyer and Crombie 1979), and of *Rhinoderma darwini*, which develop inside the father's enlarged vocal sac (Busse 1970).

14.3.2 *Provisioning of external larvae*

In some *Dendrobates* (Wells 1978) and *Chirixalus* (Ueda 1986) species, the parents continue to assist the young they have already carried to their aquatic nursery by making two or more deliveries of trophic eggs. If siblings are maximally dispersed (one per catchment), then any rivalries among them must be mediated through the care-giving parent (i.e. Triversian inter-brood conflict). If they are sometimes clumped spatially, such that multiple siblings share the same nursery, then competition for food can be much more direct and analogous to that in a typical avian nest. Such clumping has been reported for captive *D. pumilio* (Weygoldt 1980, 1987), but its occurrence under natural conditions remains unclear. For provisioning anurans, most of the critical behavioural events are undocumented, but the overall pattern and some of the details known for a few species are provocative. For example, in *Chirixalus,* the mother provides trophic eggs for her tadpoles, which congregate around her and poke their snouts into her cloaca (Ueda 1986). Furthermore, when an adult female of *D. pumilio* enters one of these small bodies of water, the resident tadpole vibrates its tail, which 'presumably signals the tadpole's presence to the female' (Duellman and Trueb 1986). Is this a form of begging (in which case, does it affect the mother's willingness to deliver food)? Is it a signal of occupancy, essentially warning transporting parents not to release tadpole rivals that, although probably doomed to lose in

---

[5] Other hylids that brood eggs on the dorsum include *Stefania, Cryptobrachus* and *Hemiphractus*. In addition, *Fritziana* spp. have open dorsal basins, while *Flectonotus* and *Amphignathodon* have true dorsal pouches.

competition with the older incumbent, might nevertheless consume part of the resident's resource base? Are there perhaps 'signature' components to the signal that allow the mother to identify her own progeny and which cue her to deposit trophic eggs?

Even for dendrobatids that do not provision their tadpoles, the placing of multiple tadpoles in a severely restricted site seems likely to result in either scramble-type or cannibalistic sibling competition, unless resources are consistently lavish. However, although *D. auratus* larvae are thought to consume siblings (Myers and Daly 1983), the frequency of such activity in nature remains unknown. Sib cannibalism may be a mere artefact of captivity or a trivial enhancement of a strategy for destroying unrelated rivals.

14.3.3 *Cannibalism in the nursery*

The dependency of amphibians with complex life cycles on aquatic nurseries for breeding or larval development, coupled with the evaporative properties of water, provide the backdrop for some truly amazing physical and behavioural traits that seem likely to have been shaped primarily by sibling competition. Furthermore, if one visualizes amphibian radiation as involving sporadic excursions into increasingly dry and unstable habitats as the more predictable niches filled (e.g. Wilbur and Collins 1973, Blair 1976, Wilbur 1980, Crump 1983), one glimpses how the multiple selective forces favouring parental over-production must have induced sibling competition repeatedly.

These confined aquatic domains may not press brood-mates into quite the forced proximity that is generated by retention inside a parent's body, but their microhabitats tend to be far harsher and more tenuous, forcing a developmental race toward the relative safety of metamorphosis. At the same time, a given rival is far more likely to be genetically unrelated than in a parent-body nursery. Depending on puddle size and the tendency of different parents to introduce cohabiting broods, truly ruthless strategies may evolve, unmuted by kinship. At present, few of these systems have been subjected to detailed studies of the relevant costs and benefits, although a wealth of natural history data exists. For example, the male African bullfrog (*Pyxicephalus adspersus*) digs channels between small drying pools so that his tadpoles can negotiate their way to deeper water; in effect, he enlarges and reshapes the nursery as needed (Kok *et al.* 1989). Furthermore, considerable research has been conducted on related topics (such as kin discrimination abilities) that may affect how sibling rivalries are played out in these animals.

The glamour taxa in this group are three species of spadefoot toad (*Scaphiopus*) plus two *Ambystoma* salamanders (the tiger salamander, *A. tigrinum*, and long-toed salamander, *A. macrodactylum columbianum*), all of which show dramatic developmental plasticity that commonly produces both omnivorous (normal-looking) and carnivorous (i.e. cannibalistic) morphs within a single brood. The toad phenotypes were apparently first observed by Arthur N. Bragg, who was so impressed by their differences that he first thought that two different species might have been mixed inadvertently, before he later worked out that they were, in fact, siblings (Bragg 1956, 1964, 1965). Dichotomous morph names aside, these tadpoles actually exhibit considerable physical and behavioural variation. Nevertheless, the cannibal types develop massive heads with hypertrophied jaws and serrated beaks, which they use when feasting voraciously on puddle-mates, many of which are siblings (Figure 14.3).

The phenotypic plasticity of spadefoot toads appears to be maintained as a balanced polymorphism by environmental uncertainty. At nearly any point during ontogeny, the ingestion of macroscopic animal food, typically fairy

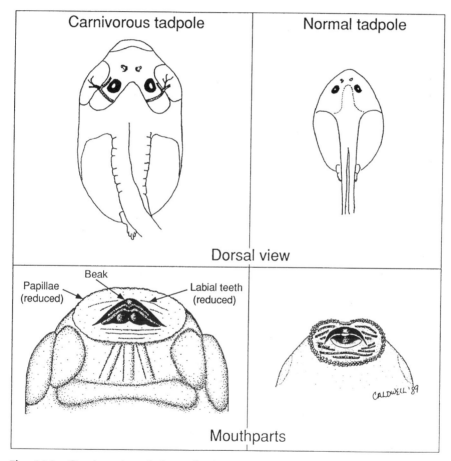

Fig. 14.3 Carnivore/cannibal morph of *Scaphiopus* larvae, drawn to same scale as a normal (omnivore) tadpole. In addition to the modifications of size, head shape and mouthparts shown here, the carnivore also develops a much shorter gut. (Original drawings by Janalee Caldwell.)

shrimp (anostracans), diverts a fraction of the tadpoles off to the carnivore developmental route (Figure 14.5); failure to ingest shrimp leads to developing as an omnivore (Pfennig 1990b, 1992). The carnivore phenotype does better, on average, in highly ephemeral ponds; specifically, it develops more rapidly, and thus is less likely to be stranded if the pond disappears (Semlitsch and Caldwell 1982, Pfennig 1992). On the other hand, omnivores fare better in longer-lasting pools because they sequester larger lipid stores (Pomeroy 1981) that enhance post-metamorphic survival (Pfennig 1992). The decision, then, as to whether a shrimp-eating individual should play Cannibal may be quite complicated, involving both its direct fitness (e.g. the accuracy with which it can predict if and when its pond will vanish, likelihood of becoming dinner for someone else, and so on) and its indirect fitness (through the loss of siblings as meals, the nutritional boost its tissues might give to a sib's future success, etc.). An ESS model for this decision is represented schematically in Figure 14.5 (from Pfennig 1992).

The question naturally arises as to whether a carnivorous tadpole can minimize its indirect fitness losses by passing up opportunities to consume its own siblings (Blaustein and O'Hara 1982), and ingesting unrelated larvae instead? The ability of anuran larvae to discriminate between sibs and non-sibs, initially suggested by Wassersug (1973: 289), has been demonstrated repeatedly (e.g. Waldman and Adler 1979, Blaustein and O'Hara 1981; for reviews see Waldman 1991, Blaustein and Waldman 1992), but the functional significance of that ability has proved to be far more elusive, especially for the tadpole stage itself. Nearly all *Scaphiopus* larvae exist in pools that contain both siblings and non-siblings (Pfennig 1990a; D.F. Pfennig, personal communication). Pfennig's first laboratory attempts to explore whether

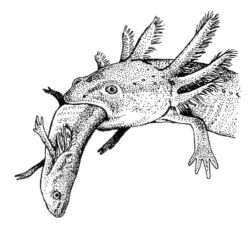

Fig. 14.4 Tiger salamander larva cannibalizing a conspecific. Recent experiments show that cannibalism is least common if all available larvae are siblings and most common if none are siblings. From Pfennig and Collins (1993).

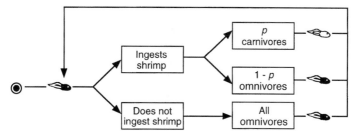

Fig. 14.5 The developmental process leading to various larval morphs of the spadefoot toad is apparently a branching pathway. Initially, all tadpoles have 'omnivore' morphology (schematically represented here as filled figures), which they retain as the 'default' condition unless they encounter some threshold amount of macroscopic animal food (especially fairy shrimp), which tends to be particularly dense when water volumes are shrinking. Only a proportion, *p*, of these shrimp-eating tadpoles switches over to the carnivore/ cannibal morph (shown here as open figures), and the remaining 1 − *p* does not change. Carnivores that fail to continue consuming large animal prey (shrimp, tadpoles, etc.) resume the omnivore morphology, so the switch is a reversible one. (From Pfennig 1992.)

discrimination affects cannibal prey-choice decisions produced negative results: sibs were consumed at the same rate as non-sibs. However, recent laboratory tests have shown that spadefoot cannibal morphs prefer to associate with non-siblings, while omnivore morphs prefer the company of siblings. Cannibals also nip at conspecifics initially, as if tasting them, before devouring them. Most intriguingly, this kin-related choosiness diminishes sharply as a function of time between meals, i.e. as the cannibal's personal condition becomes increasingly tenuous (Pfennig *et al.* 1993; see also Sadler and Elgar 1994). Whether such discrimination is routinely practised under natural conditions is not yet known.

The cannibal morphs found in long-toed salamanders and some races of tiger salamanders similarly have enlarged bodies, massively broadened heads, and teeth that exceed in both size and number those of omnivore sibs (Powers 1907, Walls *et al.* 1993a). In tiger salamanders such cannibals are usually rare (Rose and Armentrout 1976, Lannoo and Bachmann 1984). As with *Scaphiopus*, these cannibals narrow their diets to eat mostly conspecifics (Collins *et al.* 1980) and metamorphose more quickly than omnivores (Lannoo and Bachman 1984). The changeover to cannibal morphology is induced by high population densities (Collins and Cheek 1983), but the threshold for that diversion varies geographically and may be constrained by the cannibals' greater vulnerability to disease (Pfennig *et al.* 1991). In long-toed salamanders, the consumption of macroscopic animal prey (e.g. anuran tadpoles and other salamanders) instead of zooplankton accounts for some of the observed plasticity in trophic development. However, variation in diet is not sufficient to induce the fully exaggerated features of cannibal morphs (e.g. enlarged vomerine teeth), thus indicating that the addition of some other factor (e.g. conspecific density) is necessary to achieve the extreme morphological specializations (Walls *et al.* 1993b). Cannibalistic discrimination among sibs, cousins, and non-relatives has been demonstrated experimentally for captive tiger salamanders and is known to depend on olfactory cues (Pfennig *et al.* 1993). In at least one tiger salamander subspecies, sibling cannibalism is not accompanied by developmental polymorphism. Nevertheless, by eating just one sibling, a larva can achieve significantly accelerated growth and metamorphose earlier than non-cannibals (Lannoo *et al.* 1989).

The highly cannibalistic tadpoles of *Lepidobatrachus laevis* and *L. llanensis* show a comparable single-meal growth burst (Crump 1992). These obligate carnivores are not phenotypically plastic; all of them have numerous anatomical specializations that enable them to suck down whole prey nearly as large as themselves. Like *Scaphiopus* and others, *Lepidobatrachus* tadpoles inhabit temporary seasonal ponds in an otherwise dry habitat, namely the South American Chaco (Ruibal and Thomas 1988).

Overall, however, the fitness costs and benefits of eating siblings *per se* are not yet clear for these anurans and salamanders (Crump 1992). Kin recognition has now been demonstrated for tiger salamanders (Pfennig and Collins 1993) and marbled salamanders (*A. opacum*) (Walls and Roudebush 1991). The latter species' larvae are also cannibalistic and probably face sib vs. non-sib prey choices in nature. In laboratory experiments, they have shown the ability to discriminate on the basis of relatedness (rather than familiarity), although that ability may be mediated by the identity of the natal pond (see also Pfennig 1990a). Whatever the mechanism, individuals were less aggressive toward siblings than toward non-siblings (Walls and Roudebush 1991). Furthermore, the aggression studied in these experiments was more pronounced during the larval stage (when cannibalism is practised) than after metamorphosis (Walls 1991).

Although it is still not known how often marbled salamanders practise sibling cannibalism in nature, a subsequent series of laboratory experiments has produced a most paradoxical result. Whereas in the early studies, siblings were size-matched (which discourages cannibalism generally across amphibians; Crump 1992), testing of mismatched larvae showed that they feast *preferentially* on siblings (Walls and Blaustein 1995)! This illustrates the context-dependent nature of kin recognition (in this case, the context being relative body size). This was such an unexpected result that the experiment was repeated, and was successfully replicated. Assuming that the effect continues to hold up in future explorations (preferably under field conditions), such data strongly indicate that some additional factor(s) may be overriding Hamilton's rule. These might include such things as a low mean reproductive value of very young larvae (such that the indirect loss incurred by killing such a sibling is more than compensated by its nutritional value; Meffe and Crump 1987, FitzGerald 1992) and perhaps an overlay of strong pathogens with kin-correlated resistances (Pfennig *et al.* 1991). That is, selection might favour either avoiding the ingestion of close kin (if such are more likely to carry the diseases to which Self is particularly susceptible; Pfennig *et al.* 1991, 1994) or actively preferring to eat close kin (if such are more likely to carry only diseases to which Self is relatively impervious). There are other possible explanations for these findings as well, including the likelihood that these observed patterns represent non-optimal behaviour (Walls and Blaustein 1995), so further studies are eagerly awaited.

Without an obvious nursery, most free-swimming fishes have a lower potential for sibling cannibalism than amphibian larvae. Fish fry generally need not remain near siblings, or in areas of dwindling opportunity. While the eating of same-age conspecifics is apparently widespread, the evidence that it involves siblings is scarce (FitzGerald and Whoriskey 1992), some of it being based on laboratory data alone (e.g. Midas cichlids, *Cichlasoma citrinelli*; Valerio and Barlow 1986). One exception is the African sharptooth catfish (*Clarias gariepinus*) which swallows siblings tail-first when both diner and dinner are very small; survivors eventually switch styles, eating larger siblings head-first (Hecht and Appelbaum 1988). This species' cannibalism is much less common when alternative foods are available.

14.3.4 *Larval sibling scramble competition*

Diverse forms of scramble competition are likely to be important in larval nursery habitats. Amphibian metamorphosis is particularly well matched to such scenarios, and considerable research has focused on social, genetic and environmental factors affecting larval growth (Wilbur 1990). Some information is also available on how larval growth relates to post-metamorphic growth, survival and reproductive success, with chorus frogs (*Pseudacris triseriata*) at Isle Royale, Michigan, having been examined closely in this regard (D.C. Smith 1983, 1987). These anurans oviposit in small ephemeral pools scattered along the rocky shore of Lake Superior. Pools that are low down (near the lake) are ill suited to use by frogs because of ruinous wave wash, while those higher up (near the forest edge) are subject to the twin threats of evaporation and predation (especially by incumbent dragonfly larvae). As a result, chorus frogs tend to concentrate in intermediate pools, where tadpole growth is constrained by competition for space and food (D.C. Smith 1983). Faster tadpole growth translates into earlier metamorphosis, a higher level of recruitment into the breeding population 2 years later, and larger adult body size (D.C. Smith 1987). Most of the pools contain only single sibships (the mean relatedness between pool-mates has been calculated to be 0.41; D.C. Smith 1990), so the density-dependent growth is justifiably viewed as sibling rivalry. However, experimental growth studies of these same tadpoles also indicate that pure-sib poolmates may be *less* mutually harmful to their rivals than are pool-mates of mixed relatedness (D.C. Smith 1990), a result which parallels that reported for some other anurans (Jasienski 1988), salamanders (Harris 1987) and plants (Waldman 1988). If this effect is related to the facultative retention of water-soluble growth inhibitors, as was first proposed by Wassersug (1973), it could reflect kin-selected co-operation.

On the other hand, the opposite pattern—in which tadpoles are *more* inhibitory to siblings than to non-sibs—has also been reported for other anurans (see Waldman 1988, Hokit and Blaustein 1994). This effect has been interpreted in several ways, including the suggestion that some individuals altruistically refrain from consuming resources so that their larger siblings can metamorphose quickly (Waldman 1988). For that strategy to be evolutionarily stable, however, the nursery (pool) would have to be free or nearly free of non-sibs (lest they steal the benefits without sharing the costs), or some other mechanism would be needed to shunt the altruist's share preferentially toward its kin. It seems more parsimonious to postulate that the larger tadpoles' chemical releases retard the others' growth and the greater proximity or sensitivity of siblings produces the paradoxical skew. Alternatively, slower growth in pure-sib treatments could result if close kin have greater ecological overlap than unrelated larvae (Hokit and Blaustein 1994). In any case, this is at present a confusing literature, with some experimental design problems (see D.C. Smith 1990), a host of proposed causal factors (Jasienski 1988) and a paucity of field data.

14.4 Sibling competitions among adults

As described in earlier chapters, demographic conditions that keep sibships in close proximity have the potential to press individuals into protracted sibling competitions for mates or other resources. Sibling rivalry over mating opportunities (local mate competition) has been invoked to explain a significantly daughter-biased mean brood sex ratio (44:56) in 53 litters of adders belonging to an isolated Swedish population (Madsen and Shine 1992). Although rather few (8%) of the 39 witnessed bouts of ritualized combat were between known (marked) male siblings, these observations confirmed that the population size was small enough to generate sibling competition over mates. Regardless of what this implies about the skewed sex ratio, the documentation of overt sibling combat for sexual opportunity is an impressive feat in the field.

If juveniles do not disperse, they are obviously more likely to compete with litter-mates for food, hiding places and territories. Some data exist from long-term studies of individually marked reptiles that are consistent with the possibility of deferred sibling competition (e.g. Tinkle *et al.* 1993), but that point does not seem to have been addressed directly.

Summary

1. The potential for sibling rivalry in fishes, reptiles and amphibians occurs here and there among the forms that have non-shareable post-zygotic parental investment.
2. Viviparous taxa include a number of cases where siblings are likely to compete prior to parturition, although relatively little research has been undertaken on that particular problem. There is a growing list of fishes which show *in utero* cannibalism of eggs (oophagy) and/or larval siblings (adelphophagy), and there are many cases where other limited forms of parentally supplied nutrients are known to constrain offspring growth. Where larvae are simply left in a restricted site with limited food supplies, their development may be a density-dependent scramble competition; various claims have been made about how such rivalries are resolved, including some that invoke forms of nepotistic self-sacrifice.
3. Oviparous taxa in which parents provide transportation to special larval development sites such as water-filled tree-holes may have sibling competitions for riding space on the parent's back. Those that carry the larvae through metamorphosis, often inside the parent's mouth or within special brood pouches of various types, are also likely to be space-limited at times.
4. Some aquatic larvae are sequestered by parents in flooded tree-holes or leaf axils, and are then provided with 'trophic eggs'. Some of these ova are unfertilized, and hence are not viable as true siblings. Even those that are viable probably have exceedingly low probabilities of surviving, which presumably makes their consumption a trivial loss to Self's indirect fitness.
5. Many larval amphibians cannibalize siblings, and a few exhibit developmental polymorphisms (regular omnivorous features vs. dramatic specializations for cannibalism), the switch apparently being triggered by proximate cues indicating an imminent ecological catastrophe (e.g. puddle disappearance). In some salamanders and spadefoot toads, the ability to discriminate between kin and non-kin has been shown to affect prey choice, but the fact that cannibals in some taxa prefer non-relatives while in others they prefer to eat relatives complicates a general explanation at this point.
6. Adult competition among siblings is likely in many viscous and sedentary populations, but seems to have received relatively little attention to date.

15

Sibling rivalry in invertebrates

Once, when Jacob was boiling pottage, Esau came in from the field and he was famished
And Esau said to Jacob, 'Let me eat some of that red pottage for I am famished!'
Jacob said, 'First sell me your birthright.'
Esau said, 'I am about to die; of what use is a birthright to me?'

(*Genesis*, Chapter 25:29)

Invertebrates present so breathtaking an array of opportunities (and a lack of the same) for sibling rivalry that a full survey would occupy several books. At one extreme, many marine invertebrates simply shed gametes into the sea, allowing the pelagic zygotes to mix in a haphazard way with those of other nearby parents. Larvae then float in the plankton for extended periods, and for many cases it is difficult to envisage competition of any sort for larval resources, let alone any sibling competition. There is, to put it simply, no nursery and no discrete sibships. Amongst vertebrates, perhaps only certain pelagic fishes can match this *lack* of potential for sibling interaction. At the opposite extreme, certain parasitoid wasps show such profuse siblicide that only one larva from a large brood remains in the host (i.e. the nursery) after it has killed all of the others. Here, the scale of siblicide found in vertebrates, including birds that practise obligate siblicide, is dwarfed. We shall first give a synopsis of the opportunities for sibling rivalry across the invertebrate phyla, and then present some examples drawn selectively from a few groups, mainly featuring cases where detailed work has been done. In the latter, we group the cases into categories relating to the prevalent form of sib competition.

15.1 Taxonomic overview

To summarize the potential of invertebrates for sib competition we note the type of contact sibs have, the resources they might compete over, and the kinds of adaptation that might result. Regrettably, so little explicit attention has been paid to sibling competition and parent-offspring conflict across most of the invertebrates that the best we can do for most groups is to signal some possibilities (Table 15.1). Before embarking, it may be helpful to re-state the

Table 15.1 A summary chart of the general reproductive and demographic factors relevant to sibling rivalry and parent–offspring conflict across the many phyla of invertebrates.

Taxa	Reproductive mode[1]	Sib-Dispersion[2]	Parental investment[3]	Potential for sib rivalry	Candidate taxa for sib rivalry
Protozoa	A	D	None	Generally low	—
Porifera/Cnidaria	A	D	None	Generally low	Sessile hydrazoans (e.g. *Tubularia*)
Platyhelminthes	S	D (mostly)	Rare	Low, but a few exceptions	Turbellaria; triclad cocoons; some parasitic trematodes
Minor Acoelmates	A,S	D	Rare	Generally low	Parasitic nematodes
Annelida	S(H)	D (mostly)	Rare	Generally low	Cocooning Oligochaetes and Hirundinea
Arthropoda	A,S, some H	N (mostly?)	Variable but common	Many very rich pockets	Egg-retaining Branchipods and Ostracods; many insects, scorpions, and spiders.
Mollusca	A,S, some H	D (marine taxa), some N	Variable	Spotty	Prosobranchs, *Octopus*,
Minor Coelomates	A,S, and H	D	Rare	Generally low	?
Echinodermata	S	D (mostly)	Egg-retaining taxa only	Only in a few pockets	—
Protochordata	S, some H	D	None?	Low	—

[1] S = typically sexual; A = typically asexual; H = hermaphroditic
[2] D = dispersed, usually in aqueous medium; N = multiple broodmates cohabiting a nursery.
[3] U = uniparental, B = biparental

features that enhance the prospects for and intensity of sibling interactions. These include internal fertilization, prolonged retention of zygotes before release, brooding or guarding of larvae by a parent, delayed dispersal by larvae, direct development without a dispersive (e.g. planktonic) stage, oviposition in discrete clutches, and dependence of neonates on a shared and circumscribed food base. However, even in some marine organisms with external fertilization, competition between siblings can occur if there is not too much dispersal. Ultimately, following Hamilton (1964a,b) and the concepts summarized in Chapters 2–4, an important determinant of evolutionary adaptation for social interactions is the probability of sharing a given allele between two interactants. This is affected both by: (i) the levels of interaction and competition occurring between sibs rather than between unrelated offspring and (ii) the relatedness among sibs (i.e. half sibs, full sibs, or asexual products that are genetically identical).

15.2 Sib competition due to limited food resources

By far the most common context for invertebrate sibling rivalry follows the placing of multiple eggs inside a discrete food vehicle (nursery), such as a host animal, plant, or a specialized 'capsule' (e.g. Spight 1976) whose resources are in sufficiently limited supply to affect offspring fitness. Such competition usually takes the form of the model for 'scramble without parental presence' (see Section 3.3.7), which derives the ESS rate of feeding for each individual of a sibling group exploiting a shared resource. This type of model may be applicable to a wide range of invertebrates with gregarious larvae. However, some interesting cases do not obey the rules of that model. Often offspring compete by contests for limited food resources (Sections 15.2.1 and 15.2.2 below). At the extreme, acute 'ultra-siblicide' is rife, wherein only one competitor of many typically emerges from the nursery (Section 15.2.3). Such battles, which are most prevalent in cases where more than one female lays in the nursery, have also led to extreme forms of kin-selected altruism within sibships (Section 15.5.4).

In the ecological sense (Nicholson 1954), *scramble competition* was originally held to occur when all individuals share resources equally, and hence suffer equal consequences of limited food. Contest competition was characterized by opportunities to monopolize resources, which result in some individuals gaining more at the expense of others. This distinction was essentially based on predicted outcomes. By contrast, the growing use of the terms *scramble* and *contest* in behavioural ecology has tended to focus on the behavioural processes, rather than just the outcome. For instance, it is possible to have an unequal division of resources in a scramble if the competing phenotypes are unequal—the rate at which better competitors can grab resources may be higher than that for poorer rivals, even without territoriality or other types of aggressive resource monopolization (e.g. Parker 1984).

15.2.1 *Weak competition: diffuse resources and dispersed larvae*

If food resources are scattered and plentiful so that larvae normally remain apart, competition between sibs is unlikely. However, these conditions are rare. In many taxa siblings are aggregated, at least temporarily, because of synchronous emergence from the same clutch, and in many groups they may remain associated throughout much of early life, competing as a simple consequence of proximity. Finally, we mention something that might be regarded as more of a complication for researchers: the resource base may be more patchy than it initially appears. Although these general condidditions are unlikely to promote siblicide, they can certainly generate milder forms of sib competition.

One example (from many) will suffice to illustrate this case. Simmons (1987) studied competition between larvae of the field cricket (*Gryllus bimaculatus*). In nature, the larvae normally feed by scavenging omnivorously on plant material and dead animals, particularly insects. Although one would expect the food resources to be rather diffuse, larvae hatching from a given clutch are likely to remain loosely associated over the first few instars. When Simmons intensified competition for laboratory cultures by arranging different larval densities on a standardized amount of food (rodent chow pellets), he found good evidence for contest competition on the basis of both behavioural evidence and analysis of the mortality data. Successful individuals monopolized access to food and achieved higher adult weights at the expense of others. Females were more successful (they have a higher growth rate and hence gain an initial advantage in resource monopolization) and this resulted in a body size dimorphism when the larvae were at their high densities. The behavioural evidence (well-defined aggressive patterns) suggests that, in nature, larvae are adapted to compete by contests, often with siblings.

15.2.2 *Small aquatic nurseries*

In parallel with sundry arid-country anurans (see Section 14.3.3), discrete aquatic catchments often serve as sites for oviposition and larval development in many invertebrate taxa. The resource base in such nurseries is often poor and relatively closed (i.e. food replenishment is uncertain), a combination that generates considerable potential for competition among the young inhabitants. Such nursery-mates are often full or half siblings, depending on the mating system and laying habits of the mothers. For example, in the neo-tropical giant damselfly (*Megaloprepus coerulatus*) male adults frequently defend tree-hole catchments of rainwater (especially relatively large ones) and mate with all of the females that arrive to oviposit (Fincke 1992a). Each female lays up to 10 times as many eggs in the water as are likely to emerge following metamorphosis (Fincke 1994), and more than one female may use the same tree-hole, apparently being unable to detect whether a given site already contains eggs/larvae (Fincke 1992b). Thus, the offspring produced by a given female are quite likely to be full siblings and those from successive females are likely to be paternal half-sibs. Given the acute imbalance between demand and supply that follows, it is not surprising that the larvae are highly aggressive during the periods of more than 4 months required for development, with the larger individuals routinely killing (as inferred from wounding patterns in natural tree-holes plus observations in experimental pairings of larvae in artificial catchments) and frequently eating smaller nursery-mates, siblings and otherwise. In smaller tree-holes (volume < 1 litre), no more than one adult emerges eventually. In larger nurseries, the output averages approximately one emerging individual per 1-2 litres of water.

In a broader ecological context, it has been suggested that cannibalism between age/size classes can enhance the overall efficiency of resource extraction by a population from a limited resource base (Polis 1981, van den Bosch *et al.* 1988). The idea is simply that when the smallest consumers are able to exploit resources that are no longer available to the largest ones, the latter can gain indirect access to the broadest possible food base by eating small conspecifics. Cases where adult populations apparently escaped extinction by cannibalizing juveniles (that in turn were known to exploit a different diet) have been reported for copepods, scorpions, waterbugs, squid, newts, frogs, and fishes (Polis 1981). This community-level argument can easily be adapted to individual-based issues of optimal clutch size and sibling exploitation in nurseries where sib–sib cannibalism is likely (see also Section 10.1.3). An ovipositing mother that lays an inflated clutch might set in motion a competitive squeeze that leads to some rapidly developing 'haves' that later cannibalize slow-growing 'have-not' sibs (O.M. Fincke, personal communication). From the male parent's vantage point, variance in the sizes (hence ecological niches) of paternal half-sibs may be exaggerated further by the time lags between the territorial male's successive mates. The younger half-sibs can thus be viewed as serving two potential functions: (i) if they complete the ecological relay by processing inaccessible resources later acquired by their cannibalizing siblings, they serve as an active version of 'trophic offspring'; or (ii) they may inherit the nursery if some other force kills the older siblings (i.e. they also serve as insurance).

Generally similar patterns of sib cannibalism have been reported for other denizens of tree-hole nurseries. In some mosquitoes, such as *Aedes triseriatus*, different subsets of the same clutch vary in their responsiveness to external hatching stimuli, resulting in a phenomenon called 'instalment hatching'. When one subset has hatched, the others appear to delay hatching, apparently to reduce their vulnerability to cannibalistic sibs and/or to await a period of less intense local resource competition for the nursery's limited commodities (Koenekoop and Livdahl 1986).

15.2.3 *The fighting larvae of parasitoids*

In several insects, larvae compete over discrete food resources such as the body of a small host. Under such conditions, the resource can be highly limiting to larvae—there may be sufficient to ensure growth and survival of only a few individuals, and in some cases just one. Because the rivals are often (and in many cases exclusively) siblings, siblicide frequently results. The following example illustrates how severe the consequences of such limited resources can be.

Parasitic wasps typically lay their eggs on or in insect hosts. Many such parasitoid species are solitary, so that a single larva develops in each host; others are gregarious so that more than one larva—and sometimes several hundreds—develop together in a host. Although super-parasitism (when more than one female oviposits on a given host) also affects the number of larvae, the primary factors determining the number of emerging offspring per host are (i) the number of eggs a female lays and (ii) whether the larvae fight. The larvae of solitary parasitoids almost universally possess enlarged mandibles that they use when fighting, and although several eggs may be laid on a host, only one wasp ultimately emerges. In short, the progeny fight to the death until only one remains (Fisher 1961, Salt 1961).

Godfray (1987a) has examined conditions under which an allele for tolerance (non-fighting) spreads into a haplodiploid population fixed for 'fighting', and conversely conditions under which an allele for fighting invades a non-fighting population (for details, see Section 10.1.4). His main conclusions remain unchanged when diploid (rather than haplodiploid) inheritance is assumed; therefore, they should also apply to dipteran parasitoids (Tachinidae). His model predicts a dichotomy in clutch sizes, and this is found in nature. Parasitoid species tend either to lay single eggs, giving rise to fighting larvae, or relatively large clutches that produce non-fighting larvae (Le Masurier 1987). By contrast, some very recent experimental work on a solitary parasitoid wasp, *Comperiella bifasciata*, that practises an as yet unidentified form of non-aggressive obligate brood reduction (the victim larvae are believed to be somehow 'physiologically suppressed' based on the observation that they cease to develop midway through their series of five instars) strongly suggests that over-production (multiple single ovipositions by a single mother in the same scale insect host) is a widespread adaptive maternal strategy and not an accident or mistake (Rosenheim and Hongkham 1996). In *C. bifasciata*, it appears that the maternal 'decision' to remain on a host and repeat the ovipositional sequence hinges facultatively on her perception of both host availability and rival densities. The costs and benefits of such over-production have not been clearly identified, but Rosenheim and Hongkham (1996) recognized that multiple incentives from our 'replacement offspring' and 'offspring facilitation' categories (see Section 1.2) offer likely and mutually compatible incentives.

15.3 Sib competition due to limited parental care

Many invertebrate species from various phyla show advanced parental care, and some offspring compete directly for parental investment (reviewed in Clutton-Brock 1991). Zeh and Smith (1985) examined the incidence of all forms of parental care in terrestrial arthropods and many of their findings are consistent with the investment predictions of Maynard Smith (1977). In species where mating precedes the release of eggs, the female is more likely to be the sex that cares for the offspring because the male is no longer present when the zygotes are liberated. In the simplest form of care, the eggs are briefly covered by the female, and then abandoned. Here any sib competition must relate to food resources after the mother has left (Section 15.2). At the next level, some species guard unhatched eggs, sometimes until the young hatch. Guarding is much more commonly performed by females than by males, although in some orders there is biparental care, and in a few, exclusively male care. Guarding may protect offspring from physical or biotic threats (see Clutton-Brock 1991)—for example, it helps to control invasions by fungi in earwigs and some crickets (West and Alexander 1963, Wilson 1971). There seems to be little opportunity for sib competition for such parental attentions. However, PI may extend beyond mere guarding—the parent(s) may provision the young or regurgitate food to them. In these cases, there is more scope for sib competition and parent–offspring conflict following several of the models (see Chapters 3, 4, 8, and 9) involving hierarchies, scrambles or signals.

In terrestrial arthropods, although parental care by females is common, male care is relatively rare, probably because of the lag between mating and egg release (Maynard Smith 1977). The incidence of male care (relative to female care) is probably higher in marine invertebrates where there is external fertilization; the male is present when the zygotes are formed, allowing selection to favour his staying with the progeny. Clutton-Brock (1991) argues that male care has sometimes evolved because the costs of egg care to females are reduced by the fact that several successive females dump eggs in the same site, allowing the male to continue to obtain mates while guarding. In other instances (Clutton-Brock 1991), male care may have developed when a male's opportunities for additional matings is low. For instance, in sea spiders the prolonged courtship and mate-guarding (Jarvis and King 1972) may reduce the opportunity for finding second mates.

Biparental care is also relatively rare among invertebrates. Where it does occur it may, similarly, be related to limited reproductive opportunities for males (Clutton-Brock 1991). In the desert isopod, *Hemilepistus reamuri,* both sexes build, clean and alternately defend their burrow, and males show very high mortality (63–97%) during pair formation. Breeding seasons are also very brief, so a male's opportunity for mating again may be slight (Linsenmair and Linsenmair 1971; Schachak *et al.* 1976). Finally, we note that where both sexes invest in offspring, sexual conflict is expected (see Chapter 10).

We focus here on three sets of non-fatal sibling interactions with parental presence. In the first, parallels with the *in utero* competition during pregnancies of mammals, fishes and vertebrate ectotherms are found in a viviparous cockroach. Secondly, uniparental care in two groups of bugs does not appear to generate significant sibling competition, but the same phenomenon in a nursery-style crab may do so. We also mention two promising cases, burying beetles and a cockroach, in which biparental care creates clear potential for important competition. Finally, with maternal care in short-tailed crickets, we find extreme sib competition and high potential for parent-offspring conflict.

15.3.1 *Uniparental care: true bugs and bromeliad crabs*

Not all assassin bugs provide post-zygotic care, but in some there is egg-guarding by either the male or female parent that apparently reduces egg predation and vulnerability to parasitoids (Thomas 1995). In *Rhinocoris albopilosus*, a male guards the female after mating until she oviposits. At that point the male begins to brood, although he may copulate with several more females if and when they arrive to add further clutches to his egg clump. There is also male–male competition over attempted usurpation of egg masses (Odhiambo 1959, 1960, Ralston 1977). Successful usurpations necessarily reduce relatedness among clutch-mates, and thus should relax constraints on the intensity of sibling rivalries. The clutches hatch over a span of several days, so the initial cohort of first-instar nymphs may well be full sibs (e.g. if there is sperm precedence). Similar mating patterns occur in *R. tristis* and *R. albopunctatus*. By contrast, guarding is performed by the female in *R. carmelita* and involves only her own clutch, from which she strives to repel egg-dumping attempts by conspecific females.

The uniparental care in *R. tristis* and *R. carmelita* has been compared in detail by Thomas (1995). Paternal care in *R. tristis* appears to be maintained by high benefits (in terms of improved offspring survival) and low costs to the guarding male. The male circumvents some of guarding's energetic costs by eating peripheral eggs from his own egg clump (incidentally, these eggs also have the highest chance of being parasitized by wasps). In *R. carmelita*, the female makes no such sacrifice of her offspring, instead losing weight as she guards, which in turn reduces her future fecundity.

In many giant water bugs (Belostomatidae), females lay their eggs on the male's back, where they are carried continuously until hatching. In *Abedus herberti*, which has been studied extensively by R.L. Smith (1979a) and probably also in other congeners and in *Belostoma* (R.L. Smith 1980, Kight and Kruse 1992), the male requires the female to mate before he is willing to accept any of her eggs. Furthermore, the male interrupts oviposition regularly, on average after every second egg has been laid, in order to re-mate. Using a genetic marker, R.L. Smith (1979b) showed that this pattern of cyclical copulation and oviposition ensures almost total fertilization precedence by the guarding male. The male's presence while guarding probably reduces pre-dation risk and he assists in egg respiration by pumping water during hatching. Although these highly predatory adult water bugs readily cannibalize newly hatched nymphs that are unlikely to be their own, parents do not attack their own nymphs. Brooding has further costs to the encumbered male. A fully occupied egg pad may weigh twice as much as the male himself, thus reducing his mobility and increasing drag during swimming. Females reject fully loaded males, which cannot brood further eggs.

In these two groups, it is not easy to see how parent–offspring conflict could lead investment to be optimized away from parental interests, since there is little that an egg can do to inflate parental care for selfish interests. Whether sexual conflict can lead male care-givers to become overloaded is less certain, but again improbable. However, in both assassin bugs and giant water bugs, cohorts of full and/or half-sib nymphs are kept together by the brooding processes, so that the potential for eventual sib competition is present. How-ever, there is as yet no evidence for sibling cannibalism, or indeed any other manifestation of sibling selfishness in either of these systems. Similar crowding situations are found in other taxa (e.g. see review for anurans in Section 14.3.1), including scorpions where up to 95 tiny young crawl on to the mother's back and remain there until the first moult (Barnes 1987), and wolf spiders in which more than 100 spiderlings ride, several layers deep, on their mother's back for a week (Foelix 1982). In water bugs, the male is more likely to remove and discard the whole clutch (a cohesive 'egg pad') if it is well below his capacity (Kight and Kruse 1992), presumably in order to free himself for a more profitable mating cycle.

More lavish uniparental care has been reported for the Jamaican bromeliad crab (*Metopaulias depressus*) (Decapoda, Grapsidae), apparently as a result of the threadbare nursery it uses. Adults inhabit large, epiphytic bromeliads that contain substantial catchments of water in their leaf axils. During the rainy season, the female lays an average of about 50 large eggs and attaches them to her abdominal appendages for 3 months of incubation (Diesel 1989). The hatchlings spend a further 2 months in the water of a single leaf axil, which the mother has carefully prepared by removing detritus. This clearing of the nursery has the effect of doubling the dissolved oxygen content of the water during its lowest nocturnal ebb (Diesel 1992a). She also adds empty snail shells, which raise the pH and calcium ion concentration (Diesel 1992a). These are all shared forms of parental care, of course, but the mother also provides food (e.g. millipedes) which she chops into pieces for her brood, and she actively protects the larvae from predatory damselfly nymphs hatching in the same nursery (Diesel 1989, 1992b). We are unaware of any research yet on sibling rivalry (e.g. likely scramble competitions for food and even oxygen) in this system, but the potential is evident.

15.3.2 *Prenatal sibling competition in a viviparous cockroach*

Only one genus of cockroach, *Diploptera*, is known to be truly viviparous, using a maternal brood sac that generates milk-like nutrients. Ingesting this material orally, the embryos grow impressively, increasing 50-fold in mass and 60-fold in protein content from egg to the first instar stage at which they are born. Thereafter, the juveniles need only three or four additional moults to reach adulthood (cf. seven to 13 moults in oviparous cockroaches), indicating that their time in the brood sac gives them a sizeable developmental jump-start (Nalepa and Bell 1996, in press). It appears that the embryos sometimes, and perhaps routinely, compete prenatally for space and nutrients. Crowded embryos do not grow as large, presumably because of reduced contact with the structure that feeds them. Conversely, when litter size is experimentally depressed (by surgical removal of one ovary), the surviving individuals achieve greater hatching size (Roth and Hahn 1964).

15.3.3 *Biparental care in insects*

Burying beetles (*Nicrophorus* spp.) process the carcasses of small vertebrates, which are used as food for their offspring. The female beetle oviposits on or around the carcass, and either the female or both parents feed the developing larvae (Pukowski 1933, Scott 1990). In *N. tomentosus*, 22% of broods are reared by females alone (Scott 1990, Scott and Traniello 1990) and male assistance mainly involves repelling infanticidal conspecifics that might usurp the carcass.

In *N. vespilloides* a burial chamber is constructed around the corpse and eggs are scattered in the soil up to several centimetres away (Bartlett 1987, 1988, Bartlett and Ashworth 1988). Hatched larvae move to the crypt, consume tissue as they develop through three instars, then return to the surrounding soil to pupate. In addition to feeding on the corpse directly, the young are fed on regurgitate from the parents. In this species both parents are normally present, although the male typically leaves the brood 1 or 2 days earlier than the female. Parental care includes feeding the young, repairing damage to the crypt, and driving off insect trespassers.

When two adult females meet at a fresh carcass, they fight immediately, and only the victorious female buries the prize. In contrast, if two males meet with no female present, they usually co-operate in the burial of the carcass, and in signalling to attract a female (Bartlett 1998). Only upon the arrival of a female do the males fight, but the defeated (typically smaller) (Bartlett and Ashworth 1988) male nevertheless stays around the carcass and attempts to sneak matings with the female—a strategy that generates some chance of paternity (Bartlett 1988). Progeny may thus may be of mixed paternity, although most are likely to be sired by the victor. If a female is present at the time the two males meet, they usually fight immediately, with only the winner helping the female process the carcass.

When the male participates in brood care, his contribution is not influenced by the pair's relative body sizes (Bartlett 1988); he feeds larvae as often as the female does. Captive larvae that are fed exclusively by single parents are not significantly lighter in mass than those reared by both parents, nor do they differ according to which parent labours alone. (These aspects of Bartlett's results clearly do not conform to the logic of sexual conflict models outlined in Section 10.3.) Furthermore, males that process a succession of broods do not differ in median lifespan from males removed immediately after the egg-laying, and hence the mortality costs of parental care may be insignificant, although the time costs (missed reproductive opportunities) may not be. Sib competition for regurgitated food and the nature of the signals from the offspring that stimulate regurgitation have not yet been studied in this system. However, it is clear that the potential for all three dimensions of family conflict (Figure 3.1) are present in *Nicrophorus*.

An even more extraordinary example involves the woodroach (*Crypto-cercus punctulatus*), in which biparental care seems to be tied to the transfer of hindgut microfauna essential for digesting cellulose, the main dietary component (Nalepa 1990). Parental care in this single-brooded animal lasts until the adults die, which can exceed three years for the mother, with critically-limiting dietary nitrogen supplemented by ingestion of faeces, corpses, nymphal exuvia, and parental secretions (Nalepa 1994). Sibling rivalry seems to be most intense during the first few weeks after hatching, when growth is very rapid (there is an 11-fold increase in mass by the time the second instar is reached) and dependence on parental resources is absolute. The nymphs scramble for unpredictable deliveries of hindgut fluids and special faecal pellets produced by the parents, during which the closest offspring actively 'jostle' one another for improved position. Interestingly, during the first few months after hatching, when the nymphs' own hindgut protozoan faunas are not yet sufficiently established to allow independent feeding, family size falls from a mean of 73 eggs to just 35 eggs. Furthermore, hatching asynchrony is unusually pronounced, with some siblings up to 12 days older than others (Nalepa 1988). A number of nymphs are eaten by other family members (either while still alive or after being allowed to languish; Nalepa and Mullins 1992), but it is not yet known whether these tend to be the last-hatching sibs (C.A. Nalepa, personal communication).

15.3.4 *Maternal care in short-tailed crickets*

The female short-tailed cricket (*Anurogryllus muticus*) constructs a brood chamber and provisions it, lays her eggs, and then remains with her offspring until their third or fourth moult. West and Alexander (1963) documented the remarkable intra-familial conflict in this species, and made very detailed observations of one female, which mated once and then showed hyper-aggressive behaviour towards other males. She constructed and provisioned her burrow (with particles of apple, peach, and grass supplied in the cage), and after laying her eggs spent the next 2 weeks mainly in the burrow chamber, removing faecal material, maintaining the burrow, and guarding the eggs and nymphs. Two types of eggs were laid: normal-sized (all of which eventually hatched) and miniatures (none of which hatched, and all of which were eventually seized, carried off and consumed by the nymphs). The latter appear to be 'trophic' eggs (see Section 15.4.3). Normal eggs were repeatedly palpated and manipulated by the female until hatching. Nymphs continually snatched the (probably infertile) miniature eggs, and were often chased by the female, who removed the egg from the nymph's mouthparts and returned it temporarily to the egg pile. The nymphs were gregarious and followed the female, and when she laid a miniature egg, it was competed for violently by the nymphs, and either eaten immediately or carried off first. However, the only reaction of nymphs to normal eggs was occasional examination with the mouthparts. On the female's death, she was consumed, which may be the normal pattern (West and Alexander 1963). This being so, it is clear that there cannot be parent–offspring conflict over the female's future reproductive success. However, there can clearly be conflict over the number and timing of eggs produced by the female, and over the allocation of resources among sibs.

15.4 Invertebrate sib cannibalism

An excellent volume edited by Elgar and Crespi (1992) gives detailed reviews of invertebrate cannibalism (see chapters by Crespi, Baur, Elgar, Kukuk, Waddell, and Stevens). For this reason, only an introduction and selected sibling examples will be given here, and readers are referred to these more comprehensive sources. In general, adult cannibals may consume other adults, juveniles or eggs. The last two are of interest in relation to optimal clutch size and brood reduction. Commonly, however, the cannibals themselves are juveniles, usually consuming other juveniles (mainly smaller individuals and, of course, unhatched eggs). We shall focus on the sibilicidal subset of cannibalism.

Earlier we summarized our model for parent–offspring conflict over clutch size (Parker and Mock 1987), wherein the parent is expected to 'want' a larger clutch size than any given offspring (Section 10.1.3). However, because the parent may not be in a position to assess the true value of the larval food resource until some time after oviposition, it may often pay the mother to over-produce initially, counting on a secondary pruning of the clutch or brood, when more information has become available (see Section 15.4.1). In our earlier model, the benefits derived from killing a sib relate entirely to securing greater resources for Self, just as O'Connor's (1978) model had done. However, it is clear that cannibalism can also yield significant nutritional benefits through the tissue consumption *per se* (see Section 15.4.2), and this can lead to maternal provisioning of 'trophic' eggs, which are non-viable, and thus must function solely for the sustenance of the functional progeny.

15.4.1 *Egg cannibalism by newly hatched leaf beetles*

Conspecific oophagy (i.e. egg cannibalism) has been reported in more than 80 animal families, encompassing almost every group that lays eggs, including gastropods, spiders, insects (social and otherwise) and oviparous vertebrates (Polis 1981). Such consumption can involve either kin (especially sibs and parents) or conspecifics that are unrelated to the eggs. The routine ingestion of eggs by hatchlings is quite common across invertebrate taxa. It has perhaps been studied most in coccinellid beetles, where it can be extensive. For instance, in *Coccinella septempunctata* (Banks 1956) and *Coelomegilla maculata lengi* (Pienkowski 1965), over 20% of each clutch is lost by intra-clutch consumption; in *Chellomenes lunata* the level exceeds 40% (Brown 1972).

Levels are usually much lower, however, and in many species it is difficult to be certain whether sib oophagy represents a conflict (the extirpation of viable kin) or simply a mopping up of some otherwise useless failures. Some examples of the confusion that may arise involve chrysomelid leaf beetles. In green dock leaf beetles (*Gastrophysa viridula*), newly hatched larvae eat any unhatched egg that is at a sufficiently early stage of development (Kirk 1988). Clutches are laid on dock plants and spaced so as to preclude the possibility of inter-clutch cannibalism (although such occurs readily in artificially crowded laboratory conditions). Thus cannibalism typically occurs within the clutch and involves infertile eggs.[1] By presenting hatchlings with a range of ages of intra- or inter-clutch eggs, Kirk (1988) has shown that chemical defences (possibly oleic acid) against cannibalism develop gradually before hatching. (It is not certain whether this is a true 'defence' or merely a signal that the egg is fertile. That is, the chemical could serve as a warning cue to a hungry sib that this egg is both viable and dangerous to eat.) In these beetles, hatching asynchrony tends to increase with clutch size, as does cannibalism. Oophagous

---

[1] The eaten eggs are known to be generally infertile because only very young eggs are consumed. Within a single clutch, then, hatching synchrony is normally too tight for a viable egg to be undeveloped when its sibling, the cannibal, is old enough to have hatched.

hatchlings do not discriminate between intra- and inter-clutch eggs, and larvae fed exclusively on leaves survive less well than those whose folivorous diet is supplemented with *young* eggs (i.e. eggs too young to have accumulated their chemical defence).

Although only 4% of dock leaf beetle eggs are normally consumed under field conditions (of which approximately 50% would be infertile or otherwise non-viable in any case), there is evidence for much greater *potential* levels of sib cannibalism. Larvae fed on an *ad libitum* supply of very young conspecific eggs proved to be highly voracious, suggesting that they are normally deterred from cannibalism only by the presence of the chemical. Thus in *G. viridula*, although the proportional loss to the whole clutch is currently small, it is possible that this equilibrium was forced downward from a much higher level by the chemical defence. The ability to manufacture such a defence is likely to be provided to the eggs maternally. The fact that the defence is not manifest at the earliest stages of development may be related to advantages (to the parental genes) of having non-viable eggs consumed, and their nutrients recouped, by sibs.

In milkweed leaf beetles (*Labidomera clivicollis*), a second chrysomelid, a great many more eggs (55–87% of the clutch) are consumed by older siblings under field conditions. Of these, only a few (estimated at 15–17%) are non-viable (Dickinson 1992), so apparently no chemical defences are involved. Although the nutritional benefits of sib consumption have not been tested adequately in this species (Dickinson 1992), a sizeable advantage in growth and survival has been demonstrated in a third species, the willow leaf beetle (*Plagiodera versicolora*: Breden and Wade 1985).

15.4.2 *Sibling oophagy in land snails and ladybird beetles*

An almost exactly analogous situation exists in many terrestrial gastropods. In the land snail *Arianta arbustorum*, for example, emerging hatchlings first eat their own shells (as do many lepidopterans) and then consume unhatched eggs (Baur and Baur 1986). Cannibalism is performed exclusively on eggs by their own just-hatched siblings; older juveniles and adults are entirely herbivorous. Again, there is no discrimination between sib and non-sib victims and greater hatching asynchrony increases the amount of cannibalism. Under unfavourable conditions (e.g. lower ambient temperatures), hatching asynchrony increases, thus providing the first few hatchlings with the opportunity for a cannibalistic feast that can help them to survive (Baur 1992). However, unlike dock leaf beetles and some other taxa in which older embryos are avoided (for various reasons), hatchlings of *A. arbustorum* regularly cannibalize fully developed and viable sib embryos (Baur and Baur 1986).

One factor that is likely to have helped to tip the balance toward wanton sib consumption in land snails is the extraordinary nutritional benefit obtained from a cannibalistic meal. Baur (1990) found that the mean body mass of hatchlings fed entirely on conspecifics for their first 10 days was 2.6 times greater than that of siblings fed exclusively on lettuce. Furthermore, 67% of these cannibals reached adulthood, compared with only 28% of the lettuce-fed individuals. A second contributing factor concerns low levels of genetic relatedness; multiple mating is very common in female *Arianta* snails, so many clutch-mates are likely to be half-sibs (Baur 1993).

Finally, Baur (1993) tested whether cannibalistic hatchlings actively distinguish between (i) fertile vs. infertile eggs, (ii) eggs containing embryos at various stages of development and (iii) eggs of different sizes. Interestingly, there seems to be no discrimination among different stages of embryonic development or between fertile and infertile eggs. However, larger eggs were preferred to smaller ones. This last effect could result from a *passive* choosing process if, for example, large eggs are more likely to be encountered and/or broadcast stronger chemical cues. Baur (1993) suggested that the observed size preference might represent a selective mechanism restricting egg size. Because larger eggs take longer to develop, they experience longer periods of vulnerability during the critical period when their smaller nest-mates emerge and start to cannibalize, and of course it may also be the case that cannibals *actively* choose larger, more nutritious eggs.

From the parental point of view, sibling cannibalism suggests a possible conflict of interests between parent and offspring. It should generally be more energetically economical for a parent to provision an egg with twice the resources than to have a given offspring consume its same-size sib; if nothing else, the digestion process can never be 100% efficient. However, there are plausible arguments that siblicidal cannibalism can sometimes be in parental interests. For instance, if female morphology is constrained such that it is disproportionately expensive to produce fewer, larger eggs, the production of trophic eggs may be a better alternative.

In ladybird beetles (*Harmonia axyridis*), the larvae prefer to eat unfertilized eggs, but consume viable siblings as well. Whether such a practise is cost-effective on balance has been shown to depend on the availability of alternative food (Osawa 1992a, b). When aphid prey are scarce, cannibalism always pays for first-instar beetles (which are inept at capturing aphids anyway), but as aphid densities increase, the optimal number of viable eggs to eat decreases sharply.

15.4.3 *Trophic eggs*

What is the implication of sib cannibalism for parents? The alternatives are (i) that it concurs with parental interests or (ii) that it represents the outcome of conflict between parents and offspring ('offspring wins'). On the parental interest side, the long-standing icebox hypothesis was summarized succinctly by Polis (1984) as follows: '. . . intra-brood cannibalism is characterized by the use of some offspring as extended parental investment sacrificed to profit other offspring: victims essentially function as packages of live meat for their kin'. Despite the above-mentioned problem with digestive inefficiency, there are sometimes plausible arguments for the icebox hypothesis in terms of poor information at the time of egg-laying. If the ovipositing female cannot predict future conditions, over-production may be favoured, with the possibility of redress through sib cannibalism if there are too many offspring for resources to support later. As such, the icebox hypothesis converges towards O'Connor's thresholds for worsening conditions (1978), but with the difference that an extra boost to cannibal fitness is derived from consumption of sib tissues, rather than (or in addition to) via reduced competition for offspring resources.

On the other hand, even when the benefits of the consumed tissues are negligible, sib cannibalism can represent the outcome of parent–offspring conflict over clutch size. Where immediate nutritional benefits are high, as in *Arianta* snails, the potential for such conflict is clearly even greater. In some taxa (e.g. pond snails, *Lymnaea*), maternal oviposition behaviour seems to affect not only the spatial availability of siblings for consumption, but also the hatch timing. Clumping of eggs actually reduces their synchrony, apparently because the centremost eggs in a mass are less exposed to the external environment (Marois and Croll 1991).

The production of unequivocally 'trophic' eggs (i.e. those not fertilized by some kind of maternal decision) does, at first sight, seem to fit the icebox hypothesis well. However, Crespi (1992) and Kukuk (1992) suggest that trophic eggs may have arisen from oophagy, where hatchlings may or may not feed upon their unhatched siblings (as in many snails and insects; see above sections). Given that an offspring may consume sibs, it may be in the mother's interests to provide less expensive alternatives for consumption. If unfertilized eggs receive less post-zygotic care, for example, they may be less costly and hence more expendable. In short, the debate concerning the origins of trophic eggs is not easily resolved.

Trophic eggs (see reviews by Hölldobler and Wilson 1990, Crespi 1992) are common in social hymenopterans (particularly ants) and termites, and now have a complex array of functions. They can, however, be traced back to other groups, and occur in at least five non-social arthropod taxa, including the short-tailed cricket example already described (Section 15.3.3). Trivers and Hare (1976) suggested that the production of trophic eggs by worker ants and other eusocial Hymenopterans was originally a strategy for the production of males, which arise from unfertilized eggs, a means by which workers achieve limited reproductive success. Later in the evolutionary process, the queen gained greater control of male production by consuming most of these eggs, which now serve a trophic function, essentially by parental manipulation of worker–queen conflict. West-Eberhard (1981) suggested that the creation of trophic eggs may be beneficial to workers by virtue of keeping their ovaries active, should opportunities for producing viable eggs arise. Indeed, the retention of worker fecundity could be indirectly valuable to the queen if she were to die suddenly. Hölldobler and Wilson (1990) favoured the view that trophic eggs have an ecological rather than a social origin, and pointed out that food transfer in a colony may be effected via at least three means: (i) as trophic eggs, (ii) through the direct passing of prey items, or (iii) in trophallaxis (regurgitation)—the choice depending on the ecological conditions.

15.5 Sib competition for reproductive opportunity

Some of the most intriguing and compelling cases of sibling rivalry over breeding rights (both current or future) feature eusocial hymenoptera, in which mother and offspring remain together in discrete and compact social units. In the so-called 'primitively eusocial' taxa, the reproductive potential of individuals is facultatively adjusted, that is sterility is incomplete or only a transient condition. In many ants, wasps and bees nearly all of the eggs are produced by one or a very few queens (or 'gynes'). The great majority of females possess functional gonads, but seldom get to use them. Siblings often disperse together to form new colonies and/or to live together during sporadic but intense moments when breeding becomes possible (e.g. when a resident queen dies). The most obvious sibling rivalry, then, concerns which individuals achieve breeding status and which are relegated to non- or low-reproductive roles. In 'advanced eusocial' taxa, in which the sterility of non-breeders is essentially permanent, sibling rivalry appears in two forms: (i) potential reproductives (proto-queens) vie for the small number of breeding slots and (ii) even sterile females are capable of laying unfertilized eggs (from which males arise).

15.5.1 *Siblicide among sexually mature males*

In many insect species, including solitary bees (e.g. Alcock *et al.* 1976), males loiter near the emergence sites of developing females and mate as soon as the latter become accessible. In parasitoids, it is not uncommon for males to mate with the female even before she has emerged—e.g. by biting through cocoons and mating with the females within (see Godfray 1994). This often occurs between brothers and their own nursery-mate sisters. For parasitoids, Godfray (1994) notes that direct competition between males ranges '. . . from the mild to the lethal'. Some of the most dramatic examples of lethal sibling competition at emergence sites concern fig wasps (Agaonidae). In many cases, brothers never leave the natal fig and have become morphologically highly adapted for competitive mate-searching in its interior for their virgin sisters. Such conditions have led in different species either to *scrambles* (in which males compete by speed) or *contests* (in which males fight, often to the death). An example of such scrambles occurs in *Apocryptophagus*, whose pupae form large galls within its New Guinean fig host. If a mature fig is broken open, the air stimulates males to hatch immediately and to begin to search for female galls. These males are highly active and can bite into a gall, enter it, and mate with the resident sister in 48 s (Godfray 1994)! Interestingly, males do not enter galls containing either sisters that have already been inseminated or brothers.

In contest species, the males commonly emerge earlier than the females. In the most extreme forms, all of the brothers but one are killed (this victor then proceeds to mate with all of his sisters; see Godfray 1994). Some of the contest species are highly adapted for fighting, having sclerotized head capsules and huge biting mandibles (Hamilton 1979, Murray 1989, 1990). Hamilton (1979) also described fighting in a species of *Idarnes* in which even a small bite can cause death, suggesting use of a venom. Across these taxa there is substantial size variation among males, with small males attempting to avoid larger ones by selectively searching for unmated females in concealed sites. Small males sometimes hide in vacated galls and attempt to bite at the legs of victorious males. Similar behaviour has been seen in *Phylotrypesis pilosa* (Murray 1987).

Hamilton (1979) pointed out that the selective benefits of fighting would be reduced if the competing males were full siblings, and if a risk existed that their sisters might remain unmated (if all of the brothers perished). He predicted that fighting should be most common when there is an equal sex ratio of new reproductives, since female-biased sex ratios are typical when the emerging wasps are the progeny of only a few females (and hence often sibs). However, Murray (1987, 1989) failed to find evidence in support of this prediction—the absolute number of males in the fig appeared to be the most important correlate of fighting. In *Phylotrypesis pilosa,* the greatest number of injuries occurred when there were intermediate numbers of males in the fig (Murray 1987). Battles were rare when there were few males, although these appeared to be most severe. With many males, fights were common but typically less brutal.

A similar manifestation of local mate competition occurs in two ant species, where wingless, worker-like males kill one another (sons of other resident queens and brothers alike) over sexual access to the colony's females. In *Cardiocondyla wroughtoni*, these non-dispersive males (called 'ergatoid' males) also try to kill alate (= winged) males, although their slowness makes this difficult. Although ergatoid males have long sharp mandibles, apparently specialized as weapons for these battles, the killing of each rival can take several hours (Kinomura and Yamauchi 1987). The mandibles' main function is to clamp and immobilize the foe (Figure 15.1), which is usually an eclosing ergatoid pupa, after which the forelegs are used to drum rapidly on various of the opponent's body parts. Passing workers (sterile sisters) are attracted to the regions being struck, quite possibly because of pheromonal markers, and they seize them. It is the ensuing tug-of-war between that worker and the

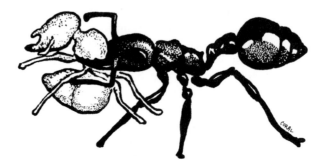

Fig. 15.1 A darkly pigmented ergatoid male *Cardiocondyla wroughtonii* (on right) holding a lethal fighting grip on a newly eclosed (less pigmented) ergatoid male. (From Stuart *et al.* 1987.)

advantaged male that causes dismemberment and eventual death (Kinomura and Yamauchi 1987, Stuart *et al.* 1987).

15.5.2 *Siblicide among sexually mature females*

There is often intense competition between foundress queens (or proto-queens) in the social insects. Because such females are frequently sisters (half or full), such killings are often siblicides. The benefits of this behaviour would seem obvious. Colony resources are restricted and the victor enjoys access to all resources for her own reproduction, rather than sharing this with others. In the most advanced eusociality (e.g. honey-bees), the eggs chosen to develop fully as future queen candidates are given very special treatment, including a larger pupal cell and a special diet ('royal jelly'). Of course, this also entitles most of them to an early death, paralleling Esau's vexation (see chapter heading).

With primary monogyny, each new colony is established by a single queen, which rears the first brood of workers on her own. In contrast, with primary polygyny (also called 'pleometrosis'), the colony is founded by several queens which co-operate in the difficult incipient-colony stage. Such co-operation appears to have evolved repeatedly because, essentially, two co-foundresses can produce more than twice the number of offspring produced by one alone, as has been demonstrated for *Polistes fuscatus* (Noonan 1981). However, this advantage quickly decreases after just a few foundresses have joined.

In more advanced social forms such as vespine wasps, apine bees, and most army ants, colonies are usually founded by primary monogyny, involving the departure of a single queen and a retinue of workers from the home colony. However, despite having elaborate nests and large colonies, most polistine wasps show primary polygyny, and typically several queens set off with workers. Thus here and in other primitively social wasps and bees, foundresses are typically related and may be full sibs. In some cases, they actively search out and collaborate with foundress sisters (reviewed in Brockmann 1984). In *Polistes*, one foundress comes to dominate the others aggressively, virtually taking over as the exclusive reproductive. The subordinate foundresses may continue to lay eggs, but their eggs are eaten by the dominant female. If the dominant female dies, the next in line takes over reproduction (West-Eberhard 1969). Noonan (1981) showed that *P. fuscatus* foundresses respond differently to sisters and non-sisters placed experimentally on their nests. As predicted, sisters were allowed to remain longer and were treated less aggressively than non-sisters. Some primary polygynous forms show more tolerance between foundresses, and in others there is siblicidal reduction to secondary monogyny.[2]

In monogynous species, the death (or departure) of the primary queen sets the stage for new proto-queens to face a cauldron of competition when they emerge from their cells. The first one to emerge usually kills all of her sisters before they pupate. If two proto-queens emerge simultaneously, they fight until one is killed. Siblicide by proto-queens to determine the secondary queen is thus common among the most advanced eusocial Hymenoptera. In some instances the reduction of proto-queens is effected by workers, which evict or kill all but one queen in order to generate secondary monogyny. Of course, these executions are siblicides, too.

In closing this section, we note that sibling mating competitions would not occur if an excess of candidate reproductives did not exist, that is, if the old queen had not over-produced in the first place. While much has been written about adaptive adjustments of brood sex ratio in the face of local mate competition, the most obvious reason for the initial over-production of honey-bee proto-queens or male fig wasps is, once again, insurance. The disaster of having no viable offspring in the key reproductive positions completely over-rides the costs of having too many in place. Second- and third-string players are worth building and the rivalry that ensues is a mere 'tidying-up'.

15.5.3 *Worker-policing*

The dilemma created by so-called 'non-breeders' that lay unfertilized eggs is far more subtle. Some of the implications for parent-offspring conflict over brood sex ratio are discussed elsewhere (see Sections 11.4 and 15.7), but the focus there was on relative numbers of sisters vs. brothers. By laying her own eggs, a worker alters the relatedness problem fundamentally; to her and her alone the choice of which male offspring deserve future investment is that of

---

[2] In contrast, there are no kin constraints in ants practising primary polygyny. Such new colonies are typically established by unrelated foundresses which aggregate after reproductive swarming, so it is unsurprising that all but one foundress is often ejected or killed, such that the colony quickly becomes secondarily monogynous. In some instances, supernumerary queens are actually ejected by workers.

brothers ($r = 0.25$) vs. sons ($r = 0.50$). An interesting puzzle follows. If laying is both inexpensive and profitable to workers, what exists to check its practise? With such an option open to every worker in the colony, as seems likely, what prevents total social collapse?

One proposed solution to the problem is that the queen controls the laying of her workers through some unknown form of pheromonal despotism, perhaps broadcasting a compound that suppresses the workers' ability to form eggs. This sounds plausible if one considers the queen's interests alone, but natural selection acting on worker interests seems likely to spoil such a mechanism. Specifically, any mutation that rendered workers insensitive to the queen's chemical mandate would almost certainly spread rapidly (being carried by sons).

A much more promising mechanism relies on sibling rivalry. While producing a son pays higher inclusive fitness dividends than raising a brother, the latter is often a better option than raising a nephew (a fellow-worker's son)[3]. It follows that each worker may be 'tempted' to lay eggs on the sly, but that most other workers in the colony fare best by destroying such eggs (Ratniecks 1988). This social force, known as worker policing, has the diffuse effectiveness of being enforced by many or all citizens acting selfishly, and requires no complicity from the worker whose attempt to procreate is being thwarted. Thus it can operate successfully in even the very largest colonies and has been shown to do so in honey-bees (Ratniecks and Visscher 1989). The effectiveness of such policing is suggested by the finding that, in colonies containing egg-laying queens, almost no surviving males are the sons of workers (Visscher 1996).

Something comparable exists in bumble-bees (*Bombus* spp.), which are obligately social, and lack the distinct morphological castes of honey-bees (so individual females function effectively either as workers or queens). Colonies are typically started by a single female that has mated just once, and this foundress rears a brood of daughters (full sisters) which behave as workers in assisting the queen to produce additional workers or reproductives. At some switch-point in the season, the queen begins to produce haploid (= male) eggs. In *B. terrestris*, the most intensively studied species, the switch-point coincides with low values in the fluctuating larva/worker ratio (i.e. when a large stock of mature workers has built up). Eventually, at the competition point, the colony begins to rear gynes (i.e. females that will mature as queens the following season). A few elite workers then begin to oviposit without having mated. Considerable oophagy ensues, by both the queen and these

---

[3] To illustrate numerically, we note that the coefficient of relatedness between a worker and her own son is 0.50, while that between her and the queen's son is only 0.25 (assuming only that Self is one of this queen's daughters). By contrast, Self's relatedness to a fellow worker can be as high as 0.75 (if they are full sisters) or as low as 0.25 (if they are maternal half-sisters). Accordingly, her relatedness to a worker's son is either 0.375 or 0.125, the probability of which depends on such factors as number of queen mating partners, sperm mixing, etc.

elite workers. Similar patterns occur in the late stages of the colony cycle in most *Bombus* species, with most mortality occurring in the egg or early larval stages. In some cases, only about a third of a queen's eggs survive to adulthood, and more than half of the eggs laid by workers are eaten by the queen (although some of the evidence suggests that a substantial fraction of males are sons of workers; see Kukuk 1992).

Crespi (1992) suggested that a *Bombus* queen's value to her workers decreases once the rearing of the next generation of queens is under way. The queen has provided the workers with sisters to rear as future queens, which is the one thing they cannot do for themselves (as they possess no sperm). It is to the worker's advantage to rear these sister proto-queens in preference to their mother's male eggs, because of the classic haplodiploid effect on r (worker to proto-queen = 0.75, worker to brother = 0.25). Furthermore, workers are more closely related to their own sons ($r = 0.50$) and to the sons of other workers ($r = 0.375$) than they are to the sons of their mother the queen ($r = 0.25$). Inescapable pheromone clues may inform workers that gynes have been created, which may trigger mutinous male egg production by workers, resulting in intense reproductive competition (manifested as widespread cannibalism of related offspring) in the late colony.

15.5.4 *Developmental polymorphism in encyrtid wasps*

We turn now to a remarkable group of parasitoid wasps that initially appeared to provide an example of extreme sibling altruism, but whose story has recently been diversified and largely reversed. Several members of the family Encyrtidae show polyembryony (many larvae deriving asexually from a single egg), and in some of these, there are special larvae that bear little resemblance to their more numerous and ordinary-looking siblings, despite the fact that they are genetically identical. These odd individuals contain no respiratory, circulatory, or excretory structures, but their digestive, muscular and nervous systems are well developed, and their growth is much accelerated (they are typically called 'precocious' larvae). Most remarkably, they lack gonadal tissue and do not survive to adulthood, for which reason they were initially regarded as abnormal 'monstrosities'. However, it was then discovered that the precocious larvae of *Pentalitomastix* spp., which are highly mobile and have advanced fighting mandibles, serve as defenders of their sibs against other, unrelated larvae. In experiments in which hosts were co-parasitized by any of three solitary parasitoids (all of which have 'fighting larvae' pheno-types), Cruz (1981, 1986) showed that the precocious larvae increased the competitive achievements of their normal sibs by selectively killing hetero-specific larvae. Precisely what role(s) the defender larvae might also play against super-parasitic competitors (i.e. conspecifics) remained uncertain.

Subsequent discoveries concerning another encyrtid, *Copidosoma flori-danum*, forced a sharp re-evaluation of the developmental polymorphism. Far from being mere guardians, the precocious larvae of this species appear to be part of a siblicidal 'hit-squad' through which the sisterhood manipulates the brood sex ratio. In two-thirds of all ovipositions, the parasite mother lays a 'mixed' combination—one male egg and one female egg—inside the host (a moth egg). The ensuing polyembrony leads to an average of 200 male clones and 1200 female clones, with 50 of the latter developing as precocious morphs. These non-reproductive sisters devote themselves to seeking and killing their own brothers (Figure 15.2), which hide as best they can in fat deposits (Grbíc *et al.* 1992). By this means, the sex ratio of reproductives decreases from 0.50 (at oviposition) to 0.17 (at hatching) and eventually to 0.09 (after the siblicides). Brother–sister matings then occur prior to the dispersal of the (winged) males. The dispersed males subsequently live for another week or two, presumably seeking opportunities for non-sib matings with virgin females that arise from single-egg (diploid) ovipositions in other hosts. For these matings, of course, the males are engaged in a scramble competition with other males arising from single-egg (haploid) ovipositions. Turning back to the mixed sibships, the incentive for precocious larvae to kill most their own brothers is presumably based on (i) the finite nature of the host as a resource base and (ii) the asymmetry in relatedness between brothers and sisters caused by haplodiploidy. In short, a precocious larva values her sisters ($r = 1.0$) far more than her brothers ($r = 0.25$), because only a few of the latter are needed for the sib matings (Grbíc *et al.* 1992). The indirect fitness pay-off available to a precocious larva through her brother's post-dispersal matings may also provide some check to the siblicidal activity, but on balance she destroys brothers in order to promote sisters.

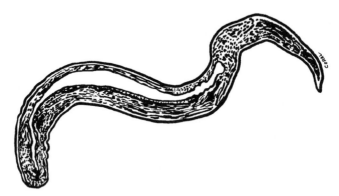

Fig. 15.2 The ingestion of male polygerm by precocious sister larvae of *Copidosoma floridana* was documented by labelling the polygerm with the vital tracer carboxy-fluorocein diacetate succinimidyl, and then injecting that material into a host containing an all-female brood. When this precocious larva was collected 24 h later, the tracer's presence in the gut (shown here as an all-white gut) was revealed by epifluourescent microscopy. (From Grbíc *et al.* 1992, © Macmillan Magazines Ltd, 1992.)

15.5.5 *Other 'defender' sibs*

Other cases in which a fraction of non-reproductive *adults* functions in the defence of the colony, and hence of their siblings, are well known in eusocial Hymenoptera and other groups. Probably the most extreme cases are the soldier castes of ants. Soldier *nymphs* also occur in termites and true bugs (Homoptera), both of which are diploid. In the aphid *Colophina clematis,* soldier nymphs within each given brood remain as first instars, and thus never reproduce. Their mouthparts and legs are apparently developed for attacking other predators (Aoki 1977). Whereas the defender larvae of the wasp *Copidosoma* are clones ($r = 1$), *Colophina* siblings may have a slightly lower degree of relatedness. They are produced endomeiotically; the maturing oöcytes undergo a modified meiotic division during which a diploid polar body is ejected and the remaining diploid chromosome elements can undergo genetic recombination by crossing-over (White 1973).

15.6 Filial cannibalism

Over and above the destructive consumption of other parents' young, invertebrate adults occasionally consume some of their own viable young, as we have already seen in the paternally investing assassin bug (*Rhinocoris tristis*) (see Section 15.3.1), in which fathers circumvent the costs of guarding by eating a few of their least valuable eggs. This must be viewed as a case of parent-offspring conflict (since not all of the consumed eggs would have been parasitized, and so are presumably still viable when eaten), albeit one of the 'parent-wins' category (see Section 11.2).

Cannibalism of one's own offspring may also occur when there are insufficient resources available for proper development and survival of the core brood. Elgar and Crespi (1992) argue that 'a parent that kills and consumes its offspring during periods of low availability effectively converts its reproductive investment back into somatic tissue, thereby postponing reproduction until conditions are improved.' Such parents effectively perform the brood reduction as a means of adjusting clutch size to fit the larval food resource when the prospects could not have been predicted at the time of laying.

Examples can be found in the burying beetles already described (Section 15.3.3). In addition to interring and processing small vertebrate carcasses, parents of *Nicrophorus vespilloides* also practise filial infanticide. They kill and eat up to half of their own first-instar larvae (Bartlett 1987), a practise less easily explained than parental consumption of trophic eggs. Of course, the number of larvae that can be sustained depends entirely on the size of the corpse. In laboratory experiments using mouse carcasses weighing a uniform 10–15 g, initial clutch size is such that complete hatching success would produce far more larvae than such a corpse can support without severely compromising individual growth in the final instars. Bartlett proposed that parents benefit from the insurance value of a large clutch, if predation on eggs and newly hatched larvae is common (a point that remains unknown). Parents kill and eat offspring on the first day of larval emergence. Crespi (1992) suggests that filial cannibalism is a simple, inexpensive mechanism enabling parents to adjust clutch size (and possibly sex ratio) optimally to carcass size, at the time when this adjustment can best be made. However, there seems to be little reason why such an assessment could not be made before, rather than after, laying the eggs, thereby avoiding the energetic losses due to digestion. Of course, these two incentives for parental over-production (insurance and resource-tracking) are often mutually compatible (Mock and Forbes 1995; see Section 1.2). By contrast, one other interpretation, that the cannibalism represents some manifestation of sexual conflict between the parents, can probably be discounted, since both parents eat progeny with approximately equal probability (Bartlett 1987).

15.7 Sex allocation and other sex conflicts in the social insects

The sex ratio of eusocial Hymenoptera has been identified as 'one of the foremost test cases of modern evolutionary biology' (Boomsma and Grafen 1990, Sundström 1994). Trivers and Hare (1976) constructed predictions about the female bias in these sex ratios from a combination of three bodies of theory, namely sex ratio (Fisher 1958), inclusive fitness (Hamilton 1964a, b, 1972) and parent-offspring conflict (Trivers 1974). The Trivers–Hare milestone (see Section 11.4) showed that in social insects relatedness asymmetries create conflicts of interest between castes over how many reproductives of each sex should be produced. Such conflicts are doubly important in the context of the present book, because they determine the potential for sib competition by altering the number and types of sibs that can interact within colonies.

The central tenet of this 'worker–queen conflict' is that disagreement of workers with their queen is driven by now familiar haplodiploid asymmetries in relatedness[4]. Because several extensive reviews of this complex problem already exist (e.g. Nonacs 1986, Bull and Charnov 1988, Clutton-Brock 1991, Seger 1991) only a brief overview will be presented here. From sex-ratio theory (Fisher 1958), the relative reproductive value of any breeder (male or

---

[4] As a quick refresher, the queen has a coefficient of relatedness, r, of 0.5 to each of her offspring, be they female (workers, proto-queens) or male. Haploid males produce identical sperm. Thus, if all sisters have the same father, the sister-sister r is 0.75 (other workers or proto-queens), but the sister-brother r is only 0.25 (because brothers have only half as much DNA). These average r-values decrease with the degree of multipaternity of the brood, and with the degree of polygyny (i.e. the number of queens) in the colony.

female) declines with the population-wide frequency of that sex and, for an autosomal 'sex-ratio allele' in a sexually reproducing diploid population, the ESS sex ratio for the population as a whole is 1:1. (This assumes only that the sexes are equally expensive to produce up to the end of parental care; when the sexes differ in their investment cost by the end of parental care, it is the *total investment ratio* that should be 1:1.) Assuming monopaternity, Trivers and Hare predicted that at the ESS under haplodiploidy, the queen would favour a 1:1 sex-ratio investment in reproductives. In contrast, workers should favour an investment ratio of 3 (into queen production) to 1 (into production of male reproductives). More strictly (see Seger 1991), workers are indifferent only when the population-wide investment ratio is 3:1, and queens are indifferent only when the population-wide investment ratio is 1:1. Whenever the population ratio is not at these values, incentives exist to produce the sex that is under-represented relative to the 3:1 or 1:1 ratio. There is clearly potential for queen–worker POC with regard to sex allocation.

By examining sex ratios at different degrees of sexual size dimorphism in 21 monogynous ant species, Trivers and Hare (1976) found that the investment ratio was in close agreement with the 3:1 prediction for the 'offspring wins' solution, although there is considerable variation. Alexander and Sherman (1977) promptly attacked the Trivers and Hare analysis, claiming that the effect could be explained by the 'local mate competition' model of Hamilton (1967), which showed that strong female biases can occur where related males compete directly with each other for matings. However, the population structure of most ants is not conducive to this interpretation, and Trivers and Hare's worker control concept is becoming accepted (Boomsma and Grafen 1990). The mechanism whereby workers exert their control is presumed to be selective oophagy by workers of diploid or haploid eggs, as needed. As predicted by Trivers and Hare (1976), polygynous ants have subsequently been shown to have only weakly female-biased investment ratios (the average relatedness between queens and workers declines rapidly as the number of egg-laying queens increases), and it approximates to 1:1 in slave-making ant species, where the worker (which, having been stolen from its natal colony, is unrelated to the brood) is not under any selection to bias their investment.

In many ant species there is a strong tendency for colonies to specialize in producing either male or female reproductives (see Nonacs 1986). Given that there is good evidence for worker control, Boomsma and Grafen (1990) have proposed that such bimodalities in sex-ratio strategy may reflect the fact that colonies vary in their degrees of relatedness asymmetry (due to variation in multiple mating, worker reproduction, and polygyny). If it is assumed that workers have the ability to assess their own relatedness asymmetry (Self to sister relative to Self to brother, i.e. 3:1 under 'ideal' monogyny) in a given colony, the best strategy may be to invest entirely in one or other sex. Workers could maximize their inclusive fitness by specializing in males if the colony has a relatedness asymmetry below the population average, and by specializing in females if it is above the population average. For example, imagine a monogynous species in which queens mate twice, on average. In the particular colonies where such a queen has mated only once it will be best for workers to produce females; conversely, in colonies where she has mated three or more times it will be best to produce males. Similarly, with polygyny, in colonies with above-average numbers of laying queens, workers should produce males, while in those with fewer queens than average, workers should produce females. Boomsma and Grafen reviewed the comparative data on colony investment ratios in ants to test their predictions (see also Seger 1991, Sundström 1994).

Finally, Hamilton's theory (1964a, b) also accounts for the facts that males (drones) contribute so little towards the life of the colony, and that relatively little attention is paid to them by workers after adult eclosion. Males have a maximum coefficient of relatedness of $r = 0.25$ with sib female reproductives. With the relationship asymmetry discussed above, workers are expected to become hostile towards males when the drones' reproductive value declines (e.g. late in the season). There is some evidence for worker-drone conflicts in nature (e.g. Wilson 1971).

Summary

1. The relatively few invertebrates that use 'nurseries' are taxonomically broad and highly diverse in morphology, ecology and life histories. Sibling rivalry ranges from non-existent to the most extreme forms of siblicide, wherein dozens or even hundreds of siblings are destroyed. A small part of the vast potential for invertebrate sibling rivalry is summarized in Table 15.1, but it must be conceded that the research attention devoted to the problem so far has been sporadic.
2. Mild scramble competition often arises from the combination of localized resources and clumped oviposition. More dramatic cases arise when the nursery is very small, as commonly occurs in tree-hole catchments of water. In giant damselflies, the mother lays up to 10 times more eggs than can survive through metamorphosis, which leads to vicious aggressiveness and rampant cannibalism among sibs. It has been suggested that asynchronous hatching in such systems might establish a cannibalistic pyramid in which tiny (later-hatching) larvae can extract the smallest food items from nursery crevices efficiently, and then be exploited as macroscopic prey by their older siblings.
3. Many parasitoids lay multiple eggs in or on a host's body, which then becomes the nursery for the developing larvae. As a discrete resource, the stage is set for diverse forms of sibling (and inter-brood) competition. In general, parasitoids that lay single eggs have fighting mandibles (to use

against non-sibs from other mothers), whereas other species use hosts that can support more than a dozen (non-fighting) offspring.

4. In taxa which show advanced parental care, some clear parallels with similar vertebrates are known. Most of these receive post-zygotic investment from only one parent and, though this is often likely to be limiting, little attention has been paid to the question of whether the offspring resolve the inadequacy themselves. An especially likely system is the Jamaican bromeliad crab, where the mother not only overhauls the nursery (a water basin formed by a leaf axil) for her approximately 50 offspring, but also delivers and processes prey for them over a period of many weeks.

5. Biparental care is found in burying beetles, where the nursery is a specially prepared vertebrate corpse that has been interred prior to oviposition. If the number of offspring proves too great for the food base, the parents may devour some of them and, in all likelihood, feed the regurgitate to the surviving siblings.

6. More commonly, the consumption of sibling tissue is accomplished directly, as sib cannibalism (or oophagy), and can be a very important feature of the life history. In green dock leaf beetles, properly developing embryos apparently arm themselves with a chemical defence; as a result, very few (c. 2%) viable offspring are cannibalized by brood-mates. In milkweed leaf beetles, by contrast, many (and perhaps most) viable siblings are eaten. The costs and benefits of sib cannibalism have been worked out quite well for a land snail, where Baur has shown that each egg consumed provides a large nutritional boost to the cannibal's fitness.

7. The question of whether sib cannibalism runs counter to parental interests or is part of an adaptive parental strategy is complex. The formation of unfertilized 'trophic' eggs, for example, seems initially like a parental adaptation, but it could be a fall-back position that evolved only because it was less costly than providing viable siblings that would be eaten. In essence, it is hard to distinguish between a gift and a bribe.

8. Many examples occur among invertebrates of siblings competing for limited reproductive opportunity, often at very young ages in highly structured social systems. With brother–sister mating in parasitoids, a number of eccentric forms of siblicide arise. In some fig wasps, for example, the males appear earlier than their sisters, possessing huge mandibles and armoured head capsules. They fight among themselves until only one remains, and he then fertilizes all of his sisters. Similarly, in other social insects there may be an equally intense competition over which females will develop as reproductives. Candidates that are being raised as proto-queens often kill all of their rivals as soon as they pupate.

9. A particularly intriguing phenomenon concerns the control of non-reproductive individuals in highly eusocial bees. Workers may not be able to mate, but they can lay unfertilized (haploid) male eggs. It appears likely

that this is prevented in several species by 'worker policing,' that is, by sibling rivalry. Workers are more closely related to their mother's sons than they are to their sister's son *if* there is a reasonable chance that the sister is not a full-sibling (i.e. if the queen mates with several males).

10. Encyrtid wasps practise polyembryony, in which an egg divides asexually to produce identical clones. In *Copidosoma floridanum*, the mother lays one male egg and one female egg in a moth host, and these transform themselves into 200 males and 1200 females. However, a few dozen of the females develop precocially as non-reproductives with no gonads but with fighting morphology. This developmental polymorphism was initially thought to represent a 'defender' caste of larvae that protected reproductive siblings from superparasitism (laying by other females). However, it turns out that, at least in *C. floridanum*, the precocious larvae are all females that hunt down and kill their own brothers, thereby preventing many of them from consuming host tissue that could go instead to reproductive sisters.

11. The broader issue of sex-ratio control has been explored across many haplodiploid species, where the different inclusive fitness values of brothers vs. sisters are likely to have shaped many aspects of brood treatment.

16

Sibling rivalry in plants

I am a tainted wether of the flock, meetest for death;
The weakest kind of fruit drops earliest to the ground.

(William Shakespeare, *Merchant of Venice*, Act IV, Scene 1)

16.1 Plant–animal differences

Plants experience sundry ecological competitions and exhibit non-behavioural
social relationships that are sculpted by their fundamental immobility. In two
general contexts, parental care and sibling interactions, Hamilton's rule
applies broadly and usually requires no assumptions about kin recognition
mechanisms (Queller 1989), the players still being in very close proximity to
one another, indeed usually physically attached. The main difference between
sibling rivalry patterns in plants and those in animals is that botanical
resolutions of kin-based conflicts have necessarily been expressed more
through a rich proliferation of adaptations in morphology, genetics and life
history. For this reason, a brief survey of plant sibling rivalry (written by non-
botanists!) may prove illuminating, although we hasten to acknowledge our
debt to the veritable cascade of excellent reviews from which we have
borrowed freely (e.g. Kress 1981, Stephenson 1981, Stephenson and Bertin
1983, Charlesworth et al. 1987, Haig 1987, 1992a, Lee 1988, Uma Shaanker *et
al.* 1988, Queller 1989, Cheplick 1992).

Our goal in this chapter, then, is to highlight both the parallels and dis-
similarities one finds in sibling rivalry and parent–offspring conflict between
the two kingdoms. The major differences range from the consequences of
monoecy (male and female functions in the same individual) and self-
fertilization—both rare in animals, but routine in plants—to the mysteries of
alternating generations. In animals, identifying what constitutes a *sibling*
seldom takes much thought; one might ponder 'sibling gametes' (from the
same individual's gonads), but would be hard-pressed to be much more
abstruse than that. In plants, however, the roles of *offspring*, *parents* and
siblings change periodically, giving the uninitiated a queasy feeling of
instability. Thus, in addition to gametes, one often encounters multiple sibling
embryos within a single seed, sibling seeds within a single fruit, sibling fruits

sharing a single branch, and so on, right up to whole adjacent individual plants, which are commonly derived from sibling seeds produced by the same maternal sporophyte.[1] (And, of course, what appear to be two neighbouring trees may, in fact, be one with underground connections.)

Some other important differences between sibling rivalry patterns in plants and animals emanate directly from the dispersal methods used by plants. In general, the reduced dispersal experienced by heavy fruits and seeds, which may simply drop from the parent by the thousand to form a 'seed shadow', simultaneously increases the likelihood that one or two offspring will eventually inherit the parental site and also the duration over which they will compete with each other for that honour. Whereas most animals move apart from their siblings within a few weeks or months (with notable exceptions such as social hymenopterans, African lions, and acorn woodpeckers, to be sure), for many plants sibling rivalry never ends, lasting for decades and even centuries. There is also the common phenomenon of *synaptospermy*, in which multiple seeds disperse together as a unit (e.g. in an apple), which also increases their potential for extended competitions (Casper and Grant 1988).

Plant parental manipulations are generally widespread and impressive. Consider *amphicarpy*, for example, which is the parental habit of producing two or more genetically dissimilar classes of offspring that are specialized to serve complex parental fitness interests in ways vaguely reminiscent of the specialized worker castes in ants. Frequently, a parent creates one batch of offspring with very high sibling relatedness coefficients (through self-fertilization), while producing another batch that is outbred and adapted for dispersal. The former progeny may arise in *cleistogamous* flowers, which have inconspicuous and non-opening reproductive organs situated at or below the soil line, sharing a common mother and a common father (and she is he!), but are locked in a permanent competition to inherit the parental site. The latter batch of offspring are outbred in *chasmogamous* flowers (often showy and open to pollination), which tend to be situated higher on the plant and whose seeds have greater dispersive potential, e.g. from wind (Schemske 1978, Cheplick 1992).

The differences in plant and animal biology are so substantial and so fascinating that one laments the tendency of their respective scientific trad-itions to maintain separate paths and distinct vocabularies (see Lovett Doust and Lovett Doust 1988). No doubt there is an interdisciplinary sibling rivalry of sorts in this, as well. On the other hand, there are important phenomena on both sides that simply cannot be accommodated by a shared language, so considerable jargon is needed for precise communication. For instance, the concept of *synaptospermy* seems unlikely to have a bright future in ornithol-

---

[1] Technically, the gametophyte stage is properly assigned to the male or female category, not the flower or whole plant (sporophyte). However, we follow recent convention in referring to the whole plant that produces ovules, seeds and fruits as the female sporophyte (e.g. Stephenson and Bertin 1983), sacrificing a small measure of precision to gain a more familiar entity as referent 'mother'.

ogy (Florida scrub jays joined at the hip?), yet it is obviously a valid and useful term for plants.

Perhaps more problematic are familiar words whose meanings differ consistently and legitimately, according to kingdom. Of special concern to us here is the split usage of *full sibling*, which we have been using happily (in naïve bliss) to connote shared maternity and shared paternity, but which is often used in the plant literature to refer to 'seeds produced via self-fertilization.' This makes a certain amount of sense in that selfing is one of the few sexual modes in which plant paternity *can* be identified easily, but it can also lead to confusion. On a genetic basis, such botanical 'full siblings' have higher coefficients-of-relatedness than zoological 'full siblings'; they are also more subject to inbreeding depression. Where the resulting homozygosity lowers the viabilities of both Self and full sib to a similar degree, the net effects on selfishness may rest mainly on the inflated r-value. However, from an ecological perspective, animal sibs are both more likely to disperse and, being more genetically different, may have less similar needs to the temporarily shared resources provided by parents.

Similarly, the term *half-sibling* is used by plant biologists to denote the relationship between two individuals that are known to have one parent in common (the mother) and known to be out-crossed. Thus $0.25 \leq r \geq 0.50$. If two sibs have derived from pollen carried by the same bee, their coefficient of relatedness could well be 0.50 (common paternity is likely); however, if they derive from wind-dispersed pollen and numerous male flowers live upwind, their coefficient of relatedness is likely to be 0.25. This imprecision is lamentable (but certainly no more so than the quaint and increasingly unfashionable ornithological assumption of common paternity and resulting 'full sibling' status for nest-mates based solely on an apparent 'pair-bond' between two care-giving adults). In any case, it is important to remember that most theoretical models applied to plants assume explicitly that siblings were fertilized by different paternal sporophytes, so $r = 0.25$.

A major conceptual obstacle to understanding sibling rivalry in plants concerns how one discriminates between a selfish offspring phenotype and a parental manipulation. As with siblicidal aggression in birds, fruit abscission that sacrifices one offspring while benefiting another might easily arise to serve the interests of the surviving offspring, the parent, or both. Even in the few cases where the proximate agent of one offspring's demise can be traced unequivocally to a sibling, one cannot assume *a priori* that the fatality was *not* also in the mother's best interest. (The fact of my holding a smoking gun over my sibling's corpse sheds no light whatever on the 'Conspiracy Hypothesis' that our parents hired me to make the hit.) Thus, one might view maternal decisions to trim family size (fruit set and/or seed set) as something that evolves because of *pending competition* (*sensu* Stinson 1979) among siblings.

The lack of persuasively conspicuous anti-sibling behaviours *per se* forces a careful consideration of what constitutes evidence for sibling rivalry in plants.

The phenomenon most readily observed in plants is the dropping of certain offspring while retaining others, which looks superficially like a maternal 'act'. By contrast, the closest analogues to overt siblicidal aggression by plant protagonists involve the relative subtlety of chemical weaponry. Furthermore, the plant equivalent of non-aggressive avian brood reduction involves no raucous pandering, but 'stealth begging' in the form of physiological sinks and silent hormonal signals. Perhaps for this reason, most botanists have tended to assume that phenomena such as fruit and seed abortion are *purely* maternal strategies (e.g. an adaptation to enhance average offspring quality by dropping only the runts and ne'er-do-wells), rather than carefully exploring possible avenues for active roles played by the surviving offspring (Uma Shaanker et al. 1988). Smoking-gun evidence not only fails to absolve the parents of any involvement, but also it distracts one from even pursuing that possibility, by lulling the mind into accepting that the case has been solved. Of course the reverse is also true. Just because fruit abscission appears most parsimoniously to be a product of parental physiology begs the question of whether other family members have played indirect roles. As discussed below, a seed or fruit that is judged to be unworthy of subsequent maternal investment may be a loser *because* other seeds/fruits outcompeted it. The maternal sporophyte may have delivered the final *coup de grâce*, but it remains unclear who contributed to the victim's demise. In short, one ought not imagine offspring as 'passive vessels into which parental investment is poured' (Trivers 1974). In this sense, the term 'abortion' is probably an unfortunate, prejudicing label, as it connotes unilateral parental decision-making. On the other hand, it is at least premature (or 'unjustified'; Lloyd 1980) to place too much stock on direct competitions between rival offspring over how limited nutrients are used. We rather suspect that the games involving plant family members are often complicated—with offspring striving against each other to achieve the attributes required by the maternal sporophyte for continued investment. The assigning of credit (or blame) for the deaths or hardships of certain offspring should be undertaken warily.

Moreover, it may all boil down to sibling rivalry in the end. Whether seed abortion is caused in the proximate sense by withering chemical assault from an adjacent pod-mate (which everyone would presumably concede to be a clear manifestation of sibling conflict) or by an entirely maternal decision to divert investment from some offspring to others, the common feature is offspring demand outstripping parental supply. This can be much less obvious in the latter scenario, because the actual point of critical resource shortages may never be reached; selection can favour a pre-emptive manoeuvre by the ablest player (here the mother) to eliminate pending competition. Thus, abscission events may parallel numerous animal parental manipulations (e.g. avian hatching asynchrony) that ameliorate the costs of imminent sibling rivalry. Nevertheless, if such competition were *not* pending, maternally-controlled abscission would presumably not evolve.

This brings us back once again to the opening gambit, wherein parents indulge in over-production of offspring. *Cui bono* hypotheses for the evolutionary significance of over-production in plants generally parallel those for animals, but differ provocatively in popularity. As a rule, plants over-produce far more extravagantly than most 'nursery' animals, in part because many are less obviously energy-limited, and in part because they face some additional uncertainties (e.g. whether pollination will suffice to fertilize all ova). Of the standard explanations for over-production (see Section 1.2), Lack's resource-tracking hypothesis finds considerable botanical support (e.g. Lloyd 1980, Stephenson 1981), but also inspires scepticism (see below), the insurance hypothesis is hardly mentioned (for reasons not entirely clear, but see Ehrlén 1991), the developmental selection (progeny-choice) argument is exceedingly well received, and the sib facilitation hypothesis is nearly invisible.[2] Some explanations for these patterns of popularity will become apparent; others must stem largely from tradition. Finally, in hermaphrodite plants (whose 'perfect' flowers produce both male and female gametophytes), extra flowers may be produced solely to increase male function, so there is a surplus of (unused) female organs that create no offspring (Sutherland and Delph 1984).

In the treatment that follows, we shall divide plant sibling competition into two convenient categories: (i) those which occur while the offspring are still physically attached to the maternal sporophyte and (ii) those which occur thereafter. Both feature numerous cases where competition takes the form of *scrambles* (e.g. physiological mechanisms whereby resources are consumed by Self at the expense of sibs), and at least a few cases where it involves chemical *interference*. Rivalries preceding separation feature the impressive and rapidly swelling literature on fruit and seed abortion (for reviews see Cook 1981, Stephenson 1981, Westoby and Rice 1982, Charlesworth et al. 1987, Lee 1988, Uma Shaanker et al. 1988, Haig 1992a). The scramble competition at this stage corresponds closely to the transplacental extraction of circulating maternal nutrients by viviparous embryos in fishes, reptiles, mammals and cockroaches, while also sharing some features with non-aggressive begging by nestling songbirds. Similarly, the use of chemical weaponry to stifle or kill siblings (e.g. Ganeshaiah and Uma Shaanker 1988a) matches reasonably closely with siblicidal attempts to control parental investment (but not sibling cannibalism).

After physical separation, offspring resting in the discrete seed shadow of the parent often vary in the timing of their germination and/or their ability to exist as stunted seedlings, ostensibly waiting for favourable conditions to arise (Cheplick 1992). In the case of rain-forest trees, century-old siblings may stand poised as gaunt saplings, awaiting an opening in the canopy that is likely

---

[2] One of the clearest botanical manifestation of 'sib facilitation' function involves creating extra inflorescences as a more powerful collective lure to pollinators, either by amplifying the visual/chemical signal, or by increasing the bribe (edible pollen and nectar rewards), or both (e.g. Willson and Rathcke 1974, Willson and Price 1977, Stephenson 1980). See review and references in Charlesworth *et al.* (1987).

to develop only with their mother's fall; in temperate beech and sugar maple forests, offspring patience can be measured in decades. Furthermore, some growing plants appear to infuse the surrounding soil with allelopathic compounds that may affect the ability of neighbours to grow (Rice 1984). To the extent that many (and often most) such neighbours are siblings, such inhibitory toxins might have evolved because of benefits conferred during sibling rivalries.

16.2 A brief primer on plant biology

We assume no botanical sophistication, and shall define terms and describe structures and cycles peculiar to plants as we go (e.g. see Figure 16.1). We follow Cheplick (1992) in excluding genetically identical offspring (among which Hamilton's rule predicts virtually no conflict in any case) from our usage of the term *sibling*.

16.2.1 *Angiosperm alternating generations*

The female reproductive function of the diploid mature sporophyte is carried out in flowers. A flower's *ovary* has one or more *ovules*, which may be thought of as the 'proto-seeds'. An ovule usually contains a single diploid *mother cell* surrounded by one or more *integuments* which communicate with the exterior via a tiny opening (*micropyle*) through which pollen tubes can invade. The mother cell undergoes meiosis I and II, thereby producing four haploid (and non-identical) *megaspores*. At this point, in most taxa, three of the four megaspores disintegrate (Haig 1990, Huang and Russell 1992). We do not know how it is determined which are to disintegrate, although that question seems worth pondering.

Next, the one surviving *functional megaspore* undergoes three mitotic divisions that are not accompanied by cytoplasmic divisions, such that a total of eight identical haploid nuclei are positioned around the cell. An uneven cytoplasmic division then follows, producing the *female gametophyte*, which typically consists of six small haploid cells (three stationed at each of the gametophyte's two opposite ends), plus one central cell containing two haploid nuclei that will become the triploid *endosperm* after it is fertilized[3].

---

[3] This is the 'standard' type of female gametophyte, characteristic of perhaps two-thirds of flowering plants. But some interesting variations on the theme offer potential research angles. At one end of the spectrum is the evening-primrose family (Onagraceae), which has a diploid endosperm that might be used for testing relatedness arguments about endosperm functions (and loyalties). Because the fertilized diploid endosperm in these plants is genetically identical to the embryo, the nurse tissue here may be an unusually aggressive advocate for its seed. In other taxa, there are polyploid endosperm arrangements ($5n$, $9n$, and so forth) derived from multiple genetically distinct meiotic products.

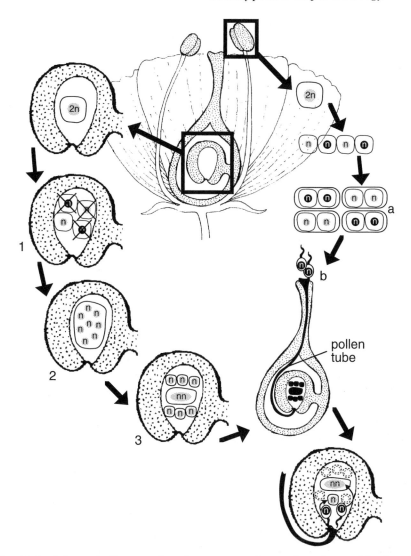

Fig. 16.1 Schematic diagram of angiosperm gametogenesis and recombination. For the female function (in the ovule, down the left side), meiotic reduction division produces four megaspores, three of which disappear (ovule **1**). That nucleus is then copied to make eight identical nuclei (**2**) that end up in only seven cells because the central cell contains a double nucleus (**3**) and extra cytoplasm. For the male function (down the right side), meiosis is followed by (**a**) each haploid cell itself dividing to produce sets of twin-celled pollen grains. Upon arrival at a female flower, one of these pollen cells develops a long tube, while the other divides one final time to give two gametes (**b**). Travelling down the tube and through the ovule's micropyle, one male gamete fertilizes the egg (producing a diploid zygote), while its partner unites with the central cell (forming the triploid endosperm).

(Note also that the pre-fertilization endosperm's version of diploidy is very strange in that its two maternally conferred nuclei are identical to each other and to each of the other six cells in the gametophyte.) The trio of haploid cells at the micropyle end includes the *egg nucleus* and her two flanking companions, the *synergids*. To summarize, our single diploid mother cell has undergone mitotic reduction division, and her haploid megaspore daughters have passed through a bottleneck that eliminated all but one, which then proliferated itself into eight copied nuclei sharing a single cell. In the peculiar cytoplasmic division that followed, two of these copies fused to form a diploid (but genetically uniform) endosperm, six became synergids, and one remained as the egg nucleus.

On the male side, simple meiotic division produces two haploid cells (microspores) encased inside a tough wall. This whole unit is the *pollen* grain, which is basically a DNA delivery crate. Upon arrival at the stigma surface of the female flower, one of the two cells grows as a *pollen tube*, while its twin splits mitotically into two identical sperm. The tube infiltrates the micropyle and female gametophyte (now called the *embryo sac*), and then ruptures and releases the sperm for a *double fertilization*. One sperm penetrates the egg nucleus while the other fuses with the doubled 2*n* nucleus of the endosperm mother, achieving triploidy.

The fertilized (3*n*) endosperm is a particularly interesting player in the intra-familial conflicts of interest, an insight due originally to Charnov (1979). It has unique relatedness patterns to the other key tissues (namely the original maternal sporophyte, the rest of the surrounding gametophyte, and the eventual embryo; Queller 1983a). Furthermore, the endosperm is often a player with clout, as it represents the nutritive tissue bank that will sustain the embryo's early development.

As embryos mature within their seeds, the surrounding ovary tissues consume circulating maternal resources, and enlarge and ripen as fruits. The fruit can exhibit many numbers and arrangements of seeds, ranging from *simple* (usually one carpel of one flower, but showing great variations, e.g. pea pod, sunflower, apple) to *aggregate* (many separate carpels of the same flower fusing as a cluster fed by a common stem, e.g. blackberry, strawberry) to *multiple* (fused ovaries plus accessory structures such as petals, e.g. fig, pineapple).

If we turn now to the task of matching up the various angiosperm players with their genetic interests, the potential for intra-familial conflict becomes apparent. Labelling a focal embryo as embryo 1 and considering how the other 'family members' should view the optimal allocation of parental investment to it as opposed to nearby alternative embryo 2, the maternal sporophyte is expected to be unbiased (following the logic of Trivers 1974) because she is mother to them both. By contrast, the paternal sporophyte of embryo 1 can usually be assumed *not* to have fathered embryo 2, so, *ceteris paribus*, it would 'prefer' an infinite skew of resources toward embryo 1. Focal embryo 1 should favour a 4:1 PI skew[4] toward itself ($r = 0.25$ between it and embryo 2). More peculiarly, the female gametophyte that gave rise to embryo 1 should do best with a 2:1 skew; embryo 1's endosperm should do best with a skew of 3:1. For the three middle players on our list (gametophyte, endosperm and embryo) the order of these differences can change position under various combinations of gene expression, inbreeding or additivity of genetic dosages and fitness effects (Queller 1989), but the potential for evolutionary games is striking. The point is that the different parties may have incongruent interests, and at least some are well situated to work 'aggressively' on behalf of embryo 1. In particular, the endosperm draws its resources directly from the maternal sporophyte's (finite) supply and may be in a position to take more than its 'share' (Haig 1992a, b).

How effectively the endosperm actually plays favourites is open to question. In one greenhouse study of wild radish (*Raphanus sativus*) only a very small amount of the variation in embryo growth (*c.* 2%) was attributable to the identity of the father. Radish embryos that had been dissected away from their endosperm and supported in petri dishes for the final third of their normal period of dependency on endosperm (nourished by an identical culture medium with constant light and temperature conditions) also showed that most size variation was due to maternal plant effects (Nakamura and Stanton 1989). (This approach seems very promising for future studies, but in this particular plant most embryo growth occurs *after* the period of endosperm dependency, making it a questionable test system for exploring endosperm power.) Another important study challenging the spectre of endosperm power involved a grass, *Anthoxanthus odoratum*, which was carefully subjected to a diallel cross among 8 genotypes (Antonovics and Schmitt 1988). Almost none of the seed size variance ($< 2\%$!) could be ascribed to paternal effects, but both maternal genotypes and environmental factors had large effects, implicating the crucial role of maternal provisioning. On the other hand, a similar diallel cross with wild radish showed strong paternity (and competition interaction) effects (Karron and Marshall 1990). It would obviously be of great interest to know the distribution of this variability in paternal effects.

The endosperm's triploid condition has itself been interpreted as a manifestation of intra-family conflicts (Charnov 1979, Westoby and Rice 1982, Queller 1983a, b, 1989, Willson and Burley 1983), especially because the maternal sporophyte's donation of extra genes and resulting vigour to the endosperm is seen as empowering an enemy. Two scenarios have developed,

---

[4] As in earlier discussions (e.g. Section 3.3.1), by 'PI skew' we mean the net effect on fitness (literally the marginal gain in fitness) resulting from uneven resoure allocation and not necessarily the indicated ratios of raw nutrients *per se*. It is also worth noting that the investment ratios predicted here are based on relatedness averaging (of maternal and paternal components). If genomic imprinting operates, these calculations would be much more complicated (e.g. see Haig 1992c for a similar case concerning eusociality).

both of which hark back to the enigmatic foreign policy of certain super-powers that provide weapons and fiscal encouragement to fringe regimes which are then branded as terrorists much in need of public humiliation.

Endosperm triploidy might have evolved from a *sexual conflict* dynamic (Charnov 1979). Here, double fertilization is seen as an event that enabled paternal genes to join the then haploid nurse tissue (making it diploid), whence they could effect increasingly aggressive favouritism toward embryo 1. 'Maternal doubling' and the resulting triploidy is seen then as an evolutionary reassertion of sporophyte control, a trumping second dosage of maternal genes that restores closer congruence between the genetic interests of nurse tissue and sporophyte (Bulmer 1986). Some tantalizing recent discussions of candidate mechanisms for this kind of sexual conflict between parents centre on 'genomic imprinting'.

As discussed earlier (Chapters 10 and 13), the logic of genomic imprinting is most easily appreciated from a gene-level perspective. In taxa where mothers are multiply inseminated, a gene derived from sperm may be unlikely to impinge on any copies of itself in the current nursery (rivals most probably having been fertilized by sperm from other males). The pay-off to such a contingent strategy hinges on two species-specific traits: (i) the mating system and (ii) the pattern of post-zygotic parental investment (which must be maternally-provided and of a magnitude worth stealing). A sperm-derived allele whose gene-product tends to extract additional resources from the mother (and, by extension, from siblings) thus receives all of the benefit and only a greatly diminished share of the cost. Genomic imprinting is there-fore well suited to viviparous animals and angiosperms[5] (Haig and Westoby 1989a, b), in which maternal investment is high and mixed paternity common[6], but rather less likely in most birds (where there is biparental care and presumably greater genetic monogamy) and pelagic-spawning fish (there is nothing to steal). In the evolution of endosperm, then, Haig and Westoby (1989a, b) proposed that endosperm was initially diploid, with one maternal and one paternal genome, and that the second maternal genome was a later addition.

An alternative scenario, the *kin conflict hypothesis*, posits a plausible sequence of coevolutionary moves stemming from the advent of double

---

[5] The direct and indirect evidence that genomic imprinting exists in endosperm is summarized nicely by Haig and Westoby (1989a, 1991), especially with reference to the maize studies of Lin. Genes on the long arm of maize chromosome 10 are active and selfish when paternally derived (producing larger kernels). Experimental manipulations of fertilization patterns have produced maize endosperms with either one or two paternal genomes plus one to eight maternal genomes, *all on the same maternal sporophyte!* Only the standard ratio of 2 maternal plus 1 paternal (2m:1p or 4m:2p) genomes produces normal-sized kernels. Crosses with a maternal overdose (e.g. 3m:1p) make small kernels; those with paternal parity (e.g. 2m:2p) abort, plausibly as an active maternal rejection.

[6] Of course, the opposite pattern of having selfish genes expressed only when they are derived from maternal gametes is to be expected in taxa where parenting roles are sharply reversed and/or maternity is likely to be mixed.

fertilization. Once that arose, the one non-zygote diploid cell in the female gametophyte was identical to embryo 1 (Willson and Burley 1983), although not yet specialized for duty as nurse tissue. It presumably shared that support task with other, unfertilized (and thus haploid) gametophyte cells. In the competition for maternal resources, nepotism by the diploid nurse on behalf of its twin (zygote 1) gradually led to withdrawal from the nurse function by the less avaricious haploid cells. During the increasing specialization toward nurse function by the endosperm, the maternal option of blocking second fertilization disappeared. Instead, the maternal response of contributing a second maternal dose may have evolved as a beneficial compromise that allowed maternal screening of variable genetic combinations (a progeny-choice pay-off) by a tissue, the endosperm, less selfish than embryo 1 itself (Westoby and Rice 1982). The sporophyte would have needed to save only enough PI (by aborting bad genetic combinations early on) to compensate for the non-optimal PI skew demanded by male-tainted endosperm. In human terms, the mother might be better off negotiating with a broker whom she had at least helped to hire, rather than negotiating with the embryo directly.

Comparative light has been shed on these evolutionary alternatives by studies of Gnetales (especially the genus *Ephedra*), the closest extant relatives of angiosperms. These plants also show double fertilization, creating 'zygote 1' and an identical diploid twin, both of which begin cell divisions. The twin lineage then aborts. In essence, there is a supernumerary embryo (usually found in the more apical position) that is lost as a future reproductive, but that may contribute altruistically to the success of its basal twin (although that key point is not clearly established; Friedman 1992). If it does contribute in some way, parental over-production in this case seems to fit the sibling facilitation category. Historically, this may well have been the ancestral condition, with endosperm evolving in the flowering plant lineage from that marginal proto-offspring (Friedman 1990, 1992), in which case the scenario of endosperm triploidy (in angiosperms) arising from a secondary maternal nuclear addition to the nutritive nurse tissue seems more parsimonious than the one featuring a paternal invasion of $2n$ maternal tissue.

16.2.2 *Gymnosperm generations*

Although less complicated than the rococo angiosperm patterns, the more ancient gymnosperm system none the less offers tantalizing food for thought. We shall focus on a few key points of divergence that are likely to affect the nature of sibling competition, namely ubiquitous polyembryony and delayed fertilization, while acknowledging the lack of litigious endosperm tissue as another vital difference.

A gymnosperm female gametophyte, derived from its megaspore, typically contains two or more *archegonia*, which can be thought of as 'proto-eggs', surrounded by various integuments. Collectively, these constitute the *ovule*.

Nearly all gymnosperms produce female gametophytes with more than one archegonium, which leads to *simple polyembryony* (the development of multiple *non-identical* embryos in a single seed). This much is 'almost universal' (Haig 1992a), and creates a very early crucible for sibling rivalry. Traditionally, this is interpreted only in terms of maternal benefits through a developmental selection/progeny-choice process (i.e. over-production of off-spring from which only the best individuals are chosen for further investment; Darwin 1876, Buchholz 1922) on a scale that is unavailable to angiosperms that lack the requisite simple polyembryony[7] (Haig 1992a).

Gymnosperms tend to dawdle between the events of *pollination* (capture of a pollen grain by the ovule) and *fertilization* (sperm invasion of the ovum). Whereas most angiosperm fertilization follows pollen arrival within just a day or two, delays of 3 or 4 months are more typical for gymnosperms, and may take a full year in some *Pinus* species. Such pre-fertilization delays allow a suite of maternal investment decisions to be made in an unhurried manner. Specifically, unpollinated ovules are jettisoned while they are still tiny and cheap. The ensuing resource deposition supports most somatic growth of the ovum prior to fertilization (which suggests that subsequent abortion might be less economical, although the most expensive nutrients are withheld and deposited later, actually after physical growth is complete (Lloyd 1980, Haig and Westoby 1989b).

Severe forms of sibling rivalry can operate on several levels, depending on who occupies the *sibling* role. If female gametophytes are regarded as sibs, they can be selectively purged (perhaps as a unilateral decision by the maternal sporophyte or as a result of actions by selfish fellow gametophytes, and quite plausibly as a collaboration between those two parties). This can operate at the level of whole seeds, whose entire polyembryonic contents are dropped selectively from cones or even at the level of whole cones that are dropped from the limb with their full complement of seeds. On a smaller scale, one can regard the polyembronic embryos themselves as siblings. Some of these individuals may die, either as a result of competition with other embryos or through some parental manipulation by the female gametophyte, which is now occupying the *mother* role. The discussions that follow embrace both angiosperms and gymnosperms and operate on all three levels of sibling competitions (embryo, seed, and fruit), with corresponding shifts in parental identity, depending on the tissue and taxon.

16.3 Pre-dispersal sibling scramble competition

The physiological reality that serves as the backdrop in this section is that some parts of a plant generate net surpluses of carbohydrate assimilate (thus

---

[7] Nevertheless, something very similar has been reported for angiosperm sugar maples (*Acer saccharum*) (Gabriel 1967).

leaves and other photosynthetic tissues are 'sources'), while other parts are net consumers (these 'sink' tissues include the reproductive organs and their dependent offspring, as well as various somatic structures). There are competitive activities engaged in by the sinks that affect how much of the total budget each will receive (Sachs and Hackett 1983) in much the same way that multiple children tapping into the same milkshake with separate straws are engaged in a sucking race (Haig 1992b). By slipping our designation of what constitutes an 'offspring' through three different scales (multiple embryos within a seed, multiple seeds within a fruit, and multiple fruits on a plant) and adjusting our attention to the appropriate 'source' entities, several possibilities emerge.

16.3.1 *Rival embryos*

Gymnosperm polyembryony leads to a typical gametophyte 'brood size' of four to seven archegonia per seed. Multiple fertilizations can follow if sufficient pollen arrive and their sperm succeed, but the situation cannot last because no more than one zygote can *germinate* per seed (Haig 1987, 1992a). Clearly, some kind of family abridgement must occur at this juncture.

These seed-mate embryos have higher average coefficients of relatedness than the normal diploid sibling maximum of 0.50. The maternal contributions are genetically identical and the paternal genes may or may not come from the same male parent (if not, the paternal contribution to sib–sib relatedness is zero; if so, it averages 0.50)[8], so embryo sibling *r*-values usually range from 0.50 to 0.75. As mentioned earlier, paternal genes favouring selfishness within gymnosperm seeds should be very active, and the maternal alleles should be indifferent (Haig and Westoby 1989a).

In fact, gymnosperm sibling rivalry may begin in some cases at the pre-zygotic (literally gametophyte) stage. Certainly, it is well known that pollen tubes engage in a growth race to reach the unfertilized ova. However, in at least one remarkable gymnosperm from Namibia, *Welwetschia mirabilis*, the four eggs grow 'prothallial tubes' that extend out to meet the elongating pollen tubes. Fertilization thus occurs at various midpoints and the zygotes then retrace the eggs' paths to return to the nutritive tissues of the female gametophyte. This supply is limiting and eventually only one embryo survives (Haig 1987).

The potential for intra-seed embryo rivalry is minimized in some conifers (e.g. *Pinus sylvestris* and *Picea abies*; Sarvas 1962, 1968, cited in Haig 1992a) by having some of the maternal pollen chambers too small to accommodate more than a single pollen grain. This maternal feature creates an automatic

---

[8] This is even something of a simplification. Many gymnosperm pollen tubes produce larger numbers of identical sperm, which may fertilize different archegonia. Under those circumstances, the overall embryo sibling *r*-values can reach 1.0.

bottleneck in zygote numbers and precludes paternally fuelled strife, at least for those ovules. Alternatively, in *Pinus monticola*, fertilization of one archegonium inhibits the fertilization of any sibling eggs (Owens and Molder 1977). A comparable blockage has been found in *Kleinhovia hospita* (see Section 16.4), where it is known to be paternally induced.

In other polyembryonic gymnosperms, there may be considerable delays between successive fertilizations within the same seed (Sweet 1973), producing effects on sibling rivalry resembling those produced by asynchronous hatching in birds. Certainly, they reduce the potential for such competition to be difficult or expensive, but they may also may influence which benefits are most likely to accrue to the maternal sporophyte for over-production. From the maternal perspective, Haig (1987, 1992a) draws an analogy between fertilization asynchrony and employee-hiring protocols. If the boss's goal is to hire the best possible job candidate, all candidates must be interviewed and tested thoroughly (i.e. the progeny-choice mechanism). Conversely, if the boss's goal is merely to hire one reasonably competent worker (and most applicants will do) while minimizing search cost, then the first applicant should be hired on a 'probationary' basis. This second argument suggests that fertilization asynchrony is a cost-cutting measure.

The common practice of 'selfing', however, greatly affects the developmental selection incentives for plants. On the plus side, inbreeding depression may greatly expand inter-embryo variation in genetic quality, increasing the value of post-zygotic choice. On the other hand, the parent is more closely related to its own selfed offspring than to its out-crossed counterpart, which complicates the assessment process. Asynchronous fertilization in gymnosperms may represent an intermediate hiring protocol, wherein one candidate is accepted on probation and then subjected to an immediate and severe test that screens for disqualifying defects. This scenario assumes that an efficient means of termination exists that allows the gametophyte to hire (fertilize) and test the next embryo. Klekowski (1982) pointed out that the fitness costs of each genetic death are trimmed by polyembryony, because a seed containing one zygote homozygous for a given recessive lethal is more likely to contain a viable embryo that is heterozygous at the same locus. (From the individual embryo's perspective, this parallels O'Connor's (1978) point that possession of siblings dampens the victim's cost of dying.)

16.3.2 *Rival seeds*

Because angiosperms lack the developmental selection inducements afforded by polyembryony, their first real opportunity for cutting losses lies at the level of the whole female gametophyte, i.e. the whole seed. The female gametophyte of angiosperms is tiny (just 8 nuclei, compared to several thousand in their gymnosperm counterparts), and therefore suited to economical culling. In addition, a maternal sporophyte faced with the task of choosing the best offspring from a large array of seeds probably has better information about the candidates' relative qualities if she is an angiosperm. Unlike a gymnosperm sporophyte, she can judge the seeds either directly or as a function of endosperm quality (Westoby and Rice 1982), without having to work through the intermediary of a large, influential and presumably contentious female gametophyte (Queller 1989, Haig 1992a). Instead, if she perceives a given seed as being unworthy of further investment, she can dispose of both it and its allied tissues.

Within multi-seeded angiosperm fruits, the potential for scramble competition among the sibling seeds themselves (i.e. not merely as the by-product of maternal seed abortion) can depend on their physical arrangement, their relative ages (developmental timing), and the means (anatomical pathways and physiological tactics) whereby parental investment reaches each seed. These variables typically show strong covariance (Lee 1988). Each seed is fed by the gametophyte's endosperm and thus by gametophytic ability to draw resources from the maternal sporophyte.

The size attained by a dependent seed is known in a few cases to affect its future performance as an independent sporophyte. Intraspecific variation in seed size was neglected in the ecological literature for nearly a century, for two main reasons (Stanton 1984). First, the test conditions used in agronomy greenhouse studies of crop plants tended to be uniform, benign and non-competitive, thus masking early advantages that greater stored reserves might provide for dealing with a stochastic environment infested with rivals. An illuminating exception involved greenhouse tests of subterranean clover (*Trifolium subterraneum*) by J.N. Black (references cited in Stanton 1984), in which seeds ranged in mass from 4 to 10 mg, but were planted in swards of either uniform size or evenly mixed large and small seeds. When such plants were grown at sufficiently high densities to induce some (25%) mortality, all victims were from small seeds, and plants from large seeds achieved twice the maximum size of small-seed survivors. Another reason for the lack of attention paid to possible impacts of seed size arose from measurement protocol. The magnitude of seed size variation itself was being under-reported because large samples were weighed *en masse* in search of mean values only (reviewed in Stanton 1984). However, when wild radish (*Raphanus raphanistrum*) seeds, ranging from 1.5 mg to 12 mg in mass, were grown under field conditions, larger seeds (i) were more likely to emerge from the soil, (ii) achieved greater early growth and (iii) produced more flowers per plant for survivors (Figure 16.2).

Similarly, a comparative study of 16 congeneric Australian angiosperm species pairs (one member of each pair having much larger seeds than the other) highlights one advantage of size. When seedlings germinated from large seeds were artificially defoliated (95% of photosynthetic tissue removed), they recovered more successfully than seedlings from smaller seeds (Armstrong and Westoby 1993), albeit presumably at some cost (e.g. reduced

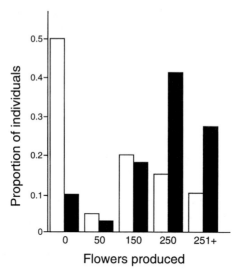

Fig. 16.2 Wild radish (*Raphanus raphanistrum*) seedlings are less likely to emerge from the soil (indicated here as zero flowers) and produce fewer total flowers when developing from small seeds (≤ 6 mg fresh mass: open bars) than large seeds (> 6mg; filled bars), both in greenhouse and field plantings. These effects may be further exaggerated when planting depth varies and/or when neighbour densities are not controlled. (Adapted from Stanton 1984a, figures 2 and 4.)

seed numbers). A parallel experimental study of 23 species also showed that larger seeds contain greater energy reserves that confer competitive advantages under conditions of protracted shade (Leishman and Westoby 1994).

Without a clear understanding of the mechanisms underlying PI allocation, we must entertain two general versions of the drama. Under the assumption that the investment decisions are entirely maternal, Haig (1990) devised an ESS model for developmental selection benefits that hinge on the *quality* of information available to the 'mother' (which could range from alive vs. dead to a continuously graded assessment of seed quality). As mentioned above, the information available to the gymnosperm sporophyte is expected to be of poor quality precisely because it gets filtered through the female gametophyte (recall that *r* between the gametophyte and her own offspring is 1.00, while her relatedness to the offspring of sibling gametophytes is only 0.50). In short, the gametophyte might be expected to 'lie' about her offspring's potential. Here a sibling competition can be envisaged as operating between gametophytes. Another way of expressing this is that the seed-abortion 'threshold' (*sensu* O'Connor 1978 and Figure 4.1) will be much higher for the gametophyte than for *her* mother, the sporophyte.

Haig's wording (1992a: 78) is clear: 'In summary, the female gametophyte of gymnosperms controls access to information about embryo quality. As a consequence, genes expressed in the female gametophyte are likely to determine which seeds are aborted in the course of developmental selection, and the sporophyte's resources will be suboptimally allocated among seeds'. If this actually works, it would seem to constitute an 'offspring wins' parent–offspring conflict.

Alternatively, individual seeds may be able to attract and retain more than their fair share of maternal resources (defined as the total parental expenditure of the whole maternal unit, or appropriate functional subunit thereof, divided by the number of viable sibling seeds affected) through their own actions, specifically by generating higher *sink intensities* than their rivals. Sink intensity is a function of each seed's hormonal activities, and these '. . . play a leading role in the mobilization of resources into the developing fruits' (Stephenson 1981). In milkweeds (*Asclepias* spp.), for example, the identity of the pollen donor (i.e. paternity) affects both the number of pods set and mean seed size, which is interpreted as being due to the embryo's resource-drawing power (Bookman 1984). Similarly, wild radish fruits containing multiply sired embryos receive more maternal resources (Marshall and Ellstrand 1986), plausibly because the selfish activities of unrelated seeds *within* such fruit are less constrained. In trumpet climber (*Campsis radicans*), flowers receiving small amounts of pollen produced fewer fruits than those receiving greater amounts, even though all dosages were from single donors and there was great variation between different donors (Bertin 1990). Large pollen loads can generate more intense pollen-tube competition (Lee 1984), which has the proximate effect of releasing auxins, gibberellins and perhaps other hormones into the ovary. These may stimulate fruit growth, both directly and indirectly (the fertilization of more seeds has positive-feedback effects on hormones produced by the seeds), making whole fruits into stronger sinks (Bertin 1990). Subsequent abscission of weaker fruit, then, could provide fitness benefits for the offspring retained and their parents (Schemske and Pautler 1984, Temme 1986). Evidence has also emerged for just such a mechanism in an understory shrub, *Lindera benzoin*, in which higher pollen load has its most acute effects on selective abortion when local resources are otherwise low (Niesenbaum and Casper 1994). One wonders in addition if 'cheater' seeds might have any means of producing extra hormones, and thus escalating their ability to attract disproportionate PI in an analogue of animal begging models.

Hamilton's rule suggests that sibling antipathy toward neighbouring seeds should be muted by anything that raises r, which is a likely result for various types of *multiple pollen units* (i.e. wads or strings of pollen gains from the same father that stay together; Cruden and Jensen 1979, Kenrick and Knox 1982, Bookman 1984) that create pockets of seeds where $r = 0.50$ ('full siblings' in the zoological sense). Regardless of whether such units evolved primarily to aid in sperm competition, for their potential in lowering 'harmful' sibling rivalry (Kress 1981, Uma Shaanker et al. 1988), or for some other reason, they may have potential for the study of parent–offspring conflict. From the maternal sporophyte's perspective, offspring sired by the same or different fathers are identical (with regard to relatedness to the sporophyte), but from the seed's perspective neighbours may differ substantially in value, depending on whether they share a common sire. In principle, one might detect which party (mother or offspring) controls PI allocation by resolving local paternity demographics in such taxa.

There appears to be little information on how nutrient limitations may affect variation in seed quality and hence selective abortion of 'lesser' seeds (Lee 1988), perhaps because so much attention has been focused on the primary issue of pollination. Obviously, an unpollinated ovule cannot be fertilized, and clearing that hurdle is problematic, but other resources can limit seed set as well. This has been demonstrated in *Cryptantha flava*, a semi-desert perennial (Boraginaceae), in which experimental ovule destruction increases the probability that other ovules will mature, an effect that can also be induced by adding water or fertilizers (Casper 1984, for similar examples see Lloyd 1980). In many plants, large numbers of healthy zygotes are abscised (20% or more in *Pinus sylvestris*, even when pollen is superabundant; Sarvas 1962, cited by Haig 1992a), which leads us to consider again the array of hypotheses put forward to explain why parents routinely over-produce, and then secondarily prune. For seeds, three of the four standard explanations for over-production look promising. The resource-tracking hypothesis can actually be applied with reference to pollen abundance: a *good year* (from the maternal sporophyte's viewpoint) is one in which pollen is abundant (Haig 1992a). That is, the parent may create numerous ovules so as to maximize seed set in low-pollen years. The insurance hypothesis (also called the 'reserve ovary hypothesis', Ehrlén 1991) may also be applied at the seed level (i) if there are threats to individual seeds that do not imperil the 'fruit-mate' seeds and (ii) if some of these marginal siblings manage to thrive specifically because of these core-brood losses. Because we cannot imagine seeds literally eating one another, the potential for 'icebox' (exploitation) benefits probably does not exist here.

The possibility of a developmental selection process has been much discussed, and requires only that seeds vary in their genetic quality to such a degree that it pays to create large numbers, and then to terminate the less promising variants. In many gymnosperms, selfed embryos are more likely to be lost early in development than are out-crossed embryos (Haig 1992a), which presumably have higher reproductive value. The very existence of selfed offspring can be viewed as insurance against low pollen availability; the policy can be trimmed or cancelled once that risk period has passed.

In many plants, direct competitive interactions between out-crossed and selfed sibling seeds may be unlikely (e.g. if they are spatially segregated on different parts of the plant). Where the potential exists, its fitness consequences will probably depend more on metabolic vigour than on relatedness *per se*. Indeed, an out-crossed embryo has the same *r*-value to a selfed sibling (0.25 from having a mother in common plus another 0.25 from the fact that the out-crossed sib's mother is also the selfed sib's *father*) as it has to an out-crossed sibling having both parents in common (also 0.50) and considerably higher than its *r*-value to an out-crossed sibling having a different father, which must often be the case. However, these factors may be easily overwhelmed if the selfed offspring's genetic constitution produces phenotypic weaknesses (Haig 1992a), not only because of the resulting competitive mismatch, but also because its reproductive value offers less to the inclusive fitness of the out-crossed sib (e.g. Eickwort 1973).

Beyond the standard list of hypotheses, a very interesting case has been made for how the loss of some seeds can influence the effectiveness of sibling dispersal. In *Cryptantha flava*, four ovules are often fertilized, but usually only one nutlet matures (Casper 1984). Aborting the other seeds reduces the weight of the survivor, which might increase its (wind-borne) dispersal distance (see Garrison and Augspurger 1983). If weight minimization is generally advantageous for wind-dispersed species, then they are likely to have lower seed-to-ovule ratios than animal-dispersed taxa, whose fruits are free to carry more seeds (Casper 1984). This prediction has received some support in reviews. For example, Uma Shaanker *et al.* (1988; Table 1) found that in 14 species of wind-dispersed plants each fruit carried a mean of 1.27 seeds (range 1–1.89 seeds), while in 13 species of animal-dispersed fruits the average was 5.47 seeds per fruit (range 1–21 seeds). Some animal-dispersed fruits apparently have just one seed in order to increase the disperser's bribe (flesh-to-seed ratio; Willson and Schemske 1980). One must be careful here to distinguish between fruits and seeds as dispersal units. Fruits that release wind-dispersed *seeds* (e.g. milkweeds) are free to have many seeds per fruit, since the seeds travel singly. In the case of *C. flava* such a dispersing advantage has not been demonstrated, and its habit of single-seed packing may be more related to local resource competition, to the extent that paired seedlings emerging from two-seeded fruits are disadvantaged (Casper 1990).

In cases where the reduction of seed numbers within fruits appears to be under offspring control, the losses have sometimes been attributed to parent–offspring conflict (e.g. Ganeshaiah and Uma Shaanker 1988a, Uma Shaanker and Ganeshaiah 1988, Uma Shaanker *et al.* 1988; see also Herrera 1990). For example, in the woody shrub *Caesalpinia pulcherrima*, a strongly bimodal distribution of seeds per pod has been interpreted as representing the differing optima of maternal sporophyte and selfish seeds (Uma Shaanker and Ganeshaiah 1988). This certainly could be the case, but it is not the only possible explanation. The authors' preferred explanation assumes that maternal interests are best served by 'economical packing' (high seed:pod ratio), with no discussion of how a mixed brood (some pods with many seeds, and others with few) might also be advantageous to the parent. One can easily imagine ecological circumstances in which the economics of seed-packing is overridden by other factors.

Conversely, where the mechanism that leads toward reduced seed packing is at least partly under the control of the maternal sporophyte (e.g. by selective abortion of multi-seeded fruits), the likelihood of parent–offspring conflict is diminished, as both players—parent and offspring—may agree that single-seededness is optimal. An example of this is provided by *Cryptantha flava*, in which one can gain an initial impression of 'offspring-wins' POC that turns out to be illusory on closer scrutiny (Figure 16.3). From the embryo's point of view, it pays to be in a singly-packed seed (it has higher nutrient stores in the pericarp, etc.), and a quick look at offspring welfare from the putative parental viewpoint would seem to indicate that double-seededness is better. Specifically, the emergence rate of seedlings is higher from fruits bearing two seeds, perhaps because there is insurance against failure by one embryo. However, various handicaps on double-seeded seedlings subsequently reverse the parental pay-off, thereby eliminating the supposed conflict (Casper 1990). The maintenance of small numbers of doubly-seeded fruits in the populations could be due either to imperfections in the abortion system or, perhaps more plausibly, to occasional conditions in which two-seededness is favoured.

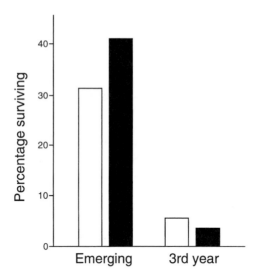

Fig. 16.3 The assumption that parental interests are best served by producing multi-seeded fruits was tested by comparing survivorship of single-seeded (open bars) vs. two-seeded (filled bars) fruits of *Cryptantha flava*. Whereas two-seeded fruits achieve a higher initial emergence rate, giving a short-term parental advantage, survival to minimum breeding age is skewed in the opposite direction, partly because singletons are 21% larger and partly because two-seeded fruits sometimes produce paired rival seedlings that impair each other. As a result, the parental optimum apparently does not differ from that of the offspring: both should favour single-seeded fruits. (From Casper 1990.)

16.3.3 *Rival fruits*

There must also be higher-level competition among whole fruits (and perhaps flowers) for maternal resources (e.g. Willson and Price 1977, Lloyd 1980). Among other things, this raises the provocative spectre of fruit-mate seed coalitions potentially co-operating, perhaps only temporarily, in order to outdo other fruits, either in direct chemical combat with sibling fruits or, more credibly, in a scramble competition between sinks (e.g. to be judged as acceptable for future investment by the maternal sporophyte). In addition, differential fruit abortion could be affected incidentally by lower-level competitions (i.e. within-fruits). For example, it appears that wild radish fruits containing multiply sired seeds obtain more maternal resources than those with single pollen-donor seeds (Marshall and Ellstrand 1986), so the un-checked 'sink' intensities of aggressive seeds might lead to the whole fruit's benefit, ironically mimicking seed–seed co-operation. The resources at issue include nutrients from soil (xylem-delivered water and inorganic compounds) and, in particular, photosynthate (phloem-delivered carbohydrates from leaves and storage organs, the 'source' tissues). Leaf-derived materials tend to be drawn from foliage near to the fruit, usually within about 1 m on a tree (Stephenson 1981) or, more generally, within what has been called the 'developmental unit' (Lloyd 1980) or 'integrated physiological unit' (IPU; Watson and Casper 1984). This is the effective nursery when fruits are being considered.

Thus one envisages at least two categories of resource budgets, two scales, that may be limiting to offspring. One budget is supplied by the whole plant and one is derived from local production. The latter may vary considerably according to each reproductive module's sunlight, number of leaves, etc. (see Watson and Casper 1984). (Academics may note a familiar parallel in their own university, which has an overall budget, and constituent departments that may have modest revenues of their own; sometimes one begs most effectively from one's chair, and at other times from the dean.)

Once the key resources have been located, an individual's position may affect the amount it receives. Although fruits cannot jockey for improved position like nestling birds, complete understanding of rivalry at this level probably requires attention to geometry. It may take take relatively little early pertubation to send even a hypothetically symmetrical competition into a 'self-organizing' positive feedback loop in which the discrepancies tend to magnify themselves. This reasoning has been applied formally to seeds within a fruit (Ganeshaiah and Uma Shaanker 1992), and applies equally well to fruits as a function of their distance from the resource supplies (e.g. Obeso 1993). Individuals closer to the lifeline will grow slightly more quickly, thereby attracting an increasing *relative* share of materials, and so on.

For example, it is known that distal inflorescences are often less richly served by vascular tissues (e.g. *Asclepias tuberosa*, Wyatt 1980; *Lupinus* spp.,

Lee 1988), arguably due to parental favouritism (or at least indifference), in that sporophyte architecture is involved that might have been shaped otherwise by natural selection. In many flowering plants, inflorescences open 'acropetally' (basal to terminal), giving a temporal advantage to the more quickly pollinated basal flowers (Stephenson 1981). Basal fruits are also situated 'upstream' with respect to water and some transported nutrients, although perhaps 'downstream' with respect to other nutrients. It is not clear what the net effect is in terms of being able to take first cut and leave less of the overall pie.

A gradual snowballing effect, reminiscent of the protracted starvation shown by the last-hatching members of avian broods, has been documented in a few cases. In currants (*Ribes nigrum*), for example, terminal inflorescences receive fewer resources than their lower counterparts, and many of their ovules degenerate prior to fertilization. The smaller resulting number of seeds per fruit fails to produce sufficient hormonal activity to draw more resources, causing still further seeds to abort within these fruits. Eventually, the affected whole fruits are abscised (Stephenson 1981). While one might regard these as the equivalent of 'damaged' fruits, the distinction is that their inadequacies are imposed by the actions of close kin, not by extrinsic factors.

From the maternal sporophyte's perspective, the costs and benefits of flower over-production and subsequent fruit abortion may well vary on different parts of its body. To the fruit as well, the optimal balance for selfishness vs. nepotism may similarly depend on location. In general, resource-tracking incentives should be more valuable for species that rely heavily on their supply of current photosynthetic assimilates for fruit maturation ('income breeders'; Stearns 1992) than for taxa with rich stores of reserves, since the latter are relatively buffered from the vicissitudes of a 'bad year' (Stephenson 1981: 271).

As reviewed above, many plants 'overproduce-and-prune' at several stages, with fruit abortion being the final and most costly correction point (Lloyd 1980, Stephenson 1981). The numbers of 'offspring' shed along the way can be staggeringly large, on both absolute and relative scales. For example, Washington navel orange trees, monitored from February to July, dropped about half of their original buds (an average of 96 343 buds!), followed by a large number(33 235) of the remaining flowers, which were followed by nearly all of the remaining fruits (68 696) while they were still very small. The final fruit crop of 419 mature oranges per tree represented 0.2% of the original buds (Kozlowski 1973)! Similarly, more than 90% of *Quercus alba* acorns drop by mid-summer (Kozlowski 1973, Sweet 1973), with most of this occurring between pollination and fertilization. Additional examples of drastic fruit reductions are provided by milkweeds (Willson and Price 1977), yucca (Udovic and Aker 1981) and many others. It is clear that both damaged and 'surplus' undamaged fruits are aborted (Stephenson 1981: 261). As usual, both can easily be interpreted as effects of sibling rivalry (current or pending, by mother or sibs themselves, etc.) as well as a means of quality control.

The economics of episodic abortions has been worked out for a few pro-fligate taxa. As expected (Hamilton 1966), early abortions are relatively cheap by most measures. For example, one can compare the smallness of the famous 'June drop' of apple fruits with the eventual size of the mature survivors. David Lloyd did this in his backyard (see Lloyd 1980), and found the abortees to be on average 7.0% and 8.2% (he had two trees) of fully ripe size. More rigorously, Bookman (1983) showed that fresh milkweed flowers contained < 3%, and a later 'young pod' stage contained < 5%, of the water present in a mature pod. These comparisons are conservative, since a growing pod also loses water via transpiration with time, but impressive figures were also obtained for key elements (N, P, K and Mg) that flux less. Furthermore, comparisons between fresh and wilted flowers suggest that the plant revokes some of the nutrients (via resorption of nectar) prior to abscission (Bookman 1983), rather in the manner of maternal cannibalism or fetal resorption in rodents.

Some of the fruit abortion issues raised thus far can be illustrated with *Catalpa speciosa*, a leguminous tree which produces long cigar-shaped fruits in which seeds overwinter before desiccation breaks them open for wind dispersal in the spring. The fruits are provisioned by photosynthate from nearby leaves, the artificial removal of which affects the likelihood that a given fruit will be jettisoned early (Figure 16.4a). The density dependence can be demonstrated even more dramatically at the consumer end. To explore how close fruits can influence each other, Stephenson (1980) pollinated artificially to give one flower per infructescence a head-start over its slightly later pollinated neighbours. Then he either removed the developing fruit of that 'senior sibling' or allowed it to remain in place. This experiment clearly revealed that the continued vigorous existence of a stronger sibling seriously jeopardized the prospects of junior neighbours, since twice as many junior sibs were aborted when the senior fruit did not predecease them (Figure 16.4b). Note that this demonstration neatly satisfies what we have termed 'O'Connor's criterion' for demonstrating true sibling competition (equation 5.1). Variation in timing of natural fertilization events within a given 'integrated physiological unit' (e.g. an infructescence) seems likely to create a sibling size hierarchy equivalent to avian hatching asynchrony.

Despite an earlier belief that fruit and seed abortions were mainly due to pollen shortages (reviewed in Stephenson 1981), there has been growing evidence that points toward resource limitations on fruit development (Mogensen 1975, Udovic and Aker 1981, Casper 1984, Stanton *et al.* 1987, Uma Shaanker *et al.* 1988, Charlesworth 1989a, b, Wiens *et al.* 1989, Obeso 1993, Niesenbaum and Casper 1994). For example, there are various negative correlations between components of seed quantity and quality (see references in Lloyd 1980). On an experimental basis, adding nitrogen fertilizer to apple trees *after the flower buds have formed* can increase the proportion of flowers that set fruit (Hill-Cottingham and Williams 1967), while restricting resources

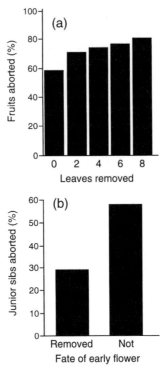

Fig. 16.4 Micro-scale economics of fruit abortion in *Catalpa speciosa*. Fruits derive most of their nutrients from leaves near them. **(a)** Accordingly, as more of these leaves are clipped off experimentally, a greater proportion of the dependent fruits are aborted. If a single flower within an inflorescence is hand-pollinated earlier than its neighbours, such that its embryo has a temporal head-start over the others using the same food base, its subsequent fate affects the younger nursery-mates. **(b)** If this early flower is allowed to develop normally, most of its junior sibs are aborted, but if it is eliminated the losses of junior sibs are halved. (From Stephenson 1980.)

increases abortion rates. Tomato leaves create essential resources and their experimental removal decreases fruit set in proportion to the number of leaves eliminated (Leopold and Scott 1952). When the viability of these shed pollinated flowers was tested by growing them on a medium of organic nutrients, 31% bore fruit (presumably a conservative figure). Similarly, when grape vines were deprived of sunlight by the application of varying layers of mosquito netting, they produced mature fruits in inverse proportion to the amount of shading (May and Antcliff 1963). Furthermore, apple trees abort fewer fruits if the competing nutrient demands of somatic growth are eliminated by the pruning of shoot tips (Quinlan and Preston 1971). The reverse experimental approach is to reduce demand on the maternal supplies by artificially thinning flowers, either on a whole-plant or within-inflorescence basis. As expected, this also reduces fruit abortion (e.g. Quinlan and Preston 1968, Stephenson 1980; see references in Lloyd 1980).

The cost-effectiveness of creating extra incipient offspring in the face of resource unpredictability may be partially buffered in many plants by the existence of storage tissues, where water and photosynthate can be stockpiled. However, the degree to which these reservoirs will be filled when needed adds another stochastic element. A high proportion of the resources eventually spent on reproduction is often generated *after* the commitment to an upper limit on family size, necessitating 'serial adjustment of maternal investment' (Lloyd 1980).

Two opposing camps have formed concerning the matter of whether pollen or nutrients limit seed and fruit numbers, as if this issue must decisively go one way or the other. This may be yet another false dichotomy, like nature–nurture, which is clouding the likelihood that different resources will surely be limiting for different taxa, at different times, in different locations, and/or under different circumstances. There is understandable concern that much of the experimental research involved with commercial crops (see above examples), which have long been under artificial selection for life-history traits intimately related to Darwinian fitness (consider seedless grapes!). Moreover, it has not escaped the notice of various sceptics that discarding the notion of pollen limitation opens the door to theoretically tasty interpretations concerning sexual selection (namely that mate choice in plants involves a kind of 'morning-after' decision to jettison zygotes fertilized by less desirable fathers) or maternal developmental selection. On the other hand, most angiosperms respond to the experimental supplementation of pollen by increasing fruit and/or seed production, in at least some years and sites (M. Burd, unpublished data). In short, it seems premature and unwise to dismiss pollen limitation in a sweeping manner at this point, although some experimental tests (e.g. Marshall and Ellstrand 1986) certainly make it look unlikely in cases.

In a few recent tests, pollen and nutrient levels were manipulated simultaneously with non-domesticated plants. These show that either or both factors of these can affect offspring number and quality. For example, Campbell and Halama (1993) supplemented wild *Ipomopsis aggregata* with doses of pollen (on toothpick analogues of hummingbird bills) and/or fertil-izer, showing that both treatments enhanced maternal fitness within the same population. Extra pollen increased seed set per flower, while fertilizer prolonged flower production. These authors also reviewed a variety of ways in which both limitations can work simultaneously.

Of course, there is no reason why maternal sporophytes should not gain in multiple ways from the one–two combination of over-production plus abortion. Decreasing to an optimal number of offspring may be enhanced further if integrated with offspring-quality decisions. Not surprisingly, fruits damaged by extrinsic factors, including the abiotic (e.g. very low tempera-tures) and biotic ones (e.g. insect attacks), tend to be aborted more often than would be predicted by (reviewed in Stephenson 1981, Fernandez and Whitham 1989), implicating insurance value for marginal offspring.

Such 'decisions' are routinely assumed to belong to the maternal sporophyte, but the same outcome could be generated by sibling competition alone (if impaired fruits cannot summon enough limited resources to sustain themselves) or as the effect of dual pressures from mother and siblings in collaboration. (By analogy, if a bot fly were to parasitize and greatly weaken a senior cattle egret nestling, such that it could no longer enforce its high rank in the brood, its eventual starvation would be ascribed to a combination of parasitism and sibling competition, *not to parental prudence*. Yet its parental investors might well gain from the elimination of their one faltering offspring.) Certainly, the elimination of damaged fruits can enhance the fitnesses of the siblings, parent(s) and even the victims themselves (O'Connor 1978). One key difference between the kinds of sibling competitive asymmetry found in angiosperm fruits and in birds is that the former results from slight and usually stochastic differences in pollination timing across adjacent sibling flowers[9] (and hence is imposed on all family members), while the latter is often effected by one party, the incubating parent.

Of course, there are also strategies available to parents for increasing seed numbers per fruit ('seed packing'). These need to be kept in mind when evaluating how final investment patterns are reached. In many taxa, fruits carrying very low numbers of seeds are selectively aborted (Janzen 1982, Bawa and Webb 1984, Ganeshaiah et al. 1986), which is typically seen as a purely parental manipulation—a maternal enforcement of a threshold tuned to her best interests. Sometimes this could be argued differently, as noted above, so caution is usually advisable in such cases. By contrast, the example of *Leucaena leucocephala* seems unambiguously parental because the maternal sporophyte has a threshold mechanism for discriminating against few-seeded fruits that operates *before the seeds even exist* (Ganeshaiah et al. 1986, Ganeshaiah and Uma Shaanker 1988b). The flowers have stigmatic pouches filled with a proteinaceous liquid pollen-inhibitor that is fully effective at its original pH. No pollen will germinate until at least 20–25 grains are present, at which time their combined buffering capacities disable the inhibitor, allowing fertilization to occur. The authors interpreted this primarily as a sexually selected trait whereby females incite mate competition (Ganeshaiah and Uma Shaanker 1988b). Be that as it may, the result is a minimum of 7 to 9 seeds per pod under moist conditions, changing to a smaller number, 6, in the dry season (presumably adjusted by a maternal alteration of

---

[9] Actually, there is a little more mystery than that. Before it drops, a juvenile fruit decreases its own production of growth hormones and shows a rise in the concentrations of abscisic acid and ethylene, which are 'growth inhibitors' (Stephenson 1981). Presumably the individual's failure to garner parental resources, for whatever reason, affects its own hormonal activity, but what about the rest? Specifically, what party is the source of these so-called 'growth inhibitors'? If maternally derived, is their function to challenge, identify, and eliminate weak or damaged offspring?

inhibitor strength). Thus, the demand level appears to be manipulated maternally in order to minimize sibling rivalry costs.

The shedding of undamaged (apparently 'perfectly good') fruits occurs in many plant species and, as with seed abortion, the 'overproduce-and-prune' combination has been shown to pay in three general ways. The experimental demonstrations of the way in which resource limitations affect fruit set described above indicate the potential importance of the resource-tracking hypothesis. A likely example is *Cassia fasciculata*, a forb that makes large numbers of small/cheap fruits, maturing only a few while the rest remain on indefinite hold, mostly withering and dying with the parent plant at the end of the growing season. Lee (1984, 1988; see also Lee and Bazzaz 1982a, b) inter- preted these offspring as being a reserve that could be used if resources prove rich. If production rate tracks resource availability, Lee and Bazzaz (1982a) reasoned that more fruits should mature when rivals are removed, especially if key resources are added. The results obtained from a field experiment (Figure 16.5) clearly show that plants normally carry many supernumerary offspring, which are capable of being ripened under conditions of bounty. They argued further that insurance did not appear to be very important because the local predators, especially a weevil, despite taking up to 43% of the fruit, typically attack older fruits, in which maternal resources have already been invested. By contrast, selective abortion appears to eliminate low-seeded fruits, as found in many other plants.

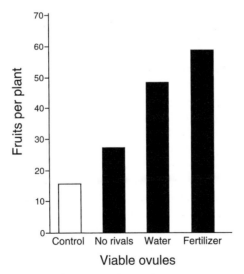

Fig. 16.5 Resource-tracking by *Cassia fasciculata*. Lee and Bazzaz (1982) weeded within 40 cm of 20 randomly chosen plants (labelled here as 'No rivals'), then compared the fruit production of those plants with 20 unweeded control plants (open bars) and two other treatment groups of 20 plants apiece that were similarly emancipated from local competition and received supplements of water or water plus fertilizer.

A closer look at progeny choice involved an experimental comparison of the offspring deemed to be 'keepers' by the parent plant with offspring resulting from a numerically equivalent trimming of family size. In each of 20 *Lotus corniculatus* weeds, one shoot was chosen after bee pollination had occurred. Alternating inflorescences were either 'hand-thinned' (half of their flowers were removed in a random manner just prior to the time of natural abortion) or allowed to remain intact (such that the sporophyte retained the chance to apply a choice criterion during abortions). The resulting fruits on the control plants averaged nearly 60% more seeds per fruit than those in the artificially thinned group (Stephenson and Winsor 1986), suggesting that proto-offspring were naturally screened for qualities likely to confer enhanced reproductive value.

An even more compelling demonstration of selective abortion on the basis of offspring quality clearly implicates the genes being screened. Sporophytes of *Mimulus guttatus* growing under copper-stressed soil conditions apparently make a choice between zygotes carrying copper-tolerant genotypes and those carrying copper-sensitive genotypes. When pollen from heterozygous donors (or mixtures of pollen from copper-tolerant and copper-sensitive donors) was given to cloned sporophytes living in copper-enriched soil, significantly more copper-resistant progeny resulted than from comparable mothers on control soils. A corresponding decrease in the seed/zygote ratio in the experimental plants indicated that the pruning occurred post-zygotically within the pistil (Searcy and Macnair 1993).

Developmental selection has acquired a large following, perhaps because of its similarities to the hot topic of mate choice in sexual selection (Charlesworth *et al.* 1987, Stearns 1987, Queller 1989). Usually, the potential value of supernumerary fruits as insurance offspring has either been overlooked (but see Casper 1984, Ehrlén 1991) or tacitly fused into the developmental selection category (Stephenson 1981, Haig 1992a, b). Both the insurance and progeny choice concepts offer some clearly testable predictions for the many puzzles still surrounding plant surplus offspring phenomena. For example, Stephenson (1981: 273) posed a closing question about why species vary in their timing of abscissions (some drop already pollinated flowers, some drop juvenile fruits, etc.). Such variation might be evaluated productively as an optimality exercise. The points at which 'extras' are jettisoned may match the age at which malevolent extrinsic forces cease to kill members of the cohort (as predicted by the insurance hypothesis) or the age at which intrinsic ontogenetic mishaps are likely to have expressed themselves (as predicted by progeny-choice). Beyond this age, whenever it occurs, the value of cheap replacement offspring should decline, so maintaining them should become less cost-effective. In a complementary way, variation in abscission timing may reflect differences in the detection timing of maternal-paternal genetic non-complementarity (*sensu* Charlesworth *et al.* 1987).

As with seed abortion, the option of self-fertilization may boost the progeny-choice hypothesis to frontline status for many plants that drop fruit. Selfed fruits in *Macadamia ternifolia* and several other species mature only when the number of fruit set from cross-pollinated flowers is low. In some squashes, the selfed flowers produce fruit only when their more valuable siblings have been removed experimentally from the parent plant (Stephenson 1981). In general, cones and fruits with *relatively* low seed set tend to be aborted (reviewed in Stephenson 1981, Lee 1988, Uma Shaanker *et al.* 1988). There are obvious questions concerning whether unfavourable genetic combinations will express themselves early enough in embryonic development to allow assessment (and rejection) by the maternal sporophyte (e.g. Charlesworth 1989a, b), but the current picture is far from clear.

One provocative demonstration of parental investment *timing* involves *Catalpa speciosa*, a tree that faces chronic and very serious seasonal herbivory from a specialized sphinx moth that oviposits large egg masses on its leaves. The caterpillars consume vast amounts of photosynthetic tissue, a factor that affects the local nutrient budget of the 'integrated physiological unit' (see Figure 16.4). The extent, locations and precise timing of such damage are presumably not predictable to the tree, but the destruction ends reliably by mid-summer, whereupon two relevant events take place. A great wave of fruit abortion occurs just after the pest vanishes, and the main maternal investment in fruit follows within several weeks (Figure 16.6).

In gymnosperms, the selective dropping of mature cones is much less common than angiosperm fruit abortion. Even cones containing mostly empty (e.g. unfertilized) seeds may be retained. This difference is attributed mainly to the timing of cone growth relative to the delayed conifer fertilization pattern. Most cones and seeds are already full-sized, or nearly so, by the time seeds are fertilized (Sweet 1973). However, in some species, especially those with the longest period between pollination and fertilization (and 2-year reproductive cycles), there may be extensive early abortion of unpollinated conelets (Haig 1992a).

16.4 Pre-dispersal sibling interference competition

Early zygotes can sometimes physically prevent the fertilization of nearby siblings by inhibiting pollen tube entry (Mogensen 1975, Sedgley 1981) or by actually crushing the embryo sacs of neighbours (Sedgley 1981). In *Kleinhovia hospita*, the first zygote to be fertilized in a multi-seeded fruit quickly synthesizes a compound that diffuses into the style to inhibit, deflect and even destroy the growth of slower pollen tubes, thereby preventing the fertilization of the remaining ovules. The full story is not yet clear, but this much is known: (i) the asynchronous nature of the initial fertilization results from maternal architecture (some pollen sites in the style are physically closer to the egg nucleus); and (ii) the production of the debilitating chemical indoleacetic acid

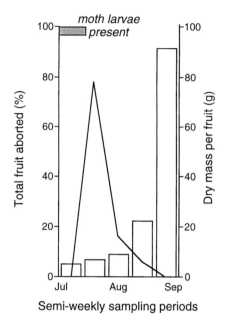

Fig. 16.6 Whole-tree investment patterns for *Catalpa speciosa*. Once the moth herbivores have completed their harvesting of leaves (open bar in upper left), the tree aborts hundreds of fruits (line) before many nutrients are packed in particular fruits (open bars). (From Stephenson 1980.)

(IAA) effectively precludes potential half-sibling seeds, which presumably benefits paternal genes. There is typically plenty of pollen on the stigma, most of which germinates, but these pollen grains are actively prevented from reaching the ovules (Uma Shaanker and Ganeshaiah 1989). Thus, the phenomenon closely parallels the vaginal sealing of certain insects (Parker 1970), and can be interpreted both as a form of sperm competition (within the proper realm of sexual selection) and as a pre-emptive strike in a pending sibling competition. The presumed benefit to the offspring (and, indirectly, to the maternal sporophyte) lies in the fact that the successful offspring is passively dispersed in a papery bladder-fruit by wind and water, so its resulting lightness may be an asset. The contributions of both parental phenotypes to the observed outcome suggests a congruence of interests with each other and with the resulting offspring.

Similarly, for over a century it has been known that the oak (*Quercus*) ovary initially contains six ovules, yet most acorns are single-seeded. The earliest proximate explanation for this (Ward 1892) was that one ovule is fertilized slightly ahead of the others, and then 'starves the rest by taking all the available nourishment to itself.' A modification of this for a different genus added the prediction that the functional ovule gains its temporal headstart from a positional advantage inside the acorn (Poole 1952, cited in Mogensen 1975), but a search for the prescribed positional correlation in *Quercus gambelii* yielded no support for this view, prompting Mogensen (1975) to propose an active 'suppression' of later ovules by the first-fertilized ovules. To our knowledge, the test to discriminate between these possibilities has not been conducted.

On an ultimate level, however, one can still ask why there is so much over-production: why make six ovules when only one embryo will march forth? Mogensen (1975) carefully microtomed 20 *Q. gambelii* acorns, cautiously discarding 'many more ovules' from ovaries in which all six ovules could not 'be interpreted with confidence.' From these data (Table 1 in Mogensen 1975), one finds that quite a few of the ovules (45% overall) were never fertilized, a further 25.8% failed to develop an embryo sac, and one ovule (0.8%) had an embryo sac that was empty for reasons that are unknown. That is, over 25% had already experienced some kind of developmental failure (suggesting the need for over-production to supply replacements), and nearly 50% apparently remained unvisited by pollen. As a result, the maximum of six viable embryos per acorn was never observed, but none of the ovaries was barren. On average, just 1.70 ± 0.92 fertilized ovules per acorn were observed, most having only one (Figure 16.7). If we assume (i) that ovules are cheap to build relative to acorns, (ii) that

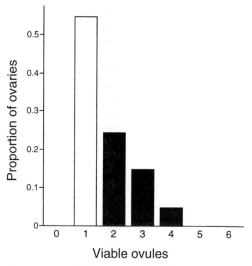

Fig. 16.7 Although each oak (*Quercus gambelii*) ovary contains six ovules, the eventual acorn contains only one embryo (with negligibly few double-seeded exceptions). Failure to be fertilized eliminates 50% of the candidate ovules and developmental irregularities account for another 25%, leaving just one zygote in the majority of acorns (open bars), and one to three redundant zygotes (filled bars) in the rest. The upshot is that none of the 20 acorns in this sample was devoid of an embryo, nor did any carry the maximum of six (or even five). The system may be initially profligate specifically because of the frequent, and cheap, losses. (Adapted from Mogensen 1975, Table 1.)

these naturally arising sources of failure eliminate most of the 'cost' of later removals of supernumerary embryos (by whatever means) and (iii) that Mogensen's small and stringently chosen sample probably *underestimated* ovule failure in the first place, it looks as if selection may have produced about the right number of extra ovules to guard against acorn-level wastage.

16.4.1 *Seed sabotage*

One of the very few studies in which the mechanisms of seed abortion have been explored systematically, with experimental tests designed to identify the perpetrators and beneficiaries, involves a tropical tree from central India, *Dalbergia sissoo* (Ganeshaiah and Uma Shaanker 1988a). The seeds are contained within flat, dry pods that also serve as sails during wind dispersal. Initial brood size can be as high as 6 seeds per pod immediately after fertilization (a mean of 4.25 ovules per ovary in week 2), but starts to fall until more than 80% of the pods contain singletons by week 6 (mean of 1.2 seeds per pod). Of the survivors, most (65%) had initially occupied the distal (stigmatic) end of the fruit.

Ganeshaiah and Uma Shaanker (1988a) showed that this is no mere statistical artefact of selective fruit abortion, that is, trees do not drop multi-seeded fruits and retain singletons. Instead, multi-seeded pods *become* single-seeded pods via progressive seed abortion (Figure 16.8) that commences with adjacent seeds and proceeds along a 'dominance gradient' that 'intensifies as the pod ages.' Basal seed failure is caused by unknown water-soluble chemicals that diffuse from the most distal sibling. In controlled laboratory experiments (developing seeds incubated on agar blocks), extract from healthy seeds induced twice the abortion rate produced by extract from aborted seeds, which had no more impact than plain water. Extract from pod tissue also had no more effect than plain water, suggesting that the maternal sporophyte is not the source of the aborting agents. Finally, the basal seeds are definitely viable offspring; when distal seeds were removed surgically (*in situ*), the junior sibs fared well.

Three results suggest fitness benefits to the dominant offspring for *Dalbergia*'s system of slow-acting chemical siblicide. (i) The distal seed achieves a mature seed weight that is about 15% higher than its counterpart in three-seeded pods, which may increase its survivorship (though we know of no data on the mass-survival relationship for this species. (ii) Despite the surviving seed's greater individual weight, after eliminating all pod-mates, its vehicle has a lower wing loading, which probably enables it to disperse further (in tests, these pods fell more slowly from a fixed height and were displaced slightly further laterally than multi-seeded pods). (iii) If it fails to develop properly, one of its viable siblings can take over. Such an insurance benefit is strongly indicated by the third of all sampled pods whose victorious seed came from a non-distal ovule.

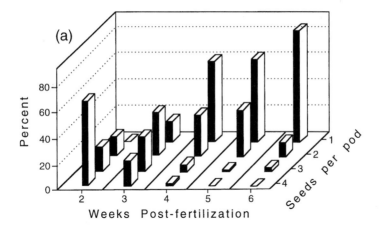

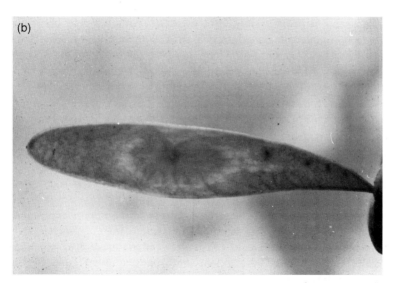

Fig. 16.8 *Dalbergia sissoo* intra-pod chemical siblicide. **(a)** Graph shows the decrease in mean seed number (i.e. brood reduction) in the weeks following fertilization. By week 6, virtually all sibships have only one seed remaining. (Modified from Figure 5 of Ganeshaiah and Uma Shaanker 1988a.) **(b)** Superior growth by the victorious distal seed. (Photograph by K.N. Ganeshaiah, University of Bangalore.)

In their discussion, Ganeshaiah and Uma Shaanker (1988a) placed most emphasis on the sibling rivalry aspects of the *Dalbergia* system, noting that it was 'likely' that the maternal parent benefits as well. However, they went on to consider the possible implications for parent–offspring conflict (see also Uma Shaanker *et al.* 1988), without considering the apparent insurance benefits accruing to both parent and offspring. We suspect parental interests may remain unharmed.

16.5 Long-term competition by sibling sporophytes

Finally, after dispersal and germination the new generation of sporophytes is often in close physical proximity to other propagules from the same maternal parent, especially in taxa with very limited dispersal (e.g. species whose propagules simply drop)[10] or whose seeds travel in sibling groups within multi-seeded fruits (*synaptospermy*) that are usually carried by animals. The effects of the ensuing competitive sibling interactions are ordinarily overlooked by plant demographers in the field, so their general importance remains obscure (Cheplick 1992). One well-studied exception is a North American prairie grass, *Sporobolus vaginiflorus*, that produces many 'full-sibling' seeds (in the botanical sense, a product of self-fertilization within closed cleistogamous flowers) that are never released for dispersal (Cheplick 1992). The parent dies and desiccates, firmly maintaining its hold of the seeds over winter. When they eventually reach the soil in the spring, the siblings germinate at very high local densities, with relatively few landing more than a few centimetres from the (now vanished) parent's centre. As a result, the reproductive scores of full siblings can be compared in relation to their degree of physical crowding in order to obtain a clear measure of sibling competition. Cheplick designated a 'high-density zone' 14 cm in diameter in the centre of the bull's-eye pattern (sibling density was 146 dm^{-2}) and a 'low-density zone' beyond (only 5 siblings dm^{-2}). The relative fitness of the seeds in these zones, expressed as mean chasmogamous (out-breeding) flowers produced by each of them, was five to seven times higher for the uncrowded siblings on the perimeter.

One cannot help wondering whether this strategy fosters unnecessarily high sibling rivalry costs through its astonishing initial maternal over-production of cleistogamous seeds and retention of them in very close quarters. Possibly these offspring are very inexpensive and the whole pattern can be interpreted as low-cost multiple insurance that virtually guarantees the probability of holding the parent's former space against incursions by non-kin.

The relatively unshackled seeds of neotropical nutmeg (*Virola surinamensis*), a canopy tree, may fall directly below the maternal sporophyte, or be carried off by birds (e.g. hornbills, trogons, and guans) and monkeys. The seeds of different individual trees vary threefold in fresh mass (1.3–4 g), a polymorphism that is apparently maintained by shifting selection pressures. Vertebrate dispersers prefer smaller seeds, which are then regurgitated or defecated in pairs or trios that then compete with each other for local space, light and nutrients. Even slight differences in seed size (e.g. 0.2 g) produce

---

[10] In the botanical literature this is sometimes referred to as *barochory*.

measurable differences in seedling shoot mass and height within 15 weeks (Howe and Richter 1982). Among the seeds that remain in the seed shadow, larger size also appears to confer a competitive advantage. Overall, it seems that trees that produce small seeds are better able to colonize distant sites, while large-seeded trees are more likely to pass the parental site on to their own offspring. Intense sibling competition is likely in both contexts (Howe and Vande Kerckhove 1980, 1981, Howe and Richter 1982).

In wild radish, 4 to 8 gravity-dispersed seedlings often emerge in close proximity, a crowding that appears to reduce their size at flowering by a substantial degree (Figure 16.9a). However, less closely related competitors exhibit a

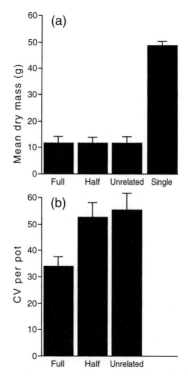

Fig. 16.9 Effects of relatedness on wild radish (*Raphanus sativa*) seedling competition. Seeds were planted in plastic greenhouse pots either in conditions of no local resource competition ('Single') or at densities comparable to that frequently seen in nature (four within a 15 cm diameter). In the group plantings, seeds were either full siblings ('Full', i.e. having both mothers and fathers in common), maternal half-siblings ('Half'), or unrelated ('Unrelated' i.e. having neither parent in common). **(a)** Mean plant size at 12 weeks (when 70% had flowered) was more than three times higher for individuals with no intraspecific competitors, but did not vary between the three competition treatments of differing degrees of relatedness. **(b)** However, variance in offspring size was dramatically influenced by relatedness. (From Karron and Marshall 1990, Figures 1 and 3.)

much greater variance in growth rate (Figure 16.9b), which arises gradually during the seedling period (Karron and Marshall 1990). It may be that the 'half-siblings' (in the zoological sense), being less constrained by inclusive fitness pay-offs, engage in an early scramble competition for local resources (due to genetic differences between them) until a positive feedback mechanism allows one or two to achieve dominance. It is conceivable that this also explains why the maternal sporophyte invests more in (or 'allows' more to be sequestered by) multiply sired fruits (Marshall and Ellstrand 1986). Following a logic akin to that of Trivers and Willard (1973), there may be conditions in which it pays to invest preferentially in high-variance offspring. Other explanations are possible as well.

Long-term sibling competitions are softened in various ways. Production of two or more different kinds of flowers and fruits ('amphicarpy') generally specializes some offspring as dispersers and others as non-dispersers (Cheplick and Quinn 1982, Cheplick 1987). The former ('diaspores') tend to be out-crossed, morphologically equipped for travel (having air-catching features for wind dispersal, fur-catching features for animal dispersal, etc.), and situated high on the plant; the latter are often self-fertilized and may be subterranean.

Seeds that are likely to compete with siblings may differ genetically in ways that affect the timing of germination. For example, buffalo grass (*Buchloe dactyloides*) of the North American plains packs its animal- and water-dispersed burs with 1–5 seeds that germinate at different times, presumably reducing the likelihood of local resource competition with siblings (Quinn and Engel 1986; see also Quinn 1987), while increasing the probability that at least one will succeed by providing variable insurance progeny. After dispersal, the bur faces unpredictable conditions—especially drought and prairie fires—and the asynchronous germination (Figure 16.10) may improve its chances of having one or more seedlings emerge under favourable conditions. The asynchrony also lessens the likelihood of siblings engaging in local resource competition, which appears to be deleterious only under conditions of very low soil moisture. Overall, the advantages of packing (protection by the hull, improved chances of having sexual partners after dispersal) more than outweigh the disadvantages of local competition, which are minimized by asynchrony (Quinn 1987).

Finally, there is a truly enormous literature on *allelopathy*, defined as the production and release into the soil of various chemical agents that affect the growth of other plants in a wide variety of ways (see review by Rice 1984). Most of that attention has been devoted to *interspecific* growth inhibitions (i.e. the so-called 'fairy rings' around mature sporophytes), but there are some very good examples of intraspecific inhibitors as well. In the light of the above arguments, we speculate that the proliferation of some of these compounds might have arisen initially in the context of discouraging sibling competition among seedlings, and we suggest that some of the elegant experimental

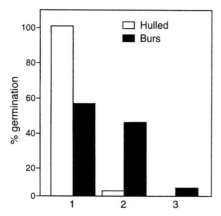

Fig. 16.10 Asynchronous germinations by buffalo-grass burs (with enclosed synaptospermic seeds; filled bars) vs. hulled seeds (open bars). The 'germination periods' are 1-month greenhouse tests, separated by comparable periods of cold temperatures. (From Quinn 1987.)

techniques already designed for exploring interspecific allelopathy be used to test these implications.

Summary

1. Plants over-produce on a scale unknown in nursery taxa of animals, apparently gaining benefits from resource-tracking and from having abundant supplies of 'replacement offspring.' The latter have been interpreted mainly as providing genetic upgrades (developmental selection, progeny choice, etc.), although they clearly also serve as insurance against numerical losses from predators and abiotic threats.
2. The roles of 'sibling', 'offspring' and 'parent' can be envisaged as changing, since there are repeated opportunities to shed tissues and complex incentives to the various players for doing so. Traditionally, flower abortion, seed abortion, and fruit abortion are viewed as purely maternal life-history decisions (as the term *abortion* strongly implies). However, offspring tissues themselves clearly can have both distinct genetic interests (from the maternal sporophyte) and some proximate power (usually as scramble competitions between 'sinks') in these processes. We urge closer attention to the matter of control and beneficiaries in cases where offspring are lost.
3. Some background on plant biology and terminology is provided to assist readers with educational handicaps similar to those of the authors.
4. In angiosperms, a triploid endosperm (nurse tissue) may function as a complex intermediary between the interests of the embryo and the maternal sporophyte. In taxa where neighbouring siblings are unlikely to

have paternal genes in common, the potential for *genomic imprinting* in this context is high.

5. In gymnosperms, ubiquitous *polyembryony* (the production of non-identical embryos within each seed) routinely leads to early sibling fatalities; only one zygote emerges.

6. The developmental selection hypothesis for over-production (defined here as a progeny choice system whereby alternate offspring with less favourable genetic combinations are either screened by the maternal sporophyte, or outcompeted by siblings, or both) is a very active topic in the plant literature. This popularity is fuelled by the paucity of pre-zygotic screening mechanisms (i.e. mate choice behaviour of animals), by the immobility of plants (to which pollen must be delivered by wind or animal vectors), and by the penchant for various forms of inbreeding (which can impose genetic loads). In short, plant zygotes seem likely to have contentious paternal and maternal genomic contributions and/or high levels of inbreeding depression, so an early purging of deleterious combinations is often cost-effective. Some elegant empirical demonstrations of 'selective abortion' are described.

7. Over-production for the purposes of resource-tracking also appears to be widespread. Embryos, seeds, fruits, and seedlings often face critical shortages of essentials, including male gametes in many cases.

8. As in other taxa, the multiple parental benefits of over-production are likely to be mutually compatible and interactive. For example, marginal offspring may prove affordable (if unusually favourable conditions develop), or the same offspring may stand in as replacements if and when various mishaps befall core brood members.

9. Offspring sometimes play highly active roles in eliminating siblings from the nursery. These can take various forms, including the prevention of other zygotes from forming and the use of chemicals to kill already formed seeds within a shared fruit.

10. After dispersal and germination, seedlings may face local resource competition with siblings lasting for many decades. In addition to simple scrambles, these may quite possibly include allelopathic interference.

17

Epilogue

And Joseph dreamed a dream, and he told it his brethren: and they hated him yet
the more.
And he said unto them, Hear, I pray you, this dream which I have dreamed:
For, behold, we were binding sheaves in the field, and, lo, my sheaf arose, and also
stood upright;
and, behold, your sheaves stood round about, and made obeisance to my sheaf.

(*Genesis*, Chapter 37: 5–7)

This book began with the observation that Hamilton's rule initially inspired a
generation of field studies on the apparent cases of altruism among close kin.
The main thrust of our efforts, by contrast, has been to explore how it may
influence the dark side of family interactions. Along the way, it will have
occurred to many readers that virtually every biological system featuring
important contact among close genetic relatives surely contains a shifting
equilibrium between the incentives for selfishness and those for altruism. As
ecological and social conditions change over time, the balance point—or
balance zone, in fact, as it will usually span a range of values—moves toward
greater or lesser co-operation among kin. The next focus, it seems to us, must
be on identifying and testing the factors that set and re-set this zone. That
search will operate on many levels, some relatively 'ultimate' (e.g. modelling
the ecological consequences of one strategy or another) and others quite
'proximate' (e.g. working out which actual cues provide the stimuli on which
behavioural changes depend or the hormonal changes that affect important
perceptual limits).

   In quite a few cases, key factors have already been identified. As we have
noted through these pages, small changes often cause major behavioural
switches. The death of a queen in primitively eusocial monogynous bee
colonies reorders worker priorities. Increases in fairy shrimp numbers and/or
correlated changes in water chemistry apparently notify spadefoot tadpoles of
an approaching crisis. Parasitoids and burying beetles adjust clutch size to
variations in corpse (nursery) size, which may have complex secondary consequences for future sibling dynamics. Plants may assess soil conditions and
discard ill-suited offspring selectively. However, in many other cases this

fuller appreciation of Hamilton's rule remains unrealized. Clearly, there is much work to be done, and there are many opportunities to be seized.

We believe many of the perspectives outlined in this book may prove useful starting points for future searches for evolutionary balance between altruism and selfishness, and we eagerly anticipate more of the accelerating progress already being shown in this whole area.

In the Biblical story, Joseph was hurled into a pit by his brothers, then sold into slavery, both for having such an obnoxious dream and, especially, for sharing it with them. Hamilton's rule is also a sweeping and pervasive vision: it is arguably amongst the most important insights of the 20th century. As we did not discover it, perhaps we are in a more appropriate position to sing its praises. And it makes a suitable parting message for us to note that its application to plant and animal systems is far from finished business.

References

Abugov, R. and R.E. Michod. 1981. On the relation of family-structured models and inclusive fitness models for kin selection. *J. Theor. Biol.* **88:** 743–54.

Aguilera, E. 1990. Parental infanticide by white spoonbills *Platalea leucorodia*. *Ibis* **132:** 124–5.

Ahnesjö, I. 1992a. Consequences of male brood care: weight and number of newborn in a sex-role-reversed pipefish. *Func. Ecol.* **6:** 274–81.

Ahnesjö, I. 1992b. Fewer newborn result in superior juveniles in the paternally brooding pipefish *Syngnathus typhle* L. *J. Fish Biol.* **41(Suppl. B):** 53–63.

Ahnesjö, I. 1996. Apparent resource competition among embryos in the brood patch of a male pipefish. *Behav. Ecol. Sociobiol.* **38:** 167–72.

Alatalo, R.V., Lundberg, A. and K. Stahlbrandt. 1982. Why do pied flycatcher females mate with already-mated males? *Anim. Behav.* **30:** 585–93.

Alatalo, R.V., Gottlander, K. and A. Lundberg. 1988. Conflict or co-operation between parents feeding nestlings in the pied flycatcher *Ficedula hypoleuca* (Pallas). *Ornis Scand.* **19:** 31–4.

Alcock, J., Jones, C.E. and S.L. Buchman. 1977. Male nesting strategies in the bee *Centris pallida* Fox (Anthophoridae: Hymenoptera). *Am. Nat.* **111:** 145–55.

Alexander, R.D. 1974. The evolution of social behavior. *Annu. Rev. Ecol. Syst.* **5:** 325–83.

Alexander, R.D. and P.W. Sherman. 1977. Local mate competition and parental investment in social insects. *Science* **196:** 494–500.

Alexander, R.McN. 1982. *Optima for animals*. Edward Arnold, London.

Allee, W.C., Emerson, A.E., Park, O., Park, T. and K.P. Schmidt. 1949. *The principles of animal ecology*. Saunders, Philadelphia, PA.

Altmann, J., Altmann, S. and G. Hausfater. 1978. Primate infant's effects on mother's future reproduction. *Science* **201:** 1028–30.

Alvarez, F., Arias de Reyna, L. and M. Segura. 1976. Experimental brood parasitism of the magpie (*Pica pica*). *Anim. Behav.* **24:** 907–16.

Amundsen, T. and J.N. Stokland. 1988. Adaptive significance of asynchronous hatching in the shag: a test of the brood reduction hypothesis. *J. Anim. Ecol.* **57:** 329–44.

Amundsen, T. and T. Slagsvold. 1991. Hatching asynchrony: facilitating adaptive or maladaptive brood reduction. *Acta XX Congr. Int. Ornithol.* III: 1707–19.

Anderson, D.J. 1989. The role of hatching asynchrony in siblicidal brood reduction of two booby species. *Behav. Ecol. Sociobiol.* **25:** 363–8.

Anderson, D.J. 1990a. Evolution of obligate siblicide in boobies. I: A test of the insurance egg hypothesis. *Am. Nat.* **135:** 334–50.

Anderson, D.J. 1990b. Evolution of obligate siblicide in boobies. II: Food limitation and parent-offspring conflict. *Evolution* **44:** 2069–82.

Anderson, D.J. 1990c. On the evolution of human brood size. *Evolution* **44:** 438–40.

Anderson, D.J. 1991. Parent blue-footed boobies are not infanticidal. *Ornis Scand.* **22:** 169–70.

Anderson, D.J., Budde, C., Apanius, V., Martinez Gomez, J.E., Bird, D.M., and W.W. Weathers. 1993. Prey size influences female competitive dominance in nestling American kestrels (*Falco sparverius*). *Ecology* **74:** 567–76.

Andersson, M. 1982. Sexual selection, natural selection and quality advertisement. *Biol. J. Linn. Soc.* **17:** 375–93.

Ankney, C.D. 1982. Sex ratio varies with egg sequence in lesser snow geese. *Auk* **99:** 662–6.

Antonovics, J. and J. Schmitt. 1988. Paternal and maternal effects on propagule size in *Anthoxanthum odoratum*. *Oecologia* **69:** 277–82.

Aoki, S. 1977. *Colophinas climatis* (Homoptera: Pemphigidae), an aphid species with 'soldiers'. *Kontyu* **45:** 276–82.

Arbrey, A.N.S. 1990. Wattled cranes. Destruction of wetlands causes decline in population. *Custos* **19:** 24–6. [cited in Simmons 1991, original not consulted].

Armstrong, D.P. and M. Westoby. 1993. Seedlings from large seeds tolerate defoliation better: a test using phylogenetically independent contrasts. *Ecology* **74:** 1092–100.

Baker-Gabb, D.J. 1982. Asynchronous hatching, fratricide and double clutches in the marsh harrier. *Corella* **6:** 83–6.

Balfour, E. 1957. Observations on the breeding of the hen harrier in Orkney. *Bird Notes* **27:** 177–83, 216–224.

Balon, E.K. 1991. Probable evolution of the coelacanth's reproductive style: lecithotrophy and orally feeding embryos in cichlid fishes and in *Latimeria chalumnae*. *Environ. Biol. Fishes* **32:** 249–65.

Bancroft, G.T. 1984. Growth and sexual dimorphism of the boat-tailed grackle. *Condor* **86:** 423–32.

Banks, C.J. 1956. Observations on the behaviour and mortality in Coccinellidae before dispersal from the egg shells. *Proc. R. Entomol. Soc. Lond. (A).* **31:** 56–60.

Barnes, R.D. 1987. *Invertebrate zoology*, 5th edn. Saunders College Publishers, New York.

Barnhart, R. 1988. *The Barnhart dictionary of etymology*. H.W. Wilson, Bronxville, NY.

Bartlett, J. 1987. Filial cannibalism in burying beetles. *Behav. Ecol. Sociobiol.* **21:** 179–83.

Bartlett, J. 1988. Male mating success and parental care in *Nicrophorus vespilloides* (Coleoptera: Silphidae). *Behav. Ecol. Sociobiol.* **23:** 297–303.

Bartlett, J. and C.M. Ashworth. 1988. Brood size and fitness in *Nicrophorus vespilloides* (Coleoptera: Silphidae). *Behav. Ecol. Sociobiol.* **22:** 429–34.

Barton, N.H. and R.J. Post. 1986. Sibling competition and the advantage of mixed families. *J. Theor. Biol.* **120:** 381–7.

Bateson, P. 1994. The dynamics of parent-offspring relationships in mammals. *Trends Ecol. Evolution* **9:** 399–403.

Baur, B. 1990. Possible benefits of egg cannibalism in the land snail *Arianta arbustorum*. *Func. Ecol.* **4:** 679–84.

Baur, B. 1992. Cannibalism in gastropods. In *Cannibalism: ecology and evolution among diverse taxa*, (ed. M.A. Elgar and B.J. Crespi Jr.), pp. 102–27. Oxford University Press, Oxford.

Baur, B. 1993. Intraclutch egg cannibalism by hatchlings of the land snail *Arianta arbustorum*: non-random consumption of eggs? *Ethol. Ecol. Evol.* **5:** 329–36.

Baur, B. and A. Baur 1986. Proximate factors influencing egg cannibalism in the land snail *Arianta arbustorum* (Pulmonata, Helicidae). *Oecologia* **70:** 283–7.

Bawa, K.S. and C.J. Webb. 1984. Flower, fruit and seed abortion in tropical forest trees: implications for the evolution of parental and maternal reproductive patterns. *Am. J. Bot.* **70:** 1031–7.

Bawa, K.S., Hedge, S.G., Ganeshaiah, K.N., and R. Uma Shaanker. 1989. Embryo and seed abortion in plants. *Nature* **342:** 625.

Baylis, J.R. 1981. The evolution of parental care in fishes, with reference to Darwin's rule of male sexual selection. *Environ. Biol. Fishes* **6:** 223–51.

Beatty, R.A. 1972. The genetics of size and shape of spermatozoan organelles. In *The genetics of the spermatozoan* (ed. R.A. Beatty and S. Gluechsohn-Wrelsh), pp. 97–115. Edinburgh University Press, Edinburgh.

Bechard, M.J. 1983. Food supply and the occurrence of brood reduction in Swainson's hawk. *Wilson Bull.* **95:** 233–42.

Bednarz, J.C. and T.J. Hayden. 1991. Skewed brood sex ratio and sex-biased hatching sequence in Harris's hawks. *Am. Nat.* **137:** 116–32.

Beecham, J.J. and M.N. Kochert. 1975. Breeding biology of the golden eagle in southwestern Idaho. *Wilson Bull.* **87:** 506–13.

Beer, J.R., MacLeod. C.F., and L.D. Frenzel. 1957. Prenatal survival and loss in some cricetid rodents. *J. Mammal.* **38:** 392–402.

Begon, M. and G.A. Parker. 1986. Should egg size and clutch size decrease with age? *Oikos* **47:** 293–302.

Beissinger, S.R. and S.H. Stoleson. 1991. Nestling mortality patterns in relation to brood size and hatching asynchrony in green-rumped parrotlets. *Acta XX Congr. Int. Ornithol.* III: 1727–33.

Bekoff, M. 1977. Mammalian dispersal and the ontogeny of individual behavioral phenotypes. *Am. Nat.* **111:** 715–32.

Bekoff, M. 1981. Mammalian sibling interactions. In *Parental care in mammals* (ed. D.J. Gubernick and P.H. Klopfer), pp. 307–46. Plenum, NY.

Bell, G. 1982. *The masterpiece of nature: the evolution and genetics of sexuality*. University of California Press, Berkeley, CA.

Bengtsson, H. and O. Rydén. 1981. Development of parent-young interaction in asynchronously-hatched broods of altricial birds. *Z. Tierpsychol.* **56:** 255–72.

Bengtsson, H. and O. Rydén. 1983. Parental feeding rate in relation to begging behavior in asynchronously hatched broods of the great tit *Parus major. Behav. Ecol. Sociobiol.* **12:** 243–51.

Bent, A.C. 1937. Life histories of north american birds of prey. Part 1. *US Natl. Mus. Bull.* No. 167. Washington, DC.

Berglund, A., Rosenqvist, G., and I. Svensson. 1986. Reversed sex roles and parental energy investment in zygotes of two pipefish (Syngnathidae) species. *Mar. Ecol. Prog. Ser.* **29:** 209–15.

Berglund, A., Magnhagen, C., Bisazza, A., König, B. and F. Huntingford. 1993. Female-female competition over reproduction. *Behav. Ecol.* **4:** 184–7.

Bertin, R.L. 1990. Effects of pollination intensity on *Campsis radicans. Am. J. Bot.* **77:** 178–87.

Birkhead, T.R. and A.P. Møller. 1992. *Sperm competition in birds.* Academic Press, London.

Birney, E.C. and D.D. Baird. 1985. Why do some mammals polyovulate to produce a litter of two? *Am. Nat.* **126:** 136–40.

Black, J.M. (ed.) 1996. *Partnerships in birds: the study of monogamy.* Oxford University Press, Oxford.

Black, J.M. and M. Owen. 1984. The importance of the family unit to barnacle goose offspring: a progress report. *Norsk Polarinstitutt Skrifer* **181:** 79–85.

Black, J.M. and M. Owen. 1987. Determinant factors of social rank in goose flocks: acquisition of social rank in young geese. *Behaviour* **102:** 129–46.

Black, J.M. and M. Owen. 1989a. Parent-offspring relationships in wintering barnacle geese. *Anim. Behav.* **37:** 187–98.

Black, J.M. and M. Owen. 1989b. Agonistic behaviour in goose flocks: assessment, investment, and reproductive success. *Anim. Behav.* **37:** 199–209.

Black, J.M., Choudhury, S., and M. Owen. 1996. Do barnacle geese benefit from life-long monogamy? In *Partnerships in birds: the study of monogamy* (ed. J.M. Black). pp. 91–117. Oxford University Press, Oxford.

Blackburn, D.G. and L.J. Vitt. 1992. Reproduction in viviparous South American lizards of the genus *Mabuya.* In *Reproductive biology of South American vertebrates* (ed. W. Hamlett), pp. 150–64. Springer-Verlag, New York.

Blackburn, D.G., Vitt, L.J., and C.A. Beuchat. 1984. Eutherian-like reproductive specializations in a viviparous reptile. *Proc. Natl. Acad. Sci. USA* **81:** 4860–3.

Blair, W.F. 1976. Adaptation of anurans to equivalent desert scrub of North and South America. In *Evolution of desert biota* (ed. D.W. Goodall), pp. 197–222. University of Texas Press, Austin.

Blaker, D. 1969. Behaviour of the cattle egret *Ardeola ibis. Ostrich* **40:** 75–129.

Blanco, D.E., Yorio, P., and P.D. Boersma. 1996. Feeding behavior, size asymmetry, and food distribution in Magellanic penguin (*Spheniscus magellanicus*) chicks. *Auk* **113:** 496–8.

Blaustein, A.R. and R.K. O'Hara. 1981. Genetic control for sibling recognition? *Nature* **290:** 246–8.

Blaustein, A.R. and R.K. O'Hara. 1982. Kin recognition in *Rana cascadae* tadpoles: maternal and paternal effects. *Anim. Behav.* **30:** 1151–7.

Blaustein, A.R. and B. Waldman. 1992. Kin recognition in anuran amphibians. *Anim. Behav.* **44:** 207–22.

Blaustein, A.R., Bekoff, M., and T.J. Daniels. 1987a. Kin recognition in vertebrates (excluding primates): empirical evidence. In *Kin recognition in animals* (ed. D.J.C. Fletcher and C.D. Michener), pp. 287–331. John Wiley, Chichester.

Blaustein, A.R., Bekoff, M. and T.J. Daniels. 1987b. Kin recognition in vertebrates (excluding primates): mechanisms, function, and future research. In *Kin recognition in animals* (ed. D.J.C. Fletcher and C.D. Michener), pp. 333–57. John Wiley, Chichester.

Blaustein, A.R., Bekoff, M. and T.J. Daniels. 1991. Kin recognition in vertebrates: what do we really know about adaptive value? *Anim. Behav.* **41:** 1079–84.

Blick, J.E. 1977. Selection for traits which lower individual reproduction. *J. Theor. Biol.* **67:** 597–601.

Blumer, L.S. 1979. Male parental care in the bony fishes. *Q. Rev. Biol.* **54:** 149–61.

Boal, C.W. and J.E. Bacorn. 1994. Siblicide and cannibalism in northern goshawk nests. *Auk* **111:** 748–50.

Boersma, P.D. 1991. Asynchronous hatching and food allocation in the Magellanic penguin *Spheniscus magellanicus*. *Acta XX Congr. Int. Ornithol.* II: 961–73.

Boersma, P.D. and P.D. Stokes. 1995. Mortality patterns, hatching asynchrony, and size asymmetry in Magellanic penguin (*Spheniscus magellanicus*) chicks. In *Penguin biology 2* (ed. P. Dann and P. Reilly), pp. 15–43. Surrey Beatty and Sons, NSW, Australia.

Bookman, S.S. 1983. Costs and benefits of flower abscission and fruit abortion in *Asclepias speciosa*. *Ecology* **64:** 264–73.

Bookman, S.S. 1984. Evidence for selective fruit production in *Asclepias*. *Evolution* **38:** 72–86.

Boomsma, J.J. 1991. Adaptive colony sex ratios in primitively eusocial bees. *Trends Ecol. Evolution* **6:** 92–5.

Boomsma, J.J. and A.Grafen. 1990. Intraspecific variation in ant sex ratios and the Trivers-Hare hypothesis. *Evolution* **44:** 1026–34.

Bortolotti, G.R. 1986a. Evolution of growth rates in eagles: sibling competition vs. energy considerations. *Ecology* **67:** 182–94.

Bortolotti, G.R. 1986b. Influence of sibling competition on nestling sex ratios of sexually dimorphic birds. *Am. Nat.* **127:** 495–507.

Bortolotti, G.R. and K.L. Wiebe. 1993. Incubation behaviour and hatching patterns in the American kestrel *Falco sparverius*. *Ornis Scand.* **24:** 41–7.

Bortolotti, G.R., Wiebe, K.L., and W.M. Iko. 1991. Cannibalism of nestling American kestrels by their parents and siblings. *Can. J. Zool.* **69:** 1447–53.

Boulenger, G.A. 1912. Observations sur l'accouplement et la pointe de l'Alyte accoucheur, *Alytes obstetricans*. *Bull. Class. Sci. Acad. Belgique* **1912:** 570–9.

Bragg, A.N. 1956. Dimorphism and cannibalism in tadpoles of *Scaphiopus bombifrons* (Amphibia, Salientia). *Southwest. Nat.* **1:** 105–8.

Bragg, A.N. 1964. Further study of predation and cannibalism in spadefoot tadpoles. *Herpetologica* **20:** 17–24.

Bragg, A.N. 1965. *Gnomes of the night*. University of Pennsylvania Press, Philadelphia, PA.

Braun, B.M. 1981. Siblicide, the mechanism of brood reduction in the black-legged kittiwake *Rissa tridactyla*. Unpublished M.S. Thesis, University of California, Irvine.

Braun, B.M. and G.L. Hunt Jr. 1983. Brood reduction in black-legged kittiwakes. *Auk* **100:** 469–76.

Breden, F. and M.J. Wade. 1985. The effect of group size and cannibalism rate on larval growth and survivorship in *Plagiodera versicolora*. *Entomography* **3:** 455–63.

Brockmann, H.J. 1984. The evolution of social behavior in insects. In *Behavioral ecology. An evolutionary approach* (ed. J.R. Krebs and N.B. Davies), pp. 340–61. Blackwell Scientific Publications, Oxford.

Brockelman, W.Y. 1975. Competition, the fitness of offspring, and optimal clutch size. *Am. Nat.* **109:** 677–99.

Bronson, F.H. 1989. *Mammalian reproductive biology*. University of Chicago Press, Chicago.

Brown, H.D. 1972. The behaviour of newly hatched coccinellid larvae (Coleoptera: Coccinellidae). *J. Entomol. Soc. S. Afr.* **35:** 149–57.

Brown, J.L. 1964. The evolution of diversity in avian territorial systems. *Wilson Bull.* **76:** 160–9

Brown, J.L. 1978. Avian communal breeding systems. *Annu. Rev. Ecol. Syst.* **9:** 123–55.

Brown, J.L. and E.R. Brown. 1981. Kin selection and individual selection in babblers. In *Natural selection and social behavior* (ed. R.D. Alexander and D.W. Tinkle), pp. 244–56. Chiron Press, NY.

Brown, K.M. and R.D. Morris. 1994. The influence of investigator disturbance on the breeding success of ring-billed gulls (*Larus delawarensis*). *Col. Waterbirds* **17:** 7–17.

Brown, L.H., Gargett, V. and P. Steyn. 1977. Breeding success in some African eagles relative to theories about sibling aggression and its effects. *Ostrich* **48:** 65–71.

Bruckhardt, D. and A. Studer-Thiersch. 1970. Über das Zugverhalten der schweizerischen Fischreiher *Ardea cinerea* aufgrund der Beringungsergebnisse. *Ornithol. Beob.* **67:** 230–55.

Bryant, D.M. 1975. Breeding biology of house martins *Delichon urbica* in relation to aerial insect abundance. *Ibis* **117:** 180–216.

Bryant, D.M. 1978a. Establishment of weight hierarchies in the broods of house martins *Delichon urbica*. *Ibis* **120:** 16–26.

Bryant, D.M. 1978b. Environmental influences on growth and survival of nestling house martins *Delichon urbica*. *Ibis* **120:** 271–83.

Bryant, D.M. 1979. Reproductive costs in the house martin (*Delichon urbica*). *J. Anim. Ecol.* **48:** 655–75.

Bryant, D.M. and K.R. Westerterp. 1983. Time and energy limits to brood size in house martins (*Delichon urbica*). *J. Anim. Ecol.* **52:** 905–25.

Bryant, D.M. and P. Tatner. 1990. Hatching asynchrony, sibling competition and siblicide in nestling birds: studies of swiftlets and bee-eaters. *Anim. Behav.* **39:** 657–71.

Buchholz, J.T. 1922. Developmental selection in vascular plants. *Bot. Gaz.* **73:** 249–86.

Bühler, P. 1981. Das Fütterungsverhalten der Schleiereule *Tyto alba*. *Ökol. Vögel* **3:** 183–202.

Bull, J.J. and E.L. Charnov. 1988. How fundamental are Fisherian sex ratios? In *Oxford surveys in evolutionary biology*. Vol. 5 (eds. P.H. Harvey and L. Partridge), pp. 96–135. Oxford University Press, New York.

Bulmer, M.G. 1986. Genetic models of endosperm evolution in higher plants. In *Evolutionary theory* (ed. S. Karlin and E. Nevo), pp. 743–63. Academic Press, London.

Burfening, P.J. 1972. Parental and prenatal competition among twin lambs. *Anim. Prod.* **15:** 61–6.

Busse, K. 1970. Care of the young by male *Rhinoderma darwini*. *Copeia* **1970:** 395.

Bustamente, J., Cuervo, J.J. and J. Moreno. 1992. The function of feeding chases in the chinstrap penguin *Pygoscelis antarctica*. *Anim. Behav.* **44:** 753–9.

Byers, J.A. 1983. Social interactions of juvenile collared peccaries, *Tayassu tajacu* (Mammalia: Artiodactyla). *J. Zool. Lond.* **201:** 83–96.

Byers, J.A. and M. Bekoff. 1991. Development, the conveniently forgotten variable in 'true kin recognition' *Anim. Behav.* **41:** 1088–90.

Caldwell, J.P. 1986. Selection of egg deposition sites: a seasonal shift in the southern leopard frog *Rana sphenocephala*. *Copeia* **1986:** 249–53.

Cameron, G.N. 1973. Effect of litter size on postnatal growth and survival in the desert woodrat. *J. Mammal.* **54:** 489–93.

Campbell, W.R. and K.J. Halama. 1993. Resource and pollen limitation to lifetime seed production in a natural plant population. *Ecology* **74:** 1043–51.

Carl, R.A. 1987. Age-class variation in foraging techniques by brown pelicans. *Condor* **89:** 525–33.

Caro, T.M. 1990. Cheetah mothers bias parental investment in favour of co-operating sons. *Ethol. Ecol. Evolution* **2:** 381–95.

Cash, K. and R.M. Evans. 1986a. Brood reduction in the American white pelican (*Pelecanus erythrorhynchos*). *Behav. Ecol. Sociobiol.* **18:** 413–8.

Cash, K. and R.M. Evans. 1986b. The occurrence, context and functional significance of aggressive begging behaviours in young American white pelicans. *Behaviour* **102:** 119–28.

Casper, B.B. 1984. On the evolution of embryo abortion in the herbaceous perennial *Cryptantha flava*. *Evolution* **38:** 1337–49.

Casper, B.B. 1990. Seedling establishment from one- and two-seeded fruits of *Cryptantha flava*: a test of parent-offspring conflict. *Am. Nat.* **136:** 167–77.

Casper, B.B. and B.W. Grant. 1988. Morphology and dispersal of one- and two-seeded diaspores of *Cryptantha flava*. *Am. J. Bot.* **75:** 859–63.

Charlesworth, B. 1978. Some models of the evolution of altruistic behaviour between siblings. *J. Theor. Biol.* **72:** 297–319.

Charlesworth, B. 1980. *Evolution in age-structured populations*. Cambridge University Press, Cambridge.

Charlesworth, B. and E.L. Charnov. 1980. Kin selection in age-structured populations. *J. Theor. Biol.* **88:** 103–19.

Charlesworth, D. 1989a. Why do plants produce so many more ovules than seeds? *Nature* **338:** 21–2.

Charlesworth, D. 1989b. Embryo and seed abortion in plants. *Nature* **342:** 625–6.

Charlesworth, D., Schemske, D.W. and V.L. Sork. 1987. The evolution of plant reproductive characters; sexual versus natural selection. In *The evolution of sex and its consequences* (ed. S.C. Stearns), pp. 317–35. Birkhauser Verlag, Basel.

Charnov, E.L. 1976. Optimal foraging, the marginal value theory. *Theor. Popul. Biol.* **9:** 129–36.

Charnov, E.L. 1977. An elementary treatment of the genetical theory of kin selection. *J. Theor. Biol.* **66:** 541–50.

Charnov, E.L. 1979. Simultaneous hermaphroditism and sexual selection. *Proc. Natl. Acad. Sci. USA* **76:** 2480–4.

Charnov, E.L. 1982. Parent-offspring conflict over reproductive effort. *Am. Nat.* **119:** 736–7.

Charnov, E.L. and J. Krebs. 1974. On clutch-size and fitness. *Ibis* **116:** 217–19.

Charnov, E.L. and S. Skinner. 1984. Evolution of host selection and clutch size in parasitoid wasps. *Florida Entomol.* **67:** 5–21.

Chase, I.D. 1980. Cooperative and noncooperative behavior in animals. *Am. Nat.* **115:** 827–57.

Cheplick, G.P. 1987. The ecology of amphicarpic plants. *Trends Ecol. Evolution* **2:** 97–101.

Cheplick, G.P. 1992. Sibling competition in plants. *J. Ecol.* **80:** 567–75.

Cheplick, G.P. and J.A. Quinn. 1982. *Amphicarpum pushii* and the 'pessimistic strategy in amphicarpic annuals with subterranean fruit. *Oecologia* **52:** 327–32.

Choi, I.-H. and G.S. Bakken. 1990. Begging response in nestling red-winged blackbirds (*Agelaius phoeniceus*): effect of body temperature. *Physiol. Zool.* **63:** 965–86.

Clark, A.B. 1978. Sex ratio and local resource competition in a prosimian primate. *Science* **201:** 163–5.

Clark, A.B. and D.S. Wilson. 1981. Avian breeding adaptations: hatching asynchrony, brood reduction and nest failure. *Q. Rev. Biol.* **56:** 253–77

Clark, A.B. and D.S. Wilson. 1985. The onset of incubation in birds. *Am. Nat.* **125:** 603–11.

Clark, M.M. and B.G. Galef Jr. 1988. Effects of uterine position on rate of sexual development in female Mongolian gerbils. *Physiol. Behav.* **42:** 15–8.

Clark, M.M., Malenfant, S.A., Winter, D.A. and B.G. Galef Jr. 1990. Fetal uterine position affects copulation and scent marking by adult male gerbils. *Physiol. Behav.* **47:** 301–5.

Clark, M.M., Tucker, L. and B.G. Galef Jr. 1992. Stud males and dud males: intra-uterine position effects on the reproductive success of male gerbils. *Anim. Behav.* **43:** 215–21.

Clark, M.M., Karpluk, P. and B.G. Galef Jr. 1993. Hormonally mediated inheritance of acquired characteristics in Mongolian gerbils. *Nature* **364:** 712.

Clutton-Brock, T.H. 1986. Sex ratio variation in birds. *Ibis* **128:** 317–29.

Clutton-Brock, T.H. 1991. *The evolution of parental care.* Princeton University Press, Princeton, NJ.

Clutton-Brock, T.H. and G.A. Parker. 1995. Punishment in animal societies. *Nature* **373:** 209–16.

Clutton-Brock, T.H., Albon, S.D. and F.E. Guinness. 1981. Parental investment in male and female offspring in polygynous mammals. *Nature* **289:** 487–9.

Clutton-Brock, T.H., Guinness, F.E. and S.D. Albon. 1982a. *Red deer. behavior and ecology of two sexes.* University of Chicago Press, Chicago.

Clutton-Brock, T.H., Albon, S.D. and F.E. Guinness. 1982b. Competition between female relatives in a matrilocal mammal. *Nature* **300:** 178–80.

Clutton-Brock, T.H., Albon, S.D. and F.E. Guinness. 1985. Parental investment and sex differences in juvenile mortality in birds and mammals. *Nature* **313:** 131–3.

Clutton-Brock, T.H., Albon, S.D. and F.E. Guinness. 1989. Fitness costs of gestation and lactation in wild mammals. *Nature* **337:** 260–2.

Cockburn, A. 1989. Adaptive patterns in marsupial reproduction. *Trends Ecol. Evolution* **4:** 126–30.

Cockburn, A. 1994. Adaptive sex allocation by brood reduction in antechinuses. *Behav. Ecol. Sociobiol.* **35:** 53–62.

Cockburn, A., Scott, M.P. and C.R. Dickman. 1985. Sex ratio and intrasexual kin competition in mammals. *Oecologia* **66:** 427–9.

Cohen, J. 1969. Is sexual reproduction wasteful? *New Sci.* **44:** 282–5.

Cohen Fernandez, E.J. 1988. La reducci wasteful? Sex ratio and intrase*Sula leuco-gaster nesiotes*, Heller and Snodgrass 1901). Tesis de Licenciatura. Universidad Nacional Auton competition in mammals. ld mammals. Collins, J.P. and J.E. Cheek. 1983. Effect of food and density on development of typical and cannibalistic salamander larvae in *Ambystoma tigrinum nebulosum. Am. Zool.* **23:** 77–84.

Collins, J.P., Mitton, J.P. and B.A. Pierce. 1980. *Ambystoma tigrinum*: a multispecies conglomerate? *Copeia* **1980:** 938–41.

Cook, R.E. 1981. Plant parenthood. *Nat. Hist.* **90:** 30–5.

Cooke, F. and R. Harmsen. 1983. Does sex ratio vary with egg sequence in lesser snow geese? *Auk* **100:** 215–17.

Cooper, J. 1980. Fatal sibling aggression in pelicans-a review. *Ostrich* **51:** 173–86.

Corben, C.J., Ingram, J.G. and M.J. Tyler. 1974. Gastric brooding: a unique form of parental care in an Australian frog. *Science* **186:** 946–7.

Cotton, P.A., Kacelnik, A. and J. Wright. 1996. Chick begging as a signal: are nestlings honest? *Behav. Ecol.* **7:** 178–82.

Cramer, C.P. and E.M. Blass. 1983. Mechanisms of control of milk intake in suckling rats. *Am. J. Physiol.* **245:** R154–9.

Cramp, S and K.E.L. Simmons. 1977. *Handbook of the birds of Europe, the Middle East, and North Africa. The birds of the Western Palearctic.* Vol. I. *Ostrich to ducks.* Oxford University Press, Oxford.

Creel, S.R. and N.M. Creel. 1990. Energetics, reproductive suppression and obligate communal breeding in carnivores. *Behav. Ecol. Sociobiol.* **28:** 263–70.

Creel, S.R. and P.M. Waser. 1991. Failure of reproductive suppression in dwarf mongooses: accident of adaptation? *Behav. Ecol.* **2:** 7–16.

Creel, S.R., Creel, N., Wildt, D.E. and S.L. Monfort. 1992. Behavioural and endocrine mechanisms of reproductive suppression in Serengeti dwarf mongooses. *Anim. Behav.* **43:** 231–45.

Creighton, J.C. and G.D. Schnell. 1996. Proximate control of siblicide in cattle egrets: a test of the food amount hypothesis. *Behav. Ecol. Sociobiol.* **38:** 371–8.

Crespi, B.J. 1992. Cannibalism and trophic eggs in subsocial and eusocial insects. In *Cannibalism: ecology and evolution among diverse taxa* (ed. M.A. Elgar and B.J. Crespi), pp. 176–213. Oxford University Press, Oxford.

Crick, H.Q.P. 1992. Load-lightening in cooperatively breeding birds and the cost of reproduction. *Ibis* **134:** 56–61.

Cronmiller, J.R. and C.F. Thompson 1980. Experimental manipulations of brood size in red-winged blackbirds. *Auk* **97:** 559–65.

Crossner, K.A. 1977. Natural selection and clutch size in the European starling. *Ecology* **58:** 885–92.

Cruden, R.W. and K.G. Jensen. 1979. Viscin threads, pollination efficiency and low pollen-ovule ratios. *Am. J. Bot.* **66:** 875–9.

Crump, M.L. 1983. Opportunistic cannibalism by amphibian larvae in temporary aquatic environments. *Am. Nat.* **121:** 281–7.

Crump, M.L. 1992. Cannibalism in amphibians. *Cannibalism: ecology and evolution among diverse taxa* (ed. M.A. Elgar and B.J. Crespi Jr.), pp. 256–76. Oxford University Press, Oxford.

Cruz, Y.P. 1981. A sterile defender morph in a polyembryonic hymenopterous parasite. *Nature* **294:** 446–7.

Cruz, Y.P. 1986. The defender role of the precocious larvae of *Copidomopsis tanytmenus* Caltagirone (Encyrtidae, Hymenoptera). *J. Exp. Zool.* **237:** 309–18.

Custer, T.W. and P.C. Frederick. 1990. Egg size and laying order of snowy egrets, great egrets, and black-crowned night-herons. *Condor* **92:** 772–5.

Cutts, J.H., Krause, W.J. and C.R. Leeson. 1978. General observations on the growth and development of the young pouch opossum, *Didelphis virginiana*. *Biol. Neonate* **33:** 264–72.

Darwin, C. 1876. *Cross and self fertilisation in the vegetable kingdom*. John Murray, London.

David, S. and M. Berrill. 1987. Siblicidal attacks by great blue heron, *Ardea herodias*, chicks in a southern Ontario heronry. *Can. Field-Natur.* **101:** 105–7.

Davies, N.B. 1976. Parental care and the transition to independent feeding in the young spotted flycatcher. *Behaviour* **59:** 280–95.

Davies N.B. 1978. Parental meanness and offspring independence: an experiment with hand-reared great tits *Parus major*. *Ibis* **120:** 509–14

Davies, S.J.J.F. 1963. Aspects of the behaviour of the magpie goose *Anseranas semipalmata*. *Ibis* **105:** 76–98.

Dawkins, R. 1976. *The selfish gene*. Oxford University Press, London.

Dawkins, R. 1979. Twelve misunderstandings of kin selection. *Zt. Tierpsychol.* **51:** 184–200.

Day, C.S.D. and B.G. Galef Jr. 1977. Pup cannibalism: one aspect of maternal behavior in golden hamsters. *J. Comp. Physiol. Psych.* **91:** 1179–89.

Desgranges, J.-L. 1981. Observations sur l'alimentation du grand héron *Ardea herodias* au Québec (Canada). *Alauda* **49:** 25–34.

Desrochers, B.A. and C. D. Ankney. 1986. Effect of brood size and age on the feeding behavior of adult and juvenile American coots (*Fulica americana*). *Can J. Zool.* **64:** 1400–6.

Dickins, D.W. and R.A. Clark. 1987. Games theory and siblicide in the kittiwake gull, *Rissa tridactyla*. *J. Theor. Biol.* **125:** 301–5

Dickinson, J.L. 1992. Egg cannibalism by larvae and adults of the milkweed leaf beetle (*Labidomera clivicollis*, Coleoptera: Chrysomelidae). *Ecol. Entomol.* **17:** 209–18.

Diesel, R. 1989. Parental care in an unusual environment: *Metopaulias depressus* (Decapoda: Grapsidae), a crab that lives in epiphytic bromeliads. *Anim. Behav.* **38:** 561–75.

Diesel, R. 1992a. Managing the offspring environment: brood care in the bromeliad crab, *Metopaulias depressus*. *Behav. Ecol. Sociobiol.* **30:** 125–34.

Diesel, R. 1992b. Maternal care in the bromeliad crab, *Metopaulias depressus*: protection of larvae from predation by damselfly larvae. *Anim. Behav.* **43:** 803–12.

Dijkstra, C., Bult, A., Bijlsma, S., Daan, S., Meijer, T. and M. Zijlstra. 1990. Brood size manipulations in the kestrel (*Falco tinnunculus*): effects on offspring and parent survival. *J. Anim. Ecol.* **59:** 269–86.

Dominey, W. and L. Blumer. 1984. Cannibalism of early life stages in fishes. In *Infanticide: comparative and evolutionary perspectives* (ed. G. Hausfater and S.B. Hrdy), pp. 43–64. Aldine, NY.

Donnelly, M.A. 1989. Demographic efects of reproductive resource supplementation in a territorial frog, *Dendrobates pumilio*. *Ecol. Monogr.* **59:** 207–21.

Dopazo, H. and P. Alberch. 1994. Preliminary results on optional viviparity and intrauterine siblicide in *Salamandra salamandra* populations from northern Spain. *Mertensiella* **4:** 125–38.

Dorward, E.F. 1962. Comparative biology of the white booby and brown booby *Sula* spp. at Ascension. *Ibis* **103b:** 174–220.

Drent, R.H. and S. Daan. 1980. The prudent parent: energetic adjustments in avian breeding. *Ardea* **68:** 225–52.

Drewett, R.F. 1983. Sucking, milk synthesis, and milk ejection in the Norway rat. In *Parental behaviour of rodents.* (ed. R.W. Elwood), pp.181–203. John Wiley, Chichester.

Drummond, H. 1987. A review of parent-offspring conflict and brood reduction in the Pelecaniformes. *Col. Waterbirds* **10:** 1–15.

Drummond, H. 1989. Parent-offspring conflict and brood reduction in boobies. *Proc. XIX Int. Ornithol. Congr.* Pp. 1244–53. University of Ottawa Press, Ottawa.

Drummond, H. 1993. Have avian parents lost control of offspring aggression? *Etología* **3:** 187–98.

Drummond, H. and C. Garcia Chavelas. 1989. Food shortage influences sibling aggression in the blue-footed booby. *Anim. Behav.* **37:** 806–19.

Drummond, H. and J.L. Osorno. 1992. Training siblings to be submissive losers: dominance between booby nestlings. *Anim. Behav.* **44:** 881–93.

Drummond, H., Gonzalez, E. and J.L. Osorno. 1986. Parent-offspring co-operation in the blue-footed booby, *Sula nebouxii. Behav. Ecol. Sociobiol.* **19:** 365–392.

Drummond, H., Osorno, J.L., Torres, R., Garcia Chavelas, C. and H.M. Larios. 1991. Sexual size dimorphism and sibling competition: implications for avian sex ratios. *Am. Nat.* **138:** 623–41

Duellman, W.E. and S.J. Maness. 1980. The reproductive behavior of some hylid marsupial frogs. *J. Herpetol.* **14:** 213–22.

Duellman, W.E. and L. Trueb. 1986. *Biology of amphibians.* McGraw-Hill, NY.

Dunbar, R.I.M. 1984. The ecology of monogamy. *New Sci.* **103:** 12–5.

Dusi, J.G. 1967. Migration in the little blue heron. *Wilson Bull.* **79:** 223–35.

Dziuk, P.J. 1977. Reproduction in pigs. In *Reproduction in domestic animals* (ed. H.H. Cole and P.T. Cupps), pp. 445–74. Academic Press, New York.

Dziuk, P.J. 1985. Effect of migration, distribution and spacing of pig embryos on pregnancy and fetal survival. *J. Reprod. Fert., Suppl.* **33:** 57–63.

Dziuk, P.J. 1987. Embryonic loss in the pig: an enigma. In *Manipulating pig production.* (ed. Australasian Pig Science Association), pp. 28–39. Proc. Conf. Australasian Pig Science Association.

Edwards, T.C. Jr 1989. Similarity in the development of foraging mechanics among sibling ospreys. *Condor* **91:** 30–6.

Edwards, T.C. Jr. and M.W. Collopy. 1983. Obligate and facultative brood reduction in eagles: an examination of factors that influence fratricide. *Auk* **100:** 630–5.

Edwards, T.C. Jr, Collopy, M.W., Steenhof, K. and M.N. Kochert. 1988. Sex ratios of fledgling golden eagles. *Auk* **105:** 793–6.

Egoscue, H.J. 1962. The bushy-tailed wood rat: a laboratory colony. *J. Mammal.* **43:** 328–36.

Ehrlén, J. 1991. Why do plants produce surplus flowers? A reserve-ovary model. *Am. Nat.* **138:** 918–933.

Eickwort, K. 1973. Cannibalism and kin selection in *Labidomera clivicollis* (Coleoptera, Chrysomelidae). *Am. Nat.* **107:** 452–3.

Eisenberg, J.F. 1981. *The mammalian radiations.* University of Chicago Press, Chicago, IL.

Elgar, M.A. and B.J. Crespi. (ed.) 1992. *Cannibalism: ecology and evolution among diverse taxa.* Oxford University Press, Oxford.

Elwood, R.W. 1992. Pup-cannibalism in rodents: causes and consequences. In *Cannibalism: ecology and evolution among diverse taxa* (ed. M.A. Elgar and B.J. Crespi), pp. 299–322. Oxford University Press

Emlen, S.T. 1991. Evolution of co-operative breeding in birds and mammals. In *Behavioural ecology. An evolutionary approach* 3rd edn (ed. J.R. Krebs and N.B. Davies), pp. 301–37. Blackwell Scientific Publications, Oxford.

Emlen, S.T. 1995. An evolutionary theory of the family. *Proc. Natl. Acad. Sci. USA* **92:** 8092–9.

Emlen, S.T. and P.H. Wrege. 1992. Parent-offspring conflict and the recruitment of helpers among bee-eaters. *Nature* **356:** 331–3.

Emlen, S.T., P.H. Wrege, and N.J. Demong. 1995. Making decisions in the family: an evolutionary perspective. *Am. Sci.* **83:** 148–57.

Enquist, M. 1985. Communication during aggressive interactions with particular reference to variation in choice of behaviour. *Anim. Behav.* **33:** 1152–61.

Enquist, M. and O. Leimar. 1983. Evolution of fighting behaviour: decision rules and assessment of relative strength. *J. Theor. Biol.* **102:** 387–410.

Enquist, M. and O. Leimar. 1990. The evolution of fatal fighting. *Anim. Behav.* **39:** 1–9.

Eshel, I. 1983. Evolutionary and continuous stability. *J. Theor. Biol.* **103:** 99–112.

Eshel, I. and M.W. Feldman. 1991. The handicap principle in parent-offspring conflict: comparison of optimality and population genetic analyses. *Am. Nat.* **137:** 167–85.

Eshel, I. and U. Motro. 1981. Kin selection and strong evolutionary stability of mutual help. *Theor. Popul. Biol.* **19:** 420–33.

Eshel, I. and U. Motro. 1988. The three brothers' problem: kin selection with more than one potential helper. 1. The case of immediate help. *Am. Nat.* **132:** 550–66.

Evans, R.M. 1984. Some causal and functional correlates of creching in young white pelicans. *Can. J. Zool.* **62:** 814–19.

Evans, R.M. 1990. The relationship between parental input and investment. *Anim. Behav.* **39:** 797–813.

Evans, R.M. 1996. Hatching asynchrony and survival of insurance offspring in an obligate brood reducing species, the American white pelican. *Behav. Ecol. Sociobiol.* **39:** 203–9.

Evans, R.M. 1997. Parental investment and quality of insurance offspring in an obligate brood reducing species, the American white pelican. *Behav. Ecol.* **8:** 378–83.

Evans, R.M. and B.F. MacMahon. 1987. Within-brood variation growth and conditions in relation to brood reduction in the American white pelican. *Wilson Bull.* **99:** 190–201.

Evans, R.M., Wiebe, M.O., Lee, S.C. and S.C. Bugden. 1995. Embryonic and parental preferences for incubation temperature in herring gulls: implications for parent-offspring conflict. *Behav. Ecol. Sociobiol.* **36:** 17–23.

Ewer, R.F. 1961. Further observations on suckling behaviour in kittens, together with some general considerations of the interrelations of innate and acquired responses. *Behaviour* **17:** 247–60.

Feldman, M.W. and I. Eshel. 1982. On the theory of parent-offspring conflict: a two-locus genetic model. *Am. Nat.* **119:** 285–92.

Fernandez, G.W. and T.G. Whitham. 1989. Selective fruit abscission by *Juniperus monosperma* as an induced defense against predators. *Am. Midl. Nat.* **121:** 389–92.

Festa-Bianchet, M. and W.J. King. 1991. Effects of litter size and population dynamics on juvenile and maternal survival in Columbian ground squirrels. *J. Anim. Ecol.* **60:** 1077–90.

Fincke, O.M. 1992a. Consequences of larval ecology for territoriality and reproductive success of a neotropical damselfly. *Ecology* **73:** 449–62.

Fincke, O.M. 1992b. Interspecific competition for tree holes: consequences for mating system and coexistence in neotropical damselflies. *Am. Nat.* **139:** 80–101.

Fincke, O.M. 1994. Population regulation of a tropical damselfly in the larval stage by food limitation, cannibalism, intraguild predation and habitat drying. *Oecologia* **100:** 118–27.

Fisher, R.A. 1930. *The genetical theory of natural selection*. Clarendon Press, Oxford.

Fisher, R.A. 1958. *The genetical theory of natural selection*. Revised second edition. Dover, London.

Fisher, R.C. 1961. A study in insect multiparasitism. II. The mechanism and control of competition for the host. *J. Exp. Biol.* **38:** 605–28.

FitzGerald, G.J. 1992. Filial cannibalism in fishes: why do parents eat their offspring? *Trends in Ecol. and Evol.* **7:** 7–10.

FitzGerald, G.J. and F.G. Whoriskey. 1992. Empirical studies of cannibalism in fish. In *Cannibalism: ecology and evolution among diverse taxa* (ed. M.A. Elgar and B.J. Crespi), pp. 238–55. Oxford University Press.

Fletcher, D.J.C. and C.D. Michener (eds.) 1987. *Kin recognition in animals*. John Wiley, New York.

Foelix, R.F. 1982. *Biology of spiders*. Harvard University Press, Cambridge, MA.

Forbes, L.S. 1990. Insurance offspring and the evolution of avian clutch size. *J. Theor. Biol.* **147:** 345–59.

Forbes, L.S. 1991a. Optimal offspring size and number in a variable environment. *J. Theor. Biol.* **150:** 299–304.

Forbes, L.S. 1991b. Burgers or brothers: Food shortage and the threshold for brood reduction. *Acta XX Congr. Int. Ornithol.* III:1720–6.

Forbes, L.S. 1991c. Insurance offspring and brood reduction in a variable environment: the costs and benefits of pessimism. *Oikos* **62:** 325–32.

Forbes, L.S. 1991d. Hunger and food allocation among nestlings of facultatively siblicidal ospreys. *Behav. Ecol. Sociobiol.* **29:** 189–95.

Forbes, L.S. 1993. Avian brood reduction and parent-offspring 'conflict'. *Am. Nat.* **142:** 82–117.

Forbes, L.S. and R.C. Ydenberg. 1992. Sibling rivalry in a variable environment. *Theor. Popul. Biol.* **41:** 335–60.

Forbes, L.S. and T.C. Lamey. 1996. Insurance, developmental accidents, and the risks of putting all your eggs in one basket. *J. Theor. Biol.* **180:** 247–56.

Forbes, L.S. and D.W. Mock. 1996. Food, information, and avian brood reduction. *Écoscience* **3:** 45–53.

Forbes, M.R.L. and C.D. Ankney. 1987. Hatching asynchrony and food allocation within broods of pied-billed grebes *Podilymbus podiceps*. *Can. J. Zool.* **65:** 2872–7.

Forester, D.C. 1979. The adaptiveness of parental care in *Desmognathus ochrophaeus* (Urodela: Plethodontidae). *Copeia* **1979:** 332–41.

Fox, L.R. 1975. Cannibalism in natural populations. *Annu. Rev. Ecol. Syst.* **6:** 87–106.

Frame, L.H. and G.W. Frame. 1976. Female African wild dogs emigrate. *Nature* **263:** 227–9.

Frame, L.H., Malcolm, J.R., Frame, G.W., and H. van Lawick. 1979. Social organization of African wild dogs (*Lycaon pictus*) on the Serengeti Plains, Tanzania, 1967–1978. *Z. Tierpsychol.* **50:** 225–49.

Frank, L.G., Glickman, S.E. and P. Licht. 1991. Fatal sibling aggrtession, precocial development, and androgens in neonatal spotted hyenas. *Science* **252:** 702–4.

Frase, B.A. and K.B. Armitage. 1984. Foraging patterns of yellow-bellied marmots: role of kinship and individual variability. *Behav. Ecol. Sociobiol.* **16:** 1–10.

Fraser, D. 1990. Behavioural perspectives on piglet survival. *J. Reprod. Fert.,* Suppl. **40:** 355–70.

Fraser, D. and B.K. Thompson. 1990. Armed sibling rivalry among piglets. *Behav. Ecol. Sociobiol.* **29:** 9–15,

Friedl, T.W.P. 1993. Intraclutch egg-mass variation in geese: a mechanism for brood reduction in precocial birds? *Auk* **110:** 129–32.

Friedman, W.E. 1990. Double fertilization in *Ephedra*, a nonflowering seed plant: its bearing on the origin of angiosperms. *Science* **247:** 951–4.

Friedman, W.E. 1992. Evidence for a pre-angiosperm origin of endosperm: implications for the evolution of flowering plants. *Science* **255:** 336–9.

Fuchs, S. 1982. Optimality of parental investment: the influence of nursing on reproductive success of mother and female young house mice. *Behav. Ecol. Sociobiol.* **10:** 39–51.

Fujimaki, Y. 1981. reproductive activity in *Cleithrionomys rufocanus bedfordiae*. 4. Number of embryos and prenatal mortality. *Jap. J. Ecol.* **31:** 247–56.

Fujioka, M. 1984. Asychronous hatching, growth, and survival of chicks in the cattle egret *Bubulcus ibis*. *Tori* **33:** 1–12.

Fujioka, M. 1985a. Food delivery and sibling competition in experimentally even-aged broods of the cattle egret. *Behav. Ecol. Sociobiol.* **17:** 67–74.

Fujioka, M. 1985b. Sibling competition and siblicide in asynchronously-hatching broods of the cattle egret *Bubulcus ibis*. *Anim. Behav.* **33:** 1228–42.

Fujioka, M. and S. Yamagishi. 1981. Extramarital and pair copulations in the cattle egret. *Auk* **98:** 134–44.

Gabriel, W.J. 1967. Reproductive behaviour in sugar maple: Self compatibility, cross compatibility, agamospermy and agamocarpy. *Silvae Genet.* **16:** 165–8.

Galef, B.G., Jr 1981. The ecology of weaning. In *Paternal care in mammals* (ed. D.J. Gubernick), pp. 211–41. Plenum Press, New York.

Galef, B.G., Jr. 1983. Costs and benefits of mammalian reproduction. In *Symbiosis in parent-offspring interactions* (ed. L.A. Rosenblum and H. Moltz), pp. 249–77. Plenum Press, New York.

Gandelman, R. and N.G. Simon. 1978. Spontaneous pup-killing by mice in response to large litters. *Dev. Psychobiol.* **11:** 235–41.

Gandelman, R., vom Saal, F.S. and J.M. Reinisch. 1977. Contiguity to male foetuses affects morphology and behaviour of female mice. *Nature* **266:** 722–4.

Ganeshaiah, K.N. and R. Uma Shaanker. 1988a. Seed abortion in wind-dispersed pods of *Dalbergia sissoo*: maternal regulation or sibling rivalry? *Oecologia* **75:** 135–9.

Ganeshaiah, K.N. and R. Uma Shaanker. 1988b. Regulation of seed number and female incitation of mate competition by a pH dependent proteinaceious inhibitor of pollen grain germination in *Leucaena leucocephala*. *Oecologia* **75:** 110–3.

Ganeshaiah, K.N. and R. Uma Shaanker. 1992. Frequency distribution of seed number per fruit in plants: A consequence of the self-organizing process? *Curr. Sci.* **63:** 359–65.

Ganeshaiah, K.N., Uma Shaanker, R. and G. Shivashanker. 1986. Stigmatic inhibition of pollen grain germination-its implication for frequency distribution of seed number in pods of *Lucaena leucocephala* (Lam) de Wit. *Oecologia* **70:** 568–72.

Gargett, V. 1967 Black eagle experiment. *Bokmakierie* **19:** 88–90

Gargett, V. 1968. Two Wahlberg's eagle chicks-a one in forty eight chance. *Honeyguide* **56:** 24.

Gargett, V. 1970a. Black eagle experiment no. 2. *Bokmakierie* **22:** 32–5.

Gargett, V. 1970b. The Cain and Abel conflict in the augur buzzard. *Ostrich* **41:** 256–7.

Gargett, V. 1977. A 13-year population study of the black eagles in the Matopos, Rhodesia, 1964–1976. *Ostrich* **48:** 17–27.

Gargett, V. 1978. Sibling aggression in the black eagle in the Matopos, Rhodesia. *Ostrich* **49:** 57–63.

Gargett, V. 1990. *The black eagle*. Acorn Books. Johannesburg.

Gargett, V. 1991. The two-egg-one-chick syndrome-again. *Ostrich* **62:** 92–3.

Garnett, M.C. 1981. Body size, its heritability and influence on juvenile survival among great tits, *Parus major. Ibis* **123:** 31–41.

Garrison, W.J. and C. K. Augspurger. 1983. Double- and single-seeded acorns of bur oak (*Quercus macrocarpa*): frequency and some ecological consequences. *Bull. Torrey Bot. Club* **110:** 154–60.

Garshelis, D.L. and J.A. Garshelis. 1987. Atypical pup rearing strategies by sea otters. *Mar. Mamm. Sci.* **3:** 263–70

Gemmell, R.T. 1982. Breeding bandicoots in Brisbane (*Isodon macrourus;* Marsupialia, Peramelidae). *Austr. Mammal.* **5:** 187–93.

Gerrard, J.M. and G.R. Bortolotti. 1988. *The bald eagle*. Smithsonian Institution Press, Washington, DC.

Gibbons, D. 1987. Hatching asynchrony reduces prental investment in the jackdaw. *J. Anim. Ecol.* **56:** 403–14.

Gilbert, A.N. 1986. Mammary number and litter size in rodentia: the 'one-half rule.' *Proc. Natl. Acad. Sci. USA* **83:** 4828–30.

Gilmore, R.G. 1993. Reproductive biology of lamnoid sharks. *Environ. Biol. Fishes* **38:** 95–114.

Gilmore, R.G., Dodrill, J.W. and P.A. Linley. 1983. Reproduction and development of the sand tiger shark, *Odontapsis taurus* (Rafinesque). *Fishery Bull.* **81:** 201–25.

Gleeson, S.K., Clark, A.B. and L.A. Dugatkin. 1994. Monozygotic twinning: an evolutionary hypothesis. *Proc. Natl. Acad. Sci. USA* **91:** 11363–7.

Godfray, H.C.J. 1986. Models for clutch size and sex ratio with sibling interaction. *Theor. Popul. Biol.* **30:** 215–31.

Godfray, H.C.J. 1987a. The evolution of clutch size in parasitic wasps. *Am. Nat.* **129:** 221–233.

Godfray, H.C.J. 1987b. The evolution of clutch size in invertebrates. In *Oxford surveys in evolutionary biology* (ed. P.H. Harvey and L. Partridge), pp. 117–54. Oxford University Press, New York.

Godfray, H.C.J. 1991. The signalling of need by offspring to their parents. *Nature* **353:** 328–30.

Godfray, H.C.J. 1994. *Parasitoids. Behavioral and evolutionary ecology*. Princeton University Press, Princeton, NJ.

Godfray, H.C.J. 1995a. Signalling of need between parents and young: Parent-offspring conflict and sibling rivalry. *Am. Nat.* **146:** 1–24.

Godfray, H.C.J. 1995b. Evolutionary theory of parent-offspring conflict. *Nature* **376:** 1133–8.

Godfray, H.C.J. and A.B. Harper. 1990. The evolution of brood reduction by siblicide in birds. *J. Theoret. Biol.* **145:** 163–75.

Godfray, H.C.J. and G.A. Parker. 1991. Clutch size, fecundity and parent-offspring conflict. *Proc. R. Soc. Lond. B* **332:** 67–79.

Godfray, H.C.J. and G.A. Parker. 1992. Sibling competition, parent-offspring conflict, and clutch size. *Anim. Behav.* **43:** 473–90.

Godfray, H.C.J., L. Partridge, and P.H. Harvey. 1991. Clutch size. *Annu. Rev. Ecol. Syst.* **22:** 409–29.

Gomendio, M. 1990. The influence of maternal rank and infant sex on maternal investment trends in rhesus macaques: birth sex ratios, inter-birth intervals and suckling patterns. *Behav. Ecol. Sociobiol.* **27:** 365–75.

Gomendio, M. 1991. Parent/offspring conflict and maternal investment in rhesus macaques. *Anim. Behav.* **42:** 993–1005.

Gomendio, M. 1993. Parent-offspring conflict. *Trends Ecol. Evolution* **8:** 218.

Gordon, S. 1955. The golden eagle. Citadel Press, New York.

Gosling, L.M. 1986. Selective abortion of entire litters in the coypu: adaptive control of offspring production in relation to quality and sex. *Am. Nat.* **127:** 772–95.

Gosling, L.M., Baker, S.J. and K.M.H. Wright. 1984. Differential investment by female coypus (*Myocastor coypus*) during lactation. *Symp. Zool. Soc. Lond.* **51:** 273–300.

Gottlander, K. 1987. Parental feeding behaviour and sibling competition in the pied flycatcher *Ficedula hypoleuca Ornis Scand.* **18:** 269–76.

Gould, S.J. 1982. The guano ring. *Nat. Hist.* **91:** 12–9.

Gould, S.J. and R.C. Lewontin. 1979. The spandrels of San Marco and the Panglossian paradigm: A critique of the adaptationist programme. *Proc. R. Soc. Lond. B* **205:** 581–98.

Gowaty, P.A. 1991. Facultative manipulation of sex ratios in birds: rare or rarely observed? *Curr. Ornithol.* **8:** 141–71.

Gowaty, P.A. 1993. Differential dispersal, local resource competition, and sex ratio variation in birds. *Am. Nat.* **141:** 263–80.

Gowaty, P.A. and A.A. Karlin. 1985. Multiple paternity and maternity in single broods of apparently monogamous Eastern bluebirds. *Behav. Ecol. Sociobiol.* **15:** 91–5.

Gowaty, P.A. and D.W. Mock. (eds.) 1985. Avian Monogamy. *Ornithol. Monogr.* **37.** American Ornithologists' Union, Washington, DC.

Grafen, A. 1984. Natural selection, kin selection, and group selection. In *Behavioural ecology. An evolutionary approach,* 2nd edn (ed. J.R. Krebs and N.B. Davies), pp. 62–84. Blackwell Scientific Publications, Oxford.

Grafen, A. 1987. The logic of divisively asymmetric contests: respect for ownership and the deseperado effect. *Anim. Behav.* **35:** 462–7.

Grafen, A. 1990a. Biological signals as handicaps. *J. Theor. Biol.* **144:** 517–46.

Grafen, A. 1990b. Do animals really recognise kin? *Anim. Behav.* **39:** 42–54.

Grafen, A. 1991a. Modelling in behavioural ecology. In *Behavioural ecology. An evolutionary approach,* 3rd edn (ed. J.R. Krebs and N.B. Davies), pp. 5–31. Blackwell Scientific Publications, Oxford.

Grafen, A. 1991b. A reply to Blaustein *et al. Anim. Behav.* **41:** 1085–7.

Grafen, A. 1991c. A reply to Byers and Bekoff. *Anim. Behav.* **41:** 1091–2.

Grafen, A. 1991d. Kin vision: a reply to Stuart. *Anim. Behav.* **41:** 1095–6.

Grafen, A. and R.M. Sibly. 1978. A model of mate desertion. *Anim. Behav.* **26:** 645–52.

Grafen, A. and R.A. Johnstone. 1993. Why we need an ESS signalling theory. *Phil. Trans. R. Soc. Lond. B* **340:** 245–50.

Graves, J., Whiten, A. and P. Henzi. 1984. Why does the herring gull lay three eggs? *Anim. Behav.* **32:** 798–805.

Grbíc, M., Ode, P.J. and M.R. Strand. 1992. Sibling rivalry and brood sex ratios in polyembryonic wasps. *Nature* **360:** 254–6.

Green, W.H.C., Rothstein, A. and J. Griswold. 1993. Weaning and parent-offspring conflict: variation relative to interbirth interval in bison. *Ethology* **95:** 105–25.

Greenwood, P.J. 1980. Mating systems, philopatry and dispersal in birds and mammals. *Anim. Behav.* **28:** 1140–62.

Gregory, K.E., Echternkamp, S.E., Dickerson, G.E., Cundiff, L.V., Koch, R.M., and L.D. Van Vleck. 1990. Twinning in cattle: III. Effects of twinning on dystocia, reproductive traits, calf survival, calf growth, and cow productivity. *J. Anim. Sci.* **68:** 3133–44.

Grosberg, R.K. and J.F. Quinn. 1986. The genetic control and consequences of kin recognition by the larvae of a colonial marine invertebrate. *Nature* **322:** 457–9.

Gross, M.R. and R. Shine. 1981. Parental care and mode of fertilization in ectothermic vertebrates. *Evolution* **35:** 775–93.

Grosvenor, C.E. and F. Mena. 1974. Neural and hormonal control of milk secretion and milk ejection. In *Lactation: a comprehensive treatise* (ed. B.L. Larson and V.R. Smith), pp. 227–76. Academic Press, New York.

Groves, S. 1984. Chick growth, sibling rivalry, and chick production in American black oystercatchers. *Auk* **101:** 525–31.

Gullion, G.W. 1954. The reproductive cycle of American coots in California. *Auk* **71:** 366–412.

Gustafsson, L. and W.J. Sutherland. 1988. The costs of reproduction in the collared flycatcher *Ficedula hypoleuca. Nature* **335:** 813–15.

Gwynn, A.M. 1953. The egg-laying and incubation periods of rockhopper, macaroni, and gentoo penguins. *Australian National Antarctic Research Expedition Reports Series* B **1:** 1–29.

Hahn, D.C. 1981. Asynchronous hatching in the laughing gull: cutting losses and reducing rivalry. *Anim. Behav.* **29:** 421–7.

Haig, D. 1987. Kin conflict in seed plants. *Trends Ecol. and Evol.* **2:** 337–40.

Haig, D. 1990. Brood reduction and optimal parental investment when offspring differ in quality. *Am. Nat.* **136:** 550–66.

Haig, D. 1992a. Brood reduction in gymnosperms. In *Cannibalism: ecology and evolution among diverse taxa* (ed. M.A. Elgar and B.J. Crespi), pp. 63–84. Oxford University Press, Oxford.

Haig, D. 1992b. Genomic imprinting and the theory of parent-offspring conflict. *Sem. Dev. Biol.* **3:** 153–60.

Haig, D. 1992c. Intragenomic conflict and the evolution of eusociality. *J. Theor. Biol.* **156:** 401–3.

Haig, D. and M. Westoby. 1989a. Parent-specific gene expression and the triploid endosperm. *Am. Nat.* **134:** 147–155.

Haig, D. and M. Westoby. 1989b. Selective forces in the emergence of the seed habit. *Biol. J. Linn. Soc.* **38:** 215–238.

Haig, D. and C. Graham. 1991. Genomic imprinting and the strange case of the Insulin-like Growth Factor II receptor. *Cell* **64:** 1045–6.

Haig, D. and M. Westoby. 1991. Genomic imprinting in endosperm: its effect on seed development in crosses between species, and between different ploidies of the same species, and its implications for the evolution of apomixis. *Proc. R. Soc. Lond.* B **333:** 1–13.

Halliday, T.R. and P.A. Verrell. 1984. Sperm competition in amphibians. In *Sperm competition and the evolution of animal mating systems* (ed. R.L. Smith), p. 487–508. Academic Press, Orlando, FL.

Hamilton, W.D. 1963. The evolution of altruistic behavior. *Am. Nat.* **97:** 354–6.

Hamilton, W.D. 1964a. The genetical evolution of social behaviour. *J. Theor. Biol.* **7:** 1–16.

Hamilton, W.D. 1964b. The genetical evolution of social behaviour. *J. Theor. Biol.* **7:** 17–52.

Hamilton, W.D. 1966. The moulding of senescence by natural selection. *J. Theor. Biol.* **12:** 12–45.

Hamilton, W.D. 1967. Extraordinary sex ratios. *Science* **156:** 477–88.

Hamilton, W.D. 1970. Selection of selfish and altruistic behavior in some extreme models. In *Man and beast: comparative social behavior* (ed. J.F. Eisenberg and W.S. Dillon), pp. 59–91. Smithsonian Institution Press, Washington, DC.

Hamilton, W.D. 1972. Altruism and related phenomena, mainly in social insects. *Annu. Rev. Ecol. Syst.* **3:** 193–232.

Hamilton, W.D. 1975. Innate social aptitudes of man: an approach from evolutionary genetics. In *Biosocial anthropology* (ed. R. Fox), pp. 133–55. John Wiley, NY.

Hamilton, W.D. 1979. Wingless and fighting males in figwasps and other insects. In *Sexual selection and reproductive competition in insects* (ed. M.S. Blum and N.A. Blum), pp. 167–220. Academic Press, New York.

Hammerstein, P. In press. Darwinian adaptation, economic behaviour, and the streetcar theory of evolution. *J. Math. Biol.*

Hammerstein, P. and G.A. Parker. 1981. The asymmetric war of attrition. *J. Theor. Biol.* **96:** 647–82.

Hanken, J. and P.W. Sherman. 1981. Multiple paternity in Belding's ground squirrel litters. *Science* **212:** 351–3.

Hanwell, A. and M. Peaker. 1977. Physiological effects of lactation on the mother. *Symp. Zool. Soc. Lond.* **41:** 297–312.

Haresign, T.W. and S.E. Schumway. 1981. Permeability of the marsupium of the pipefish *Syngnathus fuscus* to ^{14}C-alpha amino isobutyric acid. *Comp. Biochem. Physiol.* **69A:** 603–4.

Harper, A.B. 1986. The evolution of begging: sibling competition and parent-offspring conflict. *Am. Nat.* **128:** 99–114.

Harper, D.G.C. 1987. Brood division in robins. *Anim. Behav.* **33:** 466–80.

Harris, M.P. 1983. Parent-young communication in the puffin *Fratercula arctica*. *Ibis* **125:** 56–73.

Harris, M.P. and P. Rothery. 1985. The post-fledging survival of young puffins *Fratercula arctica* in relation to hatching date and growth. *Ibis* **127:** 243–50.

Harris, R.N. 1987. Density-dependent paedomorphosis in the salamander *Notophthalmus viridescens* dorsalis. *Ecology* **68:** 705–12.

Hart, J.L. 1973. Pacific fishes of Canada. *Fisheries Res. Board Can. Bull.* 180. Ottawa.

Hartman, C.G. 1929. Some excessively large litters of eggs liberated at a single ovulation in mammals. *J. Mammal.* **10:** 197–203.

Harvey, J.M., Lieff, B.C., MacInnes, C.D. and J.P. Prevett. 1968. Observations on behavior of sandhill cranes. *Wilson Bull.* **80:** 421–5.

Haskell, D. 1994. Experimental evidence that nestling begging behaviour incurs a cost due to nest predation. *Proc. R. Soc. Lond.* **B 257:** 161–4.

Haukioja, E., Lemmetyinen, R. and M. Pikkola. 1989. Why are twins so rare in *Homo sapiens*? *Am. Nat.* **133:** 572–7.

Hausfater, G. and S. Blaffer Hrdy. (eds.) 1984. *Infanticide. Comparative and evolutionary perspectives*. Aldine, New York.

Hébert, P.N. and R.M.R. Barclay. 1986. Asynchronous and synchronous hatching: effect on early growth and survivorship of herring gull, *Larus argentatus*, chicks. *Can. J. Zool.* **64:** 2357–2362.

Hébert, P.N. and S.G. Sealy. 1993. Hatching asynchrony and feeding rates in yellow warblers *Dendroica petechia*: a test of the sexual conflict hypothesis. *Am. Nat.* **142:** 881–92.

Hecht, N.B., Bower, P.A. Waters, S.H. Yelick, P.C. and R.J. Distel. 1986. Evidence for haploid expression of mouse testicular genes. *Exp. Cell. Res.* **164:** 183–90.

Hecht, T. and S. Appelbaum. 1988. Observations on intraspecific aggression and coeval sibling cannibalism by larval and juvenile *Clarius gariepinus* (Claridae: Pisces) under controlled conditions. *J. Zool. Soc. Lond.* **214:** 21–44.

Hedgren, S. 1981. Effects of fledging weight and time of fledging on survival of guillemot chicks. *Ornis Scand.* **12:** 51–54.

Heintzelman, D.S. 1966. Cannibalism at a broad-winged hawk nest. *Auk* **83:** 307.

Henderson, B. 1975. Role of the chick's begging behavior in the regulation of parental feeding behavior of *Larus glaucescens*. *Condor* **77:** 488–92

Henry, J.D. 1985a. *Red fox: the cat-like canine*. Smithsonian Institution Press, Washington, D.C.

Henry, J.D. 1985b. The little foxes. *Nat. Hist.* **94:** 46–57.

Hepper, P.G. 1986. Kin recognition: functions and mechanisms. A review. *Biol. Rev.* **61:** 63–93

Herrera, C.M. 1990. Brood size reduction in *Lavandula latifolia* (Labiateae): a test of alternative hypotheses. *Evol. Trends Plants* **4:** 99–105.

Heyer, W.R. and R.I. Crombie. 1979. Natural history notes on *Crasspedoglossa stejnegeri* and *Thoropa petrapolitana* (Amphibia; Salientia, Leptodactylidae). *J. Wash. Acad. Sci.* **69:** 17–20.

Hill, J.P. 1910. The early development of the marsupialia, with special reference to the native cat (*Dasyurus viverrinus*). *Q. J. Microsc. Sci.* **56:** 1–139.

Hill-Cottingham, D.G. and R.R. Williams. 1967. Effect of time of application of fertilizer nitrogen on the growth, flower development, and fruit set of maiden apple trees var. Lord Lambourne, and on the distribution of total nitrogen within the tree. *J. Hort. Sci.* **42:** 319–38.

Hines, G. and J. Maynard Smith. 1979. Games between relatives. *J. Theor. Biol.* **79:** 19–30.

Hochachka, W. 1992. How much should reproduction cost? *Behav. Ecol.* **3:** 42–52.

Hoeck, H. 1977. 'Teat order' in hyrax (*Procavia johnston* and *Heterohyrax brucei*). *Z. Säugetierkunde* **42:** 112–5.

Hofer, H. and M.L. East. 1993. The commuting system of Serengeti spotted hyaenas: how a predator copes with migratory prey. III. Attendance and maternal care. *Anim. Behav.* **46:** 575–589.

Högstedt, G. 1980. Evolution of clutch size in birds: adaptive variation in relation to territory quality. *Science* **210:** 1148–50.

Högstedt, G. 1981. Effect of additional food on reproductive success in the magpie *Pica pica. J. Anim. Ecol.* **50:** 219–29.

Hokit, D.G. and A.R. Blaustein. 1994. The effects of kinship on growth and development in tadpoles of *Rana cascadae*. *Evolution* **48:** 1383–8.

Höldobbler, B. and E.O. Wilson. 1990. Queen control in colonies of weaver ants. *Ann. Entomol. Soc. Amer.* **76:** 235–8.

Holmes, W.G. 1986. Kin recognition by phenotype-matching in female Belding's ground squirrels. *Anim. Behav.* **34:** 38–47.

Holmes, W.G. and P.W. Sherman. 1982. The ontogeny of kin recognition in two species of ground squirrels. *Am. Zool.* **22:** 491–517.

Holmes, W.G. and P.W. Sherman. 1983. Kin recognition in animals. *Am. Sci.* **71:** 46–55.

Holstein, V. 1927. *Fiskehejren*. Gads Forlag, Copenhagen.

Hoogland, J.L. 1983. Black-tailed prairie dog coteries are co-operatively breeding units. *Am. Nat.* **121:** 275–80.

Hoogland, J.L. 1985. Infanticide in prairie dogs: lactating females kill offspring of close kin. *Science* **230:** 1037–9.

Hoogland, J.L. 1986. Nepotism in prairie dogs (*Cynomys ludovicianus*) varies with competition but not with kinship. *Anim. Behav.* **34:** 263–70.

Hoogland, J.L., Tamarin, R.H. and C.K. Levy. 1989. Communal nursing in prairie dogs. *Behav. Ecol. Sociobiol.* **24:** 91–5.

Hõrak, P. 1995. Brood reduction facilitates female but not offspring survival in the great tit. *Oecologia* **102:** 515–9.

Horsfall, J. 1984a. Brood reduction and brood division in coots. *Anim. Behav.* **32:** 216–25.

Horsfall, J.A. 1984b. Food supply and egg mass variation in the European coot. *Ecology* **65:** 89–95.

Houston, A.I. and N.B. Davies. 1985. The evolution of co-operation and life history in the dunnock *Prunella modularis*. In *Behavioural ecology: the ecological consequences of adaptive behaviour* (ed. R. Sibly and R. Smith), pp. 471–87. Blackwell Scientific Publications, Oxford

Howard, R. 1979. Estimating reproductive success in natural populations. *Am. Nat.* **114:** 221–31.

Howe, H.F. 1976. Egg size, hatching asynchrony, sex, and brood reduction in the common grackle. *Ecology* **57:** 1195–207.

Howe, H.F. 1978. Initial investment, clutch size, and brood reduction in the common grackle (*Quiscalus quiscula* L.). *Ecology* **59:** 1109–22.

Howe, H.F. and G.A. Vande Kerckhove. 1980. Nutmeg dispersal by tropical birds. *Science* **210:** 925–7.

Howe, H.F. and G.A. Vande Kerckhove. 1981. Removal of wild nutmeg (*Virola surinamensis*) crops by birds. *Ecology* **62:** 1093–106.

Howe, H.F. and W.M. Richter. 1982. Effects of seed size on seedling size in *Virola surinamensis*: a within and between tree analysis. *Oecologia* **53**: 347–51.

Huang, B.-Q. and S.D. Russell. 1992. Female germ unit: organization, isolation, and function. In *Sexual reproduction in flowering plants* (ed. S.D. Russell and C. Dumas) *Int. Rev. Cytol.* **140**: 233–93.

Husby, M. 1986. On the adaptive value of brood reduction in birds: experiments with the magpie *Pica pica*. *J. Anim. Ecol.* **55**: 75–83.

Hussell, D.J.T. 1972. Factors affecting clutch-size in Arctic passerines. *Ecol. Monogr.* **42**: 317–64.

Hussell, D.J.T. 1985. Optimal hatching asynchrony in birds: comments on Richter's critique of Clark and Wilson's model. *Am. Nat.* **126**: 123–128.

Hussell, D.J.T. 1988. Supply and demand in tree swallow broods: a model of parent-offspring food provisioning interactions in birds. *Am. Nat.* **131**: 175–202.

Hustler, C.W. and W.W. Howells. 1986. A population study of tawny eagles in the Hwange National Park. *Ostrich* **57**: 101–6.

Hutchison, V.H., Dowling, H.G. and A. Vinegar. 1966. Thermoregulation in a brooding female Indian python, *Python molurus bivittatus*. *Science* **151**: 694–6.

Inger, R.F. 1966. The systematics and zoogeography of the amphibia of Borneo. *Fieldiana Zool.* **51**: 1–402.

Ingram, C. 1959. The importance of juvenile cannibalism in the breeding biology of certain birds of prey. *Auk* **76**: 218–26.

Ingram, C. 1962. Cannibalism by nestling short-eared owls. *Auk* **79**: 715.

Ingram, G.J., Anstis, M. and C.J. Corben. 1975. Observations on the Australian leptodactylid frog, *Assa darlingtoni*. *Herpetologica* **31**: 425–9.

Inoue, Y. 1985. The process of asynchronous hatching and sibling competition in the little egret *Egretta garzetta*. *Col. Waterbirds* **8**: 1–12.

Jamieson, I.G., Seymour, N.R., Bancroft, R.P. and R. Sullivan. 1983. Sibling aggression in nestling ospreys in Nova Scotia. *Can. J. Zool.* **61**: 466–9.

Janzen, D.H. 1977. Promising directions of study in tropical plant-animal interactions. *Ann. Mo. Bot. Gard.* **61**: 706–36.

Janzen, D.H. 1982. Variation in average seed size and fruit seediness in a fruit crop of a Guanacaste tree (*Enterolobium cyclocarpum*). *Am. J. Bot.* **69**: 1169–78.

Jarvis, M.J.F. 1974. The ecological significance of clutch size in the South African gannet (*Sula capensus*). *J. Anim. Ecol.* **43**: 1–17.

Jarvis, P.H. and P.E. King. 1975. Egg development in the reproductive cycle in the pycnogonid *Pycnogonum littorale*. *Marine Biol.* **13**: 146–54.

Jasienski, M. 1988. Kinship ecology of competition: size hierarchies in kin and nonkin laboratory cohorts of tadpoles. *Oecologia* **77**: 407–13.

Jenni, D.A. 1969. A study of the ecology of four species of herons during the breeding season at Lake Alice, Alachua County, Florida. *Ecol. Monogr.* **39**: 245–70.

Johnson, K., Bednarz, J.C. and S. Zack. 1987. Crested penguins: why are first eggs smaller? *Oikos* **49**: 347–9.

Johnston, R.D. 1993. The effect of direct supplementary feeding of nestlings on weight loss in female great tits *Parus major*. *Ibis* **135**: 311–4.

Johnstone, R.A. 1994. Honest signalling, perceptual error, and the evolution of 'all-or-nothing' displays. *Proc. Roy. Soc. Lond. B* **256**: 169–75.

Johnstone, R.A. 1996a. Multiple displays in animal communication: 'backup signals' and 'multiple messages.' *Phil. Trans. R. Soc. Lond. B* **351**: 329–38.

Johnstone, R.A. 1996b. Begging signals and parent-offspring conflict: do parents always win? *Proc. R. Soc. Lond. B* **263:** 1677–81.

Johnstone, R.A. and A. Grafen. 1992a. The continuous Sir Philip Sidney game: a simple model of biological signalling. *J. Theor. Biol.* **156:** 215–34.

Johnstone, R.A. and A. Grafen. 1992b. Error-prone signalling. *Proc. R. Soc. Lond. B* **248:** 229–33.

Johnstone, R.A. and A. Grafen. 1993. Dishonesty and the handicap principle. *Anim. Behav.* **46:** 759–64.

Kacelnik, A., Cotton, P.A., Stirling, L. and J. Wright. 1995. Food allocation among nestling starlings: sibling competition and the scope of parental choice. *Proc. R. Soc. Lond. B* **259:** 259–63.

Kahl, M.P. 1964. Food ecology of the wood stork (*Mycteria americana*) in Florida. *Ecol. Monogr.* **34:** 97–117.

Karlin, S. 1975. General 2-locus selection models: some objectives, results, and interpretations. *Theor. Popul. Biol.* **7:** 364–98.

Karron, J.D. and D.L. Marshall. 1990. Fitness consequences of multiple paternity in wild radish, *Raphanus sativus*. *Evolution* **44:** 260–8.

Keane, B., Waser, P.M., Creel, S.R., Creel, N.M., Elliott, L.F. and D.J. Minchella. 1994. Subordinate reproduction in dwarf mongooses. *Anim. Behav.* **47:** 65–75.

Kear, J. 1970. The adaptive radiation of parental care in waterfowl In *Social behaviour of birds and mammals* (ed. J.H. Crook), pp. 357–92. Academic Press, London.

Keenleyside, M.H. 1978. Parental behavior in fishes and birds. In *Contrasts in behavior* (ed. E.S. Reese and F.J. Lighter), pp. 3–30. Wiley, New York.

Keenleyside, M.H. 1980. Parental care patterns of fishes. *Am. Nat.* **117:** 1019–22.

Kight, S.L. and K.C. Kruse. 1992. Factors affecting the allocation of parental care in waterbugs (*Belostoma flumineum* Say). *Behav. Ecol. Sociobiol.* **30:** 409–14.

Kinomura, K. and K. Yamauchi. 1987. Fighting and mating behaviors in the ant *Cardiocondyla wroughtonii*. *J. Ethol.* **5:** 75–81.

Kirk, H. 1988. Cannibalism in a chrysomelid beetle, *Gastrophysa viridula*. Unpublished Ph.D. thesis, University of Liverpool.

Kenagy, G.J., Masman, D., Sharbaugh, S.M. and K.A. Nagy. 1990. Energy expenditure during lactation in relation to litter size in free-living golden-mangled ground squirrels. *J. Anim. Ecol.* **59:** 73–88.

Kenrick, J. and R.B. Knox. 1982. Function of the polyad in reproduction of *Acacia*. *Annu. Bot.* **50:** 721–7.

Khayutin, S.N., Dmitrieva, L.P. and L.I. Alexandrov. 1988. Psychobiological aspects of the acceleration of postembryonic development in the asynchronous breeder, pied flycatcher (*Ficedula hypoleuca*). *Internat. J. Comp. Psych.* **81:** 145–66.

Kleiman, D.G. 1977. Monogamy in mammals. *Q. Rev. Biol.* **52:** 39–69.

Kleiman, D.G. 1979. Parent-offspring conflict and sibling competition in a monogamous primate. *Am. Nat.* **114:** 753–75.

Kleiman, D.G. 1982. Correlations among life history characteristics of mammalian species exhibiting two extreme forms of monogamy. In *Natural selection and social behavior* (ed. R.D. Alexander and D.W. Tinkle), pp. 332–44, Chiron Press, New York.

Klekowski, E.L. Jr. 1982. Genetic load and soft selection in ferns. *Heredity* **49:** 191–7.

Klomp, H. 1970. The determination of clutch-size in birds. *Ardea* **58:** 1–124.

Knopf, F.L. 1979. Spatial and temporal aspects of colonial nesting of white pelicans. *Condor* **81:** 353–63.

Koenekoop, R.K. and T.P. Livdahl. 1986. Cannibalism among *Aedes triseriatus* larvae. *Ecol. Entom.* **11:** 111–14.

Kojima, Y. 1987. Breeding success of the grey-faced buzzard eagle *Butastur indicus. Jap. J. Ornithol.* **36:** 71–8.

Kok, D., du Preez, L.H. and A. Channing. 1989. Channel construction by the African bullfrog: another anuran parental care strategy. *J. Herpetol.* **23:** 435–7.

Konarzewski, M. 1993. The evolution of clutch size and hatching asynchrony in altricial birds: the effect of environmental variability, egg failure, and predation. *Oikos* **67:** 97–106.

König, B. 1989. Kin recognition and maternal care under restricted feeding in house mice (*Mus musculus*). *Ethology* **82:** 328–43.

König, B., Reisler, J. and H. Markl. 1988. Maternal care in house mice (*Mus musculus*). II. The energy cost of lactation as a function of litter size. *J. Zool.* **216:** 195–210.

Kozlowski, J. and S.C. Stearns. 1989. Hypotheses for the production of excess zygotes: models of bet-hedging and selective abortion. *Evolution.* **43:** 1369–77.

Kozlowski, T.T. 1973. Extent and significance of shedding of plant parts. In *Shedding of plant parts* (ed. T.T. Kozlowski), pp. 1–44. Academic Press, New York.

Krementz, D.G., Nichols, J.D. and J.E. Hines. 1989. Postfledging survival of European starlings. *Ecology* **70:** 646–55.

Kress, W.J. 1981. Sibling competition and the evolution of pollen unit, ovule number, and pollen vector in angiosperms. *Syst. Bot.* **6:** 101–12.

Krupa, J.J. 1988a. Fertilization efficiency in the Great Plains toad. *Copeia* **1988(3):** 800–3.

Krupa, J.J. 1988b. Fertilization efficiency in the Great Plains toad. *Copeia* **1988:** 1117.

Krupa, J.J. 1993. Breeding biology of the Great Plains toad in Oklahoma. *J. Herpetol.* **28:** 217–24

Kukuk, P.F. 1992. Cannibalism in social bees. In *Cannibalism: ecology and evolution among diverse taxa* (ed. M.A. Elgar and B.J. Crespi), pp. 214–37. Oxford University Press, Oxford.

Kushlan, J.A. 1978. Feeding ecology of wading birds In *Wading birds.* (ed. A. Sprunt IV, J.C. Ogden, and S. Winckler), pp. 249–97. National Audubon Society Research Report No. 7. NAS, New York.

Lack, D. 1947. The significance of clutch-size. Parts 1 and 2. *Ibis* **89:** 302–52.

Lack, D. 1948. The significance of clutch-size. Part 3. *Ibis* **90:** 25–45.

Lack, D. 1954. *The natural regulation of animal numbers.* Clarendon Press, Oxford.

Lack, D. 1956. *Swifts in a tower.* Methuen Press, London.

Lack, D. 1966. *Population studies in birds.* Clarendon Press, London.

Lack, D. 1968. *Ecological adaptations for breeding in birds.* Methuen Press, London.

Lamey, T.C. 1990. Hatch asynchrony and brood reduction in penguins. In *Penguin biology* (ed. L.S. Davis and J. Darby), pp. 399–417. Academic Press, New York.

Lamey, T.C. 1992. Egg-size differences, hatch asynchrony, and obligate brood reduction in crested penguins. Ph.D. dissertation, University of Oklahoma, Norman.

Lamey, T.C. 1993. Territorial aggression, timing of egg loss, and egg-size differences in rockhopper penguins, *Eudyptes c. chrysocome*, on New Island, Falkland Islands. *Oikos* **66:** 293–7.

Lamey, T.C. and C.S. Lamey. 1994. Hatch synchrony and bad food years. *Am. Nat.* **143:** 734–8.

Lamey, T.C., Evans, R.M. and J.D. Hunt. 1996. Insurance reproductive value and facultative brood reduction. *Oikos* **77:** 285–90.

Lamey, T.C. and D.W. Mock. 1991. Nonaggressive brood reduction in birds. *Acta XX Congr. Int. Ornithol.* III: 1741–51.

Landy, H.J., Keith, L. and D. Keith. 1982. The vanishing twin. *Acta Genet. Med. Gemellol.* **31:** 179–94.

Lannoo, M.J. and M.D. Bachmann. 1984. Aspects of cannibalistic morphs in a population of *Ambystoma tigrinum* larvae. *Am. Midl. Nat.* **112:** 103–10.

Lannoo, M.J., Lowcock, L. and J.P. Bogart. 1989. Sibling cannibalism in noncannibal morph *Ambystoma tigrinum* larvae and its correlation with high growth rates and early metamorphosis. *Can. J. Zool.* **67:** 1911–14.

Lazarus, J. 1989. The logic of mate desertion. *Anim. Behav.* **39:** 357–71.

Lazarus, J. and I. Inglis. 1986. Shared and unshared parental investment, parent-offspring conflict, and brood size. *Anim. Behav.* **34:** 1791–804.

Lee, A.K. and A. Cockburn. 1985. *Evolutionary ecology of marsupials.* Cambridge University Press, Cambridge.

Lee, S.C. 1988. Third-egg neglect in the herring gull (*Larus argentatus*). Unpublished M.Sc. thesis, University of Manitoba, Winnipeg.

Lee, S.C., Evans, R.M. and S.C. Bugden. 1993. Benign neglect of terminal eggs in herring gulls. *Condor* **95:** 507–14.

Lee, T.D. 1984. Patterns of fruit maturation: a gametophyte competition hypothesis. *Am. Nat.* **123:** 427–32.

Lee, T.D. 1988. Patterns of fruit and seed production. In *Plant reproductive ecology* (ed. J. Lovett Doust and L. Lovett Doust), pp.179–202. Oxford University Press, Oxford.

Lee, T.D. and F.A. Bazzaz. 1982a. Regulation of fruit and seed production in an annual legume, *Cassia fasciculata. Ecology* **63:** 1363–73.

Lee, T.D. and F.A. Bazzaz. 1982b. Regulation of fruit maturation patterns in an annual legume, *Cassia fasciculata. Ecology* **63:** 1374–88.

Leishman, M.R. and M. Westoby. 1994. The role of large seed size in shaded conditions: experimental evidence. *Funct. Ecol.* **8:** 205–14.

Leonard, M.L. and A.G. Horn. 1996. Provisioning rules in tree swallows. *Behav. Ecol. Sociobiol.* **38:** 341–7.

Leonard, M.L., Horn, A.G. and S.F. Eden. 1988. Parent-offspring aggression in moorhens. *Behav. Ecol. Sociobiol.* **23:** 265–70.

Leopold, A.C. and F.I. Scott. 1952. Physiological factors in tomato fruit-set. *Am. J. Bot.* **39:** 310–17.

Lessells, C.M. 1991. The evolution of life histories. In *Behavioural ecology. An evolutionary approach* (ed. J.R. Krebs and N.B. Davies), pp. 32–68. Blackwell Scientific Publications, Oxford.

Lessells, C.M. and M.I. Avery. 1989. Hatching asynchrony in European bee-eaters *Merops apiaster. J. Anim. Ecol.* **58:** 815–35

Lewontin, R.C. 1970. The units of selection. *Annu. Rev. Ecol. Syst.* **1:** 1–19.

Lieberman, M. and D. Lieberman. 1978. Lactase deficiency: a genetic mechanism which regulates the time of weaning. *Am. Nat.* **112:** 625–7.

Lindström, E.R. 1994. Placental scar counts in the red fox (*Vulpes vulpes* L.) revisited. *Z. Säugetierkunde.* **59:** 169–73.

Linsenmair, K.E. and C. Linsenmair. 1971. Paarbildung und Paarzusammenhaltung bei der monomogamen Wüstenassel *Hemilepistus reamuri* (Crustacea, Isopoda. Oniscoidea). *Z. Tierpsychol.* **29:** 134–55.

Litovitch, E. and H.W. Power. 1992. Parent-offspring conflict and its resolution in the European starling. *Ornithological Monographs* No. 47. American Ornithologists' Union, Washington, DC.

Litvinenko, N. 1982. Nesting of grey heron (*Ardea cinerea* L.) on sea islands of South Primorye. *J. Yamashina Inst. Ornithol.* **14:** 220–31.

Lloyd, D.G. 1980. Sexual strategies in plants. I. An hypothesis of serial adjustment of maternal investment during one reproductive session. *New Phytol.* **86:** 69–79.

Lloyd, D.G. 1987. Selection of offspring size at independence and other size-versus-number strategies. *Am. Nat.* **129:** 800–17.

Lloyd, D.G. 1988. A general principle for the allocation of limited resources. *Evol. Ecol.* **2:** 175–87.

Longmire, J.L., Maltbie, M., Pavelka, R.W., Smith, L.M., Witte, S.M., Ryder, O.A., Ellsworth, D.L. and R.J. Baker. 1993. Gender identification in birds using microsatellite DNA fingerprint analysis. *Auk* **110:** 378–81.

Louden, A.S.I., McNeilly, A.S. and J.A. Milne. 1983. Nutrition and lactational control of fertility in red deer. *Nature* **302:** 145–7.

Lovett Doust, J. and L. Lovett Doust. 1988. Sociobiology of plants: an emerging synthesis. In *Plant reproductive ecology* (ed. J. Lovett Doust and L. Lovett Doust), pp. 5–29. Oxford University Press, Oxford.

Lowe, F.A. 1954. *The heron*. Collins, London.

Lowe, V.T. 1966. Notes on the musk duck. *Emu* **65:** 279–90.

Lund, R. 1980. Viviparity and intrauterine feeding in a new holocephalan fish from the Lower Carboniferous of Montana. *Science* **209:** 697–9.

Lundberg, A. and R.V. Alatalo. 1992. *The pied flycatcher.* T. and A.D. Poyser, London.

Lundberg, A., Alatalo, R.V., Carlson, A. and S. Ulfstrand. 1981. Biometry, habitat distribution and breeding success in the pied flycatcher *Ficedula hypoleuca*. *Ornis Scand.* **12:** 68–79.

Lundberg, C.A. and R.A. Vaisanen. 1979. Selective correlation of egg size with chick mortality in the black-headed gull *Larus ridibundus*. *Condor* **81:** 146–56.

Lundberg, S. and H.G. Smith. 1994. Parent-offspring conflicts over reproductive effort: variations on a theme by Charnov. *J. Theor. Biol.* **171:** 215–8.

Lyon, B.E., Eadie, J.M. and L.D. Hamilton. 1994. Parental choice selects for ornamental plumage in American coot chicks. *Nature* **371:** 240–3.

Machmer, M.M. 1992. Causes and consequences of sibling aggression in nestling ospreys (*Pandion haliaetus*). M.Sc. Thesis, Simon Fraser University, Burnaby, BC.

Machmer, M.M. and R.C. Ydenberg. Submitted. Sibling aggression and brood reduction in nestling ospreys. *Can. J. Zool.*

Le Masurier, A.D. 1987. A comparative study of the relationship between host size and brood size in *Apateles* spp. (Hymenoptera: Braconidae). *Ecol. Entomol.* **12:** 383–93.

McClure, P. 1981. Sex-biased litter reduction in food-restricted wood rats, *Neotoma floridana*. *Science* **211:** 1058–60.

McDiarmid, R. 1978. Evolution of parental care in frogs. In *The development of behavior* (ed. G. Burghardt and M. Bekoff), pp. 127–47. Garland STPM Press, New York.

McGuire, B., Getz, L.L., Hofmann, J.E., Pizzuto, T. and B. Frase. 1993. Natal dispersal and philopatry in prairie voles (*Microtus ochrogaster*) in relation to population density, season, and natal social environment. *Behav. Ecol. Sociobiol.* **32:** 293–302.

McKilligan, N.G. 1990. The breeding biology of the intermediate egret. Part 1. The physical and behavioural development of the chick, with special reference to sibling aggression and food intake. *Corella* **14:** 162–9.

McKinney, D.F., Cheng, K.M. and D.J. Bruggers. 1984. Sperm competition in apparently monogamous birds. In *Sperm competition* (ed. R.L. Smith), pp. 523–46. Academic Press, Orlando, FL.

McLean, P.K. and M.A. Byrd. 1991. Feeding ecology of Chesapeake Bay ospreys and growth and behavior of their young. *Wilson Bull.* **103:** 105–11.

Macnair, M.R. 1978. An ESS for the sex ratio in animals, with particular reference to the social Hymenoptera. *J. Theor. Biol.* **70:** 449–59.

Macnair, M. and G.A. Parker. 1978. Models of parent-offspring conflict. II. Promiscuity. *Anim. Behav.* **26:** 111–22.

Macnair, M. and G.A. Parker. 1979. Models of parent-offspring conflict. III. Intra-brood conflict. *Anim. Behav.* **27:** 1202–9.

McNicholl, M.K. 1977. Usage of the terms 'cannibalism' and 'scavenging' in ecological literature. *Can. Field-Natur.* **91:** 416.

Macpherson, A.H. 1969. The dynamics of Canadian Arctic fox populations. *Can. Wildl. Serv. Rep.* Ser. 8. Queen's Printer, Ottawa.

McRae, S.B., Weatherhead, P.J. and R. Montgomerie. 1993. American robin nestlings compete by jockeying for position. *Behav. Ecol. Sociobiol.* **33:** 101–6.

McShea, W.J. and D.M. Madison. 1987. Measurements of reproductive traits in a field population of meadow voles. *J. Mammal.* **70:** 132–41.

McVittie, R. 1978. Nursing behavior of snow leopard cubs. *Appl. Anim. Ethol.* **4:** 159–68.

Madsen, T. and R. Shine. 1992. Sexual competition among brothers may influence offspring sex ratio in snakes. *Evolution* **46:** 1549–52.

Madsen, T., Shine, R., Loman, J. and T. Håkansson. 1992. Why do female adders copulate so frequently? *Nature* **355:** 440–1.

Magrath, R. 1989. Hatch asynchrony and reproductive success in the blackbird. *Nature* **339:** 536–8.

Magrath, R.D. 1990. Hatching asynchrony in altricial birds. *Biol. Rev.* **95:** 587–622.

Magrath, R.D. 1991. Nestling weight and juvenile survival in the blackbird, *Turdus merula*. *J. Anim. Ecol.* **60:** 335–51.

Magrath, R.D. 1992. The effect of egg mass on the growth and survival of blackbirds: a field experiment. *J. Zool. Lond.* **227:** 639–53.

Malcolm, J.R. and K. Marten. 1982. Natural selection and the communal rearing of pups in African wild dogs (*Lycaon pictus*). *Behav. Ecol. Sociobiol.* **10:** 1–13.

Marois, R. and R.P. Croll. 1991. Hatching asynchrony within the egg mass of the pond snail, *Lymnaea*. *Invert. Repr. Dev.* **19:** 139–46.

Marshall, D.L. and N.C. Ellstrand. 1986. Sexual selection in *Raphanus sativus*: experimental data on non-random fertilization, maternal choice, and consequences of multiple paternity. *Am. Nat.* **127:** 446–61.

Marsteller, F.A. and C.B. Lynch. 1983. Reproductive consequences of food restriction at low temperature in lines of mice divergently selected for thermoregulatory nesting. *Behav. Genet.* **13:** 397–410.

Marti, C.D. 1989. Food sharing by sibling common barn-owls. *Wilson Bull.* **101:** 132–4

Martin, T.E. 1987. Food as a limit on breeding birds: a life-history perspective. *Annu. Rev. Ecol. Syst.* **18:** 453–87.

Martíns, T.L.F. and J. Wright. 1993a. Brood reduction in response to manipulated brood sizes in the common swift (*Apus apus*). *Behav. Ecol. Sociobiol.* **32:** 61–70.

Martíns, T.L.F. and J. Wright. 1993b. On the cost of reproduction and the allocation of food between parent and young in the swift (*Apus apus*). *Behav. Ecol.* **4:** 213–23.

May, C.A., Wetton, J.H. and D.T. Parkin. 1993. Polymorphic sex-specific sequences in birds of prey. *Proc. R. Soc. Lond. B* **253:** 271–6.

May, P. and A. Antcliff. 1963. The effects of shading on fruitfulness and yield in the sultana. *J. Hort. Sci.* **38:** 85–94

Maynard Smith, J. 1977. Parental investment: a prospective analysis. *Anim. Behav.* **25:** 1–9.

Maynard Smith, J. 1978. *The evolution of sex.* Cambridge University Press, Cambridge.

Maynard Smith, J. 1980. A new theory of sexual investment. *Behav. Ecol. Sociobiol.* **7:** 247–51.

Maynard Smith, J. 1981. Will a sexual population evolve to an ESS? *Am. Nat.* **117:** 1015–8.

Maynard Smith, J. 1982a. *Evolution and the theory of games.* Cambridge University Press, Cambridge.

Maynard Smith, J. 1982b. The evolution of social behaviour-a classification of models. In *Current problems in sociobiology* (ed. King's College Sociobiology Group), pp. 29–44. Cambridge University Press, Cambridge.

Maynard Smith, J. 1992. Honest signalling: the Philip Sidney game. *Anim. Behav.* **42:** 1034–5.

Maynard Smith, J. 1994. Must reliable signals always be costly? *Anim. Behav.* **47:** 1115–20.

Maynard Smith, J. and G.A. Parker. 1976. The logic of asymmetric contests. *Anim. Behav.* **24:** 159–75.

Maynard Smith, J. and D.G.C. Harper. 1995. Animal signals: models and terminology. *J. Theor. Biol.* **177:** 305–11.

Maynard Smith, J. and E. Szathmáry. 1995. *The major transitions in evolution.* Freeman Publishers, Oxford.

Mead, P. and M. Morton. 1985. Hatching asynchrony in the mountain white-crowned sparrow (*Zonotrichia leucophrys oriantha*): A selected or incidental trait? *Auk* **102:** 781–92.

Mech, L.D. *The wolf: ecology and social behavior of an endangered species.* Natural History Press, New York.

Meffe, G.K. and R.C. Vriejenhoek. 1981. Starvation stress and intraovarian cannibalism in livebearers (Atheriniformes: Poeciliidae). *Copeia* **1981:** 702–5.

Meffe, G.K. and M.L. Crump. 1987. Possible growth and reproductive benefits of cannibalism in the mosquitofish. *Am. Nat.* **129:** 203–12.

Meisel, R.L. and I.L. Ward. 1981. Fetal female rats are masculinized by male littermates located caudally in the uterus. *Science* **216:** 239–41.

Mendl, M. 1988. The effect of litter size on variation in mother-offspring relationships and behavioural and physical development in several mammalian species (principally rodents). *J. Zool. (Lond.)* **215:** 15–34.

Metcalf, R., Stamps, J.A. and V.V. Krishnan. 1979. Parent-offspring conflict that is not limited by degree of kinship. *J. Theor. Biol.* **76:** 99–107.

Meyburg, B.-U. 1974. Sibling aggression and mortality among nestling eagles. *Ibis* **116:** 224–8.

Meyerriecks, A.J. 1962. Diversity typifies heron feeding. *Nat. Hist.* **71:** 49–59.

Michener, G.R. and J.W. Koeppl. 1985. *Spermophilus richardsonii. Mammal. Species* **243:** 1–8.

Michod, R.E. 1979. Evolution of life-histories in response to age-specific mortality factors. *Am. Nat.* **113:** 531–50.

Michod, R.E. 1982. The theory of kin selection. *Annu. Rev. Ecol. Syst.* **13:** 23–55.

Michod, R.E. and W.D. Hamilton. 1980. Coefficients of relatedness in sociobiology. *Nature* **288:** 694–7.

Milinski, M. 1978. Kin selection and reproductive value. *Z. Tierpsychol.* **47:** 328–329.

Millar, J.S. 1978. Energetics of reproduction in *Peromyscus leucopus*: the cost of lactation. *Ecology* **59:** 1055–61.

Millar, J.S. 1979. Energetics of lactation in *Peromyscus maniculatus. Can. J. Zool.* **57:** 1015–9.

Miller, R.S. 1973. The brood size of cranes. *Wilson Bull.* **85:** 436–41.

Milstein, P. LeS., Prestt, I. and A.A. Bell. 1970. The breeding cycle of the grey heron. *Ardea* **58:** 171–257.

Miskelly, C.M. and P.W. Carey. 1990. Egg-laying and egg-loss by erect-crested penguins. In *Antipodes Islands Report* (ed. C.M. Miskelly, Carey, P.W. and S. Pollard), pp. 2–11. Unpublished University of Canterbury Expedition Report.

Mock, D.W. 1976. Pair-formation displays of the great blue heron. *Wilson Bull.* **88:** 185–230.

Mock, D.W. 1984a. Infanticide, siblicide, and avian nestling mortality. In *Infanticide: comparative and evolutionary perspectives* (ed. G. Hausfater and S.B. Hrdy), pp. 3–30. Aldine Publishing Co., New York.

Mock, D.W. 1984b. Siblicidal aggression and resource monopolization in birds. *Science* **225:** 731–3.

Mock, D.W. 1985. Siblicidal brood reduction: the prey-size hypothesis. *Am. Nat.* **125:** 327–43.

Mock, D.W. 1987. Siblicide, parent-offspring conflict, and unequal parental investment by egrets and herons. *Behav. Ecol. Sociobiol.* **20:** 247–56.

Mock, D.W. 1994. Brood reduction: broad sense, narrow sense. *J. Avian Biol.* **25:** 3–7.

Mock, D.W. and K.C. Mock. 1980. Feeding behavior and ecology of the goliath heron. *Auk* **97:** 433–48.

Mock, D.W. and G.A. Parker. 1986. Advantages and disadvantages of ardeid brood reduction. *Evolution* **40:** 459–70.

Mock, D.W. and B.J. Ploger. 1987. Parental manipulation of optimal hatch asynchrony in cattle egrets: an experimental study. *Anim. Behav.* **35:** 150–60.

Mock, D.W. and M. Fujioka. 1990. Monogamy and long-term pair bonding. in vertebrates. *Trends Ecol. Evolution* **5:** 39–43.

Mock, D.W. and P.L. Schwagmeyer. 1990. The peak load reduction hypothesis for avian hatching asynchrony. *Evol. Ecol.* **4:** 249–60.

Mock, D.W. and T.C. Lamey. 1991. The role of brood size in regulating egret sibling aggression. *Am. Nat.* **138:** 1015–26.

Mock, D.W. and L.S. Forbes. 1992. Parent-offspring conflict: a case of arrested development? *Trends Ecol. Evolution* **7**: 409–13.

Mock, D.W. and L.S. Forbes. 1993. Reply to Gomendio. *Trends Ecol. Evolution* **8**: 218.

Mock, D.W. and L.S. Forbes. 1994. Life-history consequences of avian brood reduction. *Auk* **111**: 115–23.

Mock, D.W. and L.S. Forbes. 1995. The evolution of parental optimism. *Trends Ecol. Evolution* **10**: 130–4.

Mock, D.W., T.C. Lamey, and B.J. Ploger. 1987a. Proximate and ultimate roles of food amount in regulating egret sibling aggression. *Ecology* **68**: 1760–72.

Mock, D.W., T.C. Lamey, C.F. Williams, and A. Pelletier. 1987b. Flexibility in the development of heron sibling aggression: an intraspecific test of the prey-size hypothesis. *Anim. Behav.* **35**: 1386–93.

Mock, D.W., T.C. Lamey, and D.B.A. Thompson. 1988. Falsifiability and the information centre hypothesis. *Ornis Scand.* **19**: 231–48.

Mock, D.W., Drummond, H. and C.H. Stinson. 1990. Avian siblicide. *Am. Sci.* **78**: 438–49.

Mock, D.W., Schwagmeyer, P.L. and G.A. Parker. 1996. The model family. In *Partnerships in birds: the study of monogamy* (ed. J.M. Black), pp.52–69. Oxford University Press, Oxford.

Mock, D.W., Forbes, L.S. and C.C. St. Clair. In press. The study of bi-parental care in colonial waterbirds. In *Colonial breeding in waterbirds* (ed. F. Cézilly, Hafner, H. and D.N. Nettleship), Oxford University Press, Oxford.

Mogensen, H.L. 1975. Ovule abortion in *Quercus* (Fagaceae). *Am. J. Bot.* **62**: 160–5.

Moilanen, I. 1987. Dominance and submissiveness between twins. I. Perinatal and developmental aspects. *Acta Genet. Med. Gemellol.* **36**: 249–55.

Moore, T. and D. Haig. 1991. Genomic imprinting in mammalian development: a parental tug-of-war. *Trends Genet.* **7**: 45–9.

Moreno, J. 1987. Nestling growth and brood reduction in the wheatear *Oenanthe oenanthe*. *Ornis Scand.* **18**: 302–9.

Morton, S.R. 1978. An ecological study of *Smithopsis crassicaudata* (Marsupialia: Dasyuridae). III. Reproduction and life history. *Aust. Wildl. Res.* **5**: 183–211.

Morton, S.R., Recher, H.F., Thompson, S.D. and R.W. Braithwaite. 1982. Comments on the relative advantages of marsupial and eutherian reproduction. *Am. Nat.* **120**: 128–34.

Moses, R.A., Boutin, S. and T. Teferi. Submitted. Male-biased litter mortality in a sexually dimorphic rodent: evidence against maternal adjustment of sex ratios. *Anim. Behav.* [if not accepted, this becomes unpub. data]

Mossman, H.W. 1937. Comparative morphogenesis of the fetal membranes and accessory uterine structures. *Contrib. Embryol* **26**: 129–246.

Mossman, H.W. 1987. *Vertebrate fetal membranes*. Rutgers University Press, New Brunswick.

Motro, U. 1991. Avoiding inbreeding and sibling competition: the evolution of sexual dimorphism for dispersal. *Am. Nat.* **137**: 108–15.

Motro, U. and I. Eshel. 1988. The three brothers' problem: kin selection with more than one potential helper. 2. The case of delayed help. *Am. Nat.* **132**: 567–75.

Mueller, U. 1991. Haplodiploidy and the evolution of facultative sex ratios in a primitively eusocial bee. *Science* **254**: 442–4.

Muller, R.E. and D.G. Smith. 1978. Parent-offspring interactions in zebra finches. *Auk* **95:** 485–95.

Mumme, R., Koenig, W. and F. Pitelka. 1983. Reproductive competition in the communal acorn woodpecker: sisters destroy each others' eggs. *Nature* **306:** 583–4.

Murray, M.G. 1987. The closed environment of the fig receptacle and its influence on male conflict in the Old World fig wasp, *Philotrypesis pilosa*. *Anim. Behav.* **35:** 488–506.

Murray, M.G. 1988. Environmental constraints on fighting in flightless male fig wasps. *Anim. Behav.* **38:** 186–93.

Murray, M.G. 1990. Comparative morphology and mate competition of flightless male fig wasps. *Anim. Behav.* **39:** 434–43.

Myers, C.W. and J.W. Daly. 1983. Dart-poison frogs. *Sci. Am.* **248:** 120–33.

Nakamura, D., Tiersch, T.R. Douglass, M. and R.W. Chandler. 1990. Rapid identification of sex in birds by flow cytometry. *Cytogen. Cell Genet.* **53:** 201–5.

Nakamura, R.R. and M.L. Stanton. 1989. Embryo growth and seed size in *Raphanus sativus*: maternal and paternal effects *in vivo* and *in vitro*. *Evolution* **43:** 1435–43.

Nalepa, C.A. 1988. Reproduction in the woodroach *Cryptocercus punctulatus* Scudder (Dictyoptera: Cryptocercidae): mating, oviposition, and hatch. *Annu.Entomol. Soc. Am.* **81:** 637–41.

Nalepa, C.A. 1990. Early development of nymphs and establishment of the hindgut symbiosis in *Cryptocercus punctulatus* Scudder (Dictyoptera: Cryptocercidae). *Annu.Entomol. Soc. Am.* **83:** 786–9.

Nalepa, C.A. 1994. Nourishment and the origins of termite eusociality. In *Nourishment and evolution in insect societies*. (ed. J.H. Hunt and C.A. Nalepa), pp.57–104. Westview Press, Boulder, and Oxford & IBH Publishing Co., Pvt. Ltd., New Delhi.

Nalepa, C.A. and D.E. Mullins. 1992. Initial reproductive investment and parental body size in *Cryptocercus punctulatus* Scudder (Dictyoptera: Cryptocercidae). *Physiol. Entom.* **17:** 255–9.

Nalepa, C.A. and W.J. Bell. 1996. Post-ovulation parental investment and parental care in cockroaches. In *Social competition in insects and arachnids:* Vol. 2. *Evolution of sociality* (ed., J.C. Choe and B.J. Crespi). Princeton University Press, Princeton, NJ.

Nelson, J.B. 1978. *The sulidae: gannets and boobies*. Oxford University Press, Oxford.

Nelson, J.B. 1989. Cainism in the sulidae. *Ibis* **131:** 609.

Newton, I. 1977. Breeding strategies in birds of prey. *Living Bird* **16:** 51–82.

Newton, I. 1979. *Population ecology of raptors*. T. and A.D. Poyser. Berkhamsted.

Newton, I. and D. Moss. 1986. Post-fledging survival of sparrowhawks *Accipiter nisus* in relation to mass, brood size and brood composition at fledging. *Ibis* **128:** 73–80.

Nicholson, A.J. 1954. An outline of the dynamics of animal populations. *Austr. J. Zool.* **2:** 9–65.

Niesenbaum, R.A. and B.B. Casper. 1994. Pollen tube numbers and selective fruit maturation in *Lindera benzoin*. *Am. Nat.* **144:** 184–91.

Nilsson, J.-A. and M. Svensson. 1993. Fledging in altricial birds: parental manipulation or sibling competition? *Anim. Behav.* **46:** 379–86.

Nisbet, I.C.T. 1973. Courtship feeding, egg size, and breeding success in common terns. *Nature* **241:** 141–2.

Nisbet, I.C.T. and W.H. Drury. 1972. Post-fledgling survival in herring gulls in relation to brood-size and date of hatching. *Bird-Banding* **43**: 161–72.

Nonacs, P. 1986. Ant reproductive strategies and sex allocation theory. *Q. Rev. Biol.* **61**: 1–21.

Noonan, K.M. 1981. Sex ratios of parental investment in colonies of the social wasp (*Polistes fuscatus*). *Science* 199:1354–6.

Novakowski, N.S. 1966. Whooping crane population dynamics on the nesting grounds, Wood Buffalo Park, Northwest Territories, Canada. *Can. Wildl. Serv. Rept. Series* No. 1, Ottawa.

Nuechterlein, G.L. 1981. Asynchronous hatching and sibling competition in western grebes. *Can. J. Zool.* **59**: 994–8.

Nuechterlein, G.L. and A. Johnson. 1981. The downy young of the hooded grebe. *Living Bird* **19**: 69–71.

Nur, N. 1984a. The consequences of brood size for breeding blue tits. I. Adult survival, weight change, and the cost of reproduction. *J. Anim. Ecol.* **54**: 479–96.

Nur, N. 1984b. The consequences of brood size for breeding blue tits. II. Nestling weight, offspring survival and optimal brood size. *J. Anim. Ecol.* **53**: 497–517

Nur, N. 1986. Is clutch size variation in the blue tit (*Parus caeruleus*) adaptive? An experimental study. *J. Anim. Ecol.* **55**: 983–99.

Nur, N. 1987. Parents, nestlings and feeding frequency: a model of optimal parental investment and implications for avian reproductive strategies. In *Foraging behavior* (ed. A.C. Kamil, Krebs, J.R. and H.R. Pulliam), pp. 457–75. Plenus Press, NY.

Nur, N. 1988. The cost of reproduction in birds: an examination of the evidence. *Ardea* **76**: 155–68.

Nussbaum, R.A. 1985. The evolution of parental care in salamanders. *Misc. Pub. Mus. Zool. University Mich.* **169**: 1–50.

Obeso, J.R. 1993. Selective fruit and seed maturation in *Asophodelus albus* Miller (Liliaceae). *Oecologia* **93**: 564–70.

O'Connor, R.J. 1975. Initial size and subsequent growth in passerine nestlings. *Bird-Banding* **46**: 329–40.

O'Connor, R.J. 1978. Brood reduction in birds: selection for infanticide, fratricide, and suicide? *Anim. Behav.* **26**: 79–96.

O'Connor, R.J. 1984. *The growth and development of birds.* J.Wiley & Sons, Chichester.

Odhiambo, T.R. 1959. An account of parental care in *Rhinocoris albopilosus* Signoret (Hemiptera-Heteroptera: Reduviidae) with notes on its life history. *Proc. R. Entomol. Soc. Lond. A.* **34**: 175–85.

Odhiambo, T.R. 1960. Parental care in bugs and non-social insects. *New Scient.* **8**: 449–51.

O'Gara, B. 1969. Unique aspects of reproduction in the female pronghorn (*Antilocapra americana*). *Am. J. Anat.* **125**: 217–32.

Ohlsson, T. and H.G. Smith. 1994. Development and maintenance of nestling size hierarchies in the European starling. *Wilson Bull.* **106**: 448–55.

Olsen, P.D. and A. Cockburn. 1991. Female-biased sex allocation in peregrine falcons and other raptors. *Behav. Ecol. Sociobiol.* **28**: 417–23.

O'Malley, J.B.E. and R.M. Evans. 1980. Variations in measurements among white pelican eggs and their use as a hatch date predictor. *Can. J. Zool.* **58**: 603–8.

Orians, G.H. 1969. Age and hunting success in the brown pelican (*Pelecanus occidentalis*). *Anim. Behav.* **17**: 316–19.

Orlove, M.J. 1975. A model of kin selection not invoking coefficients of relationship. J. Theor. Biol. **49:** 289–310.

Orlove, M.J. and C.L. Wood 1978. Coefficients of relationship and coefficients of relatedness in kin selection: a covariance form for the rho formula. *J. Theor. Biol.* **73:** 679–86.

Osawa, N. 1992a. A life table of the ladybird beetle *Harmonia axyridis* Pallas (Coleoptera, Coccinellidae) in relation to the aphid abundance. *Jpn. J. Ent.* **60:** 575–9.

Osawa, N. 1992b. Sibling cannibalism in the ladybird beetle *Harmonia axyridis*: Fitness consequences for mother and offspring. *Res. Popul. Ecol.* **34:** 45–55.

Osorno, J.-L. and H. Drummond. 1995. The function of hatching asynchrony in the blue-footed booby. *Behav. Ecol. Sociobiol.* **37:** 265–73.

Owen, D.F. 1955. The food of the heron *Ardea cinerea* in the breeding season. *Ibis* **97:** 276–95.

Owen, D.F. 1960. The nesting success of the heron *Ardea cinerea* in relation to the availability of food. *Proc. Zool. Soc. Lond.* **113:** 597–617.

Owen, M., Wells, R.C. and J.M. Black. 1992. Energy budgets of wintering barnacle geese: the effects of declining food resources. *Ornis Scand.* **23:** 451–8.

Owens, J.N. and M. Molder. 1977. Seed-cone differentiation and sexual reproduction in western white pine (*Pinus monticola*). *Can. J. Bot.* **55:** 2574–90.

Packer, C. and A.E. Pusey. 1982. Cooperation and competition within coalitions of lions: Kin selection or game theory? *Nature* **296:** 740–2.

Packer, C. and A.E. Pusey. 1984. Infanticide in carnivores. In *Infanticide. comparative and evolutionary perspectives* (ed. G. Hausfater and S. Blaffer Hrdy), pp.31–42. Aldine, New York.

Packer, C., Gilbert, D.A., Pusey, A.E. and S.J. O'Brien. 1991. A molecular genetic analysis of kinship and co-operation in African lions. *Nature* **351:** 562–5.

Palmer, R.S. (ed.) 1962. *Handbook of North American birds*, Vol 1. Yale University Press, New Haven, CT.

Parker, G.A. 1970. Sperm competition and its evolutionary consequences in the insects. *Biol. Rev.* **45:** 525–67.

Parker, G.A. 1974. Courtship persistence and female-guarding as male time investment strategies. *Behaviour* **48:** 157–84.

Parker, G.A. 1982. Phenotype limited evolutionarily stable strategies. In *Current problems in sociobiology* (ed. B.R. Bertram, T.H. Clutton-Brock, R.I.M. Dunbar, D.I. Rubenstein, and R. Wrangham), pp 173–201. Cambridge University Press, Cambridge.

Parker, G.A. 1984. Evolutionarily stable strategies. In *Behavioural ecology. An evolutionary approach* 2nd edn. (ed. J.R. Krebs and N.B. Davies), pp. 30–61. Blackwell Scientific Publications, Oxford.

Parker, G.A. 1985. Models of parent-offspring conflict. V. Effects of the behaviour of the two parents. *Anim. Behav.* **33:** 519–33.

Parker, G.A. 1989. Hamilton's rule and conditionality. *Ethol. Ecol. Evolution* **1:** 195–211.

Parker, G.A. and R.A. Stuart. 1976. Animal behavior as a strategy optimizer: evolution of resource assesssment strategies and optimal emigration thresholds. *Am. Nat.* **110:** 1055–76.

Parker, G.A. and M. Macnair. 1978. Models of parent-offspring conflict. I. Monogamy. *Anim. Behav.* **26:** 97–111.

Parker, G.A. and M. Macnair. 1979. Models of parent-offspring conflict. IV. Suppression: evolutionary retaliation by the parent. *Anim. Behav.* **27:** 1210–35.

Parker, G.A. and S.P. Courtney. 1984. Models of clutch size in insect oviposition. *Theor. Popul. Biol.* **26:** 27–48.

Parker, G.A. and M.E. Begon. 1986. Optimal egg size and clutch size: effects of environment and maternal phenotype. *Am. Nat.* **128:** 573–92.

Parker, G.A. and D. W. Mock. 1987. Parent-offspring conflict over clutch size. *Evol. Ecol.* **1:** 161–74.

Parker, G.A. and J. Maynard Smith. 1990. Optimality theory in evolutionary biology. *Nature* **348:** 27–33.

Parker, G.A., Mock, D.W. and T.C. Lamey. 1989. How selfish should stronger sibs be? *Am. Nat.* **133:** 846–68.

Parmelee, D.F., Stephens, R.A. and R.H. Schmidt. 1967. The birds of southeastern Victoria Island and adjacent small islands. *Natl. Mus. Can. Bull.* 222. Ottawa.

Parsons, J. 1970. Relationship between egg size and post-hatching chick mortality in the herring gulls (*Larus argentatus*). *Nature* **228:** 1221–2.

Parsons, J. 1975. Asynchronous hatching and chick mortality in the herring gull *Larus argentatus*. *Ibis* **117:** 517–20.

Payne, R.B. and C.J. Risley. 1976. Systematics and evolutionary relationships among the herons (Ardeidae). *Misc. Publications Mus. Zool., University Mich.* No. 150. Museum of Zoology, Ann Arbor.

Perrins, C.M. 1963. Survival in the great tit *Parus major. Proc. Intl. Ornithol. Congr.* XIII: 717–28.

Perrins, C.M. 1965. Population fluctuations and clutch size in the great tit, *Parus major* L. *J. Anim. Ecol.* **34:** 601–47.

Perrins, C.M. and D. Moss. 1975. Reproductive rates in the great tit. *J. Anim. Ecol.* **44:** 695–706.

Perrins, C.M., M.P. Harris, and C.K. Britton. 1972. Survival of Manx shearwaters *Puffinus puffinus. Ibis* **115:** 535–48.

Petrides, G.A. 1949. Sex and age determination in the opossum. *J. Mammal.* **30:** 364–78.

Pfennig, D.F. 1990a. 'Kin recognition' among spadefoot toad tadpoles: a side-effect of habitat selection? *Evolution* **44:** 785–98.

Pfennig, D.F. 1990b. The adaptive significance of an environmentally-cued developmental switch in an anuran tadpole. *Oecologia* **85:** 101–7.

Pfennig, D.F. 1992. Polyphenism in spadefoot toad tadpoles as a locally adjusted evolutionarily stable strategy. *Evolution* **46:** 1408–20.

Pfennig, D.W. and J.P. Collins. 1993. Kinship affects morphogenesis in cannibalistic salamanders. *Nature* **362:** 836–8.

Pfennig, D.W., Loeb, M.L.G. and J.P. Collins. 1991. Pathodgens as a factor limiting the spread of cannibalism in tiger salamanders. *Oecologia* **88:** 161–6.

Pfennig, D.W., Reeve, H.K. and P.W. Sherman. 1993. Kin recognition and cannibalism in spadefoot toad tadpoles. *Anim. Behav.* **46:** 87–94.

Pfennig, D.W., Sherman, P.W. and J.P. Collins. 1994. Kin recognition and cannibalism in polyphenic salamanders. *Behav. Ecol.* **5:** 225–32.

Pianka, E.R. 1976. Natural selection and optimal reproductive tactics. *Am. Zool.* **16:** 775–84.

Pienkowski, R.L. 1965. The incidence and effect of egg cannibalism in first-instar

Coleomegilla maculata lengi (Coleoptera: Coccinellidae). *Ann. Entomol. Soc. Am.* **58:** 150–2.

Pijanowski, B.C. 1992. A revision of Lack's brood reduction hypothesis. *Am. Nat.* **139:** 1270–92.

Pilz, W.R. 1976. Possible cannibalism in Swainson's hawk. *Auk* **93:** 838.

Pinson, D. and H. Drummond. 1993. Brown pelican siblicide and the prey-size hypothesis. *Behav. Ecol. Sociobiol.* **32:** 111–18.

Platt, J.R. 1964. Strong inference. *Science* **146:** 347–53.

Ploger, B.J. In press. Effect of brood size manipulations on food deliveries and apportionment in brown pelicans (*Pelecanus occidentalis*). *Anim. Behav.*

Ploger, B.J. 1992. Proximate and ultimate causes of brood reduction in brown pelicans (*Pelecanus occidentalis*). Unpublished Ph.D. thesis, University of Florida, Gainesville.

Ploger, B.J. and D.W. Mock. 1986. Role of sibling aggression in distribution of food to nestling cattle egrets (*Bubulcus ibis*). *Auk* **103:** 768–76.

Polge, C., Rowson, L.E.A. and M.C. Chang. 1966. The effect of reducing the number of embryos during early stages of gestation on the maintenance of pregnancy in the pig. *J. Reprod. Fert.* **12:** 395–7.

Polis, G.A. 1981. The evolution and dynamics of intraspecific predation. *Annu. Rev. Ecol. Syst.* **12:** 225–251.

Polis, G.A. 1984. Intraspecific predation and 'infant killing' among invertebrates. In *Infanticide. Comparative and evolutionary perspectives* (ed. G. Hausfater and S. Blaffer Hrdy), pp.87–104, Aldine, New York.

Pomeroy, D.E. 1978. Biology of marabou storks. II. Breeding biology and general review. *Ardea* **66:** 1–23.

Pomeroy, L.V. 1981. Developmental polymorphism in the tadpoles of the spadefoot toad. Unpublished Ph.D. thesis, University of California, Riverside.

Poole, A.L. 1952. The development of *Nothofagus* seed. *Trans. R. Soc. N.Z.* **80:** 207–12.

Poole, A. 1982. Brood reduction in temperate and sub-tropical ospreys. *Oecologia* **53:** 111–9.

Powell, G.V.N. 1983. Food availability and reproduction by great white herons, *Ardea herodias*: a food addition study. *Col. Waterbirds* **6:** 139–47.

Powers, J.H. 1907. Morphological variation and its causes in *Ambystoma tigrinum* (Green). *Studies Univ. of Mich.* **7:** 197–274.

Pratt, H.M. and D.W. Winkler. 1985. Clutch size, timing of laying, and reproductive success in a colony of great blue herons and great egrets. *Auk* **102:** 49–63.

Price, G.R. 1970. Selection and covariance. *Nature* **227:** 520–1.

Price, G.R. 1972. Extension of covariance selection mathematics. *Ann. Hum. Genet.* **35:** 485–90.

Price, K. and R. Ydenberg. 1995. Begging and provisioning in broods of asynchronously-hatched yellow-headed blackbird nestlings. *Behav. Ecol. Sociobiol.* **37:** 201–8.

Procter, D.L.C. 1975. The problem of chick loss in the South Polar skua *Catharacta maccormicki*. *Ibis* **117:** 452–9.

Pukowski, E. 1933. Okologische untersuchungen an *Necrophorus*. *Z. für Morphologie und Ökologie der Tiere* 27:518–86.

Pusey, A. and C. Packer. 1994. Non-offspring nursing in social carnivores: minimizing the costs. *Behav. Ecol.* **5:** 362–74.

Quast, W.D. and N.R. Howe. 1980. The osmotic role of the brood pouch in the pipefish *Syngnathus scovelli*. *Comp. Biochem. Physiol.* **67A:** 675–8.

Queller, D.C. 1983a. Kin selection and conflict in seed maturation. *J. Theor. Biol.* **100:** 153–72.

Queller, D.C. 1983b. Sexual selection in a hermaphroditic plant. *Nature* **305:** 706–8.

Queller, D.C. 1989. Inclusive fitness in a nutshell. In *Oxford surveys in evolutionary biology*, Vol. 6 (ed. P.H. Harvey and L. Partridge), pp.73–109. Oxford University Press, New York.

Queller, D.C. 1994. Male-female conflict and parent-offspring conflict. *Am. Nat.* **114:** S84–99.

Quinlan, J.D. and A.P. Preston. 1968. Effects of thinning blossoms and fruitlets on growth and cropping of sunset apple. *J. Hort. Sci.* **43:** 373–81

Quinlan, J.D. and A.P. Preston. 1971. The influence of shoot competition on fruit retention and cropping of apple trees. *J. Hort. Sci.* **46:** 525–34.

Quinn, J.A. 1987. Relationship between synaptospermy and dioecy in the life history strategies of *Buchloe dactyloides* (Gramineae). *Am. J. Bot.* **74:** 1167–72.

Quinn, J.A. and J.L. Engel. 1986. Life-history strategies and sex ratios for a cultivar and a wild population of *Buchloe dactyloides* (Gramineae). *Am. J. Bot.* **73:** 874–81.

Quinn, J.S. and R.D. Morris. 1986. Intraclutch egg-weight apportionment and chick survival in Caspian terns. *Can. J. Zool.* **64:** 2116–22.

Rabenold, P.P., Rabenold, K.N., Piper, W.H., Haydock, J. and S.W. Zack. 1990. Shared paternity revealed by genetic analysis in co-operatively breeding tropical wrens. *Nature* **348:** 538–40.

Ralston, J.S. 1977. Egg guarding by male assassin bugs of the genus *Zelus* (Hemiptera: Reduviidae. *Psyche* **84:** 103–7.

Ratniecks, F.L.W. 1988. Reproductive harmony via mutual policing by workers in eusocial Hymenoptera. *Am. Nat.* **132:** 217–36.

Ratniecks, F.L.W. and P.K. Visscher. 1989. Worker policing in the honeybee. *Nature* **342:** 796–7.

Redondo, T. 1991. Early stages of vocal ontogeny in the magpie (*Pica pica*). *J. Ornithol.* **132:** 145–63.

Redondo, T. 1993. Exploitation of host mechanisms for parental care by avian brood parasites. *Etologia* **3:** 235–97.

Redondo, T. and F. Castro. 1992a. The increase in risk of predation with begging activity in broods of magpies, *Pica pica. Ibis* **134:** 180–7.

Redondo, T. and F. Castro. 1992b. Signalling of nutritional need by magpie nestlings. *Ethology* **92:** 193–204.

Redondo, T., M. Gomendio, and R. Medina. 1992. Sex-biased parent-offspring conflict. *Behaviour* **123:** 261–89.

Reid, W. 1987. The cost of reproduction in the glaucous-winged gull. *Oecologia* **74:** 458–67.

Rettig, N.L. 1978. Breeding behavior of the harpy eagle (*Harpia harpyja*). *Auk* **95:** 629–43.

Rice, E.L. 1984. *Allelopathy*, 2nd edn. Academic Press. Orlando, FL.

Richner, H., Schneiter, P. and H. Stirnimann. 1989. Life-history consequences of growth rate depression: an experimental study on carrion crows (*Corvus corone corone* L.) *Funct. Ecol.* **3:** 617–24.

Ricklefs, R. 1965. Brood reduction in the curve-billed thrasher. *Condor* **67:** 505–10.

Ricklefs, R.E. 1968. On the limitation of brood size in passerine birds by the ability of adults to nourish their young. *Proc. Natl. Acad. Sci. USA* **61:** 847–51

Ricklefs, R.E. 1982. Some considerations on sibling competition and avian growth rates. *Auk* **99:** 141–7.

Ricklefs, R.E. 1993. Sibling competition, hatching asynchrony, incubation period, and lifespan in altricial birds. In *Curr. Ornithol. 11* (ed. D.M. Power), pp. 199–276. Plenum Press, New York.

Ridley, M. 1978. Paternal care. *Anim. Behav.* **26:** 904–32.

Robertson, A. 1966. A mathematical model of the culling process in dairy cattle. *Anim. Prod.* **8:** 95–108.

Rodríguez-Gironés, M.A. In press. Siblicide: the evolutionary blackmail. *Am. Nat.*

Rodríguez-Gironés, M.A., P.A. Cotton, and A. Kacelnik. 1996. The evolution of begging: signaling and sibling competition. *Proc. Natl. Acad. Sci. USA* 93:14637–41.

Rodríguez-Gironés, M.A., Drummond, H. and A. Kacelnic. In press. Effect of food deprivation on dominance status in blue-footed booby (*Sula nebouxii*) broods. *Behav. Ecol.*

Roff, D. 1992. *The evolution of life histories*. Chapman and Hall, London.

Roitberg, B.D. and M. Mangel. 1993. Parent-offspring conflict and life-history consequences in herbivorous insects. *Am. Nat.* **142:** 443–56.

Rood, J.P. 1980. Mating relationships and breeding suppression in the dwarf mongoose. *Anim. Behav.* **28:** 143–50.

Rose, F.L. and D. Armentrout. 1976. Adaptive strategies of *Ambystoma tigrinum* (Green) inhabiting the Llano Estacado of west Texas. *J. Anim. Ecol.* **45:** 713–29.

Rosenblatt, J., G. Turkewitz, and T.C. Schneirla. 1961. Early socialization in the domestic cat. In *Determinants of infant behavior* (ed. B. Foss), pp.51–74. Methuen, London.

Rosenheim, J.A. and D. Hongkham. 1996. Clutch size in an obligately siblicidal parasitoid wasp. *Anim. Behav.* **51:** 841–52.

Røskaft, E. 1985. The effect of enlarged brood size on the future reproductive potential of the rook. *J. Anim. Ecol.* **54:** 255–60.

Røskaft, E. and T. Slagsvold. 1985. Differential mortality of male and female offspring in experimentally manipulated broods of the rook. *J. Anim. Ecol.* **54:** 261–6.

Roth, L.M. and W. Hahn. 1964. Size of new-born larvae of cockroaches incubating eggs internally. *J. Insect Physiol.* **10:** 65–72.

Rowe, E.G. 1947. The breeding biology of *Aquila verreauxi* Lesson. *Ibis* **89:** 576–606.

Royama, T. 1966. A re-interpretation of courtship feeding. *Bird Study* **13:** 116–29.

Ruffer, D.G. 1965. Sexual behaviour of the northern grasshopper mouse (*Onychomys leucogaster*). *Anim. Behav.* **13:** 447–52.

Ruibal, R. and E. Thomas. 1988. The obligate carnivorous larvae of the frog, *Lepidobatrachus laevis* (Leptodactylidae). *Copeia* **1988:** 591–604.

Rydén, O. and H. Bengtsson. 1980. Differential begging and locomotory behaviour by early and late hatched nestlings affecting the distribution of food in asynchronously hatched broods of altricial birds. *Z. Tierpsychol.* **53:** 209–24.

Ryder, J.P. 1983. Sex ratio and egg sequence in ring-billed gulls. *Auk* **100:** 726–8.

Rydzewski, W. 1956. The nomadic movements and migrations of the European common heron *Ardea cinerea*. *Ardea* **44:** 71–188.

Sachs, R.M. and W.P. Hackett. 1983. Source-sink relationships and flowering. In

Strategies in plant reproduction (ed. W.J. Meudt), pp.263–72, Allanheld, Osmun and Co., Totowa.

Sadler, L.M. and M.A. Elgar. 1994. Cannibalism among amphibian larvae: a case of good taste. *Trends Ecol. Evolution* **9:** 5–6.

Safriel, U.N. 1981. Social hierarchy among siblings in broods of the oystercatcher *Haematopus ostralegus. Behav. Ecol. Sociobiol.* **9:** 59–63.

Salt, G. 1961. Competition among inestc parasitoids. Mechanisms in biological competition. *Symp. Soc. Exp. Biol.* **15:** 96–119.

Salthe, S.N. and J.S. Mecham. 1974. Reproductive and courtship patterns. In *Physiology of the Amphibia* (ed. B. Lofts), pp. 309–521. Academic Press, New York.

Salter, J.H. 1904. Nesting habits of the common buzzard. *Zoologist* **8:** 96–102. [cited in Ingram 1959, original not consulted].

Salzer, D. and G. Larkin. 1990. Impact of courtship feeding on clutch and third-egg size in glaucous-winged gulls. *Anim. Behav.* **39:** 1149–62.

Sargent, R.C. 1992. Ecology of filial cannibalism in firh: theoretical perspectives. In *Cannibalism: ecology and evolution among diverse taxa* (ed. M.A. Elgar and B.J. Crespi), pp. 38–62. Oxford University Press, Oxford.

Sarvas, R. 1962. Investigations on the flowering and seedcrop of *Pinus silvestris. Commun. Inst. Forest. Fenn.* **53:** 1–198 [cited in Haig 1992a, original not consulted].

Sarvas, R. 1968. Investigations on the flowering and seedcrop of *Picea abies. Commun. Inst. Forest. Fenn.* **67:** 1–84. [cited in Haig 1992a, original not consulted].

Sato, T. 1986. A parasitic catfish of mouthbreeding cichlid fishes in Lake Tanganyika. *Nature* **323:** 58–9.

Schachak, M.E., Chapman, A., and Y. Steinberger. 1976. Feeding, energy flow and soil turnover in the desert isopod, *Hemilepistus reamuri. Oecologia* **24:** 57–69.

Schadler, M.H. and G.M. Butterstein. 1979. Reproduction in the pine vole, *Microtus pinetorum. J. Mammal.* **60:** 841–4

Scheel, D.E., Graves, H.B. and G.W. Sherritt. 1977. Nursing order, social dominance and growth in swine. *J. Anim. Sci.* **45:** 219–29.

Schemske, D.W. 1978. Evolution of reproductive characteristics in *Impatiens* (Balsaminaceae): the significance of cleistogamy and chasmogamy. *Ecology* **59:** 596–613.

Schemske, D.W. and L.P. Pautler. 1984. The effects of pollen composition on fitness components in a neotropical herb. *Oecologia* **62:** 31–6.

Schifferli, L. 1978. Experimental modification of brood size among house sparrows *Passer domesticus. Ibis* **120:** 365–9.

Schneider, L., Bessis, R. and T. Simmonet. 1979. The frequency of ovular resorption during the first trimester of twin pregnancy. *Acta. Genet. Med. Gemellol.* **28:** 271–2.

Schonecht, P.A. 1984. Growth and teat ownership in a litter of binturongs. *Zoo Biol.* **3:** 272–7.

Schüz, E. 1943. Uber die jungenaufzucht des weissen storches (*C. ciconia*). *Z. Morph. Okol. Tiere* **40:** 181–237.

Schüz, E. 1957. Das verschlingen eigener junger ('Kronismus') bei Vogeln und seine bedeutung. *Vogelwarte* **19:** 1–15.

Schwabl, H. 1993. Yolk is a source of maternal testosterone for developing birds. *Proc. Natl. Acad. Sci. USA.* **90:** 11446–50.

Schwabl, H. 1996. Maternal testosterone in the avian egg enhances postnatal growth. *Comp. Biochem. Physiol.* **114 A:** 271–6.

Schwabl, H., Mock, D.W. and J.A. Gieg. 1997. A hormonal mechanism for parental favouritism. *Nature.* **386:** 231.

Schwagmeyer, P.L. 1988. Ground squirrel kin recognition abilities: are there social and life history correlates? *Behav. Genet.* **18:** 495–510.

Schwagmeyer, P.L. and G.A. Parker. 1987. Queuing for mates in thirteen-lined ground squirrels. *Anim. Behav.* **35:** 1015–25.

Schwagmeyer, P.L. and G.A. Parker. 1990. Male mate choice as predicted by sperm competition in thirteen-lined ground squirrels. *Nature* **348:** 62–4.

Schwagmeyer, P.L., Mock, D.W., Lamey, T.C., Lamey, C.S. and M.D. Beecher. 1991. Effects of sibling contact on hatch timing in an asynchronously hatching bird. *Anim. Behav.* **41:** 887–894.

Scott, M.P. 1990. Brood guarding andd the evolution of male parental care in burying beetles. *Behav. Ecol. Sociobiol.* **26:** 31–9.

Scott, M.. and F.A. Traniello. 1990. Behaviorural and ecological correlates of male and female parental care and reproductive success in burying beetles. *Anim. Behav.* **39:** 274–83.

Scott, P.E. and R.F. Martin. 1986. Clutch size and fledging success in the turquoise-browed motmot. *Auk* **103:** 8–13.

Searcy, K.B. and M.R. Macnair. 1993. Developmental selection in response to environmental conditions of the maternal parent in *Mumulus guttatus*. *Evolution* **47:** 13–24.

Sedgley, M. 1981. Early development of the *Macadamia* ovary. *Austr. J. Bot.* **29:** 185–93.

Seger, J. 1981. Kinship and covariance. *J. Theor. Biol.* **91:** 119–213.

Seger, J. 1991. Cooperation and conflict in social insects. In *Behavioural ecology. An evolutionary approach,* 3rd edn. (ed. J.R. Krebs and N.B. Davies), pp.338–73. Blackwell Scientific Publications, Oxford.

Semlitsch, R.D. and J.P. Caldwell. 1982. Effects of density on growth, metamorphosis, and survivorship in tadpoles of *Scaphiopus holbrooki. Ecology* **63:** 905–11.

Shaw, P. 1985. Brood reduction in the blue-eyed shag *Phalacrocorax atriceps. Ibis* **127:** 476–94.

Sheldon, F.H. 1987. Rates of single-copy DNA evolution in herons. *Mol. Biol. Evol.* **4:** 56–69.

Shine, R. 1985. The evolution of viviparity in reptiles: An ecological analysis. In *Biology of the reptilia,* Vol. 15 (ed. C. Gans and F. Billett), pp.605–94. Wiley, New York.

Shine, R. 1988. Parental care in reptiles. In *Biology of the reptilia.* Vol. 16 (ed. C. Gans), pp.276–329. Alan R. Liss, New York.

Sibley, C.G. and J.E. Ahlquist. 1990. *Phylogeny and classification of birds. A study in molecular evolution.* Yale University Press, New Haven, CT.

Siegfried, W.R. 1968. Breeding season, clutch, and brood sizes in Verreaux's eagle. *Ostrich* **39:** 139–45.

Siegfried, W.R. 1972. Breeding success and reproductive output of the cattle egret. *Ostrich* **43:** 43–55.

Sikes, R.S. 1995a. Costs of lactation and optimal litter size in northern grasshopper mice (*Onychomys leucogaster*). *Anim. Behav.* **76:** 348–57.

Sikes, R.S. 1995b. Maternal response to resource limitations in eastern woodrats. *Anim. Behav.* **49:** 1551–8.

Sikes, R.S. 1996. Effects of maternal nutrition on post-weaning growth in two North American rodents. *Behav. Ecol. Sociobiol.* **38:** 303–10.

Simmons, L.W. 1987. Competition between the larvae of the field cricket, *Gryllus bimaculatus* (Orthoptera: Gryllidae) and its effects on some life-history components of fitness. *J. Anim. Ecol.* **56:** 1015–27.

Simmons, R.E. 1988. Offspring quality and the evolution of Cainism. *Ibis* **130:** 339–57.

Simmons, R.E. 1991. Offspring quality and sibling aggression in the black eagle. *Ostrich* **62:** 89–92.

Simons, L.S. and T.E. Martin. 1990. Food limitation of avian reproduction: an experiment with the cactus wren. *Ecology* **71:** 869–76.

Sivinski, J. 1984. Sperm in competition. In *Sperm competition.* (ed. R.L. Smith), pp.86–116. Academic Press. Orlando, FL.

Skagen, S.K. 1987. Hatching asynchrony in American goldfinches: an experimental study. *Ecology* **68:** 1747–59.

Skagen, S.K. 1988. Asynchronous hatching and food limitation: a test of Lack's hypothesis. *Auk* **105:** 78–88.

Slagsvold, T. 1985. Asynchronous hatching in passerine birds: influence of hatching failure and brood reduction. *Ornis Scand.* **16:** 81–7.

Slagsvold, T. 1990. Fisher's sex ratio theory may explain hatching patterns in birds. *Evolution* **44:** 1009–17.

Slagsvold, T. and J.T. Lifjeld. 1989. Hatching asynchrony in birds: the hypothesis of sexual conflict over parental investment. *Am. Nat.* **134:** 239–53.

Slagsvold, T., Amundsen, T. and S. Dale. 1994. Selection by sexual conflict for evenly spaced offspring in blue tits. *Nature* **370:** 136–8.

Slagsvold, T., Sandvik, J., Rofstad, G., Lorentsen, O., and M. Husby. 1984. On the adaptive value of intraclutch egg-size variation in birds. *Auk* **101:** 685–97.

Smith, C.C. and S.D. Fretwell. 1974. The optimal balance between size and number of offspring. *Am. Nat.* **108:** 499–506.

Smith, D.C. 1983 Factors controlling tadpole populations of the chorus frog (*Pseudacris triseriatus*). *Ecology* **64:** 501–10.

Smith, D.C. 1987. Adult recruitment in chorus frogs: effects of size and date at metamorphosis. *Ecology* **68:** 344–50.

Smith, D.C. 1990. Population structure and competition among kin in the chorus frog (*Pseudacris triseriata*). *Evolution* **44:** 1529–41.

Smith, H.G. and R. Montgomerie. 1991. Nestling American robins compete with siblings by begging. *Behav. Ecol. Sociobiol.* **29:** 307–12.

Smith, R.H. and C.M. Lessells. 1985. Oviposition, ovicide and larval competition in granivorous insects. In *Behavioural ecology: ecological consequences of adaptive behaviour* (ed. R.M. Sibley and R.H. Smith), pp. 423–48. Blackwell Scientific Publications, Oxford.

Smith, R.L. 1979a. Paternity assurance and altered roles in the mating behaviour of a giant water bug, *Abedus herberti* (Heteroptera, Belostomatidae). *Anim. Behav.* **27:** 716–23.

Smith, R.L. 1979b. Repeated copulation and sperm precedence: paternity assurance for a male brooding water bug. *Science* **205:** 1029–31.

Smith, R.L. 1980. Evolution of exclusive postcopulatory paternal care in the insects. *Fla. Entomol.* **63:** 65–78.

Smith-Gill, S.J. 1983. Developmental plasticity: developmental conversion versus phenotypic modulation. *Am. Zool.* **23:** 47–55.

Spear, L.B. and N. Nur. 1994. Brood size, hatching order and hatching date: effects on four life-history stages from hatching to recruitment in western gulls. *J. Anim. Ecol.* **63:** 283–98.

Spellerberg, I.F. 1971a. Breeding behaviour of the McCormick skua *Catharacta mccormicki* in Antarctica. *Ardea* **59:** 189–230.

Spellerberg, I.F. 1971b. Aspects of McCormick skua breeding biology. *Ibis* **113:** 357–63.

Spight, T.M. 1976. Hatching size and the distribution of nurse eggs among prosobranch embryos. *Biol. Bull.* **150:** 491–9.

Springer, S. 1948. Oviphagous embryos of the sand shark, *Carcharias taurus. Copeia* **1948:** 153–7.

St Clair, C.C. 1992. Incubation behavior, brood patch formation and obligate brood reduction in Fiordland crested penguins. *Behav. Ecol. Sociobiol.* **31:** 409–16.

St Clair, C.C. 1996. Multiple mechanisms of reversed hatching asynchrony in rockhopper penguins. *J. Anim. Ecol.* **65:** 485–94.

St Clair, C.C. and R.C. St Clair. 1996. Causes and consequences of loss in rockhopper penguins, *Eudyptes chrysocome. Oikos* **77:** 459–66.

St Clair, C.C., Waas, J.R., St Clair, R.C. and P.T. Boag. 1995. Unfit mothers? Maternal infanticide in royal penguins. *Anim. Behav.* **50:** 1177–85.

Stacey, P.B. and J.D. Ligon. 1987. Territory quality and dispersal options in the acorn woodpecker, and a challenge to the habitat-saturation model of co-operative breeding. *Am. Nat.* **130:** 654–76.

Stamps, J. 1990. When should avian parents differentially provision sons and daughters? *Am. Nat.* **135:** 671–85.

Stamps, J. 1993. Begging in birds. *Etología* **3:** 69–77.

Stamps, J. and R.A. Metcalf. 1980. Parent-offspring conflict. In *Sociobiology: beyond nature-nurture?* (ed. G. Barlow and J. Silverberg), pp. 598–618. Westview Press, Boulder, Col.

Stamps, J., Metcalf, R.A. and V.V. Krishnan. 1978. A genetic analysis of parent-offspring conflict. *Behav. Ecol. Sociobiol.* **3:** 369–92.

Stamps, J., Clark, A.B., Arrowood, P. and B. Kus. 1985. Parent-offspring conflict in budgerigars. *Behaviour* **94:** 1–40.

Stamps, J., Clark, A.B., Arrowood, P. and B. Kus. 1989. Begging behaviour in budgerigars. *Ethology* **81:** 177–92.

Stanback, M.T. 1994. Dominance within broods of the co-operatively breeding acorn woodpecker. *Anim. Behav.* **47:** 1121–6.

Stanback, M.T. and W.D. Koenig. 1992. Cannibalism in birds. In *Cannibalism: ecology and evolution among diverse taxa* (ed. M.A. Elgar and B.J. Crespi), pp. 277–98. Oxford University Press, Oxford.

Stanton, M.L. 1984. Seed variation in wild radish: effect of seed size on components of seedling and adult fitness. *Ecology* **65:** 1105–12.

Stanton, M.L. Bereczky, J.K. and H.D. Hasbrouck. 1987. Pollination thoroughness and maternal yield regulation in wild radish Raphus raphanus raphanistrum (Brassicaceae). *Oecologia* **74:** 68–76.

Stearns, S.C. 1987. The selection arena hypothesis. In *The evolution of sex and its consequences* (ed. S.C. Stearns), pp. 337–49. Birkhauser, Basel, Switzerland.

Stearns, S.C. 1992. *The evolution of life-histories.* Oxford University Press, Oxford.

Stephenson, A.G. 1980. Fruit set, herbivory, fruit reduction, and the fruiting strategy of *Catalpa speciosa* (Bignoniaceae). *Ecology* **61:** 57–64.

Stephenson, A.G. 1981. Flower and fruit abortion: proximate causes and ultimate functions. *Annu. Rev. Ecol. Syst.* **12:** 253–80.

Stephenson, A.G. and R.I. Bertin. 1983. Male competition, female choice, and sexual selection in plants. In *Pollination biology.* (ed. L. Real), pp. 109–49. Academic Press, New York.

Stephenson, A.G. and J.A. Winsor. 1986. *Lotus corniculatus* regulates offspring quality through selective fruit abortion. *Evolution* **40:** 453–8.

Steyn, P. 1980. Further observations on the tawny eagle. *Ostrich* **51:** 54–5.

Steyn, P. 1982. *Birds of prey of southern Africa.* Croom Helm, Beckenham.

Stinson, C.H. 1977. Growth and behaviour of young ospreys *Pandion haliaetus. Oikos* **28:** 299–303.

Stinson, C.H. 1979. On the selective advantage of fratricide in raptors. *Evolution* **33:** 1219–25.

Stoddart, D.M. and R.W. Braithwaite. 1979. A strategy for utilization of regenerating heathland habitat by the brown bandicoot (*Isodon obesulus*; Marsupialia, Peramelidae). *J. Anim. Ecol.* **48:** 165–79.

Stoleson, S.H. and S.R. Beissinger. 1995. Hatching asynchrony and the onset of incubation in birds, revisited: When is the critical period? *Curr. Ornithol.* **12:** 191–270.

Strassmann, J.E. 1993. Weak queen or social contract? *Nature* **363:** 502–3.

Strickland, D. 1991. Juvenile dispersal in gray jays: dominant brood member expels sibling from natal territory. *Can. J. Zool.* **69:** 2935–45.

Stuart, R.J. 1991. Kin recognition as a functional concept. *Anim. Behav.* **41:** 1093–4.

Stuart, R.J., Francoeur, A. and R. Loiselle. 1987. Lethal fighting among dimorphic males of the ant, *Cardiocondyla wroughtonii. Naturwissenschaften* **74:** 548–9.

Sundström, L. 1994. Sex ratio bias, relatedness asymmetry, and queen mating frequency in ants. *Nature* **367:** 266–8.

Sutherland, S. and L.F. Delph. 1984. On the importance of male fitness in plants: patterns of fruit set. *Ecology* **65:** 1093–104.

Sweet, G.B. 1973. Shedding of reproductive structures in forest trees. In *Shedding of plant parts* (ed. T.T. Kozlowski), pp. 341–82. Academic Press, NY.

Taborsky, M. 1994. Sneakers, satellites, and helpers: parasitic and co-operative behavior in fish reproduction. *Adv. Study Behav.* **23:** 1–99.

Taigen, T.L., F.H. Pough, and M.M. Stewart. 1984. Water balance of terrestrial anuran (*Eleuthrodactylus coqui*) eggs: importance of parental care. *Ecology* **65:** 248–55.

Tait, D.E.N. 1980. Abandonment as a reproductive tactic in grizzly bears. *Am. Nat.* **115:** 800–8.

Teather, K.L. 1992. An experimental study of competition for food between male and female nestlings of the red-winged blackbird. *Behav. Ecol. Sociobiol.* **31:** 81–7.

Teather, K.L. and P.J. Weatherhead. 1988. Sex-specific energy requirements of great-tailed grackle (*Quiscalus mexicanus*) nestlings. *J. Anim. Ecol.* **57:** 659–68.

Teicher, M.H. and E.M. Blass. 1976. Suckling in newborn rats: eliminated by nipple lavage, reinstated by pup saliva. *Science* **193:** 422–4.

Teicher, M.H. and E.M. Blass. 1977. First suckling response of the newborn albino rat: the roles of olfaction and amniotic fluid. *Science* **198:** 635–6.

Telfair, R.C. IV 1983. *The cattle egret. A Texas focus and world view.* C. Kleberg Foundation, Kingsville.

Temme, D.H. 1986. Seed size variability: a consequence of variable genetic quality among offspring? *Evolution* **40:** 414–17.

Temme, D.H. and E.L. Charnov. 1987. Brood size adjustment in birds: economical tracking in a temporally varying environment. *J. Theor. Biol.* **126:** 137–47.

Tepedino, V. and D. Frohlich. 1984. Fratricide in *Megachile rotundata*, a non-social megachilid bee: impartial treatment of sibs and nonsibs. *Behav. Ecol. Sociobiol.* **15:** 19–23.

Thomas, L. 1995. The evolution of parental care in assassin bugs. Unpublished D.Phil. thesis, University of Cambridge.

Thornhill, R. 1980. Rape in *Panorpa* scorpionflies and a general rape hypothesis. *Anim. Behav.* **28:** 52–9.

Tiersch, T.R. and R.L. Mumme. 1993. An evaluation of the use of flow cytometry to identify sex in the Florida scrub jay. *J. Field Ornithol.* **64:** 18–26.

Tiersch, T.R., Mumme, R.L., Chandler, R.W. and D. Nakamura. 1991. The use of flow cytometry for rapid identification of sex in birds. *Auk* **108:** 206–7.

Tinbergen, J.M. and M.C. Boerlijst. 1990. Nestling weight and survival in individual great tits (*Parus major*). *J. Anim. Ecol.* **59:** 1113–27.

Tilley, S.G. 1972. Aspects of parental care and embryonic development in *Desmognathus achrophaeus*. *Copeia* 1972**:** 532–40.

Tinkle, D.W., Dunham, A.E. and J.D. Congdon. 1993. Life history and demographic variation in the lizard *Sceloporus graciosus*: a long-term study. *Ecology* **74:** 2413–29.

Townsend, D.S., Stewart, M.M. and F.H. Pough. 1984. Male parental care and its adaptive significance in a neotropical frog. *Anim. Behav.* **32:** 421–31.

Tortosa, F.S. and T. Redondo. 1992. Motives for parental infanticide in white storks (*Ciconia ciconia*). *Ornis Scand.* **23:** 1859.

Trillmich, F. 1986. Maternal investment and sex-allocation in the Galapagos fur seal, *Arctocephalus galapagoensis*. *Behav. Ecol. Sociobiol.* **19:** 157–64.

Tripp, H.R.H. 1971. Reproduction in elephant-shrews (Macroscelididae) with special reference to ovulation and implantation. *J. Reprod. Fert.* **26:** 149–59.

Trivers, R.L. 1972. Parental investment and sexual selection. In *Sexual selection and the descent of man, 1871–1971* (ed. B. Campbell), pp.136–79. Aldine Atherton, Chicago.

Trivers, R.L. 1974. Parent-offspring conflict. *Am. Zool.* **14:** 249–64.

Trivers, R.L. and H. Hare. 1976. Haplodiploidy and the evolution of social insects. *Science* **191:** 249–63.

Trivers, R.L. and D.E. Willard. 1973. Natural selection of parental ability to vary the sex ratio of offspring. *Science* **179:** 90–2.

Tyler, M.J. and D.B. Carter. 1981. Oral birth of the young of the gastric brooding frog *Rheobatrachus silus*. *Anim. Behav.* **29:** 280–2.

Tyndale-Biscoe, C.H. and M.B. Renfree. 1987. *Reproductive physiology of marsupials.* Cambridge University Press, Cambridge.

Udovic, D. and C. Aker. 1981. Fruit abortion and the regulation of fruit number in *Yucca whipplei*. *Oecologia* **49:** 245–8.

Ueda, H. 1986. Reproduction of *Chirixalus eiffingeri* (Boettger). *Sci. Rep. Lab. Amphib. Biol., Hiroshima University* **8:** 109–16.

Uma Shaanker, R. and K.N. Ganeshaiah. 1988. Bimodal distribution of seeds per pod in *Caesalpinia pulcherrima*-parent-offspring conflict? *Evol. Trends Plants* **2:** 91–8.

Uma Shaanker, R. and K.N. Ganeshaiah. 1989. Stylar plugging by fertilized ovules in *Kleinhovia hospita* (Sterculeaceae)-a case of vaginal sealing in plants? *Evol. Trends Plants* **31:** 59–64.

Uma Shaanker, R., Ganeshaiah, K.N. and K.S. Bawa. 1988. Parent-offspring conflict, sibling rivalry, and brood size patterns in plants. *Annu. Rev. Ecol. Syst.* **19:** 177–205.

Urrutia, L.P. and H. Drummond. 1990. Brood reduction and parental infanticide in Heermann's gull. *Auk* **107:** 772–4.

Valerio, M. and G.W. Barlow. 1986. Ontogeny of young Midas cichlids: a study of feeding, filial cannibalism and agonism in relation to differences in size. *Biol. Behavior* **11:** 16–35.

van den Bosch, F., de Roos, A.M. and W. Gabriel. 1988. Cannibalism as a life boat mechanism. *J. Math. Biol.* **26:** 619–33.

van Heezik, Y.M. and P.J. Seddon. 1996. Scramble feeding in jackass penguins: within-brood food distribution and the maintenance of sibling asymmetries. *Anim. Behav.* **51:** 1383–90.

Vehrencamp, S.L. 1983. A model for the evolution of despotic versus egalitarian species. *Anim. Behav.* **31:** 667–82.

Verhulst, S. 1994. Supplementary food in the nestling phase affects reproductive success in pied flycatchers (*Ficedula hypoleuca*). *Auk* **111:** 714–6.

Verner, J. and M.F. Willson. 1969. Mating systems, sexual dimorphism, and the role of male North American passerine birds in the nesting cycle. *Ornithol. Monogr.* **9:** 1–76. American Ornithologists' Union, Washington, DC.

Vesey-Fitzgerald, D. 1957. The breeding of the white pelican (*Pelecanus onocrotalus*) in the Rukwa Valley, Tanganyika. *Bull. Brit. Ornithol. Club* **77:** 1279.

Vestal, B.M. and H. McCarley. 1984. Spatial and social relations of kin in thirteen-lined and other ground squirrels In *The biology of ground-dwelling squirrels.* (ed. J.O. Murie and G.R. Michener), pp.404–23. University of Nebraska Press, Lincoln.

Vince, M.A. 1969. Embryonic communication, respiration, and the synchronization of hatching. In *Bird vocalization* (ed. R.A. Hinde), pp. 233–60. Cambridge University Press, Cambridge.

Vinegar, A., Hutchison, V.H. and H.G. Dowling. 1970. Metabolism, energetics and thermoregulation during brooding of snakes in the genus *Python* (Reptilia: Boidae). *Zoologica* **55:** 19–48.

Vincent, A.C.J. 1990. Reproductive ecology of seahorses. Ph.D. dissertation, University of Cambridge, Cambridge.

Vincent, A.C.J., Ahnesjö, I., Berglund, A. and G. Rosenqvist. 1992. Pipefishes and seahorses: are they all sex role reversed? *Trends Ecol. Evolution* **7:** 237–41.

Visscher, P.K. 1989. A quantitative study of worker reproduction in honey bee colonies. *Behav. Ecol. Sociobiol.* **25:** 247–55.

Visscher, P.K. 1996. Reproductive conflict in honey bees: a stalemate of worker egg-laying and policing. *Behav. Ecol. Sociobiol.* **39:** 237–44.

Vitt, L.J. and D.G. Blackburn. 1983. Reproduction in the lizard *Mabuya heathi*

(Scincidae): Commentary on viviparity in New World *Mabuya. Can. J. Zool.* **61:** 2798–806.

vom Saal, F.S. 1984. Proximate and ultimate causes of infanticide and parental behavior in male house mice. In *Infanticide. Comparative and evolutionary perspectives* (ed. G. Hausfater and S. Blaffer Hrdy), pp. 401–24. Aldine, New York.

vom Saal, F.S. 1989. Sexual differentiation in litter-bearing mammals: influence of sex of adjacent fetuses *in utero. J. Anim. Sci.* **67:** 1824–40.

vom Saal, F.S. and F.H. Bronson. 1980. Sexual characteristics of adult female mice are correlated with their blood testosterone levels during prenatal development. *Science* **208:** 597–9.

von Haartman, L. 1953. Was reizt den Trauerfliegeschnapper (*Muscicapa hypoleuca*) zu futtern? *Vogelwarte* **16:** 157–64.

Wake, M.H. 1977. The reproductive biology of caecilians: an evolutionary perspective. In *The reproductive biology of amphibians* (ed. D.H. Taylor and S.I. Guttman), pp. 73–101. Plenum Press, New York.

Wake, M.H. 1980. Reproduction, growth, and population structure of the caecilian *Dermophis mexicanus. Herpetologica* **36:** 244–56.

Wake, M.H. 1982. Diversity within a framework of constraints: reproductive modes in the Amphibia. In *Environmental adaptation and evolution: a theoretical and empirical approach* (ed. D. Mossakowski and G. Roth), pp.87–106. Gustav Fischer Verlag, Stuttgart.

Wake, M.H. 1992. Reproduction in caecilians. In *Reproductive biology of South American* vertebrates. (ed. W.C. Hamlett), pp.112–20. Springer-Verlag, New York.

Wakerly, J.B., Clarke, G. and A.J.S. Summerlee. 1988. Milk ejection and its control In *The physiology of reproduction.* (ed. E. Knobil and J. Neil), pp. 2283–321. Raven Press, New York.

Waldman, B. 1988. The ecology of kin recognition. *Annu. Rev. Ecol. Syst.* **19:** 543–71.

Waldman, B. 1991. Kin recognition in amphibians. In *Kin recognition* (ed. P.G. Hepper), pp.162–219. Cambridge University Press, Cambridge.

Waldman, B. and K. Adler. 1979. Toad tadpoles associate preferentially with siblings. *Nature* **282:** 611–13.

Walk, R.D. 1978. Depth perception and experience In *Perception and experience.* (ed. R.D. Wald and H.L. Pick), pp. 77–103. Plenum, New York.

Walkinshaw, L.H. 1973. *Cranes of the world.* Winchester Press, New York.

Walls, S.C. 1991. Ontogenetic shifts in the recognition of siblings and neighbours by juvenile salamanders. *Anim. Behav.* **42:** 423–34.

Walls, S.C. and R. Roudebush. 1991. Reduced aggression toward siblings as evidence of kin recognition in cannibalistic salamanders. *Am. Nat.* **138:** 1027–38.

Walls, S.C. and A.R. Blaustein. 1995. Larval marbled salamanders, *Ambystoma opacum*, eat their kin. *Anim. Behav.* **50:** 537–45.

Walls, S.C., Beatty, J.J., Tissot, B.N., Hokit, D.G. and A.R. Blaustein. 1993a. Morphological variation and cannibalism in a larval salamander (*Ambystoma macrodactylum columbianum*). *Can. J. Zool.* **71:** 1543–51.

Walls, S.C., Belanger, S.S.and A.R. Blaustein. 1993b. Morphological variation in a larval salamander: dietary induction of plasticity in head shape. *Oecologia* **96:** 162–8.

Walsh, J.F. 1980. Inter-sibling conflict in the laughing dove. *Ostrich* **51:** 191.

Wandrey, R. 1975. Contribution to the study of social behaviour of golden jackals (*Canis aureus* L.). *Z. Tierpsychol.* **39:** 365–402.

Ward, H.M. 1892. *The oak.* Appleton, New York.

Warham, J. 1974. The Fiordland crested penguin *Eudyptes pachyrhynchus. Ibis* **116:** 1–27.

Warham, J. 1975. The crested penguins. In *The biology of penguins* (ed. B. Stonehouse), pp. 189–269. University Park Press, Baltimore.

Waser, P.M. 1985. Does competition drive dispersal? *Ecology* **66:** 1170–5.

Waser, P.M. 1988. Resources, philopatry, and social interactions among mammals. In *The ecology of social behavior* (ed., C.N. Slobodchikoff), pp.109–30. Academic Press, New York.

Waser, P.M. and W.T. Jones. 1983. Natal philopatry among solitary mammals. *Q. Rev. Biol.* **58:** 355–90.

Wasser, S.K. and D.P. Barash. 1983. Reproductive suppression among female mammals: implications for biomedicine and sexual selection theory. *Q. Rev. Biol.* **58:** 513–38.

Wassersug, R.J. 1973. Aspects of social behavior in anuran larvae. In *Evolutionary biology of the anurans* (ed., J.L. Vial), pp. 273–97. University of Missouri Press, Columbia.

Watson, M.A. and B.B. Casper. 1984. Morphogenetic constraints on patterns of carbon distribution in plants. *Annu. Rev. Ecol. Syst.* **15:** 233–58.

Weathers, W.W. and K.A. Sullivan. 1989. Juvenile foraging proficiency, parental effort, and avian reproductive success. *Ecol. Monogr.* **59:** 223–46.

Webel, S.K. and P.J. Dziuk. 1974. Effect of stage of gestation and uterine space on prenatal survival in the pig. *J. Anim. Sci.* **38:** 960–3.

Weir, B.J. 1971a. The reproductive physiology of the plains viscacha, *Lagostomus maximus. J. Reprod. Fert.* **25:** 355–64.

Weir, B.J. 1971b. The reproductive organs of the female plains viscacha, *Lagostomus maximus. J. Reprod. Fert.* **25:** 365–73.

Welcomme, R.L. 1967. The relationship between fecundity and fertility in the mouthbrooding cichlid fish *Tilapia lecosticta. J. Zool.* **151:** 453–68.

Wells, K.D. 1978. Courtship and parental behavior in a Panamanian poison arrow frog (*Dendrobates auratus*). *Herpetologica* **34:** 148–55.

Wells, K.D. 1980a. Behavioral ecology and social organization of a dendrobatid frog (*Colostethus inguinalis*). *Behav. Ecol. Sociobiol.* **6:** 199–209.

Wells, K.D. 1980b. Evidence for growth of tadpoles during parental transport in *Colostethus inguinalis. J. Herpetol.* **14:** 426–8.

Wells, K.D. 1981. Parental behavior in male and female frogs. In *Natural selection and social behavior* (ed. R.D. Alexander and D.w. Tinkle), pp. 184–97. Chiron Press, New York.

Werschkul, D.F. 1979. Nestling mortality and the adaptive significance of early locomotion in the little blue heron. *Auk* **96:** 116–30.

Werschkul, D.F. and J.A. Jackson. 1979. Sibling competition and avian growth rates. *Ibis* **121:** 97–102.

West, M.J. and R.D. Alexander. 1963. Sub-social behavior in a burrowing cricket *Anurogryllus muticus* (DeGeer). Orthoptera: Gryllidae. *Ohio J. Sci.* **63:** 19–24.

West Eberhard, M.J. 1969. The social biology of polistine wasps. *Misc. Publ. Mus. Zool. University of Michigan* **140:** 1–101.

West Eberhard, M.J. 1981. Intragroup selection and the evolution of insect societies. In *Natural selection and social behavior* (ed. R.D. Alexander and D.W. Tinkle), pp. 3–17, Chiron Press, New York.

Westerterp, K., Gortmaker, W. and H. Wijngaarden. 1982. An energetic optimum in brood-raising in the starling *Sturnus vulgaris*: an experimental study. *Ardea* **70:** 153–62.

Westneat, D.F., Sherman, P.W. and M.L. Morton. 1990. The ecology and evolution of extra-pair copulations in birds. *Curr. Ornithol* **7:** 331–69.

Westoby, M. and B. Rice. 1982. Evolution of the seed plants and inclusive fitness of plant tissues. *Evolution* **36:** 713–24.

Weygoldt, P. 1980. Complex brood care and reproductive behavior in captive poison-arrow frogs, *Dendrobates pumilio*. *Behav. Ecol. Sociobiol.* **7:** 329–32.

Weygoldt, P. 1987. Evolution of parental care in dart poison frogs (Amphibia: Anura: Dendrobatidae). *Z. Zool. Systematik Evolutionsforschung* **1:** 51–67.

White, M.D. 1973. *Animal cytology and evolution,* 3rd edn. Cambridge University Press, Cambridge.

Wiebe, K.L. and G.D. Bortolotti. 1992. Facultative sex ratio manipulation in American kestrels. *Behav. Ecol. Sociobiol.* **30:** 379–86.

Wiebe, K.L. and G.D. Bortolotti. 1994a. Food supply and hatching spans of birds: energy constraints or facultative manipulations? *Ecology* **75:** 813–23.

Wiebe, K.L. and G.D. Bortolotti. 1994b. Energetic efficiency of reproduction: the benefits of asynchronous hatching for American kestrels. *J. Anim. Ecol.* **63:** 551–60.

Wiens, D., Nickrent, D.L., Davern, C.I., Calvin, C.L. and N.J. Vivrette. 1989. Developmental failure and loss of reproductive capacity in the rare palaeo-endemic shrub *Dedeckera eurekensis*. *Nature* **338:** 65–7.

Wilbur, H.M. 1980. Complex life cycles. *Annu. Rev. Ecol. Syst.* **11:** 67–93.

Wilbur, H.M. 1990. Coping with chaos: toads in ephemeral ponds. *Trends Ecol. Evolution* **5:** 37.

Wilbur, H.M. and J.P. Collins. 1973. Ecological aspects of amphibian metamorphosis. *Science* **182:** 1305–14.

Williams, A.J. 1980a. Offspring reduction in macaroni and rockhopper penguins. *Auk* **97:** 754–9.

Williams, A.J. 1980b. The clutch sizes of macaroni and rockhopper penguins. *Emu* **81:** 87–90.

Williams, A.J. and A.E. Burger. 1979. Aspects of the breeding biology of the imperial cormorant *Phalacrocorax atriceps* at Marion Island, South Atlantic. *Le Gerfaut* **69:** 407–24.

Williams, G.C. 1966a. *Natural selection and adaptation*. Princeton University Press, Princeton, NJ.

Williams, G.C. 1966b. Natural selection, the costs of reproduction, and a refinement of Lack's principle. *Am. Nat.* **100:** 687–690.

Williams, G.C. and D.C. Williams. 1957. Natural selection of individually harmful social adaptations among sibs with special reference to social insects. *Evolution* **11:** 32–9.

Williams, T.D. 1994. Intraspecific variation in egg size and egg composition in birds: effects on offspring fitness. *Biol. Rev.* **68:** 35–59.

Williams, T.D., Loonen, M.J.J.E. and F. Cooke. 1994. Fitness consequences of parental behavior in relation to offspring number in a precocial species: the lesser snow goose. *Auk* **111:** 563–72.

Willson, M.F. and B.J. Rathke. 1974. Adaptive design of the floral display in *Asclepias syriaca* L.*Am. Midl. Nat.* **92:** 47–57.

Willson, M.F. and P.W. Price. 1977. The evolution of inflorescence size in *Asclepias* (Asclepiadaceae). *Evolution* **31:** 495–511.

Willson, M.F. and D.H. Schemske. 1980. Pollinator limitation, fruit production and floral display in Pawpaw (*Asimina triloba*). *Bull. Torrey Bot. Club* **107:** 401–8.

Willson, M.F. and N. Burley. 1983. *Mate choice in plants*. Princeton University Press, Princeton, NJ.

Willson, M.F., Thomas, P.A., Hoppes, W.G., Katusic-Malmborg, P.L., Goldman, D.A. and J.L. Bothwell. 1987. Sibling competition in plants: an experimental study. *Am. Nat.* **129:** 304–11.

Wilson, E.O. 1966. Behaviour of social insects. In *Symp. No. 3 of the R. Entomol. Soc. Lond: Insect Behaviour* (ed. P.T. Haskell), pp.81–96. Royal Entomological Society, London.

Wilson, E.O. 1971. *The insect societies*. Harvard University Press, Cambridge, Mass.

Wimsatt, W.A. 1945. Notes on breeding behavior, pregnancy, and parturition in some vespertilionid bats of the eastern United States. *J. Mammal.* **26:** 23–33.

Wimsatt, W.A. 1975. Some comparative aspects of implantation. *Biol. Reprod.* **12:** 1–40.

Winemiller, K.O. and K.A. Rose. 1993. Why do most fish produce so many tiny offspring? *Am. Nat.* **142:** 585–603.

Winkler, D.W. 1987. A general model for parental care. *Am. Nat.* **130:** 526–43.

Winkler, D.W. and K. Wallin. 1987. Offspring size and number: a life history model linking effort per offspring and total effort. *Am. Nat.* **129:** 708–20.

Winkler, D.W. and G.S. Wilkinson. 1988. Parental effort in birds and mammals: theory and measurement. In *Oxford surveys in evolutionary biology*. Vol 5. (ed., P.H. Harvey and L. Partridge), pp.185–214. Oxford University Press, New York.

Wooller, R.D. and K.C. Richardson. 1992. Reduction in the number of young during pouch-life in a small marsupial. *J. Zool.* **226:** 445–54.

Wourms, J.P. 1977. Reproduction and development in chondrichthyan fishes. *Am. Zool.* **17:** 379–410.

Wourms, J.P. 1981. Viviparity: the maternal-fetal relationship in fishes. *Am. Zool.* **21:** 473–515.

Wourms, J.P. and L.S. Demski. 1993. The reproduction and development of sharks, skates, rays and ratfishes: Introduction, history, overview, and future prospects. *Environ. Biol. Fishes* **38:** 7–21.

Wourms, J.P., Hamlett, W.C. and M.D. Stribling. 1981. Embryonic oophagy and adelphophagy in sharks. *Am. Zool.* **21:** 1019.

Wourms, J.P., Grove, B.D. and J. Lombardi. 1988. The maternal-embyonic relationship in viviparous fishes. In *Fish physiology*. Vol. 11B. (ed. W.S. Hoar and D.J. Randall), pp. 1–134.. Academic Press, San Diego, CA.

Wourms, J.P., Atz, J.W. and M.D. Stribling. 1991. Viviparity and the maternal-embryonic relationship in the coelacanth *Latimeria chalumnae*. *Environ. Biol. Fishes* **32:** 225–48.

Wright, J. and I. Cuthill. 1989. Manipulation of sex differences in parental care. *Behav. Ecol. Sociobiol.* **25:** 171–81.

Wright, S. 1922. Coefficients of inbreeding and relationship. *Am. Nat.* **56:** 330–8.

Wu, M.C., Chen, Z.Y., Jarrell, V.L. and P.J. Dziuk. 1989. Effects of initial length of uterus per embryo on fetal survival and development in the pig. *J. Anim. Sci.* **67:** 1767–72

Wyatt, R. 1980. The reproductive biology of *Asclepias tuberosa*: I. Flower number, arrangement, and fruit-set. *New Phytol.* **85:** 119–31.

Wynne-Edwards, V.C. 1962. *Animal dispersion in relation to social behaviour*. Oliver and Boyd, Edinburgh.

Yamamura, N. and M. Higashi. 1992. An evolutionary theory of conflict resolution between relatives: altruism, manipulation, compromise. *Evolution* **46:** 1236–9.

Yanega, D. 1988. Social plasticity and early-diapausing females in a primitively social bee. *Proc. Natl. Acad. Sci. USA* **85:** 4374–77.

Yanega, D. 1989. Caste determination and differential diapause within the first brood of *Halictus rubicundus* in New York (Hymenoptera: Halictidae). *Behav. Ecol. Sociobiol.* **24:** 97–107.

Yanagisawa, Y. and H. Ochi. 1991. Food intake by mouthbrooding females of *Cyphotilapia frontosa* (Cichlidae) to feed both themselves and their young. *Environ. Biol. Fishes* **30:** 353–8

Yanagisawa, Y. and T. Sato. 1990. Active browsing by mouthbrooding females of *Tropheus duboisi* and *Tropheus moorii* (Cichlidae) to feed both themselves and their young. *Environ. Biol. Fishes* **37:** 43–50.

Ydenberg, R.C. 1994. The behavioral ecology of provisioning in birds. *Ecoscience* **1:** 1–14.

Young, E.C. 1963. The breeding behaviour of the South Polar skua, *Catharacta maccormicki. Ibis* **105:** 203–33.

Young, J.P.W. 1981. Sibling competition can favour sex two ways. *J. Theor. Biol.* **88:** 755–6.

Zahavi, A. 1975. Mate selection-a selection for a handicap. *J. Theor. Biol.* **53:** 205–14.

Zahavi, A. 1977. The cost of honesty (further remarks on the handicap principle). *J. Theor. Biol.* **67:** 603–5.

Zahavi, A. 1987. The theory of signal selection and some of its implications. In *International symposium on biological evolution* (ed., V.P. Delfino), pp. 305–27. Adriatica Editrice, Bari.

Zahavi, A. 1990. Arabian babblers: the quest for social status in a co-operative breeder. In *Co-operative breeding in birds* (ed., P.B. Stacey and W.D. Koenig), pp.103–30. Cambridge University Press, New York.

Zeh, D.W. and R.L. Smith. 1985. Parental investment by terrestrial arthropods. *Am. Zool.* **25:** 785–805.

Zimen, E. 1975. Social dynamics of the wolf pack. In *The wild canids* (ed. M.W. Fox), pp. 336–63. Van Nostrand Reinhold, New York.

Index